APPLIED NUMERICAL METHODS IN C

Shoichiro Nakamura
Ohio State University

P T R PRENTICE HALL
Englewood Cliffs, New Jersey 07632

Nakamura, Shoichiro
    Applied numerical methods in C / Shoichiro Nakamura.
        p.   cm.
    Includes bibliographical references and index.
    ISBN 0-13-042052-2
    1. C (Computer program language) 2. Numerical analysis.
I. Title.
QA76.73.C15N35   1992
519.4'0285'5133--dc20

Editorial/production supervision: *Brendan M. Stewart*
Prepress buyer: *Mary Elizabeth McCartney*
Manufacturing buyer: *Susan Brunke* and *Margaret Rizzi*
Acquisitions editor: *Mary Franz*

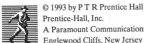

Limits of Liability and Disclaimer of Warranty of Software

Trade Names

CSL is a trademark of Computational Methods, Inc.
IBM PC is a registered trademark of International Business Machine Corporation.
ProComm is a registered trademark for a software of Datastorm Technologies, Inc.
QuickC and QuickBasic are registered trademarks of Microsoft, Inc.
Silicon Graphics and IRIS are registered trademarks of Silicon Graphics, Inc.
WATFOR-77 and WATCOM are registered trademarks of WATCOM Publications in Canada.
Unix is a registered trademark of AT&T Bell Laboratories.
VAX is a registered trademark of Digital Computer Corporation.

Printed in the United States of America

10 9 8 7 6 5 4 3

ISBN 0-13-042052-2

Prentice-Hall International (UK) Limited, *London*
Prentice-Hall of Australia Pty. Limited, *Sydney*
Prentice-Hall Canada Inc., *Toronto*
Prentice-Hall Hispanoamericana, S.A., *Mexico*
Prentice-Hall of India Private Limited, *New Delhi*
Prentice-Hall of Japan, Inc., *Tokyo*
Simon & Schuster Asia Pte. Ltd., *Singapore*
Editora Prentice-Hall do Brasil, Ltda., *Rio de Janeiro*

Contents

Programs

Preface

In recent years, the number of C language users has dramatically increased and now includes professionals who were trained in Fortran but are switching to C as well as students who are learning C as their first programming language. The surge of popularity in C is related to the increasing popularity of UNIX and computer graphics. To what extent numerical computations in the future will be programmed in C is uncertain. Nonetheless, there is no question that a need exists for a comprehensive text especially for those who want to learn, or use, numerical methods in C. This book has been written in response to this need.

Topics covered in this book range from the basics of numerical methods through to intermediate level subjects. The first nine chapters can be taught in undergraduate courses with certain omissions, while the last four chapters are recommended for an introductory graduate course in numerical methods. The mathematical content of this book is essentially identical to my previous book, *Applied Numerical Methods with Software*, except that a few new examples have been added and all the programs in this book are written in C. The material in this book has been tested many times at both undergraduate (sophomore) and graduate levels and has been adopted in many universities worldwide.

The objectives of this book include: (1) to be easily understood by undergraduate students with minimal knowledge of mathematics, (2) to enable students to practice the methods in C, (3) to provide short programs that can be easily used for scientific applications with or without modifications, and (4) to provide software that are easy to understand.

Chapter 1 covers truncation and round-off errors that are preparatory subjects for numerical computations. To explain the sources of these errors, there are brief discussions of Taylor series and how numbers are computed and saved in computers.

Chapter 2 describes Lagrange and Newton interpolations as primary subjects. Then, it covers the interpolation using the Chebyshev points, Hermite interpolation, two-dimensional interpolation, and a brief discussion on extrapolation. The causes and behavior of interpolation errors are explained with an intuitive approach.

Chapter 3 describes iterative methods used to solve nonlinear equations, including bisection method, false position method, Newton and secant methods, successive substitution method, and finally Bairstow method.

Chapter 4 covers numerical integration methods, starting with simple but fundamental integration methods such as trapezoidal and Simpson's rules. The Newton-Cotes formulas are then introduced. Chapter 4 also includes Gauss integration method and numerical method for improper integrals and double integrals. The numerical methods for improper integrals are based on trapezoidal rule and double exponential transformation.

Chapter 5 covers the basic concepts of numerical differentiation. It develops a systematic method to derive difference approximations with truncation error term. The approach is implemented in a computer program.

Chapter 6 covers basic computational methods to solve inhomogeneous linear equations. Gauss elimination and Gauss-Jordan methods are discussed first without pivoting and then with pivoting. The effect of pivoting is illustrated with both single precision and double precision. After the matrix notations are introduced, the concepts of matrix inversion, LU decomposition, and determinant are introduced. Ill-conditioned problems are explained using inverse matrix and determinant. Finally, the solution of m equations with n unknowns is described.

Chapter 7 covers selected methods of computing matrix eigenvalues. The basic aspects of homogeneous linear equations are explained first. The method of interpolation is then introduced. This approach should help students to understand the relation of eigenvalues to roots of the characteristic equation. The remainder of the chapter introduces iterative methods, Householder-tridiagonal method, and QR iteration.

Chapter 8 describes curve fitting of experimental data based on least square methods.

Chapter 9 covers numerical methods of ordinary differential equations, including Runge-Kutta method and predictor-corrector method. Applications of the methods to numerous engineering problems are illustrated.

Chapter 10 describes numerical methods for boundary value problems of ordinary differential equations, including eigenvalue problems. This chapter may serve also as a preparation for numerical methods for partial differential equations that follow.

The last three chapters cover numerical methods for partial differential equations. Chapter 11 is on elliptic partial differential equations. Its early sections cover derivation of difference equations and implementation of boundary conditions. It then describes iterative methods, including successive-over-relaxation, extrapolated Jacobi-iterative method based on 2-cyclic property, and ADI. The convergence analyses and optimization of iterative parameters are introduced using a one-dimensional model. The chapter also introduces the concept of direct solution methods.

Chapter 12 is on parabolic partial differential equations. It includes the numerical methods based on Euler explicit, implicit, and modified Euler methods. The concepts of stability analyses based on eigenfunctions as well as Fourier expansion are written.

Chapter 13 is on hyperbolic partial differential equations. Focusing on a simple first-order wave equation, it starts with methods of characteristics that are followed by fundamental finite difference methods. The analyses of truncation errors are described with the modified equation concept. A simplified formula to derive modified equations is also developed.

The book contains more than 40 programs in C. The numerical methods described in this book are not limited to microcomputers. The methods are essentially independent of the kinds of computers used. The number of programs in this book is not meant to deemphasize the importance of students' own programming practice. Developing a program is always important in learning numerical methods. The instructor should give assignments of short programs to be developed by the students. On the other hand, developing every program from scratch is time consuming and often inefficient as well as frustrating. Furthermore, it is impossible to give all necessary instructions for programming, arithmetic protections, formatting, and testing. One effective approach is to assign students frequent projects to modify parts of the programs in this book.

Comments and feedback from the reader are most welcome. Indeed, the comments pertaining to my previous book, *Applied Numerical Methods with Software*, have been very helpful in preparing this book.

I am indebted to many readers and instructors, particularly Professor Max Platzer of Naval Postgraduate School in Camel, CA, and Professors Bob Essenhigh, Terry Conlisk, and Yann Geuzennec of The Ohio State University, for valuable suggestions and assistance. The assistance of Mr. C. Y. R. Chong in developing graphic programs is acknowledged. Finally, I would like to express special appreciation to Gunawan Hadi, Greg Doench, and Mary Franz of Simon & Schuster, who suggested the idea for this book, encouraged me to pursue the project, and helped me to complete the book.

S. Nakamura

Columbus, Ohio

Before You Read and Use the Programs in This Book

All the programs except Program 3–3 are compatible with most C compilers, and were tested with Microsoft QuickC, C compilers running on VAX and IRIS (Silicon Graphics) machines. Program 3–3 which plots functions is prepared in two versions, one for IBM PC's with QuickC, and another for IRIS machines. Readers can run each program easily with or without modifications.

Although no instructions in C programming are given in this book, it is assumed that the readers of this book are beginners in C. Therefore, the programs are written as simply and clearly as possible. The style of programming may vary somewhat among similar programs. This is done intentionally to illustrate alternative approaches to programming for similar problems. Recommended reference books for C programming are listed at the end of this section.

Each program can stand alone, and there is no need to link to other programs, or to use input data files. To facilitate this approach, however, all the equations and data are hard-coded within the program. Users are expected to change these equations and data. In the lists of programs, hard-coded equations and data are highlighted. In the sample outputs, the user input data given interactively are also highlighted. The meanings of symbols are explained in the comments within the programs.

The logical contents of most programs in C are identical to those in the FORTRAN version of the book, *Applied Numerical Methods with Software*. However, the algorithms of certain programs have been modified to improve accuracy.

To those who want to link C programs with Fortran, a brief guide is provided with examples in Appendix I.

The source code of all the programs is optionally available as CSL/C (Computational Software Library in C) on a disk for the IBM PC or Macintosh. The disk also includes the programs that could not be printed here because of space limitations. To order a diskette, please use the order form at the end of this book.

Recommended books for learning the C language:

Brown, T. D., Jr., *C for Fortran Programmers*, Silicon Press, 1990.

C for Yourself (a user's manual of QuickC), Microsoft Corporation, 1988.

Darnel, P. A., and P. E. Margolis, *The Software Engineering in C*, Springer-Verlag, 1988.

Kochan, S. G., *Programming in C*, Hayden Books, 1989.

For more advanced readers, the following references are recommended:

Harbison, S. P., and G. L. Steele, *C: A Reference Manual*, 3rd ed., Prentice-Hall, 1991.

Kernighan, B., and D. M. Ritchie, *The C Programming Language*, 2nd ed., Prentice-Hall, 1989.

Plauger, P. J., and J. Brodie, *Standard C: Programmer's Quick Reference*, Microsoft Press, 1989.

The following books are recommended for mathematical and numerical programming using C:

Baker, L., *C Mathematical Function Handbook*, McGraw-Hill, 1991.

Press, W. H., B. P. Flannery, S. A. Teukolsky, and W. T. Vetterling, *Numerical Recipes in C*, Cambridge University Press, 1988.

1
Major Sources of Errors in Numerical Methods

1.1 INTRODUCTION

There are two major sources of errors in numerical computations. The first is truncation error, and the second is round-off error. Truncation error is caused by the approximations used in the mathematical formula of the scheme. The Taylor series is the most important means used to derive numerical schemes and analyze truncation errors.

Round-off errors are associated with the limited number of digits that represent numbers in a computer. To understand the nature of round-off errors, it is necessary to learn the ways numbers are stored and additions and subtractions are performed in a computer.

This chapter covers the Taylor series and numbers, the most fundamental subjects in numerical methods.

1.2 TAYLOR SERIES

Numerical solutions are mostly approximations for exact solutions. Most numerical methods are based on approximating functions by polynomials, even when this is not apparent. More advanced algorithms are constructed by combining the basic algorithms. Therefore, when an error of a numerical method is questioned, one has to investigate how accurately the polynomial is approximating the true function.

The Taylor expansion, which is an infinite power series, exactly represents a function within a certain radius about a given point. Therefore, by comparing the polynomial expansion of the numerical solution to the Taylor series of the exact solution, particularly by finding at which order the discrepancy occurs, it becomes possible to evaluate the error, called *truncation error* [Conte/de Boor; King; Hornbeck].

The Taylor series is also used in deriving numerical methods. By ignoring all the terms in the Taylor series except a few, a polynomial approximating the true function may be obtained. This is called a *truncated Taylor series* and is used as a starting point in deriving numerical methods [Morris; Cheney/Kincaid]. However, the error of the numerical method originates from the truncation.

TAYLOR EXPANSION OF ONE-DIMENSIONAL FUNCTIONS. A function $f(x)$ is said to be analytic at $x = a$ if $f(x)$ can be represented by a power series in powers of $h = x - a$ within a radius of convergence, $D > |x - a| > 0$. A necessary condition for a function to be analytic is that all its derivatives be continuous at $x = a$ and in the neighborhood.

A point at which a function $f(x)$ is not analytic is called a *singular point*. If $f(x)$ is differentiable everywhere in the neighborhood of x_0 except at x_0, then x_0 is a singular point. For example, $\tan(x)$ is analytic except at $x = \pm(n + \frac{1}{2})\pi$, $n = 0, 1, 2, \ldots, \infty$, which are singular points. Polynomials are analytic everywhere.

If f is analytic about $x = a$, $f(x)$ in the neighborhood of $x = a$ can be exactly represented in the Taylor series, which is a power series given by

$$f(x) = f(a) + hf'(a) + \frac{h^2}{2} f''(a) + \frac{h^3}{6} f'''(a) + \frac{h^4}{24} f''''(a)$$

$$+ \frac{h^5}{5!} f'''''(a) + \cdots + \frac{h^m}{m!} f^{(m)}(a) + \cdots \tag{1.2.1}$$

where

$$h = x - a$$

For example, Taylor expansions of e^{-x} and $\sin(x)$ about $x = 1$ are, respectively,

$$\exp(-x) = \exp(-1) - h \exp(-1) + \frac{h^2}{2} \exp(-1)$$

$$- \frac{h^3}{6} \exp(-1) + \frac{h^4}{24} \exp(-1) - \cdots$$

$$\sin(x) = \sin(1) + h \cos(1) - \frac{h^2}{2} \sin(1) \tag{1.2.2}$$

$$- \frac{h^3}{6} \cos(1) + \frac{h^4}{24} \sin(1) + \cdots$$

where

$$h = x - 1.$$

The Taylor series is unique. That is, there is no other power series in $h = x - a$ to represent $f(x)$.

The Taylor expansion of a function about $x = 0$ is called the *Maclaurin series*. For example, Maclaurin series of $\exp(x)$, $\sin(x)$, $\cos(x)$ and $\ln(x + 1)$ are, respectively,

$$e^x = 1 + x + \frac{x^2}{2!} + \frac{x^3}{3!} + \frac{x^4}{4!} + \cdots$$

$$\sin(x) = x - \frac{x^3}{3!} + \frac{x^5}{5!} - \frac{x^7}{7!} + \cdots$$

$$\cos(x) = 1 - \frac{x^2}{2!} + \frac{x^4}{4!} - \frac{x^6}{6!} + \cdots$$

$$\ln(x + 1) = x - \frac{x^2}{2} + \frac{x^3}{3} - \frac{x^4}{4} + \cdots$$

In practical applications, the Taylor series has to be truncated after a certain order term because it is impossible to include an infinite number of terms. If the Taylor series is truncated after the Nth order term, it is expressed as

$$f(x) = f(a) + hf'(a) + \frac{h^2}{2} f''(a) + \frac{h^3}{6} f'''(a) + \frac{h^4}{24} f''''(a)$$

$$+ \frac{h^5}{5!} f'''''(a) + \cdots + \frac{h^N}{N!} f^{(N)}(a) + 0(h^{N+1}) \qquad (1.2.3)$$

where $h = x - a$ and $0(h^{N+1})$ represents the error caused by truncating the terms of order $N + 1$ and higher. However, the whole error can be expressed by

$$0(h^{N+1}) = f^{(N+1)}(a + \xi h) \frac{h^{N+1}}{(N+1)!}, \quad 0 < \xi < 1 \qquad (1.2.4)$$

Since ξ cannot be found exactly, the error term is often approximated by setting $\xi = 0$:

$$0(h^{N+1}) \simeq f^{(N+1)}(a) \frac{h^{N+1}}{(N+1)!} \qquad (1.2.5)$$

which is the leading term of the truncated terms.

If $N = 1$, for example, the truncated Taylor series is

$$f(x) \simeq f(a) + f'(a)h, \quad h = x - a \qquad (1.2.6)$$

Including the effect of the error, it is also expressed as

$$f(x) = f(a) + f'(a)h + 0(h^2) \tag{1.2.7}$$

where

$$0(h^2) = f''(a + \xi h)\frac{h^2}{2}, \quad 0 < \xi < 1 \tag{1.2.8}$$

TAYLOR SERIES OF A TWO-DIMENSIONAL FUNCTION. The Taylor expansion of a two-dimensional function $f(x, y)$ about (a, b) is given by

$$f(x, y) = f(a, b) + hf_x + gf_y + \frac{1}{2}[h^2 f_{xx} + 2hg f_{xy} + g^2 f_{yy}]$$

$$+ \frac{1}{6}[h^3 f_{xxx} + 3h^2 g f_{xxy} + 3hg^2 f_{xyy} + g^3 f_{yyy}]$$

$$+ \frac{1}{24}[h^4 f_{xxxx} + 4h^3 g f_{xxxy} + 6h^2 g^2 f_{xxyy} + 4hg^3 f_{xyyy} + g^4 f_{yyyy}] + \cdots \tag{1.2.9}$$

where

$$h = x - a, \quad g = y - b,$$

$$f_x = \frac{\partial}{\partial x} f(x, y)\big|_{x=a, y=b}$$

$$f_y = \frac{\partial}{\partial y} f(x, y)\big|_{x=a, y=b}$$

and similar notations such as $f_{x\ldots x}$, $f_{xy\ldots}$, and $f_{yy\ldots}$ are partial derivatives of f at $x = a$ and $y = b$; each x or y in subscripts indicates one time of partial differentiation with respect to x or y, respectively.

SUMMARY OF THIS SECTION

(a) The Taylor series is the most important tool for deriving numerical methods and analyzing errors.
(b) The Taylor series expanded about $x = 0$ is called the Maclaurin series.

1.3 NUMBERS ON COMPUTERS

In solving a mathematical problem by using a hand calculator, we are well aware that the decimal numbers we calculate may not be exact. Decimal numbers are almost always rounded when we record them. Even if numbers are not rounded off intentionally, the limited number of digits of the calculator may cause round-off errors. (A hand calculator well designed for scientific calculations may have 10 or 11 digits, but a cheap one often has only six digits.)

In an electronic computer, round-off errors occur for the same reasons and affect the computed results [Wilkinson]. In some cases, the round-off errors cause very serious effects and make computed results totally meaningless. Therefore, it is important to learn some basic aspects of arithmetic operations in computers and understand in what circumstances severe round-off errors can occur. Many round-off error problems can be avoided through appropriate programming practices.

Range and accuracy of numeric constants vary among different programming languages and on different computers. Such differences sometimes cause different results or even difficulties when a program is translated from one language to another. In this section, we therefore review how numbers are treated in three different languages, namely, C, Basic, and Fortran on IBM PC and mainframe computers.

1.3.1 Base of Numbers

The number system we use daily is called the *decimal system*. The base of the decimal number system is 10. Computers, however, do not use the decimal system in computations and memory but use the binary system. The binary system is natural for computers because computer memory consists of a huge number of electronic and magnetic recording devices, of which each element has only "on" and "off" statuses.

However, if we look into the machine languages, we soon realize that other number systems, particularly the octal and hexadecimal systems, are also used [Hannula; Bartee]. The octal and hexadecimal systems are close relatives of the binary and can be translated to and from binary easily. Expressions in octal or hexadecimal are shorter than in binary, so they are easier for humans to read and understand. Hexadecimal also provides more efficient use of memory space for real numbers (as explained later).

The base of a number system is also called the *radix*. The radix for the decimal system is 10; it is 8 for the octal system, and 2 for the binary system. The radix of the hexadecimal system is 16.

The base of a number is denoted by a subscript: for example $(3.224)_{10}$ is 3.224 in base 10 (decimal), $(1001.11)_2$ is 1001.11 in base 2 (binary), and $(18C7.90)_{16}$ is 18C7.90 in base 16 (hexadecimal).*

The decimal value of a number in base r, for example,

$$(abcdefg \cdot hijk)_r$$

* In hexadecimal, each digit can range from 0 to 15. Digits from 10 to 15 are expressed by the capital letters A, B, C, D, E, and H. The decimal values of hexadecimal digits are shown next:

Decimal	0	1	2	3	4	5	6	7	8	9	10	11	12	13	14	15
Hexadecimal	0	1	2	3	4	5	6	7	8	9	A	B	C	D	E	F

is computed by

$$ar^6 + br^5 + cr^4 + dr^3 + er^2 + fr + g + hr^{-1} + ir^{-2} + jr^{-3} + kr^{-4}$$

The numbers that appear without a suffix in this book are all in base 10 unless otherwise mentioned.

1.3.2 Range of Numeric Constants

The numbers in a computer are largely classified into three categories: (a) integers, (b) real numbers, and (c) complex numbers. Complex numbers, however, are not available as a native type in C and Basic. In order to define complex variables in those languages, the real part and imaginary parts of complex variables must be treated separately, and the user has to write his or her own arithmetic functions for complex variables (see PROGRAM 3–6).

In C, integers are defined as different types, namely *unsigned*, *int*, *long*, and *short* depending on the range of the integers and signs. However, these names are used somewhat differently among different computers. Table 1.1A shows some examples of different types for signed integers.

Table 1.1A Range of integers in C

	short	int	long
QuickC (IBM PC)	N/A	-2^{15} to $2^{15} -1$	-2^{31} to $2^{31} -1$
IRIS (SG)	-2^{15} to $2^{15} -1$	-2^{31} to $2^{31} -1$	-2^{31} to $2^{31} -1$
Cray	-2^{31} to $2^{31} -1$	-2^{63} to $2^{63} -1$	-2^{63} to $2^{63} -1$

The smallest and largest magnitudes of a real number that can be represented on a computer varies depending on the design of both the hardware and software. On IBM PC (Basic) the range is approximately from 2.9×10^{-39} to 1.7×10^{38}. With QuickC on IBM PC, the range is from 1.4×10^{-45} to 3.4×10^{38}. The range on IBM 370 is from approximately 5.4×10^{-79} to 7.2×10^{75}. See Table 1.1B for similar data of other computers.

We must realize that real numbers on a computer are not continuous. If we look at the numbers near zero, the smallest positive number on IBM PC (Basic) is 2.9×10^{-39}. Therefore, no numbers between zero and 2.9×10^{-39} as well as between

Table 1.1B Range of real numbers (single precision)

IBM PC (Basic)	2.9E − 39	1.7E + 38
IBM PC (QuickC)	1.4E − 45	3.4E + 38
IBM 370	5.4E − 79	7.2E + 75
IRIS (SG)	1.4E − 45	3.4E + 38
Cray XMP	4.6E − 2476	5.4E + 2465
VAX	2.9E − 39	1.7E + 38

zero and -2.9×10^{-39} can be represented. The interval between the smallest positive number (2.9×10^{-39}) and the next smallest positive number is approximately 3.45×10^{-46}, which is much smaller than 2.9×10^{-39}.

The difference between 1 and the smallest number that is greater than 1 but distinguishable from 1 is approximately 1.19×10^{-7} with QuickC and Basic on IBM PC. This interval is named *machine epsilon* [Forsythe/Malcolm/Moler; Cheney/ Kincaid; Yakowitz/Szidarovszky]. The interval between any real number to the next real number approximately equals

$$\text{machine epsilon} \times \text{R}$$

where R is the real number. More about machine epsilon is described later.

A typical distribution of floating numbers in single precision with no extension to denormalized range is pictorially shown in Figure 1.1.

```
 ┌─  1.7x10³⁸ (largest positive floating number)
 ├─  (second largest floating number)
┌┘
│gap
┆
┆
 ├─  1.000000119 (1plus machine epsilon)
 ├─  1.0
┆
┆
 ├─  2.9x10⁻³⁹+ 3.45x10⁻⁴⁶ (second smallest floating number)
 ├─  2.9x10⁻³⁹ (smallest positive floating number)
│gap
 └─  0
```

Figure 1.1 Typical design of real numbers with normalized floating point binary format in single precision (no extension to denormalized range)

1.3.3 Numbers Inside Computer Hardware

Bit is an acronym of binary digit and represents an element of memory consisting of on and off positions such as a semiconductor device or magnetized spot on a recording surface. *Byte* is a set of bits considered as a unit that normally consists of 8 bits.

The ways bits are used for integers and floating point values vary depending on the design of a computer. The remainder of this section describes typical examples of how bits are used to save numbers.

INTEGERS.   In the binary number system, the mathematical expression of an integer is

$$\pm\, a_k a_{k-1} a_{k-2} \cdots a_2 a_1 a_0 \tag{1.3.1}$$

where a_i is a bit that is 0 or 1. Its decimal value is

$$I = \pm \left[a_k 2^k + a_{k-1} 2^{k-1} + \cdots a_2 2^2 + a_1 2 + a_0 \right] \qquad (1.3.2)$$

For example, a binary number given by

$$\pm \, 110101$$

equals

$$I = \pm \left[(1)(2^5) + (1)(2^4) + (0)(2^3) + (1)(2^2) + (0)(2) + (1) \right]$$
$$= \pm \left[32 + 16 + 0 + 4 + 0 + 1 \right] = \pm 53 \qquad (1.3.3)$$

On a computer the maximum value of k in Eq. (1.3.1) is limited because of the hardware design. On IBM PC (QuickC and Basic), 2 bytes (or equivalently 16 bits) are used to represent an integer. The first bit records the sign: positive if 0, or negative if 1. The remainder of 15 bits are used for a_i's. Therefore, the largest positive integer is

Binary:	0	1	1	1	1	1	1	1	1	1	1	1	1	1	1	1
(bit no:	0	1	2	3	4	5	6	7	8	9	10	11	12	13	14	15)	(1.3.4)

The decimal value of this is

$$\sum_{i=0}^{14} 2^i = 32767$$

One way to store a negative number is to use exactly the same digits as the real positive number of the same magnitude except that the first bit is set to 1. However, many computers use *two's complement* to store negative numbers. The two's complement for $(-32767)_{10}$, for example, is

Binary:	1	0	0	0	0	0	0	0	0	0	0	0	0	0	0	1
(bit no:	0	1	2	3	4	5	6	7	8	9	10	11	12	13	14	15)	(1.3.5)

Bits of Eq. (1.3.5) are obtained by changing 0 to 1 as well as 1 to 0 in the bits of Eq. (1.3.4) for 32767 and adding 1. In two's complement, the decimal value is found first as if the whole 16 bits express a positive number. If this number is less than 2^{15}, or 32768, then the number is interpreted as a positive number. If it is equal or greater, then it is translated to a negative number by subtracting 2^{16}. In the foregoing example of binary, the decimal equivalent of the binary in Eq. (1.3.5) equals $Z = 2^{15} + 1$, so the subtraction yields

$$32768 + 1 - 2^{16} = 32768 + 1 - 65536 = -32767 \qquad (1.3.6)$$

The negative integer smallest in magnitude is represented by

Binary: 1 1 1 1 1 1 1 1 1 1   1   1   1   1   1   1

(bit no: 0 1 2 3 4 5 6 7 8 9 10 11 12 13 14 15)

which equals -1 in decimal.

On IBM 370, 4 bytes are used for an integer. Therefore, the maximum positive integer becomes $2^{32-1} - 1 = 2147483648 - 1 = 2147483647$.

REAL NUMBERS. The format for a real number on a computer is different depending on the hardware and software design. So, we focus on IBM PC (QuickC and Basic), IRIS, VAX and IBM mainframe (Fortran 77) as major examples.

Normally in small and medium-size computers, single precision floating point numbers are saved with 32 bits in normalized binary form with or without extension to include denormalized floating numbers. Among the 32 bits, 1 bit is used for the sign, 8 bits for the exponent, and 23 bits for the mantissa (see Figure 1.2).

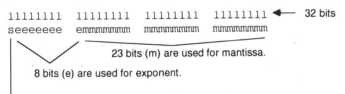

Figure 1.2 Allocation of 32 bits on IBM PC (QuickC and Basic), IRIS, and VAX

When only the normalized form is used without the extension, a floating number x in decimal form is expressed as follows:

$$\text{if } g > 0, \ x = (\text{sign}) \ (0.1bbb \ bbbb \ bbbb \ bbbb \ bbbb \ bbbb)_2 2^{(g-p)}$$
$$\text{if } g = 0, \ x = 0 \qquad (1.3.7a)$$

where g is the biased exponent (integer), p is the bias (integer), and $bbb .. b$ represent 23 bits. In the normalized format the most significant bit is always 1, so the effective number of bits for the mantissa is 24. The value of p is typically 128 (Fortran and C compilers on VAX, and Microsoft Basic on IBM PC).

Because the integer g is saved in 8-bit binary, the possible range of g is $0 \leqslant g \leqslant 255$. However, $g = 0$ and 255 are reserved. With $p = 128$ for example, the range of

$2^{(g-p)}$ in Eq. (1.3.7a) is $2^{-127} \leqslant 2^{(g-p)} \leqslant 2^{+126}$. The smallest and largest mantissa are $(0.1)_2 = 0.5$ and $(0.111\ldots)_2 = 1 - 2^{-24}$, respectively. Therefore, the smallest and largest floating values become $0.5 \times 2^{-127} = 2.938735\ldots \times 10^{-39}$ and $(1 - 2^{-24}) \times 2^{+127} \sim 1.70141\ldots \times 10^{38}$, respectively (see Table 1.1B).

The format of Eq. (1.3.7a) is by no means unique for the normalized floating value format. Indeed, some computers use the following form rather than Eq. (1.3.7a):

$$\text{if } g > 0, \ x = (\text{sign})\ (1.bbbb\ bbbb\ bbbb\ bbbb\ bbbb\ bbb)_2 2^{(g-p)}$$
$$\text{if } g = 0, \ x = (\text{sign})\ (0.bbbb\ bbbb\ bbbb\ bbbb\ bbbb\ bbb)_2 2^{(-p+1)}$$

$$(1.3.7b)$$

where the meaning of $bbb\ldots$, g and p are same as Eq. (13.7a) for a total of 32 bits for single precision. The numbers represented by the second equation in Eq. (1.3.7b) are called denormalized floating values. For the present format, p is typically 127, and g is saved in 8-bit binary. Because of the bias $p = 127$, the range of $2^{(g-p)}$ becomes $2^{-126} \leq 2^{(g-p)} < 2^{+128}$ (remember that $g = 255$ is reserved). Thus, the largest floating value becomes

$$(1.1111\ 1111\ 1111\ 1111\ 1111\ 111)_2 2^{(g-p)}$$

$$= (2 - 2^{-24}) \times 2^{+127} \sim 3.40282 \times 10^{38}$$

The smallest denormalized floating value occurs if all the bits of the mantissa are zero except the last one; that is,

$$(0.0000\ 0000\ 0000\ 0000\ 0000\ 001)_2 2^{(-127+1)}$$

$$= 2^{-149} = 1.40129\ldots \times 10^{-45}$$

This analysis explains the range of the smallest and largest floating values in QuickC on IBM PC and C on IRIS in Table 1.1B.

In Fortran 77 on IBM maintrame computers such as the IBM 370, the normalized floating point value format in hexadecimal is used, which is written as

$$x = (0.abbbbb)_{16} \times 16^k$$

$$(1.3.8)$$

where a is a non-zero hexadecimal digit and b's are hexadecimal digits including zero; k is an exponent and expressed in binary. The mantissa has 6 hexadecimal digits.

In IBM 370, a single precision floating point value uses 32 bits, among which the first bit records the sign of mantissa, the next 7 bits are for the exponent, and the last 24 bits are for the mantissa. The exponent is biased by $(64)_{10}$ and stored in the 7 bits. One hexadecimal digit is represented by 4 bits. Therefore, 6 hexadecimal digits of the mantissa need 24 bits, among which the first 4 bits represent the first hexadecimal digit, the next 4 the second hexadecimal digit, and so on.

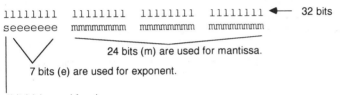

24 bits (m) are used for mantissa.

7 bits (e) are used for exponent.

1 bit (s) is used for sign.

Figure 1.3 Allocation of 32 bits for a floating value on IBM 370

Now we can figure out why the largest positive number of IBM mainframes is 7.2×10^{75}. The maximum positive mantissa that can be represented with 6 hexadecimal digits is $(0.FFFFFF)_{16}$, which equals

$$1 - (16)^{-6} \tag{1.3.9}$$

in decimal. The maximum exponent that can be represented by the 7-digit binary is

$$127 - 64 = 63$$

where 64 is the bias. Therefore, the largest positive floating value becomes

$$(1 - 16^{-6})16^{63} \simeq 7.23 \times 10^{75} \tag{1.3.10}$$

On the other hand the smallest positive number is $(0.100000)_{16} \times 16^{-64}$, which equals

$$16^{-1} \times 16^{-64} = 16^{-65} = 5.39 \times 10^{-79} \tag{1.3.11}$$

in decimal.

The machine epsilon on the IBM mainframe is

$$16^{-6} \times 16^{1} = 16^{-5} = 9.53 \times 10^{-7} \tag{1.3.12}$$

The numbers of bits used in typical computers are summarized in Table 1.2.

Table 1.2 Number of bits for floating point numbers

	Single precision		Double precision	
	Mantissa	Exponent	Mantissa	Exponent
IBM PC, AT, XT[a]	23	8	55	8
IBM 370	24	7	56	7
CDC 7600 and Cyber	48	11	96	11
VAX 11	23	8	55	8
Cray XMP	48	11	96	11

[a] Based on Microsoft Basic [IBM].

The computational results of some programs can be significantly affected by the machine epsilon. Such discrepancies are often observed with solutions of linear equations (see Example 6.4 and also Chapter 8). Therefore, the reader is encouraged to find the machine epsilon of the computer by running the program shown in the next section.

1.3.4 Round-off Errors in a Computer

ROUND-OFF ERRORS IN STORING A NUMBER IN MEMORY. The most basic source of errors in a computer is attributed to the error in representing a real number with a limited number of bits.

As already explained, machine epsilon, ε, is the interval between 1 and the next number greater than 1 that is distinguishable from 1. This means that no number between 1 and $1 + \varepsilon$ can be represented in the computer. In the case of IBM PC (QuickC and Basic), any number $1 + \alpha$ is cut off to 1 if $0 < \alpha < \varepsilon/2$, or rounded up to $1 + \varepsilon$ if $\varepsilon/2 \leqslant \alpha$. Thus $\varepsilon/2$ may be considered as the maximum possible round-off error for 1. In other words, when 1.0 is found in a memory, the original value could have been somewhere in $1 - \varepsilon/2 < x < 1 + \varepsilon/2$.

Machine epsilon can be found by the following program:

```
/*Machine Epsilon      eps.c */
#include <stdio.h>
main ()
{
float ex, g = 1, eps;
    do
    {
        g = g/2;
        ex = g*0.98 + 1;
        ex = ex - 1;
        printf ("g = %15.8e    ex = %15.8e\n", g, ex);
        if (ex>0) eps = ex;
    } while (ex>0);
    printf("\nMachine epsilon = %16.8e\n\n", eps);
}
```

The last number printed by the program equals the machine epsilon. Machine epsilon in both single and double precisions for a few computers are:

Precision	IBM PC†	IBM 370	VAX 11	Cray XMP
Single	1.19E − 7	9.53E − 7	1.19E − 7	3.55E − 15
Double	2.77E − 17	2.22E − 16	2.77E − 17	1.26E − 29

† QuickC and Basic

The round-off error involved in saving any real number R in memory is approximately equal to $\varepsilon R/2$ if the number is rounded, and εR if cut off.

EFFECTS OF ROUND-OFF ERRORS. When numbers are added or subtracted, an accurate representation of the result may require a much larger number of digits than needed for the numbers added or subtracted.

Serious amounts of round-off error occur in two types of situations: (a) when adding (or subtracting) a very small number to (or from) a large number, and (b) when a number is subtracted from another that is very close.

To test the first case on the computer, let us add 0.00001 to unity ten thousand times. A program to do this job would be:

```
/*Summation by Single Precision     sum_singl.c */
#include <stdio.h>
main ()
{
float x, sum = 1.0;
int i, k = 0;
   for (i = 1; i <= 10000; i++)
   {
      sum = sum + 0.00001;
   }
   printf("\nSum = %f\n", sum);
}
```

The result of this program on **IBM PC** is

$$SUM = 1.100136$$

Since the exact answer is 1.1, the relative error of this computation is

$$\frac{1.1 - 1.100136}{1.1} = -0.000124 \text{ or } -0.0124\% \tag{1.3.13}$$

Another annoying problem is that two numbers that ought to be mathematically identical are not always equal on computers. For example, consider the equations given by

```
y=a/b
w=y*b
z=a—w
```

where a and b are constant. Mathematically, w equals a, so z must become zero. If these equations are computed on a computer, z becomes zero or a very small nonzero value, depending on the values of a and b. Try the following program:

```
/*Equality Test     equ_tst.c */
#include <stdio.h>
#include <math.h>
main ()
{
float a, b, w, y, z, k;
   a = cos(0.3);
   printf("%f", a);
   for (k = 1; k <= 20; k++)
   {
      b = sin(0.7*k);
      z = a/b;
      w = z*b;
      y = a-w;
      printf("k=%6.2f, a=%13.4e, b=%13.4e", k, a, b);
      printf("w=%13.4e, y=%13.4e\n", w, y);
   }
}
```

What happens in the computer is that a round-off error occurs when $z = a/b$ and $w = a*b$ are computed and saved. So, $w = z*b$ in the fifth line does not become exactly a. The relative magnitude of the round-off error attributed to multiplying or dividing by a constant and saving in a memory is approximately equal to the machine epsilon.

The error of a number caused by round-off increases as the number of arithmetic operations increases [Wilkinson].

CAUSES OF ROUND-OFF ERRORS. To explain how round-off errors occur, let us consider the computation of $1 + 0.00001$ with QuickC or Basic on IBM PC. The binary representation of 1 and 0.00001 are, respectively,

$$(1)_{10} = (0.1000\ 0000\ 0000\ 0000\ 0000\ 0000)_2 \times 2^1$$

$$(0.00001)_{10} = (0.1010\ 0111\ 1100\ 0101\ 1010\ 1100)_2 \times 2^{-16}$$

$$(1.3.15)$$

The summation of these two numbers becomes

$$(1)_{10} + (0.00001)_{10}$$

$$= (0.1000\ 0000\ 0000\ 0000\ 0101\ 0011\ 1110\ 0010\ 1101\ 0110\ 0)_2 \times 2^1$$
$$*$$

$$(1.3.16)$$

However, the numbers at and after the asterisk (∗) are rounded because the mantissa has 24 bits. Therefore, the result of this calculation is stored in memory as

$$(1)_{10} + (0.00001)_{10} \simeq (0.1000\ 0000\ 0000\ 0000\ 0101\ 0100)_2 \times 2^1 \qquad (1.3.17)$$

which is equivalent to $(1.0000\ 1001\ 36)_{10}$.

Thus, whenever 0.00001 is added to 1, the result gains 0.0000000136 as an error. When addition of 0.00001 to 1 is repeated ten thousand times, an error of exactly ten thousand times 0.0000000136 is generated. Although the calculated result gains in the present example, it can lose if some digits are cut off. Loss and gain are both referred to as *round-off error*.

We next illustrate an effect of the round-off errors involved in subtracting a number. From calculus we know that

$$\lim_{\theta \to 0} \frac{f(x + \theta) - f(x)}{\theta} = f'(x) \qquad (1.3.18)$$

For illustration purpose we set $f(x) = \sin(x)$, and compute

$$d = \frac{\sin(1 + \theta) - \sin(1)}{\theta} \qquad (1.3.19)$$

with decreasing values of θ. The computed results on **IBM PC (Basic)** are shown next:

θ	d	(Exact Value $- d$)
0.1	0.49736	0.042938
0.01	0.53607	0.004224
0.001	0.53989	0.000403
0.0001	0.54061	−0.003111
0.00001	0.53644	0.003860
0.000001	0.53644	0.003860
0.0000001	0.59604	−0.055744
Exact value = cos (1) = 0.54030		

It is observed that when θ is decreased, d approaches the exact value until θ reaches 0.001, but then error starts increasing as θ is decreased further. The increase of errors with decrease of θ occurs because, as the difference between $f(1 + \theta)$ and $f(1)$ becomes small, round-off error relative to θ increases. The errors of d for larger values of θ are caused by truncation errors of the approximation, Eq. (1.3.19).

To analyze the round-off in subtraction, let us consider the computation of $1.00001 - 1$. Since we have found that 1.00001 is stored in binary as

$$(0.1000\ 0000\ 0000\ 0000\ 0101\ 0100)_2 \times 2^1$$

then, $1.00001 - 1$ becomes

$$(0.1000\ 0000\ 0000\ 000\ 0101\ 0100)_2 \times 2^1 - (0.1)_2 \times 2^1$$
$$= (0.0000\ 0000\ 0000\ 0000\ 0101\ 0100)_2 \times 2^1$$
$$= (0.1010\ 1)_2 \times 2^{-16} \tag{1.3.20}$$

Its decimal value is 1.00136×10^{-5}. Comparing this to the exact value, 0.00001, the relative error is

$$(0.00001 - 0.0000100136)/0.00001 = -0.00136$$

or -0.136%.

STRATEGIES. The effects of round off can be minimized by changing the computational algorithm although it must be devised case by case. Some useful strategies include:

- Double precision [McCracken]
- Grouping
- Taylor expansions
- Changing definition of variables
- Rewriting the equation to avoid subtractions

Applications of these approaches are illustrated in Example 1.1.

Example 1.1

Add 0.00001 ten thousand times to unity by using (a) the grouping method and (b) double precision.

⟨**Solution**⟩

(a) When the sum of many small numbers is computed, grouping them helps to reduce round-off errors. A program to add 0.00001 to unity ten thousand times may be written as

```
/*Summation by Grouping     sum_gr.c */
#include <stdio.h>
main ()
{
```

```
float x, gr_total, sum = 1;
int i, k = 0;
   for (i = 1; i <= 100; i++)
   {
       gr_total = 0;
       for (k = 1; k <= 100; k++)
       {
           gr_total = gr_total + 0.00001;
       }
       sum = sum+gr_total;
   }
   printf("\nSum = %13.8e\n", sum);
}
```

In the foregoing program, each 100 of small values are grouped, the group total is computed, and then the group totals are accumulated. The answer of this revised program is

sum=1.10000467

(b) Another approach is to use double precision for the sum as follows:

```
/*Summation by Double Precision        sum_dbl.c */
#include <stdio.h>
main ()
{
double x, sum = 1.0;
int i, k = 0;
   for (i = 1; i <= 10000; i++)
   {
       sum = sum + 0.00001;
   }
   printf("\nSum = %f\n", sum);
}
```

where sum is defined as a double precision variable and initialized to 1. The result of this version is

sum=1.10000000000

The comparison of the two approaches shows that grouping significantly increases accuracy, although using double precision gives better results in the present example. The grouping method may be implemented also with double precision.

Example 1.2

As θ approaches 0, accuracy of a numerical evaluation for

$$d \equiv \frac{\sin (1 + \theta) - \sin (1)}{\theta}$$

becomes very poor because of round-off errors. By using *Taylor expansion*, we can rewrite the equation so that the accuracy for θ is improved.

⟨**Solution**⟩

When a small difference of two functional values is to be calculated, Taylor expansion is useful. The Taylor expansion of $\sin (1 + \theta)$ is

$$\sin (1 + \theta) = \sin (1) + \theta \cos (1) - 0.5 \, \theta^2 \sin (1) \cdots$$

If we approximate $\sin (1 + \theta)$ by the first three terms, d becomes

$$d \simeq \cos (1) - 0.5\theta \sin (1) \qquad\qquad (A)$$

The computed values for various θ are

θ	d
0.1	0.49822
0.01	0.53609
0.001	0.53988
0.0001	0.54026
0.00001	0.54030
0.000001	0.54030
0.0000001	0.54030
Exact value = 0.54030	

Accuracy of the approximation increases as θ approaches 0.

Example 1.3

If the following equation is directly evaluated in a program, round-off errors occur when x approaches $+\infty$ and $-\infty$.

$$y = \frac{1}{(a - z)(b - z)} \qquad\qquad (A)$$

where

$$z = \frac{a + b + (b - a)\tanh (x)}{2} \qquad\qquad (B)$$

Rewrite the equations so that no serious round-off error occurs.

⟨**Solution**⟩

Because $-1 < \tanh(x) < 1$, the domain of z is $a < z < b$. As x approaches ∞, z approaches b; and as x approaches $-\infty$, z approaches b. Two sources of round-off errors must be considered.

If $b = 0$, Eq. (B) becomes $a[1 - \tanh(x)]$. Therefore, a severe round-off error occurs when $\tanh(x)$ becomes very close to 1. Subtraction of similar numbers in Eq. (B) can be avoided by recognizing the relation:

$$\tanh(x) = \frac{\exp(x) - \exp(-x)}{\exp(x) + \exp(-x)} \tag{C}$$

Introducing Eq. (C) into (B) and rewriting the equation yields

$$z = \frac{b \exp(x) + a \exp(-x)}{\exp(x) + \exp(-x)}$$

When z approaches a or b, a round-off error occurs in the denominator of Eq. (A). To avoid this, computation of Eq. (A) may be divided into two cases:

Case 1 $a < z < (a + b)/2$

Case 2 $(a + b)/2 \leqslant z < b$

For Case 1, $a - z$ in the denominator of Eq. (A) is written as

$$a - z = a - \frac{b \exp(x) + a \exp(-x)}{\exp(x) + \exp(-x)}$$

$$= \frac{(a - b) \exp(x)}{\exp(x) + \exp(-x)}$$

For Case 2, $b - z$ is written as

$$b - z = b - \frac{b \exp(x) + a \exp(-x)}{\exp(x) + \exp(-x)} = \frac{(b - a) \exp(-x)}{\exp(x) + \exp(-x)}$$

SUMMARY OF THIS SECTION

(a) The largest magnitude of positive integers and that of negative integers are limited by the number of bytes used. With two's complement, the latter becomes larger than the former by 1.

(b) The interval between 1 and the next real number is called *machine epsilon*. The interval between any two consecutive real numbers in a computer, except at 0, is approximately equal to the real number times the machine epsilon.

(c) Round-off errors are serious causes of errors in computations. A significant amount of round-off error occurs, particularly when a small number is added to a large number or when a number is subtracted from a similar number.

(d) Round-off error effects can be reduced by several approaches, including using double precision, rewriting equations, and expansion of a function in polynomials. However, the one that is best depends on the nature of the computation.

PROGRAM

PROGRAM 1–1 Decimal to Binary Conversion

(A) Explanations

This program converts a positive decimal value less than $1.0E + 38$ to a binary representation in the normalized floating point value format with a 24-bit mantissa. This conversion may be expressed by

$$(x)_{10} = (0.bbbb\ bbbb\ bbbb\ bbbb\ bbbb\ bbbb)_2 \times 2^{(M)_{10}}$$

where the x on the left side is the decimal input, bs on the right side are 0 or 1, and the exponent M of 2 is in decimal. (The subscripts 2 or 10 in the foregoing equation indicate the base of the number: 10 for decimal and 2 for binary.)

The program works interactively. When executed, the program asks for a decimal value for x, which is then converted to binary. The program prints out the mantissa in binary and the exponent M in decimal.

All input is given interactively. There is no need to change the program.

(B) List

```
/* CSL/c1-1.c     Decimal to Binary Conversion */
#include <stdio.h>
#include <math.h>
/*           a[k] : k-th bit in the normalized mantissa
             decimal : decimal value of input
             m_expnt : exponent (in decimal)
             pwr_of_2 : power of 2                              */
main()
{
int a[255], i, k, l, m_expnt, n;
float pwr_of_2, s, decimal, y;
printf( "\nCSL/C1-1  Decimal to Binary Conversion \n\n" );
    do {
        printf( "Decimal value input ?  " ); scanf( "%f", &decimal );
        l = 0;    k = 1;    i = log( decimal )/log( 2.0 )  + 3;
        do {
            i = i - 1;
            if( i < -200 )  exit(0);
            pwr_of_2 = pow(2.0, (float)( i-1 ));
            if( decimal >= pwr_of_2 ) {
               a[k] = 1;
               decimal = decimal - pwr_of_2;
               if( l == 0 )  {m_expnt = i;  l = 1; }
```

```
      }
      else {
          if ( k > 1 )    a[k] = 0;
      }
          if ( l > 0 ) k = k + 1;
   } while ( k < 25 );
   printf( "\n----------------------------------------\n" );
   printf( "Binary\n Mantissa = ." );
   for ( k = 1; k <= 24; k++ ) {
      printf( "%d", a[k] );
      if (k/4*4 == k) printf( " " );
   }
   printf( "\n Exponent= %d \n", m_expnt );
   printf( "----------------------------------------\n" );
   printf( "\n Type 1 to continue, 0 to stop.\n" );
   scanf( "%d", &n );
   } while (n>0);
}
```

(C) Sample Output

```
CSL/C1-1  Decimal to Binary Conversion

Decimal value input ?  0.5
----------------------------------------
Binary
 Mantissa = .1000 0000 0000 0000 0000 0000
 Exponent=  0
----------------------------------------

Decimal value input ?  64.0
----------------------------------------
Binary
 Mantissa = .1000 0000 0000 0000 0000 0000
 Exponent=  7
----------------------------------------

Decimal value input ?  0.0001751
----------------------------------------
Binary
 Mantissa = .1011 0111 1001 1011 0000 1100
 Exponent= -12
----------------------------------------
```

(D) Discussions

In the first case, the input is 0.5 in decimal. Its computed binary becomes $(0.1)_2 \times 2^0$, or equivalently $(0.1)_2$. The second conversion is

$$(64)_{10} = (0.1)_2 \times 2^7$$

The third conversion is

$$(0.0001751)_{10} = (0.1011\ 0111\ 1001\ 1011\ 0000\ 1100)_2 \times 2^{-12}$$

QUIZ. What are the normalized floating binary numbers for 5, 0.001, 2^{-7}, and 3.1415?

PROBLEMS

(1.1) If 8 bits are used to represent positive and negative integers in two's complement, what are the greatest positive and smallest negative integers (largest in magnitude) in decimal?

(1.2) Two 16-bit binary numbers in two's complement are given:

(a) Binary: 1 1 1 1 1 1 1 1 1 1 1   1   1   1   1   1
    (bit no: 0 1 2 3 4 5 6 7 8 9 10 11 12 13 14 15)

(b) Binary: 1 0 0 0 0 0 0 0 0 0   0   0   0   1   1   0
    (bit no: 0 1 2 3 4 5 6 7 8 9 10 11 12 13 14 15)

Find the decimal values of the two binary numbers.

(1.3) Find the machine epsilon of a mainframe computer you have access to.

(1.4) Express the largest floating value with QuickC in the normalized floating value format (32 bit).

(1.5) Evaluate

$$\exp(x) - 1$$

for $x = 0.0001$ by applying Taylor expansion of $\exp(x)$. Use the first three terms.

(1.6) Expand the following functions into the Maclaurin series

$$1/(1 + x^2)$$
$$\tan(x)$$
$$1/(1 - x)$$
$$\ln(1 + x)$$

(1.7) Show that the Taylor expansion of $\ln[(1 + x)/(1 - x)]$ about $x = 1$ is

$$2 \sum_{n=1}^{\infty} \frac{x^{2n-1}}{2n - 1}$$

(1.8) Using the Maclaurin expansion of e^x and e^{-x}, find the Maclaurin expansion of $\sinh(x)$ and $\cosh(x)$, where

$$\sinh(x) = \frac{1}{2}(e^x - e^{-x})$$

$$\cosh(x) = \frac{1}{2}(e^x + e^{-x})$$

(1.9) **(a)** If the following function is written in a program, in what range of x would overflow or zero divide originated from round-off error occur?

$$f(x) = \frac{1}{1 - \tanh(x)}$$

Assume that the smallest positive number is 3×10^{-39} and machine epsilon is 1.2×10^{-7}.

(b) Rewrite the equation so that no subtraction is necessary.

(1.10) In a hypothetical computer using 32 bits for single precision normalized floating numbers, assume that the exponent biased by 125 is saved in 8 bits (the biased exponents of 0 and 255 are reserved). Assume the decimal value of the most significant bit of the normalized mantissa is (0.1), which is not saved in the memory. Find the machine epsilon, the smallest and largest floating numbers.

(1.11) If the computer considered in Problem 1.10 is extended to include denormalized floating numbers, what is the smallest positive floating number?

REFERENCES

Bartee, T. C., *Digital Computer Fundamentals*, McGraw-Hill, 1981.

Cheney, W., and D. Kincaid, *Numerical Mathematics and Computing*, Brooks/Cole, 1985.

Conte, D. C., and S. D. de Boor, *Elementary Numerical Analysis: An Algorithmic Approach*, McGraw-Hill, 1980.

Forsythe, G. E., M. A. Malcolm, and C. B. Moler, *Computer Methods for Mathematical Computations*, Prentice-Hall, 1977.

Goldstein, L. J., and M. Goldstein, *IBM PC*, Prentice-Hall and Communications, 1984.

Hannula, R., *Computing and Programming*, Houghton Mifflin, 1974.

Hornbeck, W. H., *Numerical Methods*, Quantum, 1975.

IBM, "BASIC by Microsoft Corp.," IBM Personal Computer Hardware Reference Library 6025013, 1982.

International Business Machine Corporation (IBM), *IBM System/360 and System/370 Fortran IV Language*, GC28-6515-9.

King, J. T., *Introduction to Numerical Computation*, McGraw-Hill, 1984.

Kline, R. M., *Structured Digital Design Including MSI/LSI Components and Microprocessors*, Prentice-Hall, 1983.

Leventhal, L. A., *Introduction to Microprocessors*, Software, Programming, Prentice-Hall, 1978.

McCracken, D. D., *Computing for Engineers and Scientists with Fortran 77*, Wiley, 1984.

Morris, J. L., *Computational Methods in Elementary Numerical Analysis*, Wiley, 1983.

Sterbenz, P. H., *Floating-Point Computations*, Prentice-Hall, 1974.

Wilkinson, J. H., *Rounding Errors in Algebraic Processes*, Prentice-Hall, 1963.

Yakowitz, S., and F. Szidarovszky, *An Introduction to Numerical Computations*, Macmillan, 1986.

2

Polynomial Interpolation

2.1 INTRODUCTION

An *interpolating* function is the function that passes through given data points, which are typically sampled from a function table or taken directly from a known function.

Interpolation of the data points may be done with a polynomial, spline functions, a rational function, or Fourier series among other possible forms [Stoer/Burlish]. Polynomial interpolation (fitting a polynomial to given data points) is one of the most fundamental subjects in numerical methods because many other numerical schemes are based on polynomial interpolation. For example, numerical integration schemes are obtained by integrating polynomial interpolation formulas, and numerical differentiation schemes are derived by differentiating polynomial interpolations. Therefore, the objective of this chapter is to discuss the basic aspects of polynomial interpolation and its applications. Cubic spline interpolation and transfinite interpolations are described in Appendices G and H, respectively.

Measured data may be interpolated, but a direct interpolation is not desirable in most cases because of random errors involved in measurements. Therefore, fitting a curve to measured data is separately described in Chapter 8.

Table 2.1 is a brief overview of interpolation schemes described in this chapter.

2.2 LINEAR INTERPOLATION

Linear interpolation is a basis for several fundamental numerical schemes. By integrating the linear interpolation, one derives the integration scheme called the *trap-*

Table 2.1 Summary of one-dimensional interpolation schemes

Interpolation scheme	Advantages	Disadvantages
Lagrange interpolation	Convenient form. Easy to program.	Cumbersome for hand calculations.
Newton interpolation	The order of the polynomial can be changed easily. Evaluation of errors is easy.	A difference or divided difference table must be prepared.
Lagrange interpolation using Chebyshev points	Errors are more evenly distributed than with equispaced grid.	Unevenly distributed grid points.
Hermite interpolation	Accuracy is high because polynomial is fitted to derivatives also.	Needs values of derivatives.
Cubic spline (Appendix G)	Applicable to any number of data points.	Needs to solve simultaneous equations.

ezoidal rule. The gradient of the linear interpolation is an approximation for the first derivative of the function. The objective of this section is to introduce linear interpolation and then discuss its errors.

The linear interpolation is a line fitted to two data points. The linear interpolation shown in Figure 2.1 is given by

$$g(x) = \frac{b-x}{b-a} f(a) + \frac{x-a}{b-a} f(b) \qquad (2.2.1)$$

where $f(a)$ and $f(b)$ are known values of $f(x)$ at $x = a$ and $x = b$ respectively.

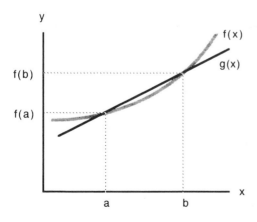

Figure 2.1 Linear interpolation

The error of the linear interpolation may be expressed in the form:

$$e(x) = \frac{1}{2}(x - a)(x - b)f''(\xi), \quad a < \xi < b \tag{2.2.2}$$

where ξ (a Greek letter named "xi") is dependent on x but somewhere between a and b (the proof for Eq. (2.2.2) is given in Appendix A). Equation (2.2.2) is a rather awkward function because we have no means to evaluate ξ exactly. However, it is possible to analyze $e(x)$ when $f''(x)$ is approximated by a constant in $a \leqslant x \leqslant b$ as discussed next.

If f'' is a slowly varying function, or if $[a, b]$ is a small interval so that f'' changes very little, we can approximate $f''(\xi)$ by $f''(x_m)$ where x_m is the midpoint between a and b: $x_m = (a + b)/2$. Equation (2.2.2) then indicates that:

(a) The maximum error occurs approximately at the midpoint between the two data points.

(b) The error increases as $b - a$ increases.

(c) The error also increases as $|f''|$ increases.

An exception for these trends is if f'' has a zero in $[a, b]$ because the assumption that f'' is approximately constant does not hold.

SUMMARY OF THIS SECTION

(a) Linear interpolation is a line fitted to two data points.

(b) If the sign of $f(x)$ does not change in $a \leqslant x \leqslant b$, the maximum error of a linear interpolation occurs at approximately midpoint and its magnitude is proportional to the second derivative of the function approximated.

2.3 LAGRANGE INTERPOLATION FORMULA

Can three or four or more data points be fitted by a curve? One of the fundamental methods to find a function that passes through given data points is to use a polynomial (see Figure 2.2).

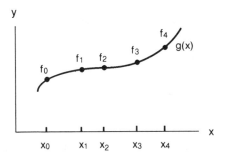

Figure 2.2 Data points fitted by a polynomial

Polynomial interpolations may be expressed in various alternative forms that can be transformed from one to another. Among them are power series, Lagrange interpolation, and Newton forward/backward interpolations.

As discussed later in more detail, a Nth order polynomial that passes through $N + 1$ data points is unique. It means, regardless of the interpolation formula, that all polynomial interpolations fitted to the same data points are mathematically identical.

Suppose $N + 1$ data points are given as

$$
\begin{array}{cccc}
x_0 & x_1 & \cdots & x_N \\
f_0 & f_1 & \cdots & f_N
\end{array}
$$

where x_0, $x_1 \ldots$ are abscissas of the data points (grid points) and assumed to be in increasing order. The spacings between grid points are arbitrary. The polynomial of order N passing through the $N + 1$ data points may be written in a power series as

$$g(x) = a_0 + a_1 x + a_2 x^2 + \cdots + a_N x^N \tag{2.3.1}$$

where the a_i are undetermined coefficients. Fitting the power series to the $N + 1$ data points gives a system of linear equations as

$$
\begin{aligned}
f_0 &= a_0 + a_1 x_0 + a_2 x_0^2 + \cdots + a_N x_0^N \\
f_1 &= a_0 + a_1 x_1 + a_2 x_1^2 + \cdots + a_N x_1^N \\
&\ \vdots \\
f_N &= a_0 + a_1 x_N + a_2 x_N^2 + \cdots + a_N x_N^N
\end{aligned}
\tag{2.3.2}
$$

Although the coefficients, a_i, may be determined by solving the simultaneous equations using a computer program, such an attempt is not desirable for two reasons. First, one needs a program to solve a set of linear equations, and second, the computer solution may not be accurate. (Indeed, powers of x_i in the equations can be very large numbers, and if so, the effects of round-off errors become serious.) Fortunately, there are better methods to determine the interpolation polynomial without solving linear equations. They include the Lagrange interpolation formula and the Newton forward/ backward interpolation formula.

To introduce the basic idea behind the Lagrange formula, consider the product of factors given by

$$\overline{V}_0(x) = (x - x_1)(x - x_2) \cdots (x - x_N)$$

which is related to the $N + 1$ data points shown earlier. The function $\overline{V}_0$ is a Nth order polynomial of x, and becomes zero at $x = x_1, x_2, \ldots, x_N$. If we divide $\overline{V}_0(x)$ by

$\bar{V}_0(x_0)$, the resulting function

$$V_0(x) = \frac{(x - x_1)(x - x_2) \cdots (x - x_N)}{(x_0 - x_1)(x_0 - x_2) \cdots (x_0 - x_N)}$$

becomes unity for $x = x_0$ and zero for $x = x_1, x = x_2, \ldots, x = x_N$. Similarly, we can write V_i by

$$V_i(x) = \frac{(x - x_0)(x - x_1) \cdots (x - x_N)}{(x_i - x_0)(x_i - x_1) \cdots (x_i - x_N)}$$

where the numerator does not include $(x - x_i)$ and the denominator does not include $(x_i - x_i)$. The function $V_i(x)$ is an Nth order polynomial and becomes unity at $x = x_i$ and zero for $x = x_j, j \neq i$. Thus if we multiply $V_0(x), V_1(x), \ldots, V_N(x)$ by $f_0, f_1, \ldots, f_N$, respectively, and add them together, the summation should become a polynomial of order N again and equals f_i for each of $i = 0$ through $i = N$.

The Lagrange interpolation formula of order N thus derived is written as follows [Conte/de Boor]:

$$\begin{aligned} g(x) = &\frac{(x - x_1)(x - x_2) \cdots (x - x_N)}{(x_0 - x_1)(x_0 - x_2) \cdots (x_0 - x_N)} f_0 \\ &+ \frac{(x - x_0)(x - x_2) \cdots (x - x_N)}{(x_1 - x_0)(x_1 - x_2) \cdots (x_1 - x_N)} f_1 \\ &\vdots \\ &+ \frac{(x - x_0)(x - x_1) \cdots (x - x_{N-1})}{(x_N - x_0)(x_N - x_1) \cdots (x_N - x_{N-1})} f_N \end{aligned} \qquad (2.3.3)$$

Equation (2.3.3) is equivalent to the power series determined by solving the linear equations. It looks lengthy, but even memorizing it is not difficult if the structure is understood.

Example 2.1

(a) Densities of sodium for three temperatures are given as follows:

i	Temperature T_i	Density ρ_i
0	94° C	929 kg/m^3
1	205	902
2	371	860

Write the Lagrange interpolation formula that fits the three data points.

(b) Find the density for $T = 251°$ C by using the Lagrange interpolation. (In computing the value of $g(x)$, do not expand the formula into a power series.)

⟨**Solution**⟩

(a) Because the number of data points is three, the order of the Lagrange formula is $N = 2$. The Lagrange interpolation becomes

$$g(T) = \frac{(T - 205)(T - 371)}{(94 - 205)(94 - 371)} \,(929)$$

$$+ \frac{(T - 94)(T - 371)}{(205 - 94)(205 - 371)} \,(902)$$

$$+ \frac{(T - 94)(T - 205)}{(371 - 94)(371 - 205)} \,(860)$$

(b) By setting $T = 251$ in the equation above, we obtain

$$g(251) = 890.5 \text{ kg/m}^3$$

(*Comments*: In evaluating $g(x)$ for a given value of x, one should not expand the Lagrange interpolation formula into a power series, because it not only is cumbersome but also increases chances of making human errors.)

Equation (2.3.3) is lengthy particularly if order N is large. However, writing it in a computer program needs only a small number of lines as follows: Looking at Eq. (2.3.3), we recognize that the first term is f_0 times a product of

$$\frac{(x - x_i)}{(x_0 - x_i)}$$

for all i except for $i = 0$. The second term is f_1 times a product of

$$\frac{(x - x_i)}{(x_1 - x_i)}$$

for all i except for $i = 1$. All other terms have the same pattern. Therefore, Eq. (2.3.3) may be programmed as follows:

```
g = 0;
for( i = 0; i <= n; i++ )
{
  z = 1.0;
  for( j = 0; j <= n; j++ )
  {
    if( i != j ) z = z*(xa - x[j])/(x[i] - x[j]);
  }
  g = g + z*f[i];
}
```

Here, meanings of notations are:

f[i], x[i], i = 0, 1, ... n : given data
g : result of interpolation
z : product of factors
xa : x

See PROGRAM 2-1 for an actual implementation.

Equation (2.3.3) is a polynomial of order N or less because each term on the right side is a polynomial of order N. The order of the polynomial becomes less than N if f_i is sampled from a polynomial $f(x)$ of an order less than N. In this case $g(x)$ becomes exactly equal to $f(x)$.

The interpolation polynomial of order N fitted to $N + 1$ data points is a unique polynomial. This is important because it indicates that all the polynomials of order N fitted to a given set of $N + 1$ data points are mathematically identical even if their forms are different.

The uniqueness of the interpolation polynomial can be proved by considering the hypothesis that the Lagrange interpolation is not an unique polynomial. If not unique, there must be another polynomial of order N, $k(x)$, that passes through the same $N + 1$ data points. The difference between the Lagrange interpolation $g(x)$ and $k(x)$ defined by

$$r(x) = g(x) - k(x) \tag{2.3.4}$$

must be a polynomial of order N or less because $g(x)$ and $k(x)$ are both polynomials of order N. On the other hand, since $g(x)$ and $k(x)$ both agree at the $N + 1$ data points, $r(x)$ must become zero at the $N + 1$ data points. It means that $r(x)$ has $N + 1$ zeroes, that is, $r(x)$ must be a polynomial of order $N + 1$. This is contradictory to the fact that $r(x)$ is a polynomial of order N or less and proves that the hypothesis is incorrect.

When a known function $f(x)$ is approximated by an interpolation polynomial, our concern is the error of the polynomial. The error is defined by

$$e(x) = f(x) - g(x) \tag{2.3.5}$$

where $f(x)$ is the function from which the data points are sampled: $f_i = f(x_i)$. Distribution and magnitude of $e(x)$ are affected by the following parameters:

(a) Distribution of abscissas of the data points.

(b) The size of the whole domain of interpolation.

(c) The order of polynomial (or equivalently the number of data points used in the interpolation minus one).

These aspects are discussed in more detail in the next few paragraphs.

The most frequently chosen distribution of x_i is the equispaced points (equally spaced intervals between two consecutive abscissas), but nonuniformly spaced x_i are also often used (see Section 2.5). We assume here that x_i are equispaced. On an equispaced grid, however, the magnitude of $e(x)$, namely $|e(x)|$, tends to be smallest in the interval nearest to the center of the domain and tends to increase rapidly toward the edges.

The size of the interpolation domain defined by

$$D = x_N - x_0 \tag{2.3.6}$$

has a significant effect on the magnitude and distribution of $e(x)$. In general, the maximum value of $|e(x)|$ approaches zero as D is decreased. On the other hand, if D is increased, the maximum value of $|e(x)|$ increases and can even dominate $|g(x)|$, particularly for a large order, N.

If D is fixed but N is increased from a small value, the maximum error tends to decrease up to a certain value of N. Then, the maximum error may start increasing. We have to understand that there is no guarantee that the interpolations $g(x)$ will converge to $f(x)$ by increasing N.

The error of the Lagrange interpolation formula is given by a formula similar to Eq. (2.2.2) for linear interpolation. Derivation is described in Appendix A, so we write it here without proof:

$$e(x) = f(x) - g(x) = L(x)f^{(N+1)}(\xi), \quad x_0 < \xi < x_N \tag{2.3.7}$$

where $N + 1$ is the number of data points, $f^{(N+1)}$ is the $(N + 1)$th derivative of $f(x)$ and

$$L(x) = \frac{(x - x_0)(x - x_1) \cdots (x - x_N)}{(N + 1)!} \tag{2.3.8}$$

In Eq. (2.3.7) ξ depends on x, but it satisfies $x_0 < \xi < x_N$. Notice that if $f(x)$ is a polynomial of order N or less, the $(N + 1)$-th derivative of $g(x)$ vanishes. Therefore, the error becomes zero. If $f(x)$ is not, then we have the same difficulty that we had for Eq. (2.2.3) because ξ is dependent on x but not known.

However, for a small interval $[a, b]$ where $f^{N+1}(\xi)$ can be approximated by a constant, Eq. (2.3.7) becomes

$$e(x) \simeq L(x)f^{(N+1)}(x_m) \tag{2.3.9}$$

where x_m is the midpoint between the two end points. Then, $e(x)$ becomes approximately proportional to $L(x)$ given by Eq. (2.3.8). An approximate value of $f^{(N+1)}$ in the interval may be estimated, if one more data point is available, by using a difference approximation. Approximating $f^{N+1}(\xi)$ by a constant is not appropriate in the following situations, however: first, if $[a, b]$ is a large interval or if $f^{(N+1)}(x)$ changes substantially, and second, if $f^{(N+1)}(x)$ changes sign in the middle of the domain.

The function $L(x)$ has significant effects on the distribution of error as follows:

(a) With an equispaced grid, the amplitude of oscillation of $L(x)$ is minimum at the center of the interpolation range but increases toward the edges. This causes an increase of errors toward end points.

(b) As the size of the interpolation range increases, the amplitude of oscillation increases rapidly.

Figure 2.3 shows $L(x)$ for the Lagrange interpolation using six grid points in the range of $0 \leqslant x \leqslant 5$. The maximum of $|L(x)|$ occurs in the leftmost and rightmost intervals, while the local maximum of $|L(x)|$ in any grid interval is smallest in the interval at the center of the domain.

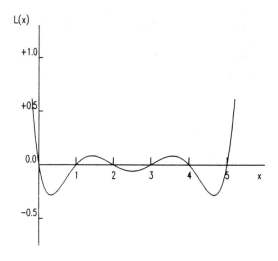

Figure 2.3 Distribution of $L(x)$

Example 2.2

A function table for $f(x) = \log_{10}(x)$ is given next:

i	x_i	$f(x_i)$
0	1	0
1	2	0.30103
2	3	0.47712
3	4	0.60206

If the function is approximated by the Lagrange interpolation fitted to the data points, estimate errors at $x = 1.5, 2.5, 3.5$.

⟨**Solution**⟩

The estimate for the error given by Eq. (2.3.9) becomes

$$e(x) \simeq \frac{(x-1)(x-2)(x-3)(x-4)f''''(2.5)}{(4!)}$$

The fourth derivative of $f(x)$ is

$$f''''(x) = \left(\frac{d}{dx}\right)^4 \log_{10}(x) = \frac{\left(\dfrac{d}{dx}\right)^4 \log_e(x)}{\log_e(10)}$$

$$= -\frac{6}{x_4 \log_e(10)}$$

Thus, $e(x)$ for $x = 1.5$, 2.5, and 3.5 becomes as follows, with comparison to exact values:

x	$e(x)$	Exact
1.5	0.0026	0.0053
2.5	−0.0015	−0.0021
3.5	0.0026	0.0026

Agreements of the estimates of errors to exact values are good. The discrepancies between estimates and the exact values arise from the approximation of $f''''(\xi)$ by $f''''(2.5)$.

It often becomes necessary to write an interpolation formula in the power series form. Unfortunately, the Lagrange interpolation formula is not suitable to derive a power series form, because expanding the Lagrange interpolation into a power series is cumbersome. A better approach, particularly if grid points are equispaced, is to use the Newton forward interpolation formula and then transform it to the power series form by using the Markov coefficients (see Sections 2.5 and 7.3 for more details).

SUMMARY OF THIS SECTION

(a) A polynomial of order N fitted to $N + 1$ data points is unique.

(b) The interpolation polynomial can be expressed in various different forms, among which we studied the power series form and the Lagrange interpolation formula. All of them are mathematically equal because of the uniqueness theorem.

(c) When abscissas of data points are equispaced, the amplitude of oscillating error of the interpolation tends to be smallest at the center or nearest to the center. The amplitude of oscillation increases toward the edges.

(d) If a function is approximated by an interpolation polynomial, there is no guarantee that the interpolation polynomial converges to the exact function when the number of data points is increased. In general, interpolation with a large-order polynomial should be avoided or used with extreme cautions.

(e) Although there is no criterion to determine the optimum order of interpolation polynomial, a general recommendation is to use a relatively low-order polynomial in a small range of x.

2.4 NEWTON FORWARD AND BACKWARD INTERPOLATIONS ON EQUISPACED POINTS

In the previous section, we studied the Lagrange interpolation formula and analyzed its error on equispaced points. The Lagrange interpolation formula itself works equally for both equispaced and nonequispaced points. The drawbacks of the Lagrange interpolation are, however, the following:

(a) The amount of computation needed for one interpolation is large.

(b) Interpolation for another value of x needs the same amount of additional computations since no part of the previous application can be used.

(c) When the number of data points has to be increased or decreased, the results of the previous computations cannot be used.

(d) Evaluation of error is not easy.

Use of the Newton interpolation formulas alleviates these difficulties.

To write a Newton interpolation for a given data set, one has to develop a difference table. Once a difference table is developed, interpolation formulas passing through different sets of consecutive data points (such as $i = 0, 1, 2, 3$ or $i = 3, 4, 5, 6$ or $i = 2, 3, 4$ etc.) can be written very easily. Therefore, the order of an interpolation polynomial can be readily increased with additional data points. Error of the Newton interpolation formula can also be easily estimated. The Newton interpolation is more suitable than the Lagrange interpolation in deriving other numerical schemes, such as difference approximations of derivatives (see Section 5.4), or in expanding an interpolation into a power series (see Section 7.3).

In the remainder of this section, two versions of the Newton interpolation—the Newton forward interpolation and the Newton backward interpolation—are discussed. Both are mathematically equivalent but different expressions. One expression can be more convenient than the other, depending on how the formula is applied. For example, the latter is preferred in deriving predictor-corrector method in Section 9.4 because data points are all in backward positions. However, in fitting data given in a table, the Newton forward interpolation is generally more convenient.

2.4.1 Forward Difference Table and Binomial Coefficients

We assume that abscissas of data points are equispaced with interval size h. The data points will be denoted by (x_i, f_i).

To evaluate a Newton forward interpolation formula, a forward difference table and binomial coefficients are necessary [Gerald/Wheatley]. Therefore, we first define forward differences by

$$\Delta^0 f_i = f_i \qquad \text{(zero-th order forward difference)} \qquad (2.4.1)$$

$$\Delta f_i = f_{i+1} - f_i \qquad \text{(first-order forward difference)} \qquad (2.4.2)$$

$$\Delta^2 f_i = \Delta f_{i+1} - \Delta f_i \qquad \text{(second-order forward difference)} \qquad (2.4.3)$$

$$\Delta^3 f_i = \Delta^2 f_{i+1} - \Delta^2 f_i \qquad \text{(third-order forward difference)} \qquad (2.4.4)$$

$$\vdots$$

$$\Delta^k f_i = \Delta^{k-1} f_{i+1} - \Delta^{k-1} f_i \quad \text{(kth-order forward difference)} \qquad (2.4.5)$$

A high-order difference may be easily derived by using the shift operator (see Appendix C for details).

The difference table, illustrated in Table 2.2, is a convenient means of evaluating the differences for a given data set. In Table 2.2, the first column is the index of the data points, the second is the ordinates of data. The third column lists the first-order differences, which are calculated from the second column. The fourth column lists the second-order differences calculated from the previous column, and so on. Each row provides a set of forward differences for the corresponding data points.

Table 2.2 Difference table

i	f_i	Δf_i	$\Delta^2 f_i$	$\Delta^3 f_i$	$\Delta^4 f_i$	$\Delta^5 f_i$
0	f_0	Δf_0	$\Delta^2 f_0$	$\Delta^3 f_0$	$\Delta^4 f_0$	$\Delta^5 f_0$
1	f_1	Δf_1	$\Delta^2 f_1$	$\Delta^3 f_1$	$\Delta^4 f_1$	
2	f_2	Δf_2	$\Delta^2 f_2$	$\Delta^3 f_2$		
3	f_3	Δf_3	$\Delta^2 f_3$			
4	f_4	Δf_4				
5	f_5					

One should know the following about the difference table. If f_i are taken as $f_i = f(x_i)$ where $f(x)$ is a polynomial of order, say L, and x_i are equispaced, then the column for the Lth order difference become entirely a constant, and the next column $((L + 1)$-th difference) becomes zero. If this happens, we know that the data set belongs to a Lth order polynomial. However, if a column of differences has one or more of abnormally large value, the chance is that there are some human errors in the process of developing the table or in the data set.

Binomial coefficients are given by

$$\binom{s}{0} = 1$$

$$\binom{s}{1} = s$$

$$\binom{s}{2} = \frac{1}{2!} s(s-1)$$

$$\binom{s}{3} = \frac{1}{3!} s(s-1)(s-2)$$

$$\vdots$$

$$\binom{s}{n} = \frac{1}{n!} s(s-1)(s-2) \cdots (s-n+1)$$

where s is a local coordinate defined by $s = (x - x_0)/h$ and h is the uniform grid interval.

Example 2.3

Develop a forward difference table for the data set given next:

i	0	1	2	3	4	5	6
x_i	0.1	0.3	0.5	0.7	0.9	1.1	1.3
$f(x_i)$	0.99750	0.97763	0.93847	0.88120	0.80752	0.71962	0.62009

⟨**Solution**⟩

The forward difference table is as follows:

i	x_i	f_i	Δf_i	$\Delta^2 f_i$	$\Delta^3 f_i$	$\Delta^4 f_i$	$\Delta^5 f_i$	$\Delta^6 f_i$
0	0.1	0.99750	−0.01987	−0.01929	0.00118	0.00052	−0.00003	−0.00006
1	0.3	0.97763	−0.03916	−0.01811	0.00170	0.00049	−0.00009	
2	0.5	0.93847	−0.05727	−0.01641	0.00219	0.00040		
3	0.7	0.88120	−0.07368	−0.01422	0.00259			
4	0.9	0.80752	−0.08790	−0.01163				
5	1.1	0.71962	−0.09953					
6	1.3	0.62009						

Comment: The higher-order differences tend to vanish but may not become exactly zero. Often the cause is due to round-off error in the data. So, this can happen even when the data belong to a low-order polynomial.

2.4.2 The Newton Forward Interpolation Formula

The Newton forward interpolation formula that passes through $k + 1$ data points, $f_0, f_1, f_2, \ldots, f_k$, is written as

$$g(x) = g(x_0 + sh) = \sum_{n=0}^{k} \binom{s}{n} \Delta^n f_0 \qquad (2.4.6)$$

When $k = 2$ for example, Eq. (2.4.6) becomes

$$g(x_0 + sh) = f_0 + s(f_1 - f_0) + \frac{s(s-1)}{2}(f_2 - 2f_1 + f_0) \qquad (2.4.7a)$$

or equivalently

$$g(x_0 + sh) = f_0 + (sh)\frac{-f_2 + 4f_1 - 3f_0}{2h} + \frac{(sh)^2}{2}\frac{f_2 - 2f_1 + f_0}{h^2} \qquad (2.4.7b)$$

Equation (2.4.6) is a polynomial of order k because the $\binom{s}{n}$ is a polynomial of order n, and its highest order is k. Equation (2.4.6) is shown to equal $f_0, f_1, f_2, \ldots, f_k$ respectively at $x = x_0, x_1, \ldots, x_k$ as follows:

$$s = 0: \quad g(x_0) = g(x_0 + 0) = f_0$$

$$s = 1: \quad g(x_1) = g(x_0 + h) = f_0 + \Delta f_0 = f_1$$

$$s = 2: \quad g(x_2) = g(x_0 + 2h) = f_0 + 2\Delta f_0 + \Delta^2 f_0 = f_2 \qquad (2.4.8)$$

$$\vdots$$

$$s = k: \quad g(x_k) = g(x_0 + kh) = f_0 + k\Delta f_0 + \frac{k(k-1)}{2}\Delta^2 f_0 + \ldots = f_k$$

The first $m + 1$ terms of Eq. (2.4.6) form an interpolation polynomial of order m fitted to $m + 1$ data points at $x_0, x_1, x_2, \ldots, x_m$. Likewise, the first $m + 2$ terms comprise an interpolation polynomial of order $m + 1$ fitted to $m + 2$ data points. Thus, the order of an interpolation polynomial can be changed easily by changing the number of differences taken from the first row of Table 2.2.

If x_0 and f_0 in Eq. (2.4.6) are replaced respectively by x_2 and f_2, for example, it becomes

$$g(x_2 + sh) = \sum_{n=0}^{k} \binom{s}{n} \Delta^n f_2 \qquad (2.4.9)$$

where s is defined by $s = (x - x_2)/h$ and is a local coordinate. The s value becomes 0 at $x = x_2$, and 1, 2, 3, $\ldots$ at $x = x_3, x_4, x_5, \ldots$, respectively. Equation (2.4.9) is a polynomial of order k fitted at $x_2, x_3, \ldots, x_{k+2}$ and uses the differences along the third row of in Table 2.2. This illustrates that, once a difference table like Table 2.2 is developed, interpolation formulas fitted to different sets of data points may be readily obtained.

Example 2.4

Derive the Newton forward interpolation polynomials fitted to the data points at (a) $i = 0, 1, 2$, (b) $i = 0, 1, 2, 3, 4$, (c) $i = 2, 3, 4$, and (d) $i = 4, 5, 6$ given in the following table:

i	0	1	2	3	4	5	6
x_i	0.1	0.3	0.5	0.7	0.9	1.1	1.3
$f(x_i)$	0.99750	0.97763	0.93847	0.88120	0.80752	0.71962	0.62009

⟨**Solution**⟩

The difference table is already developed in Example 2.3.

(a) The Newton forward interpolation passing through points $i = 0, 1, 2$ is derived by using the three values on the row corresponding to $i = 0$ in the difference table of Example 2.3 and is written as

$$y = 0.99750 - 0.01987s - \frac{0.01929}{2} s(s - 1)$$

$$s = \frac{x - x_0}{h}$$

(b) $$y = 0.99750 - 0.01987s - \frac{0.01929}{2} s(s - 1) + \frac{0.00118}{6} s(s - 1)(s - 2)$$

$$+ \frac{0.00052}{24} s(s - 1)(s - 2)(s - 3)$$

$$s = \frac{x - x_0}{h}$$

(c) $$y = 0.93847 - 0.05727s - \frac{0.01641}{2} s(s - 1)$$

$$s = \frac{x - x_2}{h}$$

(d) $$y = 0.80752 - 0.08790s - \frac{0.01163}{2} s(s - 1)$$

$$s = \frac{x - x_4}{h}$$

Because of the equivalence between the Newton interpolation formulas and the Lagrange interpolation formulas, the error of the Newton interpolation polynomial must be identical to that for the Lagrange interpolation formula. So, it can be written as

$$e(x) = f(x) - g(x) = L(x)f^{(N+1)}(\xi) \quad x_0 < \xi < x_N \tag{2.4.10}$$

where $f(x)$ is the exact function and $g(x)$ is the Newton interpolation. However, evaluation of Eq. (2.4.10) for Newton interpolation is much easier than for the Lagrange interpolation.

Consider Eq. (2.4.6) with $k = N$. If k is increased from N to $N + 1$, the additional term is

$$
\binom{s}{N+1}\Delta^{N+1}f_0 = \frac{s(s-1)(s-2)\cdots(s-N)}{(N+1)!}\Delta^{N+1}f_0
$$

$$
= \frac{(x-x_0)(x-x_1)\cdots(x-x_N)}{(N+1)!} \times \frac{\Delta^{N+1}f_0}{h^{N+1}}
$$

(2.4.11)

where $s = (x - x_0)/h$ and $x_n = x_0 + nh$ are used. The second factor of Eq. (2.4.11) may be shown to be an approximation to $f^{(N+1)}$, namely

$$
\Delta^{N+1}f_0/h^{N+1} \simeq f^{(N+1)}(x_m)
$$

where

$$
x_m = \frac{1}{2}(x_0 + x_N)
$$

Therefore, Eq. (2.4.11) is approximately equal to the right side of Eq. (2.4.10). That is, the error is represented by the next term that comes if the order of the polynomial were increased by one with an additional point, x_{N+1}.

What can we do if the next point is not available? In this case, check if an additional point is available on the other side, namely $f(x_{-1})$. If it is available, $\Delta^{N+1}f_{-1}$ can be computed and used as an approximation for $\Delta^{N+1}f_0$.

Example 2.5

Evaluate the error of Eq. (a) in Example 2.4 for $x = 0.2$.

⟨**Solution**⟩

Equation (a) in Example 2.4 is fitted at $i = 0, 1, 2$. So, the additional term that comes if the interpolation is also fitted at $i = 3$ is

$$
\frac{0.00118}{6}s(s-1)(s-2)
$$

Therefore, introducing $s = (x - x_0)/h = (x - 0.1)/0.2 = 0.5$ for $x = 0.2$, the error is

$$
e(x) \simeq \frac{0.00118}{6}s(s-1)(s-2) = 7.4 \times 10^{-5}
$$

Compare this to the actual error, 4.4×10^{-5}.

2.4.3 Newton Backward Interpolation

The Newton backward interpolation polynomial is another frequently used formula and is written in terms of backward differences and binomial coefficients. We consider equispaced points, $x_0, x_{-1}, x_{-2}, \ldots, x_{-k}$, with a constant interval of $h = x_i - x_{i-1}$.
Backward differences are defined by

$$\nabla^0 f_i = f_i \qquad \text{(zero-th order backward difference)}$$

$$\nabla f_i = f_i - f_{i-1} \qquad \text{(first-order backward difference)}$$

$$\nabla^2 f_i = \nabla f_i - \nabla f_{i-1} \qquad \text{(second-order backward difference)}$$

$$\nabla^3 f_i = \nabla^2 f_i - \nabla^2 f_{i-1} \qquad \text{(third-order backward difference)}$$

$$\vdots$$

$$\nabla^k f_i = \nabla^{k-1} f_i - \nabla^{k-1} f_{i-1} \qquad (k\text{th-order backward difference)}$$

A backward difference table may be developed as shown in Example 2.6.

Example 2.6

Develop a backward difference table for the same data set as given in Example 2.4:

i	0	1	2	3	4	5	6
x_i	0.1	0.3	0.5	0.7	0.9	1.1	1.3
$f(x_i)$	0.99750	0.97763	0.93847	0.88120	0.80752	0.71962	0.62009

⟨Solution⟩

The backward difference table is as follows:

i	x_i	f_i	∇f_i	$\nabla^2 f_i$	$\nabla^3 f_i$	$\nabla^4 f_i$	$\nabla^5 f_i$	$\nabla^6 f_i$
0	0.1	0.99750						
1	0.3	0.97763	−0.01987					
2	0.5	0.93847	−0.03916	−0.01929				
3	0.7	0.88120	−0.05727	−0.01811	0.00118			
4	0.9	0.80752	−0.07368	−0.01641	0.00170	0.00052		
5	1.1	0.71962	−0.08790	−0.01422	0.00219	0.00049	−0.00003	
6	1.3	0.62009	−0.09953	−0.01163	0.00259	0.00040	−0.00009	−0.00006

The binomial coefficients used in the Newton backward interpolations are as follows:

$$\binom{s-1}{0} = 1$$

$$\binom{s}{1} = s$$

$$\binom{s+1}{2} = \frac{1}{2!}(s+1)s$$

$$\binom{s+2}{3} = \frac{1}{3!}(s+2)(s+1)s$$

$$\vdots$$

$$\binom{s+n-1}{n} = \frac{1}{n!}(s+n-1)(s+n-2)\cdots(s+1)s$$

The Newton backward interpolation fitted to the data points at $x = x_j, x = x_{j-1}$, $x = x_{j-2}, \ldots$, and $x = x_{j-k}$ is written as

$$g(x) = g(x_j + sh) = \sum_{n=0}^{k} \binom{s+n-1}{n} \nabla^n f_j, \quad -k \leqslant s \leqslant 0 \qquad (2.4.12)$$

where s is a local coordinate defined by $s = (x - x_j)/h$; $\binom{s+n-1}{n}$ is a binomial coefficient and $\nabla^n f_j$ is the backward difference.

An equivalence relation between the forward difference and the backward difference is given by

$$\nabla^n f_j = \Delta^n f_{j-n} \qquad (2.4.13)$$

Therefore, Eq. (2.4.12) may be expressed in terms of the forward differences as

$$g(x) = \sum_{n=0}^{k} \binom{s+n-1}{n} \Delta^n f_{j-n}, \quad -k \leqslant s \leqslant 0 \qquad (2.4.14)$$

or more explicitly

$$g(x) = g(x_j + sh) = f_j + s(f_j - f_{j-1}) + \frac{1}{2}(s+1)s(f_j - 2f_{j-1} + f_{j-2})$$

$$+ \frac{1}{6}(s+2)(s+1)s(f_j - 3f_{j-1} + 3f_{j-2} - f_{j-3}) + \cdots$$

$$+ \frac{1}{k!}(s+k-1)(s+k-2)\cdots(s+1)s\Delta^k f_{j-k} \qquad (2.4.15)$$

Example 2.7

Determine the Newton backward interpolation polynomial fitted at the three points, $i = 3, 4, 5$, in the function table of Example 2.6.

⟨**Solution**⟩

Because the number of points is 3, the order of the polynomial is 2. The Newton backward interpolation polynomial given by Eq. (2.4.12) becomes, for this case,

$$g(x) = g(x_5 + sh) = \sum_{n=0}^{2} \binom{s + n - 1}{n} \nabla^n f_5$$

$$= f_5 + s\nabla f_5 + \frac{1}{2}(s + 1)s\nabla^2 f_5, \quad -2 \leqslant s \leqslant 0$$

where $s = (x - x_5)/h$. Using the values of f_5, ∇f_5, and $\nabla^2 f_5$ in the difference table of Example 2.6, the foregoing equation becomes

$$g(x) = 0.71962 - 0.08790s - \frac{0.01422}{2}(s + 1)s$$

or equivalently, by using $s = (x - x_5)/h$,

$$g(x) = 0.71962 - \frac{0.08790(x_5 - x)}{h} - \frac{0.00711(x_5 - x)(x_4 - x)}{h^2}$$

SUMMARY OF THIS SECTION

(a) Coefficients of Newton forward and backward interpolations are evaluated by developing a difference table.

(b) The Lth order difference column in the table becomes constant and the $(L + 1)$th difference column becomes zero if the data set belongs to a polynomial of order L. However, because of round-off errors in the data set, the $(L + 1)$th order difference may not vanish exactly.

(c) The Newton interpolation polynomials are equal to the Lagrange interpolation formula if the same data set is used.

(d) Error in the Newton interpolation formulas is represented by the additional term that comes with one more data point.

2.5 NEWTON INTERPOLATION ON NONUNIFORMLY SPACED POINTS

The Newton interpolation formulas described in the previous section are restricted to equispaced points. The need to write an interpolation polynomial for nonuniformly spaced points occurs often, however. The Newton interpolation scheme may be extended to nonuniformly spaced points by using the divided differences [Isaacson/ Keller; Carnahan/Luther/Wilkes]. So, the Lagrange interpolation polynomial on a nonuniformly spaced grid may be equivalently expressed in the form of a Newton interpolation polynomial.

Let us denote the Lagrange interpolation polynomial fitted at $x_0, x_1, x_2, \ldots, x_m$ as

$$P_{0,1,2,\ldots,m}(x)$$

and the one fitted at $x_1, x_2, \ldots, x_{m+1}$ as

$$P_{1,2,3,\ldots,m+1}(x)$$

The number of subscripts of $P_{a,b,c,\ldots,j}$ less one is the order of the interpolating polynomial, so the two polynomials just given are both of order m. Then, it is easily seen that the polynomial fitted at $x_0, x_1, \ldots, x_{m+1}$ is given by

$$P_{0,1,2,\ldots,m+1}(x) = \frac{(x - x_0)P_{1,2,\ldots,m+1}(x) + (x_{m+1} - x)P_{0,1,2,\ldots,m}(x)}{x_{m+1} - x_0} \quad (2.5.1)$$

If Eq. (2.5.1) is expanded into a power series, the coefficient of the highest order term is called the *leading coefficient*. The leading coefficient of $P_{a,b,c,\ldots,j}$ is denoted by $f_{a,b,c,\ldots,j}$ using the same subscripts. Applying this rule, the leading coefficient for $P_{0,1,2,\ldots,m+1}$ is $f_{0,1,2,\ldots,m+1}$. Likewise, the leading coefficients for $P_{0,1,2,\ldots,m}$ and $P_{1,2,\ldots,m+1}$ are respectively $f_{0,1,2,\ldots,m}$ and $f_{1,2,\ldots,m+1}$. By inspection of Eq. (2.5.1), its leading coefficient for the left side is related to those for the two interpolation polynomials on the right side by

$$f_{0,1,2,\ldots,m+1} = \frac{f_{1,2,\ldots,m+1} - f_{0,1,2,\ldots,m}}{x_{m+1} - x_0} \quad (2.5.2)$$

Equation (2.5.2) is a "divided difference" of order $m + 1$ because it is given by the difference of the leading coefficients of order m divided by the distance between the outermost points. Using Eq. (2.5.2), the leading coefficients may be recursively calculated starting with a function table, as symbolically shown in Table 2.3.

Using the difference table, an interpolation polynomial—$P_{a,b,c,\ldots,j}$ for example—can be derived as

$$P_{a,b,c,\ldots,j}(x) = f_a + f_{a,b}(x - x_a) + f_{a,b,c}(x - x_a)(x - x_b) + \cdots$$
$$+ f_{a,b,c,\ldots,j}(x - x_a)(x - x_b)(x - x_c) \cdots (x - x_{j-1}) \quad (2.5.3)$$

Error of the interpolation is evaluated in exactly the same manner as is the Newton interpolation on an equally spaced grid. Indeed, the error of Eq. (2.5.3) approximately

Table 2.3 Divided difference table

x_0	f_0	$f_{0,1} = \dfrac{f_1 - f_0}{x_1 - x_0}$	$f_{0,1,2} = \dfrac{f_{1,2} - f_{0,1}}{x_2 - x_0}$	$f_{0,1,2,\ldots,m+1} = \dfrac{f_{1,2,\ldots,m+1} - f_{0,1,\ldots,m}}{x_{m+1} - x_0}$
x_1	f_1	$f_{1,2} = \dfrac{f_2 - f_1}{x_2 - x_1}$	$f_{1,2,3} = \dfrac{f_{2,3} - f_{1,2}}{x_3 - x_1}$	
x_2	f_2	$f_{2,3} = \dfrac{f_3 - f_2}{x_3 - x_2}$	$f_{2,3,4} = \dfrac{f_{3,4} - f_{2,3}}{x_4 - x_2}$	
x_m	f_m	$f_{m,m+1} = \dfrac{f_{m+1} - f_m}{x_{m+1} - x_m}$		

equals the term that would be added to Eq. (2.5.3) if the interpolation were extended to fit to one more point, $j + 1$: that is, the error is

$$e(x) \simeq f_{a,b,c,\ldots,j,j+1}(x - x_a)(x - x_b)(x - x_c)\cdots(x - x_j)$$

Example 2.8

(a) Develop a divided difference table for the following data:

i	x_i	f_i
0	0.1	0.99750
1	0.2	0.99002
2	0.4	0.96040
3	0.7	0.88120
4	1.0	0.76520
5	1.2	0.67113
6	1.3	0.62009

(b) Write the interpolation formula using the divided differences fitted at points (1) $i = 0$ through 6, and (2) $i = 2$ through 4,

(c) Derive an estimate for the error of the interpolations.

(d) Evaluate the interpolating polynomials for $x = 0.3$ and $x = 0.55$.

(e) Estimate errors of the interpolation for the two values in item (d), and compare them with the exact values of errors. (The exact values are $f(0.3) = 0.97763$ and $f(0.55) = 0.92579$ respectively.)

⟨**Solution**⟩

The divided difference table is developed first.

i	x_i	f_i	$f_{i,i+1}$	$f_{i,\ldots,i+2}$	$f_{i,\ldots,i+3}$	$f_{i,\ldots,i+4}$	$f_{i,\ldots,i+5}$	$f_{i,\ldots,i+6}$
0	0.1	0.99750	−0.07480	−0.24433	0.02088	0.01478	−0.00236	0.00122
1	0.2	0.99002	−0.14810	−0.23180	0.03418	0.01218	−0.00090	
2	0.4	0.96040	−0.26400	−0.20445	0.04636	0.01119		
3	0.7	0.88120	−0.38667	−0.16736	0.05643			
4	1.0	0.76520	−0.47035	−0.13350				
5	1.2	0.67113	−0.51040					
6	1.3	0.62009						

(i) The polynomial fitted at $i = 0$ through 5 is then

$$P_{0,1,...,5}(x) = 0.99750 - 0.07480(x - 0.1) - 0.24433(x - 0.1)(x - 0.2)$$
$$+ 0.02088(x - 0.1)(x - 0.2)(x - 0.4)$$
$$+ 0.01478(x - 0.1)(x - 0.2)(x - 0.4)(x - 0.7)$$
$$- 0.00236(x - 0.1)(x - 0.2)(x - 0.4)(x - 0.7)(x - 1)$$

An estimate for the error of this interpolation is

$$e(x) = 0.00122(x - 0.1)(x - 0.2)(x - 0.4)(x - 0.7)(x - 1)(x - 1.2)$$

Calculated results are summarized below:

	$P_{0,1,...,5}$	Error (Estimate)	Error (Exact)
$x = 0.3$	0.97762	$6.17E - 7$	$5.2E - 6$
$x = 0.55$	0.92580	$-1.26E - 7$	$-5.2E - 6$

The errors shown above need some consideration. The exact values for $f(0.3)$ and $f(0.55)$ are given only to the fifth decimal places, so they are subject to round-off errors of at most $\pm 5.0E - 6$. The exact errors shown above, therefore, have no meaning except as the effect of round-off error of a subtraction. On the other hand, estimated errors are smaller than these values. Thus, we conclude that the results of interpolation are exact within the round-off errors.

(ii) The polynomial fitted at $i = 2$ through 4 is

$$P_{1,2,3,4}(x) = 0.96040 - 0.26400(x - 0.4) - 0.20445(x - 0.4)(x - 0.7)$$

Estimated error of this interpolation is approximately

$$e(x) = 0.04636(x - 0.4)(x - 0.7)(x - 1)$$

Calculated values are summarized below:

	$P_{1,2,3,4}$	Error (Estimated)	Error (Exact)
$x = 0.3$	0.97862	-0.00130	-0.00099
$x = 0.55$	0.92540	0.00047	0.00039

The estimated errors agree well with the exact errors.

SUMMARY OF THIS SECTION

(a) The Newton interpolation formula with divided differences is a variation of the Newton forward interpolation for equispaced data points.

(b) It is applicable with both nonequispaced data points and equispaced data points.

(c) Error of the interpolation formula is represented by the additional term that comes with one more data point.

2.6 INTERPOLATION WITH CHEBYSHEV ROOTS

As mentioned in the previous section, the polynomial interpolation using equally spaced data points, whether expressed in the Lagrange interpolation formula or a

Newton interpolation polynomial, is most accurate in the middle range of the interpolation domain but the error of the interpolation increases toward the edges. This is attributable to the behavior of $L(x)$ in Eq. (2.3.7).

The scheme described in this section determines the data points by a Chebyshev polynomial [Carnahan/et al.; Abramowitz/Stegun]. The spacings determined by a Chebyshev polynomial are largest at the center of the interpolation domain and decrease toward the edges. As a result, errors become more evenly distributed throughout the domain, and their magnitudes become less than with equally spaced points. Interpolation with Chebyshev points is widely used in mathematical subroutines as well as in general numerical computations.

The Chebyshev polynomials may be expressed in two different but equivalent ways: one using cosine functions and the other in the power series. In the first expression, the normalized Chebyshev polynomial of order K is defined by

$$T_K(x) = \cos(K \cos^{-1}(x)), \quad -1 \leqslant x \leqslant 1 \tag{2.6.1}$$

The Chebyshev polynomials in the power series are given by

$$
\begin{aligned}
T_0(x) &= 1 \\
T_1(x) &= x \\
T_2(x) &= 2x^2 - 1 \\
T_3(x) &= 4x^3 - 3x \\
T_4(x) &= 8x^4 - 8x^2 + 1 \\
T_5(x) &= 16x^5 - 20x^3 + 5x \\
T_6(x) &= 32x^6 - 48x^4 + 18x^2 - 1
\end{aligned}
\tag{2.6.2}
$$

The Chebyshev polynomials of any higher order in the power series can be generated by using the recursive relation,

$$T_j(x) = 2x T_{j-1}(x) - T_{j-2}(x) \tag{2.6.3}$$

The cosine form of Chebyshev polynomials in Eq. (2.6.1) indicates that their local minimums and maximums in $-1 \leqslant x \leqslant 1$ are -1 and 1, respectively. Notice also that all the Chebyshev polynomials become 1 at $x = 1$, and $+1$ or -1 at $x = -1$ as illustrated in Figure 2.4. Since the cosine function becomes zero at $\pm \pi/2, \pm 3\pi/2, \ldots$, the zeroes of a Chebyshev polynomial of order K satisfy

$$K \cos^{-1}(x_n) = \left(K + \frac{1}{2} - n\right)\pi, \quad n = 1, 2, \ldots, K \tag{2.6.4}$$

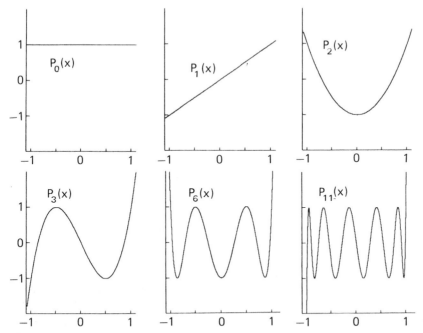

Figure 2.4 Chebyshev polynomials

or more explicitly,

$$x_n = \cos\left(\frac{K + 1/2 - n}{K}\pi\right), \quad n = 1, 2, \ldots, K \tag{2.6.5}$$

If $K = 3$ for example, x_n for $n = 1$, 2, and 3 are -0.86602, 0, $+0.86602$, respectively.

Assuming that the range of an interpolation is $[-1, 1]$, the K zeroes x_n, $i = 1$, 2, ..., K, may be used as the abscissas of points in the Lagrange interpolation instead of using equally spaced points. Notice, however, that the numbering of points in obtaining Chebyshev points and those in the Lagrange interpolation formula of Eq. (2.3.3) are different. If the three Chebyshev points of $K = 3$ as illustrated in the previous paragraph are used, the order of the Lagrange interpolation formula is $N = 2$, and x_i in Eq. (2.3.3) are $x_0 = -0.86602$, $x_1 = 0$, and $x_2 = +0.86602$. The ordinates of end points, namely at $x = -1$ and $x = +1$, are not used. Therefore, the Lagrange interpolation formula will be used as "extrapolation" in $[-1, -0.86602]$, as well as $[+0.86602, 1]$.

The Chebyshev polynomial interpolation can be applied to any range other than $[-1, 1]$ by mapping $[-1, 1]$ onto the range of interest. Writing the range of interpolation as $[a, b]$, the mapping is given by

$$x = \frac{2z - a - b}{b - a} \tag{2.6.6}$$

or equivalently

$$z = \frac{(b - a)x + a + b}{2} \tag{2.6.7}$$

where

$$-1 \leqslant x \leqslant 1 \quad \text{and} \quad a \leqslant z \leqslant b.$$

Therefore, substituting the Chebyshev points x_n in $[-1, 1]$ given by Eq. (2.6.5) into Eq. (2.6.7), the Chebyshev points z_n in $[a, b]$ become

$$z_n = \frac{1}{2}\left[(b - a)\cos\left(\frac{K + \frac{1}{2} - n}{K}\pi\right) + a + b\right], \quad n = 1, 2, \ldots, K \tag{2.6.8}$$

The error of an interpolation using Chebyshev roots is also given by Eq. (2.3.7). The behavior of $L(x)$, however, becomes different from that with equispaced points. Indeed, $L(x)$ itself becomes a Chebyshev polynomial because it passes through the roots of the Chebyshev polynomial. Consequently, the error of the interpolation with Chebyshev roots is more evenly distributed than with equispaced points. The actual error distribution $e(x)$ deviates from the Chebyshev polynomial, however, since ξ depends on x.

Example 2.9

    (a) Find the three Chebyshev points in $2 \leqslant z \leqslant 4$.
    (b) Using the three Chebyshev points, write the Lagrange interpolation formula fitted to ln (z).

⟨Solution⟩

    (a) By introducing $a = 2$, $b = 4$ and $K = 3$ into Eq. (2.6.8) and setting n to 1, 2, and 3, the Chebyshev points are found as

$$z_1 = 2.13397$$
$$z_2 = 3$$
$$z_3 = 3.86602$$

(b) We now develop a function table with the Chebyshev points as follows:

z	y = ln (z)
2.13397	0.757984
3	1.098612
3.86602	1.352226

The Lagrange interpolation formula fitted to the data set is

$$g(z) = \frac{(z - 3)(z - 3.86602)}{(2.13397 - 3)(2.13397 - 3.86602)} (0.757984)$$

$$+ \frac{(z - 2.13397)(z - 3.86602)}{(3 - 2.13397)(3 - 3.86602)} (1.098612)$$

$$+ \frac{(z - 2.13397)(z - 3)}{(3.86602 - 2.13397)(3.86602 - 3)} (1.352226)$$

SUMMARY OF THIS SECTION

(a) Chebyshev points are roots of a Chebyshev polynomial.

(b) A Kth order Chebyshev polynomial provides K Chebyshev points. The Lagrange interpolation formula using K points is a polynomial of order $K - 1$.

(c) The function $L(x)$ in representation of error given by Eq. (2.3.9) then becomes a Chebyshev polynomial of order K.

(d) Using Chebyshev points in Lagrange interpolation, error is more evenly distributed than with equispaced points.

2.7 HERMITE INTERPOLATION POLYNOMIALS

The polynomial interpolation schemes discussed earlier in this chapter do not use information of the derivative of the function fitted. However, a polynomial can be fitted not only to functional values but also to the derivatives given at data points. The polynomials fitted to functional and derivative values are named *Hermite interpolation polynomials*, or *osculating polynomials* [Isaacson/Keller].

Suppose data points are $x_0, x_1, \ldots, x_N$, and functional values and all derivatives up to pth order $(f_i, f_i', \ldots, f_i^{(p)}, i = 0, 1, \ldots, N)$ are given. The total number of data is $K = (p + 1)(N + 1)$. A polynomial of order $K - 1$, namely,

$$g(x) = \sum_{j=0}^{K-1} a_j x^j \qquad (2.7.1)$$

may be fitted to the K data, where a_j is a coefficient. By equating Eq. (2.7.1) to the data, we get a set of $K = (p + 1)(N + 1)$ equations

$$g(x_i) = f_{i'} \qquad i = 0, 1, \ldots, N$$

$$g'(x_i) = f_{i'}' \qquad i = 0, 1, \ldots, N$$

$$\vdots \qquad\qquad\qquad\qquad (2.7.2)$$

$$g^{(p)}(x_i) = f_i^{(p)} \qquad i = 0, 1, \ldots, N$$

The coefficients may be determined by solving Eq. (2.7.2) in a closed form if K is small.

An alternative expression similar to the Lagrange interpolation formula may be written as

$$g(x) = \sum_{i=0}^{N} \alpha_i(x)f_i + \sum_{i=0}^{N} \beta_i(x)f'_i + \cdots + \sum_{i=0}^{N} \theta_i(x)f_i^{(p)} \qquad (2.7.3)$$

Here,

$$\alpha_i(x_j) = \delta_{i,j} \qquad (2.7.4)$$

and all derivatives of $\alpha_i(x)$ are zero for every $x = x_i$; $\beta_i(x)$ and all its derivatives are zero for every $x = x_i$ except

$$\left[\frac{d}{dx}\beta_i(x)\right]_{x=x_j} = \delta_{i,j} \qquad (2.7.5)$$

Similarly $\theta_i(x)$ and all of its derivatives are zero for every $x = x_j$ except

$$\left[\frac{d^p}{dx^p}\theta_i(x)\right]_{x=x_j} = \delta_{i,j} \qquad (2.7.6)$$

Equation (2.7.3) is, indeed, an extension of the Lagrange interpolation formula. It reduces to the Lagrange interpolation if no derivative is fitted.

Example 2.10

Suppose a function table has the functional values and the first derivative values. Derive a polynomial for each interval that is fitted to the functional values and the first derivatives at the end points of the interval.

⟨Solution⟩

For each interval, the total number of data is four so the order of the polynomial is three. The polynomial is called a *cubic Hermite polynomial*.

Consider an interval between x_{i-1} and x_i as shown in Figure E2.10. The cubic polynomial that fits to f_{i-1}, f'_{i-1}, f_i, and f'_i, is written as

$$y(t) = a + bt + ct^2 + et^3 \qquad (A)$$

where a local coordinate $t = x - x_{i-1}$ is used. Fitting Eq. (A) to the given data

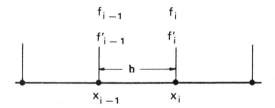

Figure E2.10 An interval for Hermite Interpolation

yields

$$f_{i-1} = a \qquad (B)$$

$$f'_{i-1} = b \qquad (C)$$

$$f_i = a + bh + ch^2 + eh^3 \qquad (D)$$

$$f'_i = b + 2ch + 3eh^2 \qquad (E)$$

where $h = x_i - x_{i-1}$. Introducing Eqs. (B) and (C) into Eqs. (D) and (E), and then solving for c and e, yield

$$c = \frac{3(f_i - f_{i-1}) - (f'_i + 2f'_{i-1})h}{h^2} \qquad (F)$$

$$e = \frac{-2(f_i - f_{i-1}) + (f'_i + f'_{i-1})h}{h^3}$$

Thus, the cubic Hermite interpolation polynomial becomes

$$y(t) = f_{i-1} + f'_{i-1}t + [3(f_i - f_{i-1}) - (f'_i + 2f'_{i-1})h]\left(\frac{t}{h}\right)^2 \qquad (G)$$

$$+ [-2(f_i - f_{i-1}) + (f'_i + f'_{i-1})h]\left(\frac{t}{h}\right)^3$$

Equation (G) may be alternatively expressed by

$$y(t) = \alpha_{i-1}f_{i-1} + \alpha_i f_i + \beta_{i-1}f'_{i-1} + \beta_i f'_i \qquad (H)$$

where

$$\alpha_{i-1} = 3(1-s)^2 - 2(1-s)^3$$
$$\alpha_i = 3s^2 - 2s^3$$
$$\beta_{i-1} = h[(1-s)^2 - (1-s)^3] \qquad (I)$$
$$\beta_i = -h[s^2 - s^3]$$

where

$$s = \frac{t}{h} = \frac{x - x_{i-1}}{h}$$

It can be shown easily that $\alpha_i(x)$ is unity for $x = x_i$ but zero for x_{i-1}, and its first derivative is zero at both x_i and x_{i-1}. Likewise, $\beta_i(x)$ is zero for x_i and x_{i-1}, its first derivative is unity at $x = x_i$ but zero for $x = x_{i-1}$.

SUMMARY OF THIS SECTION:

(a) A polynomial interpolation fitted to both functional values and the derivatives is called *Hermite interpolation.*

(b) A cubic Hermite interpolation is applied in an interval in which both functional value and the first derivative are specified at each end point. If the whole domain is divided into intervals and the scheme is applied to each interval, the interpolation scheme is called the *piecewise cubic Hermite interpolation.*

(c) The functional value and the first derivative in the piecewise cubic Hermite interpolation are continuous in the entire domain.

2.8 TWO-DIMENSIONAL INTERPOLATION

Two-dimensional interpolation schemes may be classified into two types. The first uses one-dimensional interpolation twice; it is called *double interpolation.* The second uses piecewise two-dimensional interpolating polynomials. The former is suitable for interpolating a table of functions on equispaced points. The latter is used in the finite element methods [Becker/Carey/Oden].

To explain the first type, suppose the values of a function $f(x, y)$ are given on a rectangular grid of (x, y) namely (x_i, y_j). We denote the value on the point (x_i, y_j) as $f_{i,j} = f(x_i, y_j)$. A double interpolation proceeds in two steps, in each of which a one-dimensional interpolation is used. Suppose one has to estimate the functional value at a point located in the box defined by $x_{i-1} \leqslant x \leqslant x_i$ and $y_{j-1} \leqslant y \leqslant y_j$, as shown in Figure 2.5. For simplicity of explanation, let us assume linear interpolation is used in both steps. The first step is to interpolate the table in the y-direction and find the values at E and F, respectively, as

$$f_E = \frac{y_j - y}{y_j - y_{j-1}} f_{i-1,j-1} + \frac{y - y_{j-1}}{y_j - y_{j-1}} f_{i-1,j}$$

$$f_F = \frac{y_j - y}{y_j - y_{j-1}} f_{i,j-1} + \frac{y - y_{j-1}}{y_j - y_{j-1}} f_{i,j} \tag{2.8.1}$$

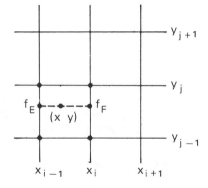

Figure 2.5 Bilinear interpolation on a two-dimensional domain

The second step is to interpolate between f_E and f_F by linear interpolation, as

$$g(x, y) = \frac{x_i - x}{x_i - x_{i-1}} f_E + \frac{x - x_{i-1}}{x_i - x_{i-1}} f_F \qquad (2.8.2)$$

Combining the two steps into one equation, we can write

$$\begin{aligned} g(x, y) = [&(x_i - x)(y_j - y)f_{i-1, j-1} + (x_i - x)(y - y_{j-1})f_{i-1, j} \\ &+ (x - x_{i-1})(y_j - y)f_{i, j-1} + (x - x_{i-1})(y - y_{j-1})f_{i, j}] / \\ &[(x_i - x_{i-1})(y_j - y_{j-1})] \end{aligned} \qquad (2.8.3)$$

The order of the linear interpolation steps can be exchanged. Namely, f_G and f_H are first found by linear interpolation in the x direction, and then a linear interpolation in the y direction is applied to calculate $g(x, y)$. The change does not affect the result.

See Appendix H for another two-dimensional interpolation named transfinite interpolation.

SUMMARY OF THIS SECTION

(a) Two-dimensional interpolation may be performed in two ways. In the first approach, one-dimensional interpolation is applied twice. In the second approach, a two-dimensional polynomial may be directly fitted to the functional values.

(b) In the present section, the first approach is illustrated using linear interpolation. Linear interpolation may be replaced by any other one-dimensional interpolation.

2.9 EXTRAPOLATIONS

Polynomial extrapolation is exactly the same as polynomial interpolation except the fitted polynomial is used outside the two end points of data.

In the domain where the function is not known but is believed to be well represented, extrapolation is used by extending the use of an interpolation formula.

In using an extrapolation, one has to decide the order of polynomial to be used and how far the extrapolation is extended. The extrapolation works most reliably if a theoretical analysis of the function to be extrapolated indicates a particular order to be used.

In general, error of extrapolation increases as the point of interest goes further from the data points. If a higher order interpolation is used for extrapolation without theoretical basis, errors may increase rapidly as the order of polynomial increases. Analysis of errors of extrapolation is described in Appendix A.

Application of extrapolations may be seen in various parts of this book: for example, see the Newton-Cotes open integration formula (Section 4.5), the Romberg integration method (Section 4.2), the predictor-corrector method (Section 9.4), and the iterative acceleration parameters (Section 12.5). For more discussion see Stoer/Burlish.

PROGRAMS

PROGRAM 2-1 Lagrange Interpolation

(A) Explanations

This program evaluates a Lagrange interpolation formula

$$g(x) = \sum_{i=0}^{N} \eta_i(x) f_i$$

with

$$\eta_i(x) = \prod_{\substack{j=0 \\ i \neq j}}^{N} \frac{x - x_j}{x_i - x_j}$$

for a value of x given by input, where (x_i, f_i), $i = 0, 1, 2, \ldots, N$, are the data of the functional table, $N + 1$ is the total number of the data.

The user defines the number of data points in the data statements. When executed, the computer asks interrogatively for the value of x.

Although the program is designed for interpolation, it may be used for extrapolation. In this case, however, a message "X is in the extrapolation range" is printed out.

(B) List

```
/* CSL/c2-1.c      Lagrange Interpolation */
#include <stdio.h>
#include <math.h>
#define TRUE 1
/*            n    : last point number (n+1 = number of data points)
              f[i], x[i] : given data ( i starts at 0)
              yans : final result
              xa : x value of input for which interpolation is
computed*/
main()
{
int i, j, kk;
float xa, yans, z;
static n = 3;          /* n+1 is the number of data points. */
/* Next two lines define data points to be used for interpolation. */
static float x[11]={1.   , 2.   , 3.   , 4.   };
static float f[11]={.671,  .620,  .567,  .512};
    printf( "\nCSL/C2-1    Lagrange Interpolation\n\n" );
    printf( "Function table used:\n" );
    printf( "i           x(i)          f(i) \n" );
    for( i = 0; i <= n; i++ )
         printf( "%d          %g           %g \n", i, x[i], f[i] );
    while(TRUE) {
       printf( "\nInput x ?   " );      scanf( "%f", &xa );
       if ( !(x[0] <= xa && xa <= x[n]) )
       printf("(WARNING: x is in the extrapolation range.)\n" );
       yans = 0;
```

```
        for( i = 0; i <= n; i++ ) {
           z = 1.0;
           for( j = 0; j <= n; j++ ) {
              if( i != j )   z = z*(xa - x[j])/(x[i] - x[j]);
           }
           yans = yans + z*f[i];
        }
        printf( "Answer:   g( %g ) = %g \n", xa, yans );
        printf("\nType 1 to continue, 0 to stop:");scanf("%d",&kk);
        if( kk != 1 ) exit(0);
   }
}
```

(C) Sample Output

```
CSL/C2-1      Lagrange Interpolation

Function table used:
i             x(i)          f(i)
0             1             0.671
1             2             0.62
2             3             0.567
3             4             0.512

Input x ?  1
Answer:   g( 1 ) = 0.671

Input x ?  3.66
Answer:   g( 3.66 ) = 0.530924

Input x ?  4.5
(WARNING: x is in the extrapolation range.)
Answer:   g( 4.5 ) = 0.48375

Input x ?  0.1
(WARNING: x is in the extrapolation range.)
Answer:   g( 0.1 ) = 0.71519
```

(D) Discussions

The program prints out the function table used in the interpolation formula. The interpolation for $x = 1$ (input) is then printed out. In the third case, the value of the interpolation for $x = 4.5$ is printed out with the warning that x is in the extrapolation range because x is beyond the end points in the function table.

QUIZ. The derivative of a function may be approximated by the derivative of a polynomial interpolation formula such as the Lagrange interpolation formula. What do you call this approximation at a data point? (See Section 5.5)

PROGRAM 2–2 Forward Difference Table

(A) Algorithm and Program Structure

Suppose a set of data, (f_i, x_i), $i = 0, 1, 2, \ldots, N$, is given in a function table, where x_i is the abscissa of a data point and f_i is the corresponding functional value. We assume

here that the abscissas are equispaced. The forward difference of order k for f_i is
defined by Eq. (2.4.5). The output of the computed difference table is in the same
format as Table 2.2 in the text.

This program computes the forward differences of all orders that can be computed for the given function table. The data with equispaced abscissas are initialized in the program.

(B) List

```
/* CSL/c2-2.c     Forward Difference Table with Equispaced Points */
#include <stdio.h>
#include <math.h>
/*          itemp0[], itemp1[] : data definition
            i_max : maximum point number
            f[i][0], x[i] : given data (i starts at 0)
            f[i][k] : k-th difference for point i          */
main()
{
int i, j, k, _r;
static float f[11][11], x[11];
static i_max = 6;
static float _itmp0[] = {1,3,5,7,9,11,13};
static float _itmp1[] =
        {1.0,0.5,0.3333333,0.25,0.2,0.1666666,0.14285714};
   for (i = 0, _r = 0; i <= i_max; i++) {
      x[i] = _itmp0[_r];    f[i][0] = _itmp1[_r++];
   }
   printf( "\nCSL/C2-2  Foward Difference Table  \n" );
   for (k = 1; k <= i_max; k++) {
      j = i_max - k;
      for (i = 0; i <= j; i++) f[i][k] = f[i + 1][k - 1] - f[i][k - 1];
   }
   printf( "\n i    x(i)     f(i)      1st odr. 2nd odr.    ... \n" );
   printf( "                             diff.    diff. \n" ) ;
   for (i = 0; i <= i_max; i++){
      printf( "%2d %8.5f ", i, x[i] );
      for (k = 0; k <= i_max - i; k++) printf( "%8.5f ", f[i][k] );
      printf( "\n" );
   }
   printf( "\n\n" );     exit(0);
}
```

(C) Sample Output

```
CSL/C2-2  Foward Difference Table

i    x(i)      f(i)      1st odr. 2nd odr.     ...
                            diff.    diff.
0   1.0000   1.00000 -0.50000  0.33333 -0.25000  0.20000 -0.16667  0.14286
1   3.0000   0.50000 -0.16667  0.08333 -0.05000  0.03333 -0.02381
2   5.0000   0.33333 -0.08333  0.03333 -0.01667  0.00952
3   7.0000   0.25000 -0.05000  0.01667 -0.00714
4   9.0000   0.20000 -0.03333  0.00952
5  11.0000   0.16667 -0.02381
6  13.0000   0.14286
```

(D) Discussions

The forward difference table is useful in developing a Newton forward (as well as backward) interpolation polynomial using equispaced data points. For example, the interpolation polynomial passing through points, $i = 2, 3$, and 4, can be written using the difference values along the third line as

$$g(x) = 0.33333 - 0.08333s + 0.03333\,\frac{s(s-1)}{2}$$

where

$$s = \frac{x - 5.0}{2}$$

The forward difference table may also be used to estimate a Nth-order derivative of the function. The Nth-order difference divided by the Nth power of the interval size of x approximately equals the Nth-order derivative (see Section 5.4). For example, the third derivative of the function at $i = 2$ approximately equals

$$-0.01667/2^3 = -0.00208$$

Quiz. Can the values of backward differences be found in a forward difference table?

PROGRAM 2–3 Divided Difference Table

(A) Explanations

Suppose a function table is given by (f_i, x_i), $i = 0, 1, 2, \ldots, N$, where x_i is the abscissa of a data point with nonequispaced intervals, and f_i is the functional value. The divided difference of order m is defined by Eq. (2.5.2).

This program develops a divided difference table and prints out in the form of Table 2.3. All the data of a function table are initialized in the beginning of the program. Abscissas of the data can be variably spaced. There is no input to be given interactively.

(B) List

```
/* CSL/c2-3.c  Divided Difference Table with Non-equispaced Data Points */
#include <stdio.h>
#include <math.h>
/*          i_max : last point number
            f[i][0],x[i] : data given
            f[i][k] : k-th divided difference for point i   */
```

```
main()
{
int i, j, k, _r;
static float _f[11][11], x[11];
static i_max = 6;
static float _itmp0[] = {0.1,0.2,0.4,0.7,1.0,1.2,1.3};
static float _itmp1[] =
         {.99750,.99002,.96040,.88120,.76520,.67113,.62009};
   for (i=0, _r=0; i <= i_max; i++) {
      x[i] = _itmp0[_r ];
      f[i][0] = _itmp1[_r++];
   }
   printf( "\n\nCSL/C2-3     Divided Difference Table \n" );
   for( k = 1; k <= i_max; k++ ) {
      for( i = 0; i <= i_max - k ; i++ )
        f[i][k] = (f[i + 1][k - 1] - f[i][k - 1])/(x[i + k] - x[i]);
   }
   printf( "\n i    x(i)      f(i)     f(i,i+1) f(i,.i+2),..\n" );
   for( i = 0; i <= i_max; i++ ) {
      j = i_max - i;
      printf( "%2d %8.5f ", i, x[i] );
      for( k = 0; k <= j; k++ ) printf( "%8.5f ", f[i][k] );
      printf( "\n" );
   }
   printf( "\n" ); exit(0);
}
```

(C) Sample Output

```
CSL/C2-3      Divided Difference Table

i  x(i)      f(i)      f(i,i+1) f(i,.i+2),..
0  0.10000  0.99750 -0.07480 -0.24433  0.02089  0.01479 -0.00239  0.00128
1  0.20000  0.99002 -0.14810 -0.23180  0.03419  0.01215 -0.00085
2  0.40000  0.96040 -0.26400 -0.20444  0.04635  0.01122
3  0.70000  0.88120 -0.38667 -0.16737  0.05644
4  1.00000  0.76520 -0.47035 -0.13350
5  1.20000  0.67113 -0.51040
6  1.30000  0.62009
```

(D) Discussions

Using the divided difference table, an interpolation polynomial fitted to the data of consecutive points can be developed easily. For example, the interpolation formula passing through $i = 1, 2,$ and 3 is written as

$$g(x) = f_1 + (x - x_1)f_{1,2} + (x - x_1)(x - x_2)f_{1,2,3}$$

$$= 0.99002 - 0.14810(x - 0.2) - 0.23180(x - 0.2)(x - 0.4)$$

where $x_1 = 0.2$, $x_2 = 0.4$, and the divided difference values along the line of $i = 1$ are used.

PROBLEMS

(2.1) The following pairs of data are sampled from $y = \cos(x)$:

(a) $x = 0,\quad y = 1$
$x = 0.1,\quad y = 0.99500$

(b) $x = 0,\quad y = 1$
$x = 0.2,\quad y = 0.98007$

(c) $x = 0,\quad y = 1$
$x = 0.5,\quad y = 0.87758$

Approximate the y value at midpoint by linear interpolation, and estimate the error by using Eq. (2.2.2). Compare the estimated error to the exact error evaluated by comparison to $\cos(x)$.

(2.2) If a data set consists of (f_i, x_i), with $i = 1, 2, \ldots, N$, the linear interpolation may be fitted to each pair of consecutive two data points, namely (f_i, x_i) and (f_{i+1}, x_{i+1}). The linear interpolation may be written for each data interval. However, the interpolation formulas for all the intervals can be written in a single equation.

(a) Show that the piecewise linear interpolation formulas for the range $[x_0, x_N]$ may be compactly expressed by

$$g(x) = \sum_{i=0}^{N} f_i \eta_i(x) \tag{a}$$

where $\eta_i(x)$ is defined by

$$\eta_i(x) = \frac{x - x_{i-1}}{x_i - x_{i-1}} \quad \text{for } x_{i-1} \leqslant x \leqslant x_i$$

$$= \frac{x_{i+1} - x}{x_{i+1} - x_i} \quad \text{for } x_i \leqslant x \leqslant x_{i+1}$$

$$= 0 \qquad \qquad \text{otherwise}$$

(b) Plot $\eta_i(x)$ and its derivative.

(2.3) (a) Write the Lagrange interpolation formula fitted to the data points, $i = 2, 3$ and 4, given in the following table.

i	x_i	$f(x_i)$
1	0	0.9162
2	0.25	0.8109
3	0.5	0.6931
4	0.75	0.5596
5	1.0	0.4055

(b) If the third derivative of the function at $i = 3$ is $f''' = -0.26$, estimate the error of the Lagrange interpolation derived in (a) at $x = 0.6$.

(2.4) A Lagrange interpolation of order N (using $N + 1$ points) for a function $f(x)$ becomes exact if $f(x)$ is a polynomial of order N or less. Explain the reason in two different ways.

(2.5) (a) Write the Lagrange interpolation that passes through the following data points:

x	0	0.4	0.8	1.2
f	1.0	1.49182	2.22554	3.32011

(b) Knowing $f''''(0.6) = 1.822$, estimate the error at $x = 0.2$, 0.6, and 1.0 by using Eq. (2.3.9) with $\xi = x_m$. (In case f'''' is not known, an approximation for f'''' may be calculated by a difference approximation if one more data point is available from the functional table.)

(c) Evaluate exact error of the interpolation formula at $x = 0.2$, 0.6, and 1.0 by $e(x) = f(x) - g(x) = \exp(x) - g(x)$.

(2.6) Fit $x \sin(x)$ in $[0, \pi/2]$ with the Lagrange interpolation polynomial of order 4 using equispaced points. Calculate the error of each interpolation formula at every increment of $\pi/16$, and plot.

(2.7) (a) Write a program to evaluate the Lagrange interpolation for $\sqrt{x} \cos(x)$ in $[0, 2]$ with six equally spaced grid points with $h = 0.4$. **(b)** Calculate the error of the interpolating polynomial at each increment of 0.1 of x. Plot the error distribution.

(2.8) Fit $\sin(x)$ in $[0, 2\pi]$ by the Lagrange interpolation polynomial of order 4 and 8 using equispaced points (5 and 9 points, respectively). Plot the interpolating polynomials together with $\sin(x)$, and the error distributions.

(2.9) (a) Develop a Lagrange interpolation for $\log_e(x)$ in $1 \leqslant x \leqslant 2$ using four equispaced points. **(b)** Estimate error of the approximation, using Eq. (2.3.9) at $x = 1, 1.2, 1.3, \ldots, 1.9$ and 2.0. **(c)** Calculate the exact error by $e(x) = \log_e(x) - g(x)$.

(2.10) Approximate

$$y = \frac{1 + x}{1 + 2x + 3x^2}$$

in $[0, 5]$ by the Lagrange interpolation of order 4, and evaluate exact error by $e(x) = y - g(x)$. Work according to the following steps: **(a)** determine the points, **(b)** write the Lagrange interpolation, **(c)** calculate the error for each increment of 0.2 in x, and **(d)** plot the error distribution.

(2.11) If a Lagrange interpolation is fitted to four data points at $x_i = 1, 2, 3$, and 4, the following cubic polynomials appear in the interpolation formula:

(a) $\dfrac{(x - 2)(x - 3)(x - 4)}{(1 - 2)(1 - 3)(1 - 4)}$

(b) $\dfrac{(x - 1)(x - 3)(x - 4)}{(2 - 1)(2 - 3)(2 - 4)}$

(c) $\dfrac{(x - 1)(x - 2)(x - 4)}{(3 - 1)(3 - 2)(3 - 4)}$

(d) $\dfrac{(x - 1)(x - 2)(x - 3)}{(4 - 1)(4 - 2)(4 - 3)}$

Plot the foregoing four functions and discuss implications of the shape of each.

(2.12) The Lagrange interpolation formula may be compactly written as

$$g(x) = \sum_{i=0}^{N} f_i \eta_i(x)$$

where $\eta_i(x)$ is a shape function defined by

$$\eta_i(x) = \prod_{\substack{j=0 \\ j \neq i}}^{N} \frac{x - x_j}{x_i - x_j}$$

Sketch the shape function in a graphic form.

(2.13) Derive the Newton forward interpolation polynomial fitted to the following data points in Example 2.4:

(a) $i = 1, 2, 3$

(b) $i = 2, 3, 4, 5$

Also estimate the error of the foregoing interpolations at $x = 0.75$, using the method shown in Example 2.5.

(2.14) Analytically prove $\Delta^n x^{n-1} = 0$.

(2.15) If $f(x)$ is a polynomial of order N or less, a Newton forward interpolation of order N will become exactly equal to $f(x)$ regardless of the interval size h. Explain why.

(2.16) **(a)** Derive the Newton forward interpolation polynomial passing through the data points, $i = 2, 3, 4$ given in the following table:

i	x_i	$f(x_i)$
1	0	0.9162
2	0.25	0.8109
3	0.5	0.6931
4	0.75	0.5596
5	1.0	0.4055

(b) Estimate the error of the interpolation formula at $x = 0.6$.

(2.17) The following function table is sampled from the polynomial:

$$y = 2x^3 + 3x + 1$$

x	y
0.1	1.302
0.2	1.616
0.3	1.954
0.4	2.328
0.5	2.750

(a) Develop a forward difference table, and show that the fourth order difference becomes zero. **(b)** Explain why the fourth order difference becomes zero.

(2.18) Develop the forward difference table from the following function table:

i	x	$f(x)$
1	0.5	1.143
2	1.0	1.000
3	1.5	0.828
4	2.0	0.667
5	2.5	0.533
6	3.0	0.428

Using the Newton forward formulas, write the interpolation polynomials fitted to:

    (a) $i = 1, 2, 3$

    (b) $i = 4, 5, 6$

    (c) $i = 2, 3, 4, 5$

(2.19) Derive an approximate expression for the error of each interpolation formula obtained in the previous problem.

(2.20) Prove that, if $k = 3$ in Eq. (2.4.12), $g(x)$ becomes equal to f_0, f_{-1}, f_{-2}, and f_{-3} for $s = 0$, -1, -2, and -3 respectively.

(2.21) The Newton backward interpolation polynomial fitted to data points at x_0, x_1, and x_2 is written by

$$g(x) = f_2 + s\nabla f_2 + \tfrac{1}{2}s(s + 1)\nabla^2 f_2, \quad -2 \leqslant s \leqslant 0$$

where

$$s = (x - x_2)/h$$

On the other hand, the Newton forward interpolation polynomial fitted to the same data points is

where

$$s = (x - x_0)/h$$

Verify the equivalence of the equations.

(2.22) Is it possible to write a Newton backward interpolation using the forward difference table? Explain how.

(2.23) Develop a divided difference table for the following function table:

Viscosity of water
$(Ns/m^2) \times 10^{-3}$

$T(C)$	μ
0	1.792
10	1.308
30	0.801
50	0.549
70	0.406
90	0.317
100	0.284

(2.24) Derive the interpolation polynomial fitted to the following data points of Example 2.8.

(a) $i = 2, 3, 4$

(b) $i = 0, 1, 2, 3$

Estimate the error of the interpolations derived at $x = 0.3$ and $x = 0.55$.

(2.25) Prove that, if $P_{0, 1, 2, \ldots, N}$ given by Eq. (2.5.3) is expanded in a power series of x, the coefficient of the highest order term x^N is given by

$$a_N = \frac{f_N - P_{0, 1, \ldots, N-1}}{\prod\limits_{i=0}^{N-1} (x_N - x_i)}$$

where f_N is the functional value at x_N.

(2.26) Examine the validity of Eq. (2.6.3) by introducing the following Chebyshev polynomials of Eq. (2.6.2):

(a) T_1, T_2, T_3

(b) T_3, T_4, T_5

(2.27) **(a)** Develop a Lagrange interpolation approximation for $\log_e(x)$ in $1 \leqslant x \leqslant 2$ using four Chebyshev points. **(b)** Estimate error of the approximation using Eq. (2.3.9) at $x = 1$, 1.2, 1.3, ..., 1.9, and 2.0. **(c)** Calculate actual error by $e(x) = \log_e(x) - g(x)$.

(2.28) Find a quadratic interpolation formula for $\log_e(x)$ in the interval of $1 < x < 3$ using three Chebyshev points. Write the equation to estimate the error and evaluate it for $x = 2.5$.

(2.29) Develop a Lagrange interpolation polynomial fitted to

$$y = \frac{x + 1}{1 + 2x + 3x^2}$$

in $[1, 3]$ using three Chebyshev points.

(2.30) Repeat the problem of Example 2.2 with the Chebyshev points (use 3 and 5 points, respectively).

(2.31) Approximate e^x by piecewise cubic Hermite interpolations in $[0, 2]$ with two intervals. Calculate the error of the piecewise Hermite interpolations for each increment of 0.2 of x, and plot the error.

(2.32) Determine the polynomial fitted to

$$f(0) = 1, \quad f(1) = 2, \quad f'(0) = 0.5$$

(2.33) Determine the polynomial fitted to

$$f(0) = 1, \quad f(1) = 2, \quad f'(0) = 0, \quad f'(1) = 1$$

(2.34) Verify that Eq. (G) of Example 2.10 is the polynomial that fits to f_{i-1}, f'_{i-1}, f_i, and f'_i.

(2.35) Calculate the values of the four functions in Eq. (I) of Example 2.10 at $s = 0$ and $s = h$.

(2.36) Determine the second-order (parabolic) Hermite interpolation polynomial fitted to

$$f(1) = 2, \quad f(2) = 3, \quad f'(2) = 1.2$$

(2.37) Determine the cubic Hermite interpolation polynomial fitted to

$$f(1) = 2, \quad f(2) = 3, \quad f'(2) = 0.5, \quad f(3) = 0$$

REFERENCES

Abramowitz, M., and I. A. Stegun, eds., *Handbook of Mathematical Functions*, National Bureau of Standards, 1970.

Becker, E. B., G. F. Carey, and J. T. Oden, *Finite Elements*: An *Introduction*, Prentice-Hall, 1981.

Carnahan, B., H. A. Luther, and J. O. Wilkes, *Applied Numerical Methods*, Wiley, 1969.

Conte, S. D., and C. de Boor, *Elementary Numerical Analysis*, 3rd ed., McGraw-Hill, 1980.

Gerald, C. F., and P. O. Wheatley, *Applied Numerical Analysis*, 3rd ed., Addison-Wesley, 1984.

Isaacson, E., and H. B. Keller, *Analysis of Numerical Methods*, Wiley, 1966.

Stoer, J., and R. Burlish, *Introduction to Numerical Analysis*, Springer-Verlag, 1980.

3

Solution of Nonlinear Equations

3.1 INTRODUCTION

Solutions of a nonlinear equation are called *roots*, or *zeroes*. A few examples of nonlinear equations follow:

(a) $1 + 4x - 16x^2 + 3x^3 + 3x^4 = 0$

(b) $f(x) - \alpha = 0, \quad a < x < b$

(c) $\dfrac{x(2.1 - 0.5x)^{1/2}}{(1 - x)(1.1 - 0.5x)^{1/2}} - 3.69 = 0, \quad 0 < x < 1$

(d) $\tan(x) = \tanh(2x)$

The first is an example of a polynomial equation, which may be encountered as a characteristic equation for a linear ordinary differential equation, among numerous other problems. The second example is equivalent to evaluating $f^{-1}(\alpha)$, where $f(x)$ is any function and f^{-1} is its inverse function. The third example is a special case of (b). The fourth example is a transcendental equation.

The primary reason why we solve nonlinear equations by using computer methods is that nonlinear equations have no closed-form solution except for very few problems. Analytical solution for polynomial equations exists up to the fourth order [Abramowitz/Stegun, p. 17], but there are no closed-form solutions for higher orders. Therefore, roots of those nonlinear equations are found by computer methods based on iterative procedures.

The numerical methods designed to find roots are powerful, although each has its own limitations and pitfalls. Therefore, students should learn the pros and cons

Table 3.1 Summary of the schemes to find roots

Name	Necessity of Specifying an Interval Containing the Root	Neccesity of Continuity of f'	Types of Equations	Other Special Features
Bisection	yes	no	any	Robust; applicable to nonanalytic functions.
False position	yes	yes	any	Slow to converge for a large interval.
Modified false position	yes	yes	any	Faster than false position method.
Newton's method	no	yes	any	Fast; needs to compute f'; applicable to complex roots.
Secant method	no	yes	any	Fast; no need to compute f'.
Successive substitution	no	yes	any	May not converge.
Bairstow's method	no	yes	Polynomial	Quadratic factors.

of each method, particularly its difficulties, and become familiar with the methods through practice on a computer.

The key features of the numerical methods for nonlinear equations described in this chapter are summarized in Table 3.1. The first three methods in Table 3.1 (namely, the bisection method, false position method, and modified false position method) share a common feature. That is, these schemes can find a root if an interval of x that contains the root is known. Therefore, all these methods require a preliminary effort to estimate an appropriate interval that contains the desired root. The Newton's and secant methods need an initial guess to find the root, but estimating an interval is not necessary. The successive substitution method is a simple iterative algorithm, although its drawback is that the iteration may not always converge. Bairstow's method is limited to polynomials. By multiple applications of Bairstow's method, however, all the roots including complex roots can be found without prior knowledge of any kind, although sometimes the iteration may not converge at all.

3.2 BISECTION METHOD

The bisection method is the simplest yet the safest and most robust method for finding one root in a given interval where the root is known to exist. Its unique advantage is that it works even for nonanalytic functions.

Suppose the interval between $x = a$ and $x = c$ denoted by $[a, c]$, or equivalently $a \leqslant x \leqslant c$, has only one root, as shown in Figure 3.1. The bisection method is based on the fact that, when an interval $[a, c]$ has a root, the sign of $y(x)$ at the two ends are opposite, or one of $f(a)$ and $f(c)$ is zero, namely $f(a)f(c) \leqslant 0$.

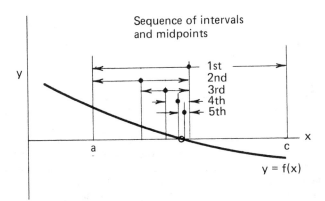

Figure 3.1 Bisection method

The first step in using the present method is to bisect the interval $[a, c]$ into two halves, namely, $[a, b]$ and $[b, c]$ where $b = (a + c)/2$. By checking the signs of $f(a)f(b)$ and $f(b)f(c)$, the half interval with the root can be found. Indeed, if $f(a)f(b) \leqslant 0$, the interval $[a, b]$ including $x = a$ and $x = b$ has the root, otherwise the other interval $[b, c]$ has the root. The new interval containing the root is bisected again. As this procedure is repeated, the size of the interval with the root becomes smaller and smaller. At each step, the midpoint of the interval is taken as the most updated approximation for the root. The iteration is stopped when the half interval is within a given tolerance ε. PROGRAM 3–1 is designed to find a root by the bisection method.

The interval size after n iteration steps becomes

$$\frac{(c - a)_0}{2^n}$$

where the numerator is the initial interval size. This also represents the maximum possible error when the root is approximated by the nth midpoint. Therefore, if the tolerance for the error is given by ε, the number of iteration steps required is the smallest integer satisfying

$$\frac{(c - a)_0}{2^n} < \varepsilon \tag{3.2.1}$$

or equivalently

$$n \geqslant \ln \frac{(c - a)_0}{\varepsilon} \Big/ 0.6931 \tag{3.2.2}$$

For example, if $(c - a)_0 = 1$ and $\varepsilon = 0.0001$, then $n = 14$.

Example 3.1

The root of

$$e^x - 2 = 0$$

is known to exist in [0, 2]. Find an approximate value of the root within a tolerance of $\varepsilon = 0.01$ by using the bisection method.

⟨**Solution**⟩

The hand calculation for the bisection method may be carried out by developing a table as shown below. When the first iteration is started, the values of $a = 0$ and $c = 2$ and the midpoint $b = (0 + 1)/2 = 1$ are written in the table for the row of $i = 1$. Also $f(a)$, $f(b)$, and $f(c)$ are computed and written in the same line. Examining the signs of these three f values, we find that the root is located between a and b. Therefore, a and b of $i = 1$ will become respectively a and c for $i = 2$. Thus $f(a)$ and $f(b)$ of $i = 1$ are copied to $f(a)$ and $f(c)$ for $i = 2$. The b for $i = 2$ is $b = (a + c)/2 = 0.5$ and $f(b)$ is computed and written in the table. The iteration for the remainder proceeds in a similar way until the tolerance is satisfied. The last value of b is the final answer.

Iteration Number, i	a	b	c	$f(a)$	$f(b)$	$f(c)$	Error Bound
1	0	1	2	−1	0.7182	5.3890	1
2	0	0.5	1	−1	−0.3512	0.7182	0.5
3	0.5	0.75	1	−0.3512	0.1170	0.7182	0.25
4	0.5	0.625	0.75	−0.3512	−0.1317	0.1170	0.125
5	0.625	0.6875	0.75	−0.1317	−0.0112	0.1170	0.0625
6	0.6875	0.7187	0.75	−0.0112	0.0518	0.1170	0.03125
7	0.6875	0.7031	0.7187	−0.0112	0.0200	0.0518	0.015625
8	0.6875	0.6953	0.7031	−0.0112	0.0043	0.0200	0.0078125

The eighth approximation for the root is $b = 0.6953$. Its error bound (maximum possible error) is 0.0078, which is within the specified tolerance.

We have assumed that the initial interval has only one root and $f(a)f(b) \leqslant 0$. However, $f(a)f(b) \leqslant 0$ is satisfied whenever the interval has an odd number of roots as illustrated in Figure 3.2. In this case, the bisection method will find one of the separate roots in the given interval. The bisection method cannot find a pair of double roots because the function touches the x axis tangentially at the double roots as shown in Figure 3.3.

One pitfall of the bisection method is that the method can catch a singularity as if it were a root because the bisection method does not recognize the difference between root and singularity. A singular point is a point at which the functional value goes to infinity as illustrated in Figure 3.4. To avoid this trouble, the program should check if $|f(c) - f(a)|$ is converging to zero as the bisection method proceeds. If this quantity diverges, the program is chasing a singularity rather than a root.

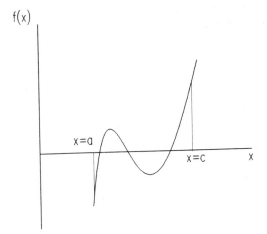

Figure 3.2 An odd number of roots in a given interval

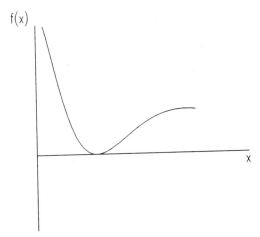

Figure 3.3 A function touching the x axis at a point

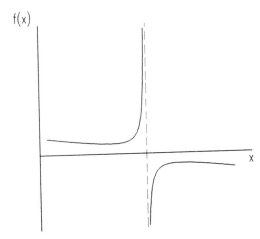

Figure 3.4 A function with singularity

When there is no prior information on approximate values of the roots, an easy way of finding intervals of x containing a root is to print out a table of the function for equally spaced values of x (see PROGRAM 3–2), or plot the function by computer graphics (see PROGRAM 3–3). If the sign of the functional value changes across an interval, at least one root exists in that interval. The graphic approach is useful in locating intervals containing a root, particularly when the equation has multiple roots.

Example 3.2

(a) Find intervals of size 1.0, each containing one or more (odd) roots of

$$y = -19(x - 0.5)(x - 1) + \exp(x) - \exp(-2x), \quad -10 < x < 10$$

(b) Repeat by using PROGRAM 3–2 with interval size 0.1.

⟨**Solution**⟩

(a) We compute y for $x = -10, -9, -8 \ldots, 10$ and make a function table. Then, mark the interval in which the function changes sign as illustrated in Table E3.2.

Table E3.2 Function table

x	y
−10.0	−48.517E + 07
−9.0	−65.662E + 06
−8.0	−88.876E + 05
−7.0	−12.037E + 05
−6.0	−16.362E + 04
−5.0	−22.653E + 03
−4.0	−34.084E + 02
−3.0	−66.938E + 01
−2.0	−19.696E + 01
−1.0	−64.021E + 00
0.0	−95.000E − 01
1.0	25.829E − 01
2.0	−21.129E + 00
3.0	−74.917E + 00
4.0	−14.490E + 01
5.0	−19.359E + 01
6.0	−11.907E + 01
7.0	35.563E + 01
8.0	19.835E + 02
9.0	68.111E + 02
10.0	20.402E + 03

Thus, three intervals, [0, 1], [1, 2] and [6, 7], each containing at least one root, are found.

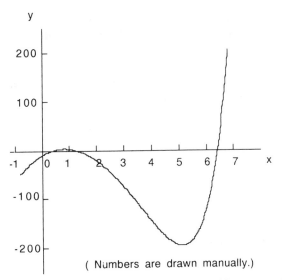

Figure E3.2 A sample of the function plotted by PROGRAM 3–3

(b) By running PROGRAM 3–2 for the range $[-10, 10]$ with the interval size $h = 1$, three intervals, [0, 1], [1, 2] and [6, 7], each containing at least one root, are found as expected from (a). By executing PROGRAM 3–2 again for each of these intervals with $h = 0.1$, the interval sizes are reduced to [0.4, 0.5], [1.2, 1.3], and [6.4, 6.5]. (PROGRAM 3–2 may be executed for a large range with a small interval h in a single run, but computing time increases.) Plotting the function using PROGRAM 3–3 is also illustrated in Figure E3.2.

SUMMARY OF THIS SECTION

(a) The bisection method finds a root of a function if the root is known to exist in a given interval.

(b) The bisection method finds a root even when the function is not analytic.

(c) On the other hand, singularity may be caught as if it were a root because the method does not distinguish roots from singularities.

(d) To find an interval containing the root is an important task to be done before applying the bisection method. The search for roots may be performed by listing a function table or plotting the function on the screen.

3.3 FALSE POSITION METHOD AND MODIFIED FALSE POSITION METHOD

The false position method that is based on linear interpolation is similar to the bisection method because the size of an interval containing a root is reduced through iteration. However, instead of monotonically bisecting the interval, a linear interpolation fitted at two end points is used to find an approximation for the root. So, if the function is well approximated by the linear interpolation, then the estimated roots should have good accuracy, and consequently the iteration would converge faster than when using the bisection method.

Given an interval $[a, c]$ containing the root, the linear function that passes through $(a, f(a))$ and $(c, f(c))$ is written as

$$y = f(a) + \frac{f(c) - f(a)}{c - a}(x - a) \tag{3.3.1}$$

or, solving for x,

$$x = a + \frac{c - a}{f(c) - f(a)}(y - f(a)) \tag{3.3.2}$$

The x coordinate at which the line intersects the x axis is found by setting $y = 0$ in Eq. (3.3.2) namely,

$$b = a - \frac{c - a}{f(c) - f(a)}f(a) = \frac{af(c) - cf(a)}{f(c) - f(a)} \tag{3.3.3}$$

After finding b, the interval $[a, c]$ is divided into $[a, b]$ and $[b, c]$. If $f(a)f(b) \leqslant 0$, the root is in $[a, b]$; and otherwise the root is in $[b, c]$. The two ends of the new interval having the root are renamed a and c. The interpolating procedure is repeated until the estimated roots converge.

The drawback of the present method is stagnation of an end point as graphically illustrated in Figure 3.5, which shows that one end of successive intervals does not move from the original end point, so that the approximations for the root denoted by $b_1, b_2, b_3, \ldots$, converge to the exact root from one side only. Stagnation is not desirable because it slows down convergence, particularly when the initial interval is very large or when the function deviates significantly from a straight line in the interval. The modified false position method explained next eliminates this difficulty.

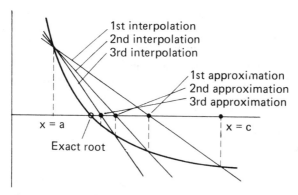

Figure 3.5 False position method

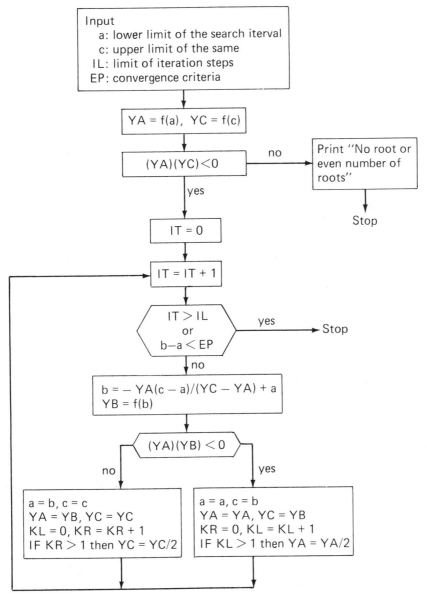

Figure 3.6 Flowchart of modified false position method

In the modified false position method, the f value of a stagnant end point is halved if that point has repeated twice or more. The end point that repeats is called a *stagnant point*. The exception to this rule is that for $i = 2$ the f value at one end is divided by 2 immediately if it does not move.

The algorithm is schematically illustrated in Figure 3.6. The effect of halving

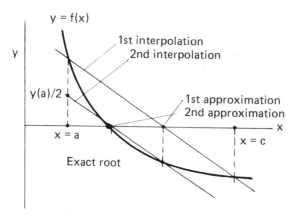

Figure 3.7 Modified false position method

the y value is that the solution of the linear interpolation becomes closer to the true root as illustrated in Figure 3.7.

If the present method is used with a hand calculator, it is suggested that you work with a table format such as that illustrated in Example 3.3.

Example 3.3

By using the false position method, find the smallest positive root of

$$f(x) = \tan(x) - x - 0.5 = 0$$

which is known to be in $0.1 < x < 1.4$.

⟨**Solution**⟩

The computations are shown in Figure E3.3. In the line for the first iteration ($i = 1$), the values of $a = 0.1$, $c = 1.4$, and the computed values of $f(a)$ and $f(c)$ are typed. The value b is found by linear interpolation,

$$b = a - \frac{c - a}{f(c) - f(a)} f(a)$$

and $f(b)$ is calculated subsequently. These two numbers are typed in the same line. By checking the signs of $f(a)$, $f(b)$, and $f(c)$, the root is located in between b and c. Therefore, b and c in the first iteration are copied to a and c for $i = 2$, respectively. The value of $f(b)$ for $i = 1$ is copied to $f(a)$ for $i = 2$, but $f(c)/2$ of $i = 1$ is copied to $f(c)$ for $i = 2$. The value of $f(c)$ that has been copied after dividing by 2 will be denoted by $f^*(c)$. The value of b for $i = 2$ is found by Eq. (3.3.3) in the same way as for $i = 1$ except $f^*(c)$ is used:

$$b = a - \frac{c - a}{f^*(c) - f(a)} f(a) \tag{3.3.4}$$

Program 3.4 Modified Linear Scheme

```
LOWER BOUND A=? 0.1
UPPER BOUND C=? 1.4
TOLERANCE   EP=? 0.00001
ITERATION LIMIT=? 13
INPUT: A= .1 , C= 1.4 , EP= .00001 , IL= 13
```

It.No	a	b	c	f(a)	f(b)	f(c)
1	1.0000E-01	2.4771E-01	1.4000E+00	-4.9967E-01	-4.9481E-01	3.8979E+00 *0.5
2	2.4771E-01	4.8102E-01	1.4000E+00	-4.9481E-01	-4.5911E-01	1.9489E+00 *0.5
3	4.8102E-01	7.7533E-01	1.4000E+00	-4.5911E-01	-2.9527E-01	9.7447E-01 *0.5
4	7.7533E-01	1.0110E+00	1.4000E+00	-2.9527E-01	8.4850E-02	4.8724E-01
5	7.7533E-01	9.5842E-01	1.0110E+00	-2.9527E-01	-3.4845E-02	8.4850E-02
6	9.5842E-01	9.7374E-01	1.0110E+00	-3.4845E-02	-2.7664E-03	8.4850E-02 *0.5
7	9.7374E-01	9.7603E-01	1.0110E+00	-2.7664E-03	2.1981E-03	4.2425E-02
8	9.7374E-01	9.7501E-01	9.7603E-01	-2.7664E-03	-6.1393E-06	2.1981E-03
9	9.7501E-01	9.7502E-01	9.7603E-01	-6.1393E-06	0.0000E+00	2.1981E-03

```
Tolerance satisfied
        Approximate root= .9750172
```

Figure E3.3 Computational procedure of the false position method

After completing the line for $i = 2$, the signs of a, b, and c are examined: again the root is located in between b and c. Therefore, the same procedure is repeated for $i = 3$, and even for $i = 4$.

In the line for $i = 4$, the root is located in a and b, so these values and $f(a)$ and $f(b)$ are copied to the next line, which is for $i = 5$. However, $f(a)$ is not divided by 2 because this is the first time a becomes stagnant. After completing the calculation of b and $f(b)$ for $i = 6$, b and c are copied for $i = 7$, and $f(c)$ is divided by 2 before being copied to the line for $i = 7$.

After $f(a)$ is modified, the b value is calculated by

$$b = a - \frac{c - a}{f(c) - f^*(a)} f^*(a) \tag{3.3.5}$$

where $f^*(a)$ is the modified value of $f(a)$.

SUMMARY OF THIS SECTION

(a) The false position method is essentially the same as the bisection method except that the bisection method is replaced by linear interpolation.

(b) The false position method is not necessarily faster than the bisection method because of stagnation of an end point.

(c) The modified false position method eliminates stagnation of an end point by halving the y values of the stagnant end points.

3.4 NEWTON'S METHOD

Newton's method (known also as the Newton-Raphson method) finds a root if an initial estimate for the desired root is known. It uses the tangential lines analytically evaluated. Newton's method may be applied on the complex domain to find complex roots. It can also be extended to simultaneous nonlinear equations (see an application in Section 3.7).

Newton's method is derived by the Taylor expansion [Press/Flannery/Teukolsky/Vetterling; Cheney/Kincaid]. Suppose the problem is to find a root of $f(x) = 0$. Using the Taylor expansion of $f(x)$ about an estimate x_0, the equation can be written as

$$f(x) = 0 = f(x_0) + f'(x_0)(x - x_0) + 0(h^2) \tag{3.4.1}$$

where $h = x - x_0$. Solving the Eq. (3.4.1) for x does not give the exact value because of the truncation error, but the solution becomes closer to the exact x than the estimate, x_0. Therefore, by repeating the solution using the updated value as a new estimate, the accuracy is successively improved.

The algorithm is graphically illustrated in Figure 3.8. The value x_0 is an initial guess for the root. The linear function passing through (x_0, y_0) tangentially is then obtained. The intersection of the tangential line with the x axis is denoted by x_1 and considered as an approximation for the root. The same procedure is repeated, using the most updated value as a guess for the next iteration cycle.

The tangential line passing through $(x_0, f(x_0))$ is

$$g(x) = f'(x_0)(x - x_0) + f(x_0) \tag{3.4.2}$$

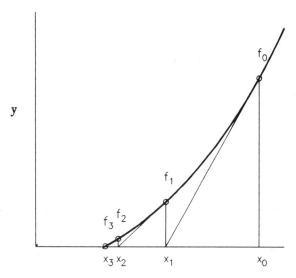

Figure 3.8 Newton's method

The root of $g(x) = 0$ denoted by x_1 satisfies

$$f'(x_0)(x_1 - x_0) + f(x_0) = 0$$

Solving the above equation yields

$$x_1 = x_0 - \frac{f(x_0)}{f'(x_0)} \qquad (3.4.3)$$

The successive approximations for the root are written as

$$x_i = x_{i-1} - \frac{f(x_{i-1})}{f'(x_{i-1})} \qquad (3.4.4)$$

Deriving the first derivative of a given function could be cumbersome or impossible. In such a case $f'(x_i)$ in Eq. (3.4.4) may be evaluated by a difference approximation rather than analytically. For example, $f'(x_{i-1})$ may be approximated by the forward difference approximation,

$$f'(x_{i-1}) \simeq \frac{f(x_{i-1} + h) - f(x_{i-1})}{h} \qquad (3.4.5)$$

where h is a small value such as $h = 0.001$, or by the backward difference approximation given by

$$f'(x_{i-1}) \simeq \frac{f(x_{i-1}) - f(x_{i-1} - h)}{h} \qquad (3.4.6)$$

Small errors in the difference approximation have no noticeable effect on the convergence rate of Newton's method. The accuracy of the final result is not affected by the difference approximation. If the function has no singularity in the neighborhood of the root, both difference approximations work well. However, we have to choose one or the other if a singularity is nearby. (More about numerical approximations for derivatives is described in Chapter 5.)

As indicated earlier, Newton's method is applicable to finding complex roots. If the programming language permits complex variables, a computer program originally designed only for real roots can be easily applied for complex roots. PROGRAM 3–7 is written in FORTRAN and can find complex roots. Newton's method is applicable to a set of nonlinear equations. An example is found in Section 3.7.

Newton's second-order methods have been proposed to increase the rate of convergence [Cheney/Kincaid; James/Smith/Wolford]. The second-order methods use the second derivative in addition to the first derivative. The method described in the present section is a first-order method because it uses only the first derivative. The number of iteration steps of second-order methods becomes smaller than the first-order method but the computational cost to evaluate the second derivatives often offsets the effect of reduction of interation times. Note also that the convergence rate of the second-order methods is higher than the first-order method only in an early stage of iteration or when the iterative solution is far from the exact solution. However, the advantage of using the second derivative disappears in the later stage of iteration because the convergence rate of the second-order method approaches that of the first-order method as the iterative solution converges to the exact root. Another disadvantage of the second-order methods is that applications to a set of equations is difficult. The second-order methods are recommended, therefore, only if any reduction of iteration times is desired in spite of the cost of computing the second derivative. In general, the first order method is best recommended for its simplicity and high efficiency.

Example 3.4

   Derive an iterative scheme to find the cubic root of a number based on Newton's method. Find the cubic root of $a = 155$ by the scheme derived.

⟨**Solution**⟩

   Suppose we desire to compute $x = \sqrt[3]{a}$. This problem can be restated as finding the zero of the function given by

$$f(x) = x^3 - a$$

Using Newton's method, an iterative scheme is written as

$$\begin{aligned}
x_{n+1} &= x_n - \frac{f(x_n)}{f'(x_n)} \\
&= x_n - \frac{x_n^3 - a}{3x_n^2} \\
&= \frac{2}{3}x_n + \frac{a}{3x_n^2}
\end{aligned}$$

   To compute the cubic root of 155, we set $a = 155$ and the initial guess to $x_0 = 5$. The iterative results become

n	x
0	5
1	5.4
2	5.371834
3	5.371686 (exact)

The exact solution is obtained after only three iteration steps. We try again with a much poorer estimate of $x_0 = 10$:

n	x
0	10
1	7.183334
2	5.790176
3	5.401203
4	5.371847
5	5.371686 (exact)

The exact value of the cubic root is obtained with five iteration steps.

Example 3.5

Find the first positive root of $y = \tan(x) - 0.5x$ by Newton's method.

⟨Solution⟩

The plotting of y in Figure E3.5 shows the smallest positive root in the neighborhood of 4.5, or between 4 and $3\pi/2$. Although analytical expression of the first derivative of $\tan(x)$ can be easily derived, we use Eq. (3.4.6) with $h = 0.001$. Thus, y' for $\tan(x) - 0.5x$ is approximated by

$$y'(x) \simeq [\tan(x) - \tan(x - 0.001)]/0.001 - 0.5$$

Newton's method with an initial guess of 4.0 is shown in Table E3.5a.

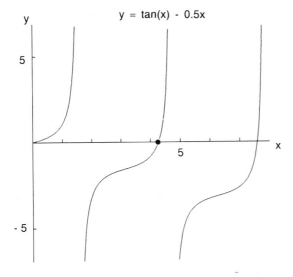

● : the smallest positive root in search

Figure E3.5 Graph of $y = \tan(x) - 0.5x$

```
CSL/F3-5       NEWTON SCHEME

TOLERANCE ?
0.0001
INITIAL GUESS FOR THE ROOT ?
4
 IT.NO.    X(N-1)          Y(N-1)         X(N)
   1     4.000000E+00  -8.421787E-01  4.458280E+00
   2     4.458280E+00   1.621111E+00  4.352068E+00
   3     4.352068E+00   4.781129E-01  4.288511E+00
   4     4.288511E+00   7.190108E-02  4.275191E+00
   5     4.275191E+00   2.075195E-03  4.274782E+00
   6     4.274782E+00  -2.861023E-06  4.274782E+00
--------------------------------
 FINAL SOLUTION=   4.274782
--------------------------------
```
Table E3.5a

This problem is very sensitive to the selection of an initial guess. If the initial
guess is set to, for example, 3.6, the iteration converges to an irrelevant value
after the x values erratically change, as shown in Table E3.5b.

```
CSL/F3-5       NEWTON SCHEME

TOLERANCE ?
0.0001
INITIAL GUESS FOR THE ROOT ?
3.6
 IT.NO. N   X(N-1)          Y(N-1)         X(N)
   1     3.600000E+00  -1.306533E+00  5.358891E+00
   2     5.358891E+00  -4.004476E+00  7.131396E+00
   3     7.131396E+00  -2.431464E+00  8.494651E+00
   4     8.494651E+00  -5.588555E+00  1.092057E+01
   5     1.092057E+01   7.847680E+00  1.087581E+01
   6     1.087581E+01   2.872113E+00  1.083419E+01
   7     1.083419E+01   7.255301E-01  1.081511E+01
   8     1.081511E+01   7.328224E-02  1.081269E+01
   9     1.081269E+01   6.022453E-04  1.081267E+01
--------------------------------
 FINAL SOLUTION=   10.81267
--------------------------------
```
Table E3.5b

SUMMARY OF THIS SECTION

(a) Newton's method iteratively uses tangential lines that pass through the consecutive approximations for the root.

(b) The method needs a good initial guess. Otherwise the iterative solution may diverge or converge to an irrelevant solution.

(c) Iterative convergence rate of Newton's method is high if it works.

(d) Newton's method can find complex roots if the variables are defined as complex variables.

3.5 SECANT METHOD

The secant method is closely related to Newton's method. The major difference from Newton's method is that f' is approximated by using the two consecutive iterative values of f. This eliminates the need to evaluate both f and f' in each iteration. Therefore, the secant method is more efficient, particularly when f is a time-consuming function to evaluate. The secant method is closely connected to the false position method also, because both are based on the linear interpolation formula, but the former uses extrapolations whereas the latter uses interpolations only [Press/et al.].

The successive approximations for the root by the secant method are given by

$$x_n = x_{n-1} - y_{n-1}\frac{x_{n-1} - x_{n-2}}{y_{n-1} - y_{n-2}}, \quad n = 2, 3 \dots \tag{3.5.1}$$

where x_0 and x_1 are two initial guesses to start the iteration.

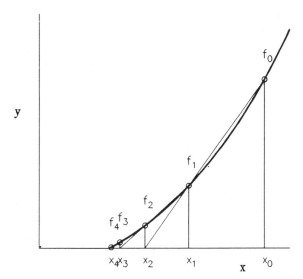

Figure 3.9 Secant method

If consecutive x_{n-1} and x_n become very close, y_{n-1} and y_n become close also, so a significant round-off error in Eq. (3.5.1) occurs. This trouble can be avoided in one of two ways: (a) when $|y_n|$ becomes less than a prescribed value, x_{n-2} and y_{n-2} in Eq. (3.5.1) are fixed (or frozen) thereafter, or (b) x_{n-2} and y_{n-2} are replaced by $x_{n-2} + \xi$ and $y(x_{n-2} + \xi)$ where ξ is a small prescribed number but large enough to avoid serious round-off errors. The secant method may converge to an unintended root or may not converge at all if the initial guess is not good.

Example 3.6

A bullet of $M = 2$ gm has been shot vertically into the air and is descending at its terminal speed [Shames, p. 417]. The terminal speed is determined by $gM = F_{drag}$ where g is gravity and M is the mass, and the whole equation may be written after evaluating the constants as

$$\frac{(2)(9.81)}{1000} = 1.4 \times 10^{-5}v^{1.5} + 1.15 \times 10^{-5}v^2$$

where v is the terminal velocity, m/sec. The first term on the right side represents the friction drag, and the second term represents the pressure drag. Determine the terminal velocity by the secant method. A crude guess is given by $v \simeq 30$ m/sec.

⟨**Solution**⟩

The problem is defined as finding the root of

$$y = f(v) = \frac{(2)(9.81)}{1000} - 1.4 \times 10^{-5}v^{1.5} - 1.15 \times 10^{-5}v^2 \qquad \text{(A)}$$

We set $v_0 = 30$ and $v_1 = 30.1$ based on the crude guess given, for which y_0 and y_1 are computed by Eq. (A). The iterative solution in accordance with Eq. (3.5.1) follows:

n	v_n	y_n
0	30.00000	1.9620001E − 02
1	30.10000	6.8889391E − 03
2	30.15411	6.8452079E − 03
3	38.62414	−8.9657493E − 04
4	37.64323	9.0962276E − 05
5	37.73358	9.9465251E − 07
6	37.73458	−1.8626451E − 09

Thus, the terminal velocity is $v = 37.7$ m/sec.

SUMMARY OF THIS SECTION

(a) The secant method is a variation of Newton's method. Computationally, it is more efficient than Newton's method.

(b) However, if two consecutive approximations are too close, round-off error can occur. Two ways to prevent the round-off error problem have been suggested.

3.6 SUCCESSIVE SUBSTITUTION METHOD

If the equation, $f(x) = 0$ is rearranged in the form

$$x = \overline{f}(x) \qquad (3.6.1)$$

then an iterative method may be written as

$$x^{(t)} = \bar{f}(x^{(t-1)}) \tag{3.6.2}$$

where the superscript t is the number of iteration steps and $x^{(0)}$ is an initial guess. This method is called the *successive substitution method*, or *fixed-point iteration* [Conte/de Boor].

The advantage of this method is in its great simplicity and the flexibility in choosing the form of $\bar{f}$. The disadvantage is, however, that the iteration does not always converge with an arbitrarily chosen form of $\bar{f}(x)$. To insure convergence of the iteration, the following condition must be satisfied:

$$|\bar{f}'(x)| < 1 \text{ in the neighborhood of the root} \tag{3.6.3}$$

Figure 3.10 illustrates how $\bar{f}'(x)$ affects the convergence of the iterative method. It can be observed that the convergence is asymptotic if $0 < \bar{f}' < 1$, and oscillatory if $-1 < \bar{f}' < 0$. Furthermore, it can be shown easily that the convergence rate becomes fastest as $\bar{f}'$ approaches 0 in the neighborhood of the root.

Figure 3.10 Convergence of successive substitution method

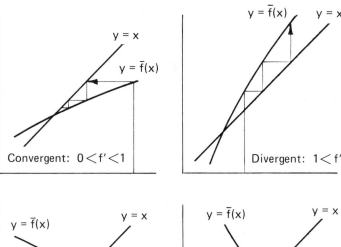

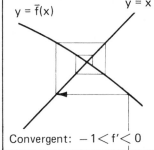

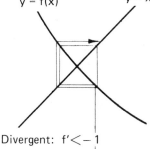

Example 3.7

The function $y = x^2 - 3x + e^x - 2$ is known to have two roots, one negative and one positive. Find the smaller root by the successive substitution method.

⟨**Solution**⟩

By checking the sign of y at $x = -1$ and $x = 0$ (namely $y(-1) = 2.367$ and $y(0) = -1$), we locate the smaller root in $[-1, 0]$. By rewriting the given equation as

$$x = \bar{f}(x) = \frac{x^2 + e^x - 2}{3} \tag{A}$$

an iterative method may be written as

$$x^{(t)} = \bar{f}(x^{(t-1)}) \tag{B}$$

The first derivative of $\bar{f}(x)$ satisfies Eq. (3.6.3) in the range of $[-1, 0]$, so the foregoing method is convergent. Numerical values of the iteration are shown next:

Iteration Count n	Successive Approximation x_n
0	0 (initial guess)
1	−0.333333
2	−0.390786
3	−0.390254
4	−0.390272
5	−0.390272

Alternative equations are

$$x = -\sqrt{3x - e^x + 2} \tag{C}$$

and

$$x = \sqrt{3x - e^x + 2} \tag{D}$$

However, the above equations have discontinuities in the vicinity of the smaller root. Furthermore, the first derivatives of both equations violate the condition of Eq. (3.6.3) in the vicinity of the root. Therefore, neither equation works.

One systematic way of finding a form of $\bar{f}(x)$ is to set

$$\bar{f}(x) = x - \alpha f(x) \tag{3.6.4}$$

so the iterative scheme becomes

$$x_n = x_{n-1} - \alpha f(x_{n-1}) \tag{3.6.5}$$

where α is a constant. If the iteration converges, x obtained from the above scheme satisfies $f(x) = 0$. The constant α may be determined as follows. By substituting Eq. (3.6.4) into Eq. (3.6.3), it is seen that iteration converges when

$$-1 < 1 - \alpha f'(x) < 1 \qquad (3.6.6)$$

or equivalently

$$0 < \alpha f'(x) < 2 \qquad (3.6.7)$$

Equation (3.6.7) indicates that first, α must have the same sign as f', and second, Eq. (3.6.5) will always converge as α approaches 0. The convergence rate is optimal when $\alpha \simeq 1/f'$.

The present scheme reduces to Newton's method if α is set to $1/f'(x_{n-1})$ for each iteration.

Example 3.8

The critical size of a nuclear reactor is determined by solving a criticality equation [Lamarsh]. Suppose a simple version of the criticality equation is given by

$$\tan(0.1x) = 9.2e^{-x} \qquad (A)$$

The solution that is physically meaningful is the smallest positive root and known to exist in [3, 4] for Eq. (A). Determine the smallest positive root.

⟨**Solution**⟩

We use the iterative scheme of Eq. (3.6.5) by writing

$$f(x) = \tan(0.1x) - 9.2e^{-x}$$

An approximate value of f' in [3, 4] is estimated by

$$f' = \frac{[f(4) - f(3)]}{(4 - 3)} = 0.40299$$

Using the preceding approximation, the parameter α is set to $1/f' = 1/0.40299 = 2.4814$.

The iteration of Eq. (3.6.5) converges as follows:

Iteration Count, n	x_n
0	4
1	3.36899
2	3.28574
3	3.29384
4	3.28280
5	3.29293
6	3.29292
7	3.29292

(a) Successive substitution is a broad class of iterative schemes to find a root of a function. Newton's method and the secant method described in the previous sections are special cases of the successive substitution.

(b) A criterion for convergence has been discussed.

3.7 BAIRSTOW'S METHOD

A few methods are specialized to find roots of polynomials almost automatically, including the quotient difference method [Gerald/Wheatley], Bairstow's method, and the application of QR iteration. Although the quotient difference method is simple and easy to use, it fails to work very often, unfortunately. Use of QR iteration, explained in Section 7.5 is the best among the three, but it cannot be used without understanding the eigenvalues of a matrix. Bairstow's method has an accuracy problem and fails to work sometimes, but it is more reliable than the quotient difference method.

Bairstow's method is an iterative scheme to find one quadratic factor of a polynomial in each application without any prior knowledge. By repeated applications of Bairstow's method to deflated polynomials, all the quadratic factors of a polynomial can be calculated.

Complex roots of a polynomial with real coefficients always exist in complex conjugate pairs. If a quadratic factor, $x^2 + \bar{p}x + \bar{q}$, which has a pair of complex conjugate roots, is factored out from the polynomial, the pair of complex roots can be computed by solving $x^2 + \bar{p}x + \bar{q} = 0$. Thus, all the roots of a polynomial can be calculated without using complex variables. One disadvantage of Bairstow's method is that accuracy of the results is often poor, so the accuracy of the computed roots must be verified or improved by some other means such as Newton's method.

Any given polynomial of order N written as

$$y = a_0 + a_1 x + a_2 x^2 + \cdots + a_N x^N \tag{3.7.1}$$

may be rewritten in the form

$$y = (x^2 + px + q)G(x) + R(x) \tag{3.7.2}$$

where p and q are arbitrary values, $G(x)$ is a polynomial of order $N - 2$, and $R(x)$ is the remainder, which is a polynomial of order one, namely a linear function at most. If p and q are chosen in such a way that the remainder $R(x)$ becomes zero, then $(x^2 + px + q)$ is a quadratic factor. The roots of a quadratic factor are given by the well-known formula

$$\frac{-p \pm \sqrt{p^2 - 4q}}{2}$$

We write the polynomial of order $N - 2$ and the remainder respectively as

$$G(x) = b_2 + b_3 x + b_4 x^2 + \cdots + b_N x^{N-2} \tag{3.7.3}$$

$$R(x) = b_0 + b_1 x \tag{3.7.4}$$

The values of b_0 and b_1 depend on the selected values of p and q, so they can be considered as functions of p and q:

$$b_0 = b_0(p, q)$$
$$b_1 = b_1(p, q) \tag{3.7.5}$$

Our aim is to find $p = \bar{p}$ and $q = \bar{q}$ such that $b_0(\bar{p}, \bar{q}) = b_1(\bar{p}, \bar{q}) = 0$, so $R(x) = 0$. Then $(x^2 + \bar{p}x + \bar{q})$ will become a quadratic factor. The iterative scheme adopted in the Bairstow method is based on the Newton's method applied to simultaneous nonlinear equations $b_0(p, q) = b_1(p, q) = 0$. When the Newton's method is referred to in the remainder of this section, however, it is as the method to find a single root of a polynomial as described in Section 3.5.

To derive an explicit form of Eq. (3.7.5), we introduce (3.7.3) and (3.7.4) into (3.7.2), and rewrite the resulting equation in a power series. Since the equation thus obtained must equal Eq. (3.7.1), the coefficients for the same power of x in the two equations must be equal. By equating the coefficients for the same order, we find the relations:

$$a_N = b_N$$
$$a_{N-1} = b_{N-1} + p b_N$$
$$a_{N-2} = b_{N-2} + p b_{N-1} + q b_N$$
$$\vdots \tag{3.7.6}$$
$$a_2 = b_2 + p b_3 + q b_4$$
$$a_1 = b_1 + p b_2 + q b_3$$
$$a_0 = b_0 \qquad\quad + q b_2$$

By rewriting Eq. (3.7.6), the coefficients b_N through b_0 can be calculated in descending order as

$$b_N = a_N$$
$$b_{N-1} = a_{N-1} - p b_N$$
$$b_{N-2} = a_{N-2} - p b_{N-1} - q b_N$$
$$\vdots \tag{3.7.7}$$
$$b_2 = a_2 - p b_3 - q b_4$$
$$b_1 = a_1 - p b_2 - q b_3$$
$$b_0 = a_0 \qquad\quad - q b_2$$

Let us now consider p and q in Eqs. (3.7.5) as arbitrary estimates for the exact values $\bar{p}$ and $\bar{q}$. The terms $b_0(\bar{p}, \bar{q})$ and $b_1(\bar{p}, \bar{q})$ may be expanded in a Taylor series about p and q:

$$b_0(\bar{p}, \bar{q}) = b_0(p, q) + \Delta p\left(\frac{\partial b_0}{\partial p}\right) + \Delta q\left(\frac{\partial b_0}{\partial q}\right) + \cdots \qquad (3.7.8a)$$

$$b_1(\bar{p}, \bar{q}) = b_1(p, q) + \Delta p\left(\frac{\partial b_1}{\partial p}\right) + \Delta q\left(\frac{\partial b_1}{\partial q}\right) + \cdots \qquad (3.7.8b)$$

where

$$\Delta p = \bar{p} - p, \quad \Delta q = \bar{q} - q$$

and the partial derivatives are evaluated at p and q. Notice that the left sides of Eqs. (3.7.8a) and (3.7.8b) are zero because $\bar{p}$ and $\bar{q}$ are exact values. Truncating the right sides of Eqs. (3.7.8a) and (3.7.8b) after the first-order derivative terms yields

$$\Delta p\left(\frac{\partial b_0}{\partial p}\right) + \Delta q\left(\frac{\partial b_0}{\partial q}\right) = -b_0(p, q) \qquad (3.7.9a)$$

$$\Delta p\left(\frac{\partial b_1}{\partial p}\right) + \Delta q\left(\frac{\partial b_1}{\partial q}\right) = -b_1(p, q) \qquad (3.7.9b)$$

Numerical values of the right sides of the Eqs. (3.7.9a) and (3.7.9b) are evaluated by the last two equations in Eq. (3.7.7). If the partial derivatives are known, Eqs. (3.7.9a) and (3.7.9b) can be solved for Δp and Δq.

The partial derivatives in Eqs. (3.7.9a) and (3.7.9b) are evaluated by recursively calculating partial derivatives of all the equations in Eq. (3.7.7):

$$
\begin{aligned}
(b_N)_p &= 0 \\
(b_{N-1})_p &= -b_N - p(b_N)_p \\
(b_{N-2})_p &= -b_{N-1} - p(b_{N-1})_p - q(b_N)_p \\
&\ \ \vdots \\
(b_2)_p &= -b_3 - p(b_3)_p - q(b_4)_p \\
(b_1)_p &= -b_2 - p(b_2)_p - q(b_3)_p \\
(b_0)_p &= \qquad\qquad\quad - q(b_2)_p
\end{aligned}
\qquad (3.7.10)
$$

and

$$(b_N)_q = 0$$

$$(b_{N-1})_q = 0$$

$$(b_{N-2})_q = -b_N$$

$$\vdots$$ (3.7.11)

$$(b_2)_q = -p(b_3)_q - b_4 - q(b_4)_q$$

$$(b_1)_q = -p(b_2)_q - b_3 - q(b_3)_q$$

$$(b_0)_q = \qquad\quad -b_2 - q(b_2)_q$$

where subscripts p and q denote partial derivatives with respect to p and q respectively. The last two equations in Eq. (3.7.10) and the last two equations in Eq. (3.7.11) give the values of the partial derivatives in Eq. (3.7.9).

An implementation of Bairstow's method proceeds as follows:

(a) Set an initial guess for p and q, and calculate b_0 and b_1 by Eq. (3.7.7).

(b) Calculate $(b_0)_p$, $(b_1)_p$, $(b_0)_q$, and $(b_1)_q$ by Eqs. (3.7.10) and (3.7.11) (all the equations must be recursively evaluated).

(c) Solve Eq. (3.7.9) for Δp and Δq.

(d) Obtain $\bar{p}$ and $\bar{q}$ by $\bar{p} = p + \Delta p$ and $\bar{q} = q + \Delta q$ respectively.

The whole procedure of (a) through (d) is iterated by using $\bar{p}$ and $\bar{q}$ from the previous step as updated estimates for p and q.

One significant advantage of Bairstow's method is that the iteration converges to one of the quadratic factors regardless of the initial guess for p and q for most problems, although sometimes iteration does not converge at all. The coefficients of the reduced polynomial $G(x)$ are also automatically obtained. So, to find another quadratic factor, Bairstow's method may be applied again to the deflated polynomial $G(x)$. By repeating this until the order of the deflated polynomial becomes less than 2, all the quadratic factors can be found. On the other hand, a disadvantage is that the accuracy of the roots found by the method may not be good. Therefore, it is desirable to improve accuracy by running Newton's method for each root. Accuracy becomes poor particularly if the polynomial has multiple roots. The Bairstow method is available as PROGRAM 3–7. See Isaacson/Keller; Shoup; Gerald/Wheatley for more information.

The QR iteration explained in Section 7.4 may also be used to find roots of a polynomial.

Example 3.9

By using PROGRAM 3–7, find quadratic factors of the equation:

$$y = 3.3 + 0.5x + 2.3x^2 - 1.1x^3 + x^4$$

⟨**Solution**⟩

The output of PROGRAM 3–7 for the present problem is shown in Figure E3.9. The output shows that (1) $P = 0.9$ and $Q = 1.1$ so the quadratic factor is $x^2 + 0.9x + 1.1$, and (2) the roots of the quadratic factor are $-0.45 \pm 0.94736i$ and (3) the deflated polynomial is

$$x^2 - 2x + 3$$

which is another quadratic factor.

The exact factors are ($x^2 + 0.9x + 1.1$) and ($x^2 - 2x + 3$). Thus, both quadratic factors found are exact. In general, however, the first quadratic factor found is the most accurate, and the deflated polynomials become less and less accurate.

```
CSL/C3-7      Bairstow Scheme

Should zero-divide occur, alter the values for
p_ini and q_ini and rerun the program.

The order of polynomial ?   4
a(0)?    3.3
a(1)?    0.5
a(2)?    2.3
a(3)?   -1.1
a(4)?    1
Tolerance ?    0.00001
-------------------------------------------------
 p =0.9     q =1.1
Quadratic coefficient = x^2 + (0.9 x) + (1.1 )
Roots of the quadratic factor:
   -0.45 + 0.947365i
   -0.45 - 0.947365i
Coefficients of deflated polynomial
      order    coefficients
        0        3.00000
        1       -2.00000
        2        1.00000
-------------------------------------------------
```

Figure E3.9 Sample output of PROGRAM 3–7

SUMMARY FOR THIS SECTION

(a) Bairstow's method finds a quadratic factor of a polynomial, from which a pair of roots is calculated.

(b) Since complex roots always appear in a pair of complex conjugates (when all the coefficients of a polynomial are real), complex roots can be calculated without complex algebra.

(c) By repeating application of the method to the deflated polynomial, one can find all the quadratic factors.

(d) Errors of deflated polynomials and quadratic factors increase as the method is repeatedly applied.

(e) Accuracy of the roots found can be poor, so accuracy must be improved by another method.

(f) The iteration may not converge at all for certain problems.

PROGRAMS

PROGRAM 3–1 Bisection Method

(A) Explanations

PROGRAM 3–1 finds a root of a nonlinear equation in the form

$$f(x) = 0$$

by the bisection method. The function $f(x)$ is defined in fun_f(x), in which currently a sample equation is set to

$$f(x) = x^3 - 3x^2 - x + 3$$

Some input data must be given interactively.

(B) List

```
/* CSL/c3-1.c    Bisection Scheme    */
#include <stdio.h>
#include <stdlib.h>
#include <math.h>
#define TRUE 1
/*          a, c : current end points
            epsilon : telerance
            it_limit : limit of iteration number
            Y_a, Y_c :/* CSL/c3-1.c    Bisection Scheme    */
#include <stdio.h>
#include <stdlib.h>
#include <math.h>
#define TRUE 1
/*          a, c : current end points
            epsilon : telerance
            it_limit : limit of iteration number
            Y_a, Y_c :   while( TRUE ) {
```

```
        printf( "Lower bound: a ?      ");  scanf( "%f", &a );
        printf( "Upper bound: c ?      " ); scanf( "%f", &c );
        printf( "Tolerance: epsilon ? " ); scanf( "%f", &epsilon );
        printf( "Iteration limit ?     " ); scanf( "%d", &it_limit );
        printf( " It.  a            b              c             f(a)    ");
        printf( "    f(c)    abs(f(c)-f(a))/2\n" );
        it = 0;
        Y_a = fun_f( a );
        Y_c = fun_f( c );
        if( Y_a*Y_c > 0 ) {
           printf( "    f(a)f(c) > 0 \n" );
        }
        else {
           while( TRUE ) {
              it = it + 1;
              b = (a + c)/2;  Y_b = fun_f( b );
              printf("%3d %10.6f, %10.6f, %10.6f,  %10.6f, %10.6f",
                                    it, a, b, c, Y_a, Y_c );
              printf( " %12.3e\n", fabs((Y_c - Y_a)/2));
              if( it > it_limit )              break;
              if( fabs( b - a ) < epsilon )  break;
              if( Y_a*Y_b <= 0 )  {
                 c = b;     Y_c = Y_b;
              }
              else {
                 a = b;     Y_a = Y_b;
              }
           }
           if ( it<=it_limit ) printf( " Tolerance is satisfied. \n" );
           if ( it>it_limit )  printf( "Iteration limit exceeded.\n" );
           printf( "Final result:  Approximate root = %g \n", b );
        }
        printf("\nType 1 to continue, or 0 to stop.\n"); scanf("%d",&ks);
        if( ks != 1 )   exit(0);
     }
  }


  float fun_f(x)     /*  Definition of Function */
  float x;
  {
  float f;
     f = x*x*x - 3*x*x - x + 3;
     return( f );
  }
```

(C) Sample Output

```
CSL/C3-1      Bisection Scheme

Lower bound: a ?     0
Upper bound: c ?     3
Tolerance: epsilon ? 0.0001
Iteration limit ?    20
 It.  a          b          c          f(a)        f(c)      abs(f(c)-f(a))/2
  1   0.000000,  1.500000,  3.000000,  3.000000,  0.000000  1.500e+00
  2   0.000000,  0.750000,  1.500000,  3.000000, -1.875000  2.438e+00
  3   0.750000,  1.125000,  1.500000,  0.984375, -1.875000  1.430e+00
```

4	0.750000,	0.937500,	1.125000,	0.984375,	-0.498047	7.412e-01
5	0.937500,	1.031250,	1.125000,	0.249756,	-0.498047	3.739e-01
6	0.937500,	0.984375,	1.031250,	0.249756,	-0.124969	1.874e-01
7	0.984375,	1.007813,	1.031250,	0.062496,	-0.124969	9.373e-02
8	0.984375,	0.996094,	1.007813,	0.062496,	-0.031250	4.687e-02
9	0.996094,	1.001953,	1.007813,	0.015625,	-0.031250	2.344e-02
10	0.996094,	0.999023,	1.001953,	0.015625,	-0.007812	1.172e-02
11	0.999023,	1.000488,	1.001953,	0.003906,	-0.007812	5.859e-03
12	0.999023,	0.999756,	1.000488,	0.003906,	-0.001953	2.930e-03
13	0.999756,	1.000122,	1.000488,	0.000977,	-0.001953	1.465e-03
14	0.999756,	0.999939,	1.000122,	0.000977,	-0.000488	7.324e-04
15	0.999939,	1.000031,	1.000122,	0.000244,	-0.000488	3.662e-04

```
Tolerance is satisfied.
Final result:  Approximate root = 1.00003
```

(D) Discussions

The initial interval specified by the input is $[0, 3]$. A root found in this interval equals 1.000031. To make sure the computed result is a valid root, it is important to check that the values of the last three columns of the output converge to zero because the bisection program may capture a pole instead. If a pole is captured, the functional values diverge, so it can be detected easily by checking the three columns.

Quiz: (1) Suppose that a huge number of files are stored in a cabinet. Many files are missing, but fortunately, the files are placed in increasing order of the file numbers. If it would take hours to go through the files one by one, how can you find the file with a minimal effort?

PROGRAM 3-2 Root Search

(A) Explanations

The bisection method and some other methods need information on an interval in which the desired root can be found. One way of finding such an interval is to print out the functional values for abscissas at a constant increment. If the sign of the function changes in an interval, one or an odd number of roots, or poles, or both, may be contained in that interval.

This program searches for intervals of a fixed size containing roots or poles of a function, $f(x)$. The interactive input consists of the lower and upper limits of x for the search, and an interval size h. The intervals in which the sign of the function changes (containing one or an odd number of roots and poles) are printed out. The func() defines the equation to be solved, which can be changed by the user. Currently a sample equation in func() is given by

$$y = -19(x - 0.5)(x - 1) + e^x - e^{-2x}$$

(B) List

```
/* CSL/c3-2.c      Root Search  */
#include <stdio.h>
#include <math.h>
#define TRUE 1
/*          a : starting value of x for search
            b : ending value of x for search
            x_intvl : interval of x
            func_v, func(x) : functional value at x           */
main()
{
int i, ks;
float a, b, x_intvl, x, y, yb;
float f();
   printf( "\n\nCSL/C3-2        Root Search \n\n" );
   while( TRUE ) {
      printf( "First x ? " );          scanf( "%f", &a );
      printf( "Last x ? " );           scanf( "%f", &b );
      printf( "Increment of x ? " ); scanf( "%f", &x_intvl );
      i = 0; x = a;
      while( TRUE ) {
         i = i + 1;
         if( x > b )  break;
         y = f( x );
         if( !(i == 1 || y*yb > 0) )
             printf(  "Interval that may contain a root: [%g,   %g] \n",
                                                x - x_intvl, x );
             yb = y;
             x = x + x_intvl;
      }
      printf("\nType 1 to continue, or 0 to stop.\n");scanf("%d",&ks);
      if( ks != 1 )  exit(0);
   }
}

float f(x)     /* Definition of function */
float x;
{
   return( -19*(x  -  0.5)*(x-1)  +  exp(x)  -  exp( -2*x));
}
```

(C) Sample Output

```
CSL/C3-2        Root Search

First x ? -10
Last x ? 10
Increment of x ? 1
Interval that may contain a root: [0,   1]
Interval that may contain a root: [1,   2]
Interval that may contain a root: [6,   7]
```

```
First x ? 0
Last x ? 1
Increment of x ? 0.001
Interval that may contain a root: [0.405998,  0.406998]
```

(D) Discussions

In the first case of the run, intervals of width 1 are searched between $x = -10$ and $x = 10$. For this range of x, three intervals, each including at least one root or pole are found. In the second case, intervals of width 0.001 are searched between $x = 0$ and $x = 1$. An interval of $[0.405998, 0.406998]$ is found, which contains a root or a pole. Because obviously Eq. (A) has no pole, one root is indicated to exist in this interval and it is appproximately 0.4060.

PROGRAM 3–3Q Function Plotting in C (QuickC Version)

(A) Explanations

Plotting a function is a useful means to understand the function and find an approximate value for the solution. This program plots a function on the screen of IBM PC using QuickC. Another version named C3-3S.C that is a Silicon Graphic IRIS version is described after this program.

To run the program the user defines the function in function_to_plot(). When executed the program asks for the input data as follows:

Xmin: minimum of x of the figure?

Xmax: maximum of x of the figure?

Ymin: minimum of y of the figure?

Ymax: maximum of y of the figure?

   m: number of intervals in plotting the curve?

The curve is plotted by connecting two consecutive data points by a straight line. If m is small, the plotted curve may not be smooth, but if the number is too large it will waste computing time. Nonetheless, a very large number may be sometimes necessary, particularly when a portion of the curve has a very large slope. In most cases $m = 200$ is more than enough.

The following figure illustrates the relation among the three coordinate systems considered in the graphics. The first is the pixel coordinate on the screen, the second is the normalized domain of the graphics representing the screen, and the third is the x-y domain of the actual graph. The actual graphic domain is mapped to the dotted area on the normalized domain, and then the coordinates on the normalized domain are mapped onto the pixel domain. Assuming the x-y graphic domain is bounded by

Xmin $< x <$ Xmax and Ymin $< y <$ Ymax, the coordinates on the normalized domain are found by

$$\xi = (1 - \Delta\xi_{left} - \Delta\xi_{right}) \frac{x - Xmin}{Xmax - Xmin} + \Delta\xi_{left}$$

$$\eta = (1 - \Delta\eta_{bottom} - \Delta\eta_{top}) \frac{y - Ymin}{Ymax - Ymin} + \Delta\eta_{bottom}$$

The pixel coordinates corresponding to the preceding point are found by

pix_x $= \xi$ times [number of pixels in the hozizontal direction -1] $+1$

pix_y $= (1 - \eta)$ times [number of pixels in the hozizontal direction -1] $+1$

In the program, the preceding equations are reorganized to increase computational efficiency.

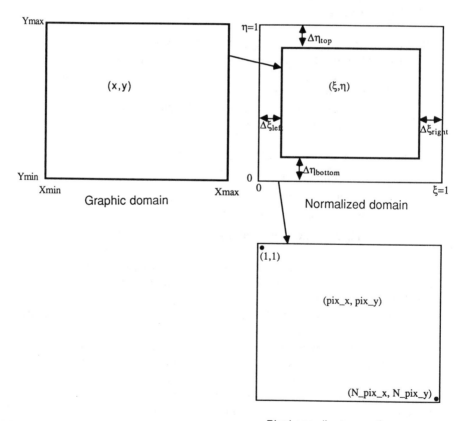

Figure 3.11 Relation among three coordinate systems

(B) List

```
/*CSL/c3-3q.c   Function Plotting (Graphics) with QuickC */
#include <stdio.h>
#include <stdlib.h>
#include <graph.h>
#include <float.h>
#include <math.h>
#include <conio.h>
/*   N_pix_x : number of pixels to the horizontal direction
             on the screen
     N_pix_y : same for the vertical direction
     mg_left, mg_rite, mg_bot, mg_top : left, right, bottom, and top
         margin on the normalized graphic domain.  The mornalized
         domain is in 0<x<1, and 0<y<1.
     Xmin, Xmax, Ymin, Ymax : limits of graphic coordinates
*/
int N_pix_x, N_pix_y, ix1, ix2, istp, x_len, iy1, iy2, y_len;
float   Xmin, Xmax, Ymin, Ymax, Ax, Bx, Ay, By;
static float mg_left=0.05, mg_rite=0.05, mg_top=0.05, mg_bot=0.05;
void begin_graphics( void );
void move_to(float x, float y);
void line_to(float x, float y);
void end_graphics( void );
float function_to_plot(float v);
main()
{
int  i, m;  float dx, x, y, x0, x1, y0, y1, l_xtic, l_ytic;
    printf( "CSL/C3-3    Function Plotting with QuickC    \n");
    printf( "Input the limits of the graphic domain:\n");
    printf( "Xmin = " );   scanf( "%f", &Xmin );
    printf( "Xmax = " );   scanf( "%f", &Xmax );
    printf( "Ymin = " );   scanf( "%f", &Ymin );
    printf( "Ymax = " );   scanf( "%f", &Ymax );
    printf( "Number of intervals = " );   scanf( "%d", &m );
    begin_graphics();
    printf( "Xmin = %4.2f, Xmax = %4.2f\n", Xmin, Xmax );
    printf( "Ymin = %4.2f, Ymax = %4.2f\n", Ymin, Ymax );
    l_xtic = (Xmax - Xmin)/100;  l_ytic = (Ymax - Ymin)/50;
    /* Plots the function */
    dx = (Xmax - Xmin)/m;
    x = Xmin;    y = function_to_plot(x);   move_to(x,y);
    for ( x = (Xmin+dx); x <= Xmax ; x = x+dx ){
       y = function_to_plot(x);   line_to(x,y);
    }
    /* Draw coordinate lines */
    move_to(0,Ymin); line_to(0,Ymax);move_to(Xmin,0); line_to(Xmax,0);
    /* draw tic marks */
    printf("Tic mark interval:\n" );
    ix1 = Xmin; ix2 = Xmax; istp = 1;if (Xmax - Xmin > 20) istp = 10;
    for (i=ix1; i<=ix2; i=i+istp) {
       move_to(i,0); line_to(i,l_ytic);
    }
    printf( "x:%d", istp );
    iy1 = Ymin; iy2 = Ymax; istp = 1; if (Ymax-Ymin > 20 ) istp = 10;
    for (i=iy1; i<=iy2; i=i+istp) {
       move_to(0, i);line_to(l_xtic,i);
```

```
   }
   printf( "   y:%d", istp );
   end_graphics();
}

void move_to( float X, float Y )
{
   int pix_x, pix_y;
   pix_x = Ax*X + Bx ; pix_y = Ay*Y + By; _moveto( pix_x, pix_y );
}

void line_to( float X, float Y )
{
   int pix_x, pix_y;
   pix_x = Ax*X + Bx; pix_y = Ay*Y + By; _lineto( pix_x, pix_y );
}

float function_to_plot (float  x)
{
   float  f;     f = sin(x)*exp(0.2*x);        return f;
}

void begin_graphics( void )
{
   struct  videoconfig myscreen;
   _getvideoconfig( &myscreen );
   switch( myscreen.adapter ){
          case _CGA:  case _OCGA:
                 _setvideomode( _HRESBW ); break;
          case _EGA:  case _VGA:  case _OVGA: case _HGC:
                 _setvideomode( _HERCMONO ); break;
          default: printf( "This program requires a CGA, EGA, \
                 VGA or Hercules card\n" );
          exit(0);
   }
   _getvideoconfig( &myscreen );
   N_pix_x = myscreen.numxpixels-1;  N_pix_y = myscreen.numypixels-1;
   Ax=(1.0 - mg_left - mg_rite)/(Xmax-Xmin);
   Bx=mg_left - Ax*Xmin; Ax=Ax*N_pix_x;  Bx=Bx*N_pix_x;
   Ay=(1.0 - mg_bot - mg_top)/(Ymax - Ymin);
   By=mg_bot - Ay*Ymin;  Ay = -Ay*N_pix_y; By=(1.0 - By)*N_pix_y;
   _setvieworg( 0, 0 );
}

void end_graphics( void )
{
   getch();_setvideomode( _DEFAULTMODE );
}
```

(C) Sample Output

```
CSL/C3-3     Function Plotting with QuickC

Input the limits of the graphic domain:
Xmin = -10
Xmax = 10
Ymin = -10
Ymax = 10
Number of intervals = 100
```

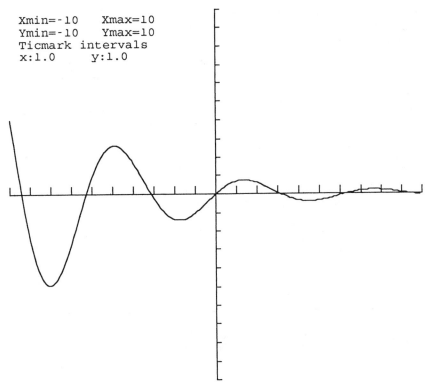

```
Xmin=-10    Xmax=10
Ymin=-10    Ymax=10
Ticmark intervals
x:1.0       y:1.0
```

Figure 3.12 Graph plotted on the screen

(D) Discussions

The function currently in function_to_plot is

$$y = e^{-0.2x} \sin(x)$$

which can be easily changed by the user.

Although the current program is written for QuickC, it can be converted to other graphic environments with appropriate conversion of the graphic commands.

PROGRAM 3–3S Function Plotting in C (IRIS Version)

(A) Explanations

This program does exactly the same plotting job on Silicon Graphics IRIS computer. The programming is, however, different from C3-3Q.C because of the difference in graphic commands. In the graphic library of IRIS, positions of the pixels are defined as shown in Fig. 3.13. The pixel at the bottom left corner is (1, 1), and the top right corner is (Nx, Ny) where Nx and Ny are the number of pixels in the horizontal and vertical directions. The command prefposition($nx1$, $nx2$, $ny1$, $ny2$) specifies the domain of the pixels to be used for the graphic plotting. In the list shown next, the horizontal pixel number 100 to 800 and the vertical pixel 100 to 800 will be used as the graphic area. By changing these numbers, the size and location of the graph on the screen may be changed. Command ortho2($x1$, $x2$, $y1$, $y2$) determines the correspondence between the graphic area on the x-y graphic plane and the pixel domain for the graphics on the screen. Namely, the corner ($x1$, $y1$) is mapped to the pixel ($nx1$, $ny1$), while the corner ($x2$, $y2$) to ($nx2$, $ny2$). Command move2(x, y) moves the pen to the pixel corresponding to (x, y) on the x-y coordinates, $v2f$(vert1) starts plotting a line in its first appearance in the subroutine, and $v2f$(vert2) terminates the line drawing in the second appearance. The function to be plotted is defined in subroutine function_to_plot(). The interactive input procedure is the same as C3-3Q.C.

To compile and link the program, type

```
>cc c3-3s.c -lgl_s -lc_s -lm -o c3-3s
```

To run the program, type

```
>c3-3s
```

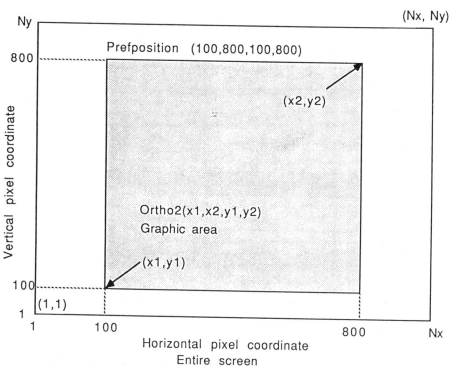

Figure 3.13 Relation between graphic coordinates and pixel coordinates on IRIS

The graph will stay on the screen for 30 seconds and disappear.

(B) List

```
/* CSL/c3-3iris.c    Function Plotting (Graphics) with C */

#include <stdio.h>
#include <string.h>
#include <math.h>
#include <gl/gl.h>
#define     N_pix_x   1000
#define     N_pix_y   1000

/* N_pix_x : number of pixels to the horizontal direction on the screen
   N_pix_y : same for the vertical direction
   Xmin, Xmax, Ymin, Ymax : limits of graphic coordinates           */


int  ix1, ix2, istpx, istpy, x_len, iy1, iy2, y_len;
float Xmin, Xmax, Ymin, Ymax, del_x, del_y, Ax, Bx, Ay, By;
float x_mgnl, x_mgnr, y_mgnb, y_mgnt, x[N_pix_x], y[N_pix_y];
char tic_x[100], tic_y[100];
char Xmin_str[100], Xmax_str[100], Ymin_str[100], Ymax_str[100];
void draw_lines( float X1, float Y1, float X2, float Y2 );
float function_to_plot( float p );


main()
{
int   i, m;
float   dx, x0, x1, y0, y1, l_xtic, l_ytic;
    printf( "CSL/C3-3 Function Plotting with C  \n");
    printf( "Input the limits of the graphic domain:\n");
    printf( "Xmin = " ); scanf( "%f", &Xmin );
    printf( "Xmax = " ); scanf( "%f", &Xmax );
    printf( "Ymin = " ); scanf( "%f", &Ymin );
    printf( "Ymax = " ); scanf( "%f", &Ymax );
    printf( "Number of intervals = " ); scanf( "%d", &m );


    x_mgnl = Xmin-(Xmax - Xmin)*0.05; x_mgnr = Xmax+(Xmax - Xmin)*0.05;
    y_mgnt = Ymax+(Ymax - Ymin)*0.05; y_mgnb = Ymin-(Ymax - Ymin)*0.05;


    prefposition(100, 800, 100, 800); winopen("C3-3");
    ortho2(x_mgnl, x_mgnr, y_mgnb, y_mgnt);
    color(WHITE);
    clear(); color(BLUE);

    l_xtic = (Xmax - Xmin)/100; l_ytic = (Ymax - Ymin)/100;
                                              /* Plot the function */
    dx = (Xmax - Xmin)/m;
    x[0] = Xmin;     y[0] = function_to_plot(x[0]);

    color(RED);
    for ( i = 0; i < m; i++ ){
        x[i+1] = x[i] + dx;
        y[i+1] = function_to_plot(x[i+1]);
        draw_lines(x[i],y[i],x[i+1],y[i+1]);
    }
                                         /* Draw coordinate lines */
```

```
    color(BLUE);
    draw_lines(0,Ymin,0,Ymax); draw_lines(Xmin,0,Xmax,0);
                                        /* Draw tic marks */
    ix1 = Xmin; ix2 = Xmax; istpx = 1; if (Xmax - Xmin > 20) istpx = 10;
    for (i=ix1; i<=ix2; i=i+istpx) draw_lines(i,0,i,l_ytic);
    iy1 = Ymin; iy2 = Ymax; istpy = 1; if (Ymax - Ymin > 20) istpy = 10;
    for (i=iy1; i<=iy2; i=i+istpy)  draw_lines(0,i,l_xtic,i);
                                        /* Print texts on the window */
    color(BLACK);
    del_x = (Xmax - Xmin)/40.0; del_y = (Ymax - Ymin)/40.0;
    sprintf( tic_x, "x: %d", istpx);
    sprintf( tic_y, "  y: %d", istpy);
    sprintf( Xmin_str, "Xmin = %4.2f", Xmin);
    sprintf( Xmax_str, "  Xmax = %4.2f", Xmax);
    sprintf( Ymin_str, "Ymin = %4.2f", Ymin);
    sprintf( Ymax_str, "  Ymax = %4.2f", Ymax);
    cmov2(Xmin, Ymax-del_y);      charstr("Tic mark interval:");
    cmov2(Xmin, Ymax-2*del_y); charstr(tic_x);      charstr(tic_y);
    cmov2(Xmin, Ymax-3*del_y); charstr(Xmin_str);charstr( Xmax_str );
    cmov2(Xmin, Ymax-4*del_y); charstr(Ymin_str);charstr( Ymax_str );
    sleep(30); gexit(); exit(0);
}

void draw_lines( float X1, float Y1, float X2, float Y2 )
{
float vert1[2], vert2[2];
    vert1[0] = X1; vert1[1] = Y1;
    vert2[0] = X2; vert2[1] = Y2;
    bgnline(); v2f(vert1); v2f(vert2); endline();
}

float function_to_plot (float p)
{
float f, p_c;
    f = sin(p)*exp(   -0.3*p);
    return f;
}
```

(C) Sample Output and Discussions

The graph plotted is the same as PROGRAM C3-3Q, except it is colored.

PROGRAM 3-4 Modified False Position Method

(A) Explanation

This program finds a root of

$$f(x) = 0$$

by the modified false position method. The user must define the equation to be solved in func(x), which is currently set to $f(x) = \tan(x) - x - 0.5$.

(B) List

```
/* CSL/c3-4.c   Modified False Postion Method     */
#include <stdio.h>
#include <stdlib.h>
#include <math.h>
#define TRUE 1
/*            a, c : current end points
              kl : counter of stagnation at the left end point
              kr : counter of stagnation at the right end point
              ep : tolerance
              Y_a, Y_c : y value at the current end points
              func(x) : function
              it_limit : iteration limit                                    */
main()
{
int it_limit, it, kl, kr;
float a, b, bb, c, ep, gr, Y_a, Y_b, Y_c;
double func();
    printf( "\nCSL/C3-4   Modified False Position Method \n\n" );
    printf( "Lower bound: a ?   " );     scanf( "%f", &a );
    printf( "Upper bound: c ?   " );     scanf( "%f", &c );
    printf( "Tolerance: ep ?    " );     scanf( "%f", &ep );
    printf( "Iteration limit ?  " );     scanf( "%d", &it_limit );
    printf(
      " It.   a              b              c          f(a)         f(c)      " );
    printf(" abs(f(a)-f(c))");
    Y_a = func( a );
    Y_c = func( c );     printf(  "\n" );
    it = 0;
    kl = 1;
    kr = 1;
    while( TRUE ) {
        it = it + 1;
        if( it > it_limit ) {
            printf( "Iteration limit exceeded\n" );
            break;
        }
        gr = (Y_c - Y_a)/(c - a);
        bb = b;
        b = -Y_a/gr + a;
        Y_b = func( b );
        printf(  "%3d, %9.5f, %9.5f, %9.5f, %9.5f, %9.5f, %12.4e\n",
                          it, a, b, c, Y_a, Y_c, (Y_c-Y_a)/2 );
        if( fabs( bb - b ) < ep ) {
            printf( "                         Tolerance satisfied\n" );
            break;
        }
        if( Y_a*Y_b <= 0 ) {
            Y_c = Y_b;
            c = b;
            kl = kl + 1;   kr = 0;
            if( kl > 1 ) Y_a = Y_a/2;
        }
        else {
            Y_a = Y_b;
            a = b;
```

```
          kr = kr + 1;   kl = 0;
          if ( kr > 1 )     Y_c = Y_c/2;
       }
   }
   printf( " ---------------------------------------------- \n" );
   printf( " Approximate root = %g \n", b );
   printf( " ---------------------------------------------- \n" );
   exit(0);
}

double func(x)     /* Definition of function */
float x;
{
float func_v;
   func_v = tan( x ) - x - .5;
   return( func_v );
}
```

(C) Sample Output

```
CSL/C3-4     Modified False Position Method

Lower bound: a ?    0
Upper bound: c ?    1.5
Tolerance: ep   ?   0.0001
Iteration limit ?   20
 It.    a           b           c         f(a)        f(c)       abs(f(a)-f(c))
  1,   0.00000,    0.05952,    1.50000,  -0.50000,  12.10142,    6.3007e+00
  2,   0.05952,    0.16945,    1.50000,  -0.49993,   6.05071,    3.2753e+00
  3,   0.16945,    0.35763,    1.50000,  -0.49836,   3.02536,    1.7619e+00
  4,   0.35763,    0.63451,    1.50000,  -0.48393,   1.51268,    9.9830e-01
  5,   0.63451,    0.93315,    1.50000,  -0.39846,   0.75634,    5.7740e-01
  6,   0.93315,    1.03560,    1.50000,  -0.08343,   0.37817,    2.3080e-01
  7,   0.93315,    0.96961,    1.03560,  -0.08343,   0.15097,    1.1720e-01
  8,   0.96961,    0.97433,    1.03560,  -0.01162,   0.15097,    8.1296e-02
  9,   0.97433,    0.97552,    1.03560,  -0.00149,   0.07549,    3.8489e-02
 10,   0.97433,    0.97502,    0.97552,  -0.00149,   0.00109,    1.2925e-03
 11,   0.97502,    0.97502,    0.97552,   0.00000,   0.00109,    5.4661e-04
                       Tolerance satisfied
------------------------------------------------------
Approximate root = 0.975017
------------------------------------------------------
```

(D) Discussions

In the foregoing output, a solution for the equation

$$f(x) = \tan(x) - x - 0.5 = 0$$

is found. The interval of $[0, 1.5]$ is specified by the input interactively. In the output, it is seen that the c value of $c = 1.5$ was repeated 7 times including the initial value. Then $f(c)$ value is divided by 2 each time c is repeated in accordance with the modified false position method. Also in the 7th to 9th iteration steps, $c = 1.0356$ stays the same, so the $f(c)$ is divided by 2. If $f(c)$ value is not halved, the convergence would become much slower.

PROGRAM 3-5 Newton's Method

(A) Explanations

This program finds a solution of

$$f(x) = 0$$

by Newton's method. The equation to be solved and its first derivative are defined in func(). The sample function defined there currently is

$$y = x^3 - 5x^2 + 6x$$

with its derivative

$$y' = 3x^2 - 10x + 6$$

The first arguments in the func() is a numerical value of x, while the second and third arguments are pointers, because they allow the values of y and $y\_derivative$ inside the function to be changed.

Often, analytical derivation of the first derivative is found to be cumbersome. To simplify the work, one can use a finite difference approximation as suggested in Section 3.4. The convergence rate and accuracy are not affected by the difference approximation.

(B) List

```
/* CSL/c3-5.c          Newton's Scheme */
#include <stdio.h>
#include <stdlib.h>
#include <math.h>
#define TRUE 1
/*            x : current x value (approximation for the root)
              xb: previous value of x
              y : y value for x
              y_derivative : derivative of y
              n : iteration counter                                 */
main()
{
int i, k, n;
float epsilon, x, xb, y, y_derivative, error;
void func();
   printf( "\nCSL/C3-5      Newton's Scheme  \n" );
   printf( "\nTolerance ?    " );  scanf( "%f", &epsilon );
   while( TRUE) {
      printf( "Initial guess ?    " );     scanf( "%f", &x );
      error=1.0e10;   xb = x;
      n = 0;
      printf( "\n It.No.     x(n-1)        y(n-1)         x(n)" );
      while( error>epsilon ) {
```

```
        n = n + 1;
        func( x, &y, &y_derivative );
        x = x - y/y_derivative;          /* finds new x. */
        printf(  "\n %3d    %12.5e    %12.5e    %12.5e ", n, xb, y, x );
        error=fabs(x-xb);       xb = x;
    }
    printf( "\n-------------------------------\n" );
    printf( " Final solution = %g \n", x );
    printf( "-------------------------------\n" );
    printf( "\nType 1 to continue, or 0 to stop.");scanf( "%d", &k );
    if( k != 1 )  exit(0);
  }
}

void func(x, y, y_derivative)   /*Computes y and y_derivative */
float x, *y, *y_derivative;
{
    *y = pow(x,  3)  -  5.0*pow(x,  2)  +  6.*x;
    *y_derivative = 3.0*pow(x,  2)  -  10.0*x  +  6.;
    return;
}
```

(C) Sample Output

```
    CSL/C3-5       Newton's Scheme

    Tolerance ?   0.00001
    Initial guess ?   4.0
     It.No.      x(n-1)           y(n-1)            x(n)
         1     4.00000e+00      8.00000e+00       3.42857e+00
         2     3.42857e+00      2.09913e+00       3.12782e+00
         3     3.12782e+00      4.50898e-01       3.01708e+00
         4     3.01708e+00      5.24009e-02       3.00038e+00
         5     3.00038e+00      1.12566e-03       3.00000e+00
         6     3.00000e+00      7.15256e-07       3.00000e+00
    -------------------------------
     Final solution = 3
    -------------------------------

    Initial guess ?   1.4
     It.No.      x(n-1)           y(n-1)            x(n)
         1     1.40000e+00      1.34400e+00       2.03396e+00
         2     2.03396e+00     -6.67319e-02       1.99936e+00
         3     1.99936e+00      1.27785e-03       2.00000e+00
         4     2.00000e+00      4.76837e-07       2.00000e+00
    -------------------------------
     Final solution = 2
    -------------------------------
```

(D) Discussions

In the foregoing output, two zeroes of

$$f(x) = x^3 - 5x^2 + 6x = 0$$

are found. First, with the initial guess of $x = 4$, the solution $x = 3$ is found. Second, with the initial guess of 1.4, the solution $x = 2$ is found.

PROGRAM 3-6 Newton's Method for Complex Roots

(A) Explanations

Newton's method explained in PROGRAM 3-5, namely

$$x_i = x_{i-1} - f(x_{i-1})/f'(x_{i-1})$$

can be applied in the complex domain to find a complex root. In the foregoing equation, x_is are interpreted as complex variables and f' is defined by

$$f'(x) = \lim_{\Delta \to 0} \frac{f(x + \Delta) - f(x)}{\Delta}$$

where x and Δ are both complex variables. This program finds a complex root of a polynomial by the Newton's method. The coefficients of the polynomial are defined in the beginning of the program. The sample polynomial currently defined is given by

$$y = x^5 - 0.2x^4 + 7x^3 + x^2 - 3.5x + 2.0$$

Roots of the same equations are discussed also in Example 7.6.

Unlike Fortran, the C language does not have complex variables. Therefore the program needs to have its own subroutines to perform complex arithmetics. In the present program, addition/subtraction, multiplication, and division are performed by add(), mult(), and div(), respectively. The arguments in the functions here are all pointers.

(B) List

```
/* CSL/c3-6.c          Newton's Method for Complex Roots */
#include <stdio.h>
#include <stdlib.h>
#include <math.h>
#define TRUE 1
/*        (xr,xi) : initial guess, iteratives, and final result (complex)
          (fr,fi) : functional value (complex)
          (fdr,fdi) : derivative of the function (complex)
          n : order of polynomial
          a[i] : coefficient for the ith power                     */
main()
{
```

```
int k, m, i;
float fdi, fdr, fi, fr, ti, tim, tr, trm, xi, xib, xr, xrb;
void add(),mult(), divide(), fdd();
static int n = 5;                                    /* order of polynomial*/
static float a[11]={2.0, -3.5, 1, 7., -0.2, 1}; /*coefficients*/
static float eps = 0.000001;
   printf("\nCSL/C3-6     Newton's Method for Complex Roots \n\n");
   while( TRUE ) {
       printf("Type real part of initial guess:\n"); scanf("%f",&xr);
       printf( "Type imaginary part of initial guess:\n");
       scanf( "%f",&xi );
       printf(    "\nA poloynomial of order: %d\n",n);
       printf(    "Power      Coefficient     \n");
       for (i=0; i<=n; i++){
          printf( " %d        %12.6f \n", i, a[i]);
       }
       printf( "\nIt. No.  Iterative Approximation (complex)\n" );
       for( k = 1; k <= 50; k++ ) {
          xrb = xr;    xib = xi;
          fdd( &n, a, &xr, &xi, &fr, &fi, &fdr, &fdi );
          divide( &fr, &fi, &fdr, &fdi, &tr, &ti );
          trm = -tr;  tim = -ti;
          add( &xr, &xi, &trm, &tim, &xr, &xi );
          printf( "%2d        (%12.5e) + (%12.5e)i \n",k, xr, xi );
          if( fabs(xr - xrb) + fabs(xi - xib) <eps) break;
       }
       printf("\nConverged solution:(%12.5e) + (%12.5e)i \n",xr,xi);
       printf("\nType 1 to continue, 0 otherwise\n");scanf("%d", &m);
       printf(" \n"); if( m != 1 )  exit(0);
   }
}

void fdd(n, a, xr, xi, fr, fi, fdr, fdi) /* Polynomial and derivative */
int *n;   float a[], *xr, *xi, *fr, *fi, *fdr, *fdi;
{
int i, i_;   float b, ti, tr, zero, zi, zr;
void add(), mult();
   zero = 0.0;
   *fr = a[0];     *fi = 0.0;     zr = 1.0;      zi = 0.0;
   *fdr = 0.0;     *fdi = 0.0;
   for( i = 1; i <= *n; i++ ) {
      b = a[i]*i;
      mult( &zr, &zi, &b, &zero, &tr, &ti );
      add( fdr, fdi, &tr, &ti, fdr, fdi );
      mult( xr, xi, &zr, &zi, &zr, &zi );
      mult( &zr, &zi, &a[i], &zero, &tr, &ti );
      add( fr, fi, &tr, &ti, fr, fi );
   }
   return;
}

void add(xr, xi, yr, yi, zr, zi)  /* Addition */
float *xr, *xi, *yr, *yi, *zr, *zi;
{
   *zr = *xr + *yr;  *zi = *xi + *yi;
   return;
}
```

```
void mult(xr, xi, yr, yi, zr, zi) /* Multiplication */
float *xr, *xi, *yr, *yi, *zr, *zi;
{
float ai, ar;
   ar = *xr**yr - *xi**yi;     ai = *xr**yi + *xi**yr;
   *zr = ar;      *zi = ai;      return;
}


void   divide(xr, xi, yr, yi, zr, zi)   /* Division */
float *xr, *xi, *yr, *yi, *zr, *zi;
{
float den, deni, yim;
void mult();
   yim = -*yi;      mult( yr, yi, yr, &yim, &den, &deni );
                    mult( xr, xi, yr, &yim, zr, zi );
   *zr = *zr/den;      *zi = *zi/den;      return;
}
```

(C) Sample Output

```
CSL/C3-6     Newton's Method for Complex Roots

Type real part of initial guess:  1
Type imaginary part of initial guess:  1
It. No.   Iterative Approximation (complex)
   1      ( 6.66086e-01) + ( 7.15681e-01)i
   2      ( 4.92989e-01) + ( 4.93482e-01)i
   3      ( 4.27748e-01) + ( 3.80124e-01)i
   4      ( 4.14529e-01) + ( 3.52202e-01)i
   5      ( 4.13905e-01) + ( 3.50649e-01)i
   6      ( 4.13904e-01) + ( 3.50645e-01)i
   7      ( 4.13904e-01) + ( 3.50645e-01)i

Converged solution:( 4.13904e-01) + ( 3.50645e-01)i

Type real part of initial guess:  -4
Type imaginary part of initial guess:  9
It. No.   Iterative Approximation (complex)
   1      (-3.16434e+00) + ( 7.25070e+00)i
   2      (-2.48683e+00) + ( 5.86379e+00)i
   3      (-1.93177e+00) + ( 4.76997e+00)i
   4      (-1.46788e+00) + ( 3.91480e+00)i
   5      (-1.06500e+00) + ( 3.25625e+00)i
   6      (-6.88271e-01) + ( 2.76417e+00)i
   7      (-2.83118e-01) + ( 2.42766e+00)i
   8      ( 2.92584e-01) + ( 2.34942e+00)i
   9      (-2.81054e-01) + ( 2.82589e+00)i
  10      (-2.12660e-03) + ( 2.62604e+00)i
  11      ( 1.79943e-01) + ( 2.70906e+00)i
  12      ( 1.38263e-01) + ( 2.72986e+00)i
  13      ( 1.40386e-01) + ( 2.73145e+00)i
  14      ( 1.40378e-01) + ( 2.73145e+00)i
  15      ( 1.40378e-01) + ( 2.73145e+00)i

Converged solution:( 1.40378e-01) + ( 2.73145e+00)i
```

(D) Discussions

In the foregoing run, two roots among five of the fifth order polynomial are found with randomly selected initial guesses. To find all the roots, it is important to know an initial guess of each root. Bairstow's method, or PROGRAM 3–7, may be used for this purpose. The initial guess for a complex root must be a complex value. (The roots of the same polynomial are also computed by the QR iteration in Example 7.6.)

PROGRAM 3–7 Bairstow's Method

(A) Explanations

This program finds a pair of roots including complex conjugate pair for a polynomial by Bairstow's method. The order of polynomial and its coefficients are given interactively as input.

Initial guesses for p and q are set to zero in the program. If zero division (a rare but unpredictable incidence) occurs, rerun the program with a different set of initial guesses for p and q. There is no need to change the program, except when a zero division occurs.

(B) List

```
/* CSL/c3-7.c    Bairstow Scheme  */
#include <stdio.h>
#include <stdlib.h>
#include <math.h>
#define TRUE 1
/*            a[i] : coefficients of the polynomial
              b[i] : coefficients in Eq. (3.7.3) and Eq. (3.7.4)
              eps : tolerance of convergence
              n : order of polynomial
              p, q : p and q
              p_ini, q_ini: initial guess for p and q (arbitrary)
              dp, dq : Greek "delta" p, and Greek "delta" q, respectively
*/
main()
{
int i, i_, k, n,  iter;
float a[11], b[11], bp[11], bq[11], dn, dp, dq, p, q, rt, eps, zz;
static float p_ini=0, q_ini=0;        /* any arbitray initial guess */
   printf( "\n\nCSL/C3-7    Bairstow Scheme \n" );
   printf( "\nShould zero-divide occur, alter the values for");
   printf( "\np_ini and q_ini and rerun the program.\n");
   while( TRUE ) {
      for( i = 0; i <= 10; i++ ) {
         a[i] = 0;              b[i] = 0;
         bp[i] = 0;            bq[i] = 0;
      }
      printf( "\nThe order of polynomial ?  " );  scanf( "%d", &n );
      for( i = 0; i <= n; i++ ){
         printf( "a(%d)?   ", i );   scanf( "%f", &a[i] );
      }
```

```
printf( "Tolerance ?    " );        scanf(  "%f", &eps );
p = p_ini;  q = q_ini;
for (iter=1; iter< 100; iter++){
   for( i=n; i >= 1; i--) b[i] = a[i] - p*b[i + 1] - q*b[i + 2];
   b[0] = a[0] - q*b[2];
   for( i = n; i >= 1; i-- ){
      bp[i] = -b[i + 1] - p*bp[i + 1] - q*bp[i + 2];
      bq[i] = -p*bq[i + 1] - b[i + 2] - q*bq[i + 2];
   }
   bp[0] = -q*bp[2];
   bq[0] = -q*bq[2] - b[2];
   dn = bp[0]*bq[1] - bp[1]*bq[0];
   dp = (b[0]*bq[1] - b[1]*bq[0])/dn;  p = p - dp;
   dq = (b[1]*bp[0] - b[0]*bp[1])/dn;  q = q - dq;
   if( fabs( dq ) + fabs( dp ) <= eps ) break;
}             /* Convergence test passed. */
printf( "--------------------------------------------------\n" );
printf( " p =%g      q =%g \n", p, q );
printf( "Quadratic coefficient = x^2 + (%g x) + (%g )\n",p, q );
printf( "Roots of the quadratic factor:\n" );
zz = p*p - 4*q;
if( zz >= 0 ){
   rt = sqrt( zz );                 /* Prints a pair of real roots */
      printf( "   %g %g \n", (-p + rt)/2, (-p - rt)/2 );
}
else if( zz < 0 ) {
   rt = sqrt( -zz );        /* Prints complex conjugate roots*/
   printf( "   %g + %gi\n", -p/2, rt/2 );
   printf( "   %g - %gi\n", -p/2, rt/2 );
}
printf( "Coefficients of deflated polynomial\n" );
printf( "       order    coefficients\n" );
for( i = 2; i <= n; i++ ){
   printf( "       %3d     %13.5f \n", i-2, b[i] );
}
printf( "--------------------------------------------------\n" );
printf( "Type 1 to continue, or 0 to stop.\n" );
scanf( "%d", &k );  if( k != 1 ) exit(0);
   }
}
```

(C) Sample Output

See Example 3.9

(D) Discussions

The polynomial defined by the input is

$$y = x^4 - 1.1x^3 + 2.3x^2 + 0.5x + 3.3$$

for which a quadratic factor, $Q(x) = x^2 + 0.9x + 1.1$, is found. The roots of the quadratic factor, as well as for the polynomial, are $x = -0.45 + 0.9473648i$ and

$x = 0.45 - 0.9473648i$. The deflated polynomial is

$$G(x) = x^2 - 2x + 3$$

Quiz: To find a quadratic factor for double roots by Bairstow's method is often difficult. If you don't believe it, try for

$$y = x^5 \pm 5x^4 + 10x^3 \pm 10x^2 + 5x \pm 1$$

both of which has a quintuple root that equals $x = 1$.

PROBLEMS

(3.1) Find the positive root for $x^2 - 0.9x - 1.52 = 0$ in the interval $[1, 2]$ by the bisection method with a tolerance of 0.001.

(3.2) Find the root of

$$x \sin (x) - 0.1 = 0, \quad 0 < x < 1.0$$

by the bisection method with a tolerance of 0.001.

(3.3) Calculate the root of $\tan (x) = 3.5$ in the interval $[0, \pi]$ by the bisection method with a tolerance of 0.005.

(3.4) (a) Determine an interval of width 0.5 for each positive root of the following equations by using PROGRAM 3–2:

(i) $f(x) = 0.5e^{x/3} - \sin x \quad = 0, \quad x > 0$
(ii) $f(x) = \log_e(1 + x) - x^2 = 0$

(b) Plot the functions defined above on the x-y plane by using PROGRAM 3–3 and verify the results in (a).

(3.5) (a) Determine an interval of width 0.5 for each root of the following equations by using PROGRAM 3–2 with a modification to the subprogram:

(i) $f(x) = e^x - 5x^2 = 0$
(ii) $f(x) = x^3 - 2x - 1 = 0$
(iii) $f(x) = \sqrt{x} + 2 - x = 0$

(b) Plot the functions defined above on the x-y plane by using PROGRAM 3–3 and verify the results in (a).

(c) Calculate the largest root of each of the problems in (a) by the bisection method with a tolerance of 0.0001.

(3.6) Find the root of

$$\frac{x(2.1 - 0.5x)^{1/2}}{(1 - x)(1.1 - 0.5x)^{1/2}} = 3.69, \quad 0 < x < 1$$

in the interval $[0, 1]$ using PROGRAM 3–1 with a change to the subprogram and with a tolerance of 0.001.

(3.7) Find all the positive roots of the following equations by the bisection method with a tolerance of 0.001. (First, determine an appropriate interval for each root by PROGRAM 3–3 or by listing the functional values for selected values of x.)

(a) $\tan(x) - x + 1 = 0, \quad 0 < x < 3\pi$

(b) $\sin(x) - 0.3e^x = 0, \quad x > 0$

(c) $-x^3 + x + 1 = 0$

(d) $16x^5 - 20x^3 + x^2 + 5x - 0.5 = 0$

(3.8) Find appropriate intervals for the roots of the following equations, and then determine the roots by the bisection method (use PROGRAM 3–1) with a tolerance of 0.001:

(a) $0.1x^3 - 5x^2 - x + 4 + e^{-x} = 0$

(b) $\log_e(x) - 0.2x^2 + 1 = 0$

(c) $x + \dfrac{1}{(x+3)x} = 0$

(3.9) A bullet of $M = 2$ gm has been shot vertically into the air and is descending at its terminal speed. The terminal speed is determined by $gM = D_{\text{drag}}$ where g is gravity and M is the mass, and may be written after evaluating the constants as

$$\frac{(2)(9.81)}{1000} = 1.4 \times 10^{-5} v^{1.5} + 1.15 \times 10^{-5} v^2$$

where v is the terminal velocity, m/sec. The first term on the right side represents the friction drag, and the second term represents the pressure drag. Determine the terminal velocity by the bisection method with a tolerance of 0.01.

(3.10) The surface configuration of the NACA 0012 airfoil of chord length 1 m and maximum thickness of 0.2 m is given by

$$y(x) = \pm[0.2969\sqrt{x} - 0.126x - 0.3516x^2 + 0.2843x^3 - 0.1015x^4]$$

where plus and minus signs refer to upper and lower surfaces respectively. Determine x where the thickness of the airfoil is 0.1 m by using the bisection method. Set tolerance to 0.00001. (There are two solutions.)

(3.11) One Kg mole of CO is contained in a vessel at $T = 215°$ K and $p = 70$ bars. Calculate the volume of the gas by using the van der Waals equation of state for a nonideal gas given [Moran/Shapiro] by

$$\left(P + \frac{a}{v^2}\right)(v - b) = RT$$

where $R = 0.08314$ bar m³/(Kg mol °K), $a = 1.463$ bar m⁶/(Kg mol)² and $b = 0.0394$ m³/Kg. Determine the specific volume v(m³/Kg) and compare the result to the volume calculated by the ideal gas equation, $Pv = RT$.

(3.12) Find the root of $f(x) = \sin(x) - x + 1$ that is known in $1 < x < 3$ by the modified false position method. Stop calculations after four iterations.

(3.13) Find the roots of the following equations by the modified false position method:

(a) $f(x) = 0.5 \exp(x/3) - \sin(x) = 0, \quad x > 0$

(b) $f(x) = \log(1 + x) - x^2 = 0$

(c) $f(x) = \exp(x) - 5x^2 = 0$

(d) $f(x) = x^3 + 2x - 1 = 0$

(e) $f(x) = \sqrt{x + 2} - x = 0$

(3.14) The transfer function for a system is given by

$$F(s) = \frac{H(s)}{1 + G(s)H(s)}$$

where

$$G(s) = \frac{1}{s}\exp(-0.1s), \quad H(s) = K$$

Investigate the roots of the characteristic equation $1 + G(s)H(s) = 0$ for K = 1, 2, and 3 by the graphic method, and then evaluate them by the modified false position method.

(3.15) Find the root of

$$\tan(x) - 0.1x = 0$$

in $\pi < x < 1.5\pi$ by Newton's method with a handheld calculator (tolerance is 0.0001).

(3.16) Find the roots of the equations in Problem 3.7 by Newton's method with a tolerance of 0.0001.

(3.17) The natural frequencies of vibration of a uniform beam clamped at one end and free at the other end [Thomson] are solutions of

$$\cos(\beta l)\cosh(\beta l) + 1 = 0 \tag{A}$$

where

$\beta = \rho\omega^2/EI$
$l = 1$ (length of the beam, m)
$\omega = $ frequency, sec^{-1}
$EI = $ flexural rigidity [Byars/Snyder/Plants]
$\rho = $ density of the beam material

Investigate the roots of Equation (A) first by the graphic method, and then determine the lowest three values of β satisfying Equation (A) by Newton's method.

(3.18) Natural frequencies of the vibration of a beam clamped at two ends satisfy

$$\tan(\beta l) = \tanh(\beta l), \quad \beta > 0$$

where l is assumed to be 1 m as given in Problem (3.17). By Newton's method based on a difference approximation to evaluate the derivative, determine the lowest three values of $\beta > 0$ satisfying the equation above. Don't include $\beta = 0$ as an answer. (*Hint:* $\tanh(x) = [\exp(x) - \exp(-x)]/[\exp(x) + \exp(-x)]$)

(3.19) Repeat Problem (3.12) by Newton's method.

(3.20) Find all the roots of the equations in Problem 3.13 by Newton's method.

(3.21) Two complex roots of

$$y = 2 - x + 2x^2 + x^4$$

are approximately $-0.5 + 1.5i$ and $0.5 - 0.7i$. Using those values as initial guesses, find exact values of the two complex roots by Newton's method (use PROGRAM 3-6).

(**3.22**) An equimolar mixture of carbon monoxide and oxygen attains equilibrium at $300°$ K and 5 atm pressure. The theoretical reaction is

$$CO + \frac{1}{2}O_2 \rightleftharpoons CO_2$$

The actual chemical reaction is written as

$$CO + O_2 \longrightarrow x\,CO + \frac{1}{2}(1 + x)O_2 + (1 - x)\,CO_2$$

The chemical equilibrium equation to determine the fraction of the remaining CO, namely x, is written as

$$K_p = \frac{(1 - x)(3 + x)^{1/2}}{x(x + 1)^{1/2}P^{1/2}}, \quad 0 < x < 1$$

where $K_p = 3.06$ is the equilibrium constant for $CO + \frac{1}{2}O^2 = CO_2$ at $3000°$ K, and $P = 5$ atm is the pressure [Wark, p. 608]. Determine the value of x by Newton's method.

(**3.23**) Consider the same chemical reaction as in the previous problem except that it occurs with the existence of N_2 at atmospheric pressure. The actual reaction is

$$CO + O_2 + 3.76\,N_2 \longrightarrow x\,CO + \frac{1}{2}(1 + x)O_2 + (1 - x)CO_2 + 3.76\,N_2$$

The equation of equilibrium is

$$3.06 = \frac{(1 - x)(10.52 + x)^{1/2}}{x(1 + x)^{1/2}}$$

Determine the value of x using Newton's method.

(**3.24**) Repeat PROBLEM (3.7) by the secant method.

(**3.25**) Repeat PROBLEM (3.8) by the secant method.

(**3.26**) The equation $x^2 - 2x - 3 = 0$ may be reformulated for the successive substitution method as follows

(**a**) $x = \dfrac{(x^2 - 3)}{2}$

(**b**) $x = \sqrt{2x + 3}$

(**c**) $x = \dfrac{(2x + 3)}{\sqrt{x}}$

(**d**) $x = x - 0.2(x^2 - 2x - 3)$

The solutions of the equation are $x = 3$ and $x = -1$. Graphically determine which formulas above converge when used for successive substitution to find the root, $x = -1$. Verify the results of the graphic approach by using the criterion given by Eq. (3.7.3). Repeat the same analysis for $x = 3$.

(3.27) Find all the solutions of the equations in Problem 3.4 using successive substitution in the form

$$x = x - \alpha f(x)$$

Hint: Determine α by using the gradient of the linear interpolation fitted to the two end points of the interval found in problem 3.4.

(3.28) The friction factor f for turbulent flow in a pipe is given by

$$\frac{1}{\sqrt{f}} = 1.14 - 2.0 \log_{10}\left(\frac{e}{D} + \frac{9.35}{R_e\sqrt{f}}\right)$$

(Colebrook correlation)

where R_e is the Reynolds number, e is the roughness of the pipe surface, and D is the pipe diameter [Shames]. **(a)** Write a computer program to solve the above equation for f using the successive substitution method. **(b)** Evaluate f by running the program for the following cases:

(i) $D = 0.1\text{m}, \quad e = 0.0025\text{m}, \quad R_{ey} = 3 \times 10^4$

(ii) $D = 0.1\text{m}, \quad e = 0.0001\text{m}, \quad R_{ey} = 5 \times 10^6$

(*Hint:* Rewrite the equation to the following form first:

$$f = \left(1.14 - 2.0 \log_{10}\left[\frac{e}{D} + \frac{9.35}{R_e\sqrt{f}}\right]\right)^{-2}$$

Introduce an initial guess to f on the right side. Reintroduce f calculated to the right side again, and repeat this iteration until f converges. The initial guess may be set to zero. The results of these calculations can be checked with a Moody's chart found in any standard fluid mechanics text.)

(3.29) Using Bairstow's method, find all the quadratic factors of the following:

(a) $x^4 - 5x^2 + 4 = 0$

(b) $2x^3 + x^2 - x - 7 = 0$

(c) $-x^4 - 4x^3 - 7x^2 + x - 3 = 0$

(d) $-x^3 + 9x^3 - 18x + 16 = 0$

(e) $x^4 - 16x^3 + 72x^2 - 96x + 24 = 0$

(f) $x^6 - 6x^5 + 14x^4 - 18x^3 + 14x^2 - 6x + 1 = 0$

(3.30) As multiplicity of a root increases, convergence of Bairstow's method becomes poorer and the results become less accurate. The following polynomial has sextuple roots of $x = 1$:

$$x^6 - 6x^5 + 15x^4 - 20x^3 + 15x^2 - 6x + 1 = 0$$

Try to find all the quadratic factors by the Bairstow's method.

(3.31) Find the roots of the following equations by Bairstow's method:

(a) $y = 2 - x + 2x^2 + x^4$

(b) $y = 2 + 4x + 3x^2 + x^3$

(c) $y = 1.1 - 1.6x - 1.7x^2 + x^3$

(d) $y = 11.55 + 0.325x - 9.25x^2 + 1.1x^3 + x^4$

(e) $y = 6 - 12x + 13x^2 - 6x^3 + x^4$

(3.32) The following polynomials have double roots:

(a) $y = -8 + 12x - 2x^2 - 3x^3 + x^4$

(b) $y = 4 - 12x + 13x^2 - 6x^3 + x^4$

Find approximate values of the roots of the above polynomials by Bairstow's method.

(3.33) The block diagram for a dynamic system follows:

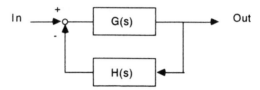

where the transfer functions are given as

$$G(s) = \frac{s + 2}{s(s + 3)}$$

$$H(s) = \frac{K}{s^2 + 2s + 2}$$

Then, the overall transfer function is given by

$$Y(s) = \frac{G(s)}{1 + G(s)H(s)}$$

The transient response of the system is characterized by the poles of $Y(s)$, namely the zeroes of the characteristic equation

$$1 + G(s)H(s) = 0$$

Find all the poles of $Y(s)$ for $K = 0$, 1 and 10 by using Bairstow's method.

(3.34) Modify PROGRAM 3–7 so that all the quadratic factors are automatically found.

REFERENCES

Abramowitz, M., and L. A. Stegun, eds., *Handbook of Mathematical Functions*, National Bureau of Standards, 1970.

Byars, E. F., R. D. Snyder, and H. L. Plants, *Engineering Mechanics of Deformable Bodies*, Harper & Row, 1983.

Cheney, W., and D. Kincaid, *Numerical Mathematics and Computing*, Brooks/Cole, 1985.

Conte, S. D., and C. de Boor, *Elementary Numerical Analysis*, 3rd ed., McGraw-Hill, 1980.

Gerald, C. F., and P. O. Wheatley, *Applied Numerical Analysis*, 3rd ed., Addison-Wesley, 1984.

Isaacson, E., and H. B. Keller, *Analysis of Numerical Methods*, Wiley, 1966.

James, M. L., G. M. Smith, and J. C. Wolford, *Applied Numerical Methods for Digital Computation*, 3rd ed., Harper & Row, 1985.

Lamarsh, J. R., *Introduction to Nuclear Reactor Theory*, Addison-Wesley, 1966.

Moran, M., and H. N. Shapiro, *Fundamentals of Engineering Thermodynamics*, Wiley, 1988.

Press, W. H. B. P. Flannery, S. A. Teukolsky, W. T. Vetterling, *Numerical Recipes*, Cambridge Univ. Press, 1986.

Shames, I. H., *Mechanics of Fluids*, McGraw-Hill, 1982.

Shoup, T. E., *Applied Numerical Methods for the Micro-Computer*, Prentice-Hall, 1984.

Thomson, W. T., *Theory of Vibration with Applications*, Prentice-Hall, 1981.

Wark, K, Jr., *Thermodynamics*, McGraw-Hill, 1988.

4

Numerical Integration

4.1 INTRODUCTION

Numerical integration methods may be used to integrate functions given in both tabular forms and analytical forms. Even when analytical integration is possible, numerical integration may save time and effort if only the numerical value of the integral is desired.

This chapter describes the numerical methods used to evaluate single integrals:

$$I = \int_a^b f(x)\, dx$$

as well as double integrals:

$$I = \int_a^b \int_{u(x)}^{v(x)} f(x, y)\, dy\, dx$$

where the functions $f(x)$ and $f(x, y)$ may be given in either an analytical form or a tabular form.

Numerical integration methods are derived by integrating interpolation polynomials. Therefore, different interpolation formulas will lead to different numerical integration methods. The methods discussed in Section 4.2 through 4.5 belong to the Newton-Cotes formulas. They are based on interpolation formulas with equally spaced data points, and they are derived by integrating the Newton forward, Newton backward, or Lagrange interpolation formulas. Newton-Cotes formulas are further subdivided into the closed type and the open type. The trapezoidal rule and the two

Simpson's rules belong to the Newton-Cotes closed type. The Gauss quadratures discussed in Section 4.6 are based on polynomial interpolation using the roots of an orthogonal polynomial such as Legendre polynomials. The integration methods discussed in Section 4.7 apply to integrals with infinite limits or those for singular functions. The last section describes numerical integration for double integrals.

The advantages and disadvantages of the numerical integration methods described in this chapter are summarized in Table 4.1.

Table 4.1 Summary of numerical integration methods

Method	Advantages	Disadvantages
Trapezoidal rule	Simplicity. Optimal for improper integrals.	Needs a large number of subintervals for good accuracy.
Simpson's 1/3 rule	Simplicity. Higher accuracy than the trapezoidal rule.	Even number of intervals only.
Simpson's 3/8 rule	Same order of accuracy as the 1/3 rule.	Intervals in multiples of three only.
Newton-Cotes formulas	Use equispaced points. Open and closed type formulas are available.	Higher-order formulas are not necessarily more accurate.
Gauss quadratures	More accurate than Newton-Cotes formulas. Functional data at two end points are not used.	Data points are not equispaced.
Double exponential transformation	Good accuracy for improper integrals.	Needs proper care to prevent overflow or underflow.

4.2 TRAPEZOIDAL RULE

The trapezoidal rule is a numerical integration method derived by integrating the linear interpolation formula. It is written as

$$I = \int_a^b f(x)\,dx = \frac{b-a}{2}\left[f(a) + f(b)\right] + E \tag{4.2.1}$$

where the first term on the right side is the trapezoidal rule (integration formula) and E represents its error. The numerical integration by Eq. (4.2.1) is graphically shown in Figure 4.1. The shadowed area under the line interpolation (which may be denoted by $g(x)$) equals the integral computed by the trapezoidal rule, while the area under the curve, $f(x)$, is the exact value. The error of Eq. (4.2.1) is, therefore, equal to the area between $g(x)$ and $f(x)$.

Equation (4.2.1) can be extended to multiple intervals. For N intervals with equal spacing h, Eq. (4.2.1) may be applied N times (as shown in Figure 4.2) to yield the extended trapezoidal rule:

$$I = \int_a^b f(x) = \frac{h}{2}\left[f(a) + 2\sum_{j=1}^{N-1} f(a + jh) + f(b)\right] + E$$

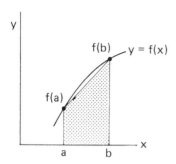

Figure 4.1 Trapezoidal rule

where $h = (b - a)/N$. The preceeding equation may be written equivalently as

$$I = \frac{h}{2}(f_0 + 2f_1 + 2f_2 + \cdots + 2f_{N-1} + f_N) + E \tag{4.2.2}$$

where $f_0 = f(a)$, $f_1 = f(a + h)$, and $f_i = f(a + ih)$.

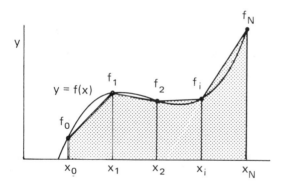

Figure 4.2 Extended trapezoidal rule

Example 4.1

   A body of revolution, shown in Figure E4.1, is obtained by rotating the curve given by $y = 1 + (x/2)^2$, $0 \leqslant x \leqslant 2$, about the x-axis. Calculate the volume

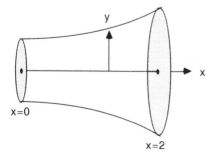

Figure E4.1 A body of revolution

using the extended trapezoidal rule with $N = 2, 4, 8, 16, 32, 64$, and 128. The exact value is $I = 11.7286$. Evaluate the error for each N.

⟨Solution⟩

The volume is given by

$$I = \int_0^2 f(x)\,dx$$

where

$$f(x) = \pi\left(1 + \left(\frac{x}{2}\right)^2\right)^2$$

The calculations for $N = 2$ and 4 are shown below:

$N = 2$: $h = 2/2 = 1$

$$I \simeq \frac{1}{2}[f(0) + 2f(1) + f(2)] = 0.5\pi[1 + 2(1.5625) + 4]$$

$$= 12.7627$$

$N = 4$: $h = 2/4 = 0.5$

$$I \simeq (0.5/2)[f(0) + 2f(0.5) + 2f(1) + 2f(1.5) + f(2)] = 11.9895$$

The integrations with other values of N are evaluated by using PROGRAM 4–1. The results are summarized in Table E4.2.

Table E4.2

N	h	I_h	E_h
2	1.	12.7627	−1.0341
4	0.5	11.9895	−0.2609
8	0.25	11.7940	−0.0654
16	0.125	11.7449	−0.0163
32	0.0625	11.7326	−0.0040
64	0.03125	11.7296	−0.0010
128	0.015625	11.7288	−0.0002
	Exact	11.7286	

It is observed that the error decreases in proportion to h^2.

The error of the trapezoidal rule is defined by

$$E = \int_a^b f(x)\,dx - \frac{b - a}{2}\left[f(a) + f(b)\right] \tag{4.2.3}$$

where the first term is the exact integral and the second term is the trapezoidal rule. To analyze Eq. (4.2.3) we will use the Taylor expansions for $f(x)$, $f(a)$, and $f(b)$ about $\bar{x} = (a + b)/2$, assuming that $f(x)$ is analytical in $a \leqslant x \leqslant b$.

The Taylor expansion for $f(x)$ is written as

$$f(x) = f(\overline{x}) + zf'(\overline{x}) + \frac{z^2}{2} f''(\overline{x}) + \cdots \tag{4.2.4}$$

where

$z = x - \overline{x}$

So, the first term of Eq. (4.2.3) becomes

$$\int_a^b f(x)\,dx = \int_{-h/2}^{h/2} \left[f(\overline{x}) + zf'(\overline{x}) + \frac{z^2}{2} f''(\overline{x}) + \cdots \right] dz \tag{4.2.5}$$

where $z = -h/2$ for $x = a$ and $z = h/2$ for $x = b$ are used. Carrying out the integration yields the following:

$$\int_a^b f(x)\,dx = hf(\overline{x}) + \frac{1}{24} h^3 f''(\overline{x}) + \cdots \tag{4.2.6}$$

On the other hand, the second term of Eq. (4.2.3) becomes

$$\frac{b-a}{2} [f(a) + f(b)] = \frac{h}{2} \left[f(\overline{x}) - \frac{h}{2} f'(\overline{x}) + \frac{1}{2} \left(\frac{h}{2} \right)^2 f''(\overline{x}) - \cdots \right.$$

$$\left. + f(\overline{x}) + \frac{h}{2} f'(\overline{x}) + \frac{1}{2} \left(\frac{h}{2} \right)^2 f''(\overline{x}) + \cdots \right]$$

$$= hf(\overline{x}) + \frac{1}{8} h^3 f''(\overline{x}) + \cdots \tag{4.2.7}$$

So, introducing Eq. (4.2.6) and Eq. (4.2.7) into Eq. (4.2.3) yields

$$E = \int_a^b f(x)\,dx - \frac{b-a}{2} [f(a) + f(b)]$$

$$\simeq -\frac{1}{12} h^3 f''(\overline{x}) \tag{4.2.8}$$

where higher-order terms have been truncated. This equation indicates that the error of the trapezoidal rule is proportional to f'' and decreases proportionally to h^3 when $h = b - a$ is reduced. As written earlier, this discussion is based on the assumption that $f(x)$ is analytic in the interval. If not, the error is not proportional to h^3.

The error of the extended trapezoidal rule is the summation of the errors for all the intervals. Suppose the extended trapezoidal rule is applied to an interval $[a, b]$, which is divided into N intervals with the $N + 1$ points $x_0, x_1, x_2, \ldots, x_N$, where $x_0 = a$ and $x_N = b$. Since the error for each interval is given by Eq. (4.2.5), the error of the extended trapezoidal rule becomes

$$E \simeq -\frac{1}{12} \frac{(b-a)^3}{N^3} \sum_{i=1}^N f''(\overline{x}_i) \tag{4.2.9}$$

where $h = (b - a)/N$ and $\overline{x}_i$ is the midpoint between x_i and x_{i+1}. If we define $\overline{f}''$ as the average of f'', namely

$$\overline{f}'' = \sum_{i=1}^{N} f''(\overline{x}_i)/N$$

Eq. (4.2.9) may be rewritten as

$$E \simeq -\frac{1}{12}(b - a)h^2\overline{f}'' \tag{4.2.10}$$

Equation (4.2.10) shows that the error of the extended trapezoidal rule is proportional to h^2 for a fixed interval $b - a$. (This agrees with the observation made in Example 4.1).

It is assumed in the error analysis for the extended trapezoidal rule that $f(x)$ is analytic in $[a, b]$. Otherwise, the error is not proportional to h^2. For example, when $f(x) = \sqrt{x}$ is integrated in $[0, 1]$ by the extended trapezoidal rule, the error decreases very slowly as the size of intervals is decreased. This is because $f = \sqrt{x}$ is not analytic at $x = 0$. For functions with a singularity, use of the double exponential transformation (Section 4.7) is recommended.

An important application of the error analysis for the trapezoidal rule is the Romberg integration. Suppose the result of the extended trapezoidal rule with point interval $h = (b - a)/N$ is I_h, and another calculation with $h' = 2h$ gives I_{2h}. Since the error of the extended trapezoidal rule is proportional to h^2, the errors with intervals h and $2h$ may be written respectively as

$$E_h \simeq Ch^2 \quad \text{and} \quad E_{2h} \simeq C(2h)^2 = 4Ch^2 \tag{4.2.11}$$

where C is a constant. On the other hand, the exact integral may be written as $I = I_h + E_h = I_{2h} + E_{2h}$, from which we can write

$$E_h - E_{2h} = I_{2h} - I_h \tag{4.2.12}$$

Introducing Eq. (4.2.11) into Eq. (4.2.12) and solving for C yield

$$C = \frac{1}{3}h^{-2}(I_h - I_{2h})$$

Thus, the first equation of Eq. (4.2.8) gives an approximate value of E_h as

$$E_h \simeq \frac{1}{3}(I_h - I_{2h}) \tag{4.2.13}$$

Knowing the values of I_h and I_{2h} from the actual calculations, a more accurate integral is calculated by

$$I = I_h + E_h \simeq I_h + \frac{1}{3}(I_h - I_{2h}) \tag{4.2.14}$$

The foregoing I is not exact because Eq. (4.2.11) is not, but the error of Eq. (4.2.14) is proportional to h^4 that is two orders higher than that of I_h. Therefore, Eq. (4.2.14) gives a more accurate result than I_h or I_{2h}. This technique is named the Romberg integration. For more applications of the Romberg integration see James/Smith/ Wolford; Ferziger; and Gerald/Wheatley.

Example 4.2

In Example 4.1, the extended trapezoidal rule gives $I_{0.5} = 11.9895$ and $I_{0.25} = 11.7940$. Find a more accurate value by using the Romberg integration.

⟨**Solution**⟩

By defining $h = 0.25$ for Eqs. (4.2.11) through (4.2.14), $E_{0.25}$ given by Eq. (4.2.13) becomes

$$E_{0.25} \simeq \frac{1}{3}(11.7940 - 11.9895) = -0.0652$$

Therefore, a more accurate value for I, given by Eq. (4.2.14), is

$$I = I_{0.25} + E_{0.25} \simeq 11.7940 - 0.0652 = 11.7288$$

This result agrees with the result for $N = 128$ (with $h = 0.0156$) in Example 4.1.

Another important application of the extended trapezoidal rule is in integrating a function from $-\infty$ to ∞. The optimum method for this type of problem is the extended trapezoidal rule. By mapping a finite interval to the infinite space, any function can be accurately integrated by the extended trapezoidal rule. This approach is explained in more detail in Section 4.7.

SUMMARY OF THIS SECTION

(a) The trapezoidal rule is based on integration of the linear interpolation.

(b) The extended trapezoidal rule is derived by repeating the trapezoidal rule.

(c) For a given domain of integration, error of the extended trapezoidal rule is proportional to h^2.

(d) The Romberg integration is based on the fact of item (c). Using the results of extended trapezoidal rule with two different data spacings, a more accurate integral is evaluated.

(e) See Section 4.7 for a more advanced integration method based on the extended trapezoidal rule.

4.3 SIMPSON'S 1/3 RULE

Simpson's 1/3 rule is based on quadratic (second-order) polynomial interpolation (see Figure 4.3). The Newton forward polynomial fitted at three points, x_0, x_1, x_2, is given

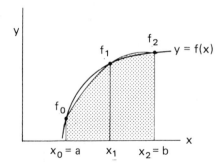

Figure 4.3 Simpson's rule

by Eq. (2.4.7). Integrating Eq. (2.4.7) from $x_0 = a$ to $x_2 = b$ yields Simpson's 1/3 rule:

$$I = \int_a^b f(x)\,dx = \frac{h}{3}\left[f(a) + 4f(\bar{x}) + f(b)\right] + E \qquad (4.3.1)$$

where $h = (b-a)/2$ and $\bar{x} = (a+b)/2$. Equation (4.3.1) may be written equivalently as

$$I = \frac{h}{3}\left[f_0 + 4f_1 + f_2\right] + E \qquad (4.3.2)$$

where $f_i = f(x_i) = f(a + ih)$. It will be shown later that the error is

$$E \simeq -\frac{h^5}{90} f^{iv}(\bar{x}) \qquad (4.3.3)$$

The error is zero if $f(x)$ is a polynomial of order 3 or less. Simpson's 1/3 rule is easy to apply with a hand calculator. Its accuracy is high enough for many applications, as illustrated in Example 4.3.

The extended Simpson's 1/3 rule is a repeated application of Eq. (4.3.2) for a domain divided into an even number of intervals. Denoting the total number of intervals by N(even), the extended Simpson's 1/3 rule is written as

$$I = \frac{h}{3}\left[f(a) + 4\sum_{\substack{i=1 \\ \text{odd } i}}^{N-1} f(a + ih) + 2\sum_{\substack{i=2 \\ \text{even } i}}^{N-2} f(a + ih) + f(b)\right] + E \qquad (4.3.4)$$

where $h = (b-a)/N$; the first summation is over odd i only and the second summation over even i only. Equation (4.3.4) may be written equivalently as

$$I = \int_a^b f(x)\,dx$$

$$= \frac{h}{3}\left[f_0 + 4f_1 + 2f_2 + 4f_3 + 2f_4 + \cdots + 2f_{N-2} + 4f_{N-1} + f_N\right] + E \quad (4.3.5)$$

The error term is given by

$$E \simeq -\frac{N}{2}\frac{h^5}{90}\bar{f}'''' = -(b-a)\frac{h^4}{180}\bar{f}''''$$ (4.3.6)

where

$$\bar{f}'''' = \sum_{i=1}^{N} f''''(\bar{x}_i)/N$$

For a fixed domain $[a, b]$, the error is proportional to h^4.

Example 4.3

Repeat the problem in Example 4.1 by using the extended Simpson's 1/3 rule with $N = 2, 4, 8, 16, 32$.

⟨Solution⟩

The interval size is $h = 2/N$. The calculations for $N = 2$ and 4 are as follows:

$$N = 2: \quad I = \frac{1}{3}[f(0) + 4f(1) + f(2)]$$

$$= \frac{1}{3}\pi[1 + (4)(1.25^2) + 2^2] = 11.7809$$

$$N = 4: \quad I = \frac{0.5}{3}[f(0) + 4f(0.5) + 2f(1) + 4f(1.5) + f(2)]$$

$$= \frac{0.5}{3}\pi[1 + 4(1.0625) + 2(1.25)^2 + 4(1.5625)^2 + 2^2]$$

$$= 11.7318$$

The calculations for larger N's may be performed similarly. The results and error evaluations are summarized next.

N	h	I_h	E_h
2	1.	11.7809	−0.0523
4	0.5	11.7318	−0.0032
8	0.25	11.7288	−0.0002
16	0.125	11.7286	0
32	0.0625	11.7286	0
64	0.03125	11.7286	0

A comparison of the preceding results to those of Example 4.1 reveals that the extended Simpson's rule is significantly more accurate than the extended trapezoidal rule using the same number of intervals. For example, the accuracy of the extended trapezoidal rule using 32 intervals is equivalent to that of the

> extended Simpson's rule using only 4 intervals. The error of the extended
> Simpson's rule is proportional to h^4 and thus two orders higher than that of the
> extended trapezoidal rule. Because of the higher order of error, the extended
> Simpson's rule approaches the exact solution faster than the extended trap-
> ezoidal rule when h is reduced.

We might try to derive the error term by integrating the error of the quadratic
interpolation given in the form of Eq. (2.3.7), but the result of integration becomes
zero and does not represent the true error. The reason for this misleading consequence
comes from the approximation $\xi = \bar{x}$. Therefore, we need a more accurate approach
to derive the error term.

For the Simpson's 1/3 rule we use the Taylor expansion. The Taylor expansions
of f_0 and f_2 about x_1, or equivalently $\bar{x} = (a + b)/2$, are written as

$$f_0 = f_1 - hf_1' + \frac{1}{2}h^2 f_1'' - \frac{1}{6}h^3 f_1''' + \frac{1}{24}h^4 f_1'''' - \cdots$$

$$f_2 = f_1 + hf_1' + \frac{1}{2}h^2 f_1'' + \frac{1}{6}h^3 f_1''' + \frac{1}{24}h^4 f_1'''' + \cdots$$

Introducing the foregoing two expansions into Eq. (4.3.2) yields

$$I = 2hf_1 + \frac{1}{3}h^3 f_1'' + \frac{1}{36}h^5 f_1'''' + \cdots + E \tag{4.3.7}$$

On the other hand, the Taylor expansion of $f(x)$ about x_1 is

$$f(x) = f_1 + zf_1' + \frac{1}{2}z^2 f_1'' + \frac{1}{6}z^3 f_1''' + \frac{1}{24}z^4 f_1'''' + \cdots \tag{4.3.8}$$

where $x = x_1 + z$ or equivalently $z = x - x_1$.

The analytical integration of this expansion over $[a, b]$ yields

$$\int_a^b f(x)\,dx = 2hf_1 + \frac{1}{3}h^3 f_1'' + \frac{1}{60}h^5 f_1'''' + \cdots \tag{4.3.9}$$

which is regarded as the exact integral in the Taylor expansion form. By subtracting Eq. (4.3.7) from Eq. (4.3.9) and truncating after the leading term, the error of
Eq. (4.3.7) is approximately given by

$$E \simeq -\frac{1}{90}h^5 f_1'''' \tag{4.3.10}$$

where $f_1'''' = f''''(x_1)$. Since $x_1 = \bar{x} = (a + b)/2$, this result has already been presented
by Eq. (4.3.3).

One drawback of the extended Simpson's 1/3 rule is that the total number of
intervals must be an even number. On the other hand, the extended Simpson's 3/8

rule described in the next section is applicable only to multiples of three intervals. Therefore, by combining the 1/3 and 3/8 rules, any even or odd number of intervals can be treated.

SUMMARY OF THIS SECTION

(a) Simpson's 1/3 rule is obtained by integrating a quadratic interpolation formula.

(b) By repeated application of the 1/3 rule to a multiple of two intervals, the extended Simpson's 1/3 rule is derived. Its error is proportional to h^4.

4.4 SIMPSON'S 3/8 RULE

Simpson's 3/8 rule is derived by integrating a third-order polynomial interpolation formula. For a domain $[a, b]$ divided into three intervals, it is written as

$$I = \int_a^b f(x)\,dx = \frac{3}{8}h[f_0 + 3f_1 + 3f_2 + f_3] + E \qquad (4.4.1)$$

where

$h = (b - a)/3, f_i = f(a + ih)$

and E represents error. The error term is given by

$$E \simeq -\frac{3}{80}h^5 f''''(\overline{x}) \qquad (4.4.2)$$

where

$\overline{x} = (a + b)/2$

This expression for the error may be derived by using the Taylor expansion in a similar manner as described for Simpson's 1/3 rule.

As explained before, the extended 1/3 rule is applicable to an even number of intervals, whereas the extended 3/8 rule is applicable to a multiple of three intervals. When the number of intervals is an odd number, but not a multiple of three, one may use the 3/8 rule for the first three or the last three intervals, and then use the 1/3 rule for the remaining even number of intervals. Since the order of error of the 3/8 rule is the same as that of the 1/3 rule, there is no gain in accuracy over the 1/3 rule when one has a free choice between the two rules.

SUMMARY OF THIS SECTION

(a) Simpson's 3/8 rule is obtained by integrating a cubic interpolation polynomial. The order of its error is the same as the 1/3 rule.

(b) This rule can be extended to a multiple of three intervals.

(c) Simpson's 3/8 rule is important in combination with the extended 1/3 rule.

4.5 NEWTON-COTES FORMULAS

The numerical integration methods that are derived by integrating the Newton interpolation formulas are the Newton-Cotes integration formulas. The trapezoidal rule and the two Simpson's rules are members of the Newton-Cotes formulas. The Newton-Cotes formulas consist of closed formulas and open formulas.

We write Newton-Cotes closed formulas in the form:

$$\int_a^b f(x)\,dx = \alpha h[w_0 f_0 + w_1 f_1 + w_2 f_2 + \cdots + w_N f_N] + E \qquad (4.5.1)$$

where α and the w's are the constants listed in Table 4.2, and

$$f_n = f(x_n), \quad x_n = a + nh, \quad \text{and} \quad h = (b - a)/N$$

Equation (4.5.1) is called a *closed formula* because the domain of integration is closed by the first and last data points.

On the other hand, integration of Eq. (4.5.1) may be extended beyond the two end points of the given data. The Newton-Cotes open formulas are obtained by

Table 4.2 Constants for Newton-Cotes closed formulas

N	α	$w_i, i = 0, 1, 2, \ldots, N$	E
1	1/2	1 1	$-\dfrac{1}{12}h^3 f''$
2	1/3	1 4 1	$-\dfrac{1}{90}h^5 f^{iv}$
3	3/8	1 3 3 1	$-\dfrac{3}{80}h^5 f^{iv}$
4	2/45	7 32 12 32 7	$-\dfrac{8}{945}h^7 f^{vi}$
5	5/288	19 75 50 50 75 19	$-\dfrac{275}{12096}h^7 f^{vi}$
6	1/140	41 216 27 272 27 216 41	$-\dfrac{9}{1400}h^9 f^{viii}$
7	7/17280	751 3577 1323 2989 2989 1323 3577 751	$-\dfrac{8183}{518400}h^9 f^{viii}$
8	8/14175	989 5888 −928 10496 −4540 10496 −928 5888 989	$-\dfrac{2368}{467775}h^{11} f^{x}$
9	9/89600	2857 15741 1080 19344 5788 5788 19344 1080 15741 2857	$-\dfrac{173}{14620}h^{11} f^{x}$
10	5/299376	16067 106300 −48525 272400 −260550 427368 −260550 272400 −48525 106300 16067	$-\dfrac{1346350}{326918592}h^{13} f^{xii}$

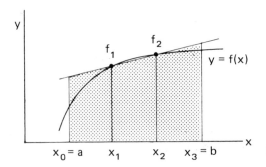

Figure 4.4 Newton-Cotes open formula ($N = 1$)

extending the integration over one interval to the left of the first data point and one interval to the right of the last data point (see Figure 4.4). They are written as

$$\int_a^b f(x) = \alpha h[w_0 f_0 + w_1 f_1 + \cdots + w_{N+2} f_{N+2}] + E \qquad (4.5.2)$$

where $h = (b - a)/(N + 2)$. The constants α and w_i are listed in Table 4.3, in which w_0 and w_{N+2} are both set to zero because these correspond to the end points of the domain. Since w_0 and w_{N+2} are zero, f_0 and f_{N+2} are dummy data, which are not actually required.

Table 4.3 Constants for Newton-Cotes open formulas

N	α	$w_i, i = 0, 1, \ldots, N + 2$	E
1	3/2	0 1 1 0	$\frac{1}{4} h^3 f''$
2	4/3	0 2 -1 2 0	$\frac{28}{90} h^5 f^{iv}$
3	5/24	0 11 1 1 11 0	$\frac{95}{144} h^5 f^{iv}$
4	6/20	0 11 -14 26 -14 11 0	$\frac{41}{140} h^7 f^{vi}$
5	7/1440	0 611 -453 562 562 -453 611 0	$\frac{5257}{8640} h^7 f^{vi}$
6	8/945	0 460 -954 2196 -2459 2196 -954 460 0	$\frac{3956}{14175} h^9 f^{viii}$

If we compare an open formula to a closed formula using the same number of data points N, the error of the open formula is significantly greater than that of the closed formula. On the other hand, the open formulas can be used when functional values at the integrating limits are not available.

It is observed in both Table 4.2 and Table 4.3 that the values of w for a large N are very large and change sign. Subtractions of large numbers cause round-off errors. For this reason, higher-order Newton-Cotes formulas are undesirable. This trend is illustrated in Example 4.4.

Example 4.4

The arc length of a curve in polar coordinates (see Figure E4.4) is given by

$$L = \int_a^b \sqrt{r^2 + \left(\frac{dr}{d\theta}\right)^2}\, d\theta$$

Calculate the arc length of the curve given by $r = 2(1 + \cos\theta)$, $0 < \theta < \pi$, by using each of the Newton-Cotes closed integration formulas listed in Table 4.2.

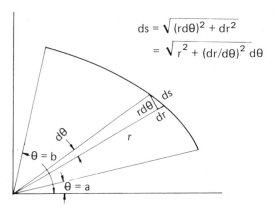

$$ds = \sqrt{(rd\theta)^2 + dr^2}$$
$$= \sqrt{r^2 + (dr/d\theta)^2}\, d\theta$$

Figure E4.4 Line segment of a curve

⟨**Solution**⟩

We set $a = 0$ and $b = \pi$, and use PROGRAM 4–2. The computational results are shown below:

Order N	Integral L
2	8.01823
3	8.00803
4	7.99993
5	7.99996
6	8.00000
7	8.00000
8	8.00000
9	8.00197
10	7.99201
Exact	8.00000

These results illustrate the effect of round-off errors. That is, as N increases, the results approach the exact value, 8.00000. After $N = 8$, however, the error gradually increases again. The increase of errors is attributed to round-off errors in additions and subtractions of the very large numbers in the formulas.

There is another reason why higher-order formulas may not be desirable. In the foregoing discussions, the order of a formula is indicative of the accuracy only if the domain of integration is small enough that derivatives of the function are nearly constant in the domain. This implies that a higher-order formula is not necessarily more accurate than a lower-order formula when the derivatives change significantly in the domain.

The aforementioned fact may be easily understood also if we recall that the Newton-Cotes formulas are obtained by integrating an interpolating polynomial. As stated in Section 2.3, there is no guarantee that a higher-order interpolation polynomial is more accurate than a lower-order interpolation polynomial.

A repeated application of a low-order formula such as the extended trapezoidal rule or extended Simpson's rule is far more desirable than using a single high-order formula.

One extreme example, although uncommon, is that the extended trapezoidal rule with two intervals can be more accurate than the Simpson's rule for the same two intervals. This occurs when one of the integrating limit is close to a singularity, or when the range of integration is large. Indeed, the extended trapezoidal rule is chosen to the extended Simpson's rule for an improper integral or in integrating a singular function as described in more details in Section 4.7. For proper integrals, the extended Simpson's rule is more accurate than the extended trapezoidal rule as described in Sections 4.3 and 4.4.

SUMMARY OF THIS SECTION

(a) The Newton-Cotes closed and open formulas are obtained by integrating polynomials fitted to equispaced data points.

(b) The trapezoidal rule and the two Simpson's rules belong to the closed formulas. The Newton-Cotes formulas of high orders are not accurate because of round-off errors caused by large coefficients of varying signs.

4.6 GAUSS QUADRATURES

4.6.1 Gauss-Legendre Quadratures

Gauss-Legendre (or simply Gauss) quadratures are numerical integration methods using Legendre points (roots of Legendre polynomials). Gauss quadratures cannot be used to integrate a function given in a tabular form with equispaced intervals because the Legendre points are not equispaced, but they are rather suitable for integrating analytical functions. The advantage of the Gauss quadratures is that their accuracy is significantly higher than the Newton-Cotes formulas.

Before discussing the general expression of the Gauss quadratures, we review error terms of the Newton-Cotes formulas. Equation (4.2.8) indicates that the error of the trapezoidal rule is proportional to f''. If the trapezoidal rule is used to integrate each of $f = 1, x, x^2, x^3, \ldots$, then the results have no error for $f = 1$ and $f = x$, but there are errors for x^2 and higher powers of x. The error of the Simpson's rule given by Eq. (4.3.6) is proportional to f^{iv}, so it is accurate if $f = 1, x, x^2$, and x^3 are integrated. In more general terms, the Newton-Cotes closed formula of odd order N is exact if the integrand is a polynomial of order N or less, whereas that of an even N is exact when the integrand is a polynomial of order $N + 1$ or less.

The next example shows, however, that a numerical integration using two points can be made exact for third-order polynomials by optimizing the x values of the data points.

Example 4.5

The integration formula using two points can be made exact when a polynomial of order 3 is integrated. Determine the points.

⟨**Solution**⟩

We consider

$$I = \int_{-1}^{1} f(x) \, dx \tag{A}$$

and write an integration formula using only two points as

$$I = w_1 f(x_1) + w_2 f(x_2) + E \tag{B}$$

where w_k, $k = 1, 2$, are weights and x_k are undetermined points, and E is the error term.

Since w_k and x_k are both undetermined, we require that $E = 0$ (namely, I becomes exact) for $f(x) = 1, x, x^2$, and x^3. Introducing each of $f(x) = 1, x, x^2$, and x^3 into Eq. (B) yields four equations:

$$2 = w_1 + w_2$$

$$0 = w_1 x_1 + w_2 x_2$$

$$\frac{2}{3} = w_1 x_1^2 + w_2 x_2^2$$

$$0 = w_1 x_1^3 + w_2 x_2^3$$

where the left sides are the exact values.

The limits of integration are -1 and 1 and symmetric about $x = 0$, so we set $x_2 = -x_1$ and require that points be located symmetrically. From the first and second equations above, we get

$$w_1 = w_2 = 1$$

With these values, the fourth equation is automatically satisfied. The third equation becomes

$$\frac{1}{3} = x_1^2$$

which yields

$$x_1 = \frac{1}{\sqrt{3}} = 0.577350269$$

and

$$x_2 = -x_1 = -0.577350269$$

With these weights and points, Eq. (B) is exact for a polynomial of order 3 or less. Although we considered the interval of $[-1, 1]$ for simplicity, it can be changed to any arbitrary interval by a transformation of the coordinate.

The integrating formula derived in Example 4.5 is the simplest member of the Gauss quadratures.

It is not easy to derive the Gauss quadratures using more than two points by extending the approach in this example. Therefore, in the remainder of this section, a general formula of the Gauss quadratures is given, and then it will be shown that the Gauss quadrature of order N is exact when any polynomial of order $2N - 1$ or less is integrated.

The Gauss quadratures differ significantly from the Newton-Cotes formulas because N grid points (called Gauss points) are set to the roots of $P_N(x) = 0$, where $P_N(x)$ is the Legendre polynomial of order N. Indeed, x_1 and x_2 determined in Example 4.5 are roots of $P_2(x) = 0$. More details of Legendre polynomials are in Appendix B.

The Gauss quadrature extended over the interval $[-1, 1]$ is given by

$$\int_{-1}^{1} f(x)\, dx \simeq \sum_{k=1}^{N} w_k f(x_k) \tag{4.6.1}$$

where N is the number of Gauss points, the w_i are the weights, and the x_i are the Gauss points given in Table 4.4. The $\pm$ signs in the table mean that the x values of the Gauss points appear in a pair, one of which is plus and the other minus. If $N = 4$ for example, Eq. (4.6.1) becomes

$$\int_{-1}^{1} f(x)\, dx \simeq 0.34785f(-0.86113) + 0.65214f(-0.33998)$$

$$+ 0.65214f(0.33998) + 0.34785f(0.86113) \tag{4.6.2}$$

The Gauss integration formula may be applied to any arbitrary interval $[a, b]$ with the transformation

$$x = \frac{2z - a - b}{b - a} \tag{4.6.3}$$

where z is the original coordinate in $a < z < b$, and x is the normalized coordinate

Table 4.4 Gauss points and weights[a]

	$\pm x_i$	w_i
$N = 2$	0.577350269	1.000000000
$N = 3$	0	0.888888889
	0.774596669	0.555555556
$N = 4$	0.339981043	0.652145155
	0.861136312	0.347854845
$N = 5$	0	0.568888889
	0.538469310	0.478628670
	0.906179846	0.236926885
$N = 6$	0.238619186	0.467913935
	0.661209387	0.360761573
	0.932469514	0.171324492
$N = 8$	0.183434642	0.362683783
	0.525532410	0.313706646
	0.796666478	0.222381034
	0.960289857	0.101228536
$N = 10$	0.148874339	0.295524225
	0.433395394	0.269266719
	0.679409568	0.219086363
	0.865063367	0.149451349
	0.973906528	0.066671344

[a] See Abramowitz and Stegun for more data.

in $-1 \leqslant x \leqslant 1$. The transformation from x to z is

$$z = \frac{(b - a)x + a + b}{2} \tag{4.6.4}$$

Using this transformation, the integral may be written as

$$\int_a^b f(z)\,dz = \int_{-1}^1 f(z)(dz/dx)\,dx \simeq \frac{b - a}{2} \sum_{k=1}^N w_k f(z_k) \tag{4.6.5}$$

where $dz/dx = (b - a)/2$ is used. The values of z_k are obtained by substituting x in Eq. (4.6.4) by the Gauss points, namely

$$z_k = \frac{(b - a)x_k + a + b}{2} \tag{4.6.6}$$

For example, suppose $N = 2$, $a = 0$, and $b = 2$. Since the Gauss points x_k for $N = 2$ on the normalized coordinate x, $-1 \leqslant x \leqslant 1$, are ± 0.57735 (from Table 4.4), the corresponding points on z are

$$z_1 = \frac{1}{2}[(2 - 0)(-0.57735) + 0 + 2] = 0.42265$$

$$z_2 = \frac{1}{2}[(2 - 0)(0.57735) + 0 + 2] = 1.57735$$

(4.6.7)

The derivative is $dz/dx = (b - a)/2 = 1$. So the Gauss quadrature becomes

$$\int_0^2 f(z)\,dz = \int_{-1}^1 f(z)(dz/dx)\,dx \simeq (1)[(1)f(0.42264) + (1)f(1.57735)] \quad (4.6.8)$$

We will show that if $f(x)$ is a polynomial of order $2N - 1$ or less, the Gauss quadrature of order N becomes exact.

Suppose $f(x)$ in Eq. (4.6.1) is a polynomial of order $2N - 1$ or less and is to be integrated in $[-1, 1]$. By using the Legendre polynomial of order N, namely $P_N(x)$, $f(x)$ may be written as

$$f(x) = c(x)P_N(x) + r(x) \qquad (4.6.9)$$

where $c(x)$ and $r(x)$ are both polynomials of order $N - 1$ or less. The following two features of this expression will become very important. First, integrating Eq. (4.6.9) in $[-1, 1]$ yields

$$\int_{-1}^1 f(x)\,dx = \int_{-1}^1 r(x)\,dx \qquad (4.6.10)$$

where the first term of Eq. (4.6.9) vanishes upon integration because $P_N(x)$ is orthogonal to any polynomial of order $N - 1$ or less (see Appendix B). Second, if x is set to one of the roots of $P_N(x) = 0$, the first term of Eq. (4.6.9) vanishes, and Eq. (4.6.9) reduces to $f(x_i) = r(x_i)$.

The integrand $r(x)$ in Eq. (4.6.10) is a polynomial of order $N - 1$ or less, so it can be exactly expressed by the Lagrange interpolation of order $N - 1$ (see Eq. (2.3.3))

$$r(x) = \sum_{i=1}^N \left[\prod_{\substack{j=1 \\ j \ne i}}^N \frac{x - x_j}{x_i - x_j} \right] r(x_i) \qquad (4.6.11)$$

where $\prod$ denotes a multiple product taken over the subscript i or j. Since x_i, $i = 1, 2, \ldots, N$, are the roots of $P_N(x)$, and $r(x_i) = f(x_i)$ from Eq. (4.6.9), Eq. (4.6.11) can be written as

$$r(x) = \sum_{i=1}^N \left[\prod_{\substack{j=1 \\ j \ne i}}^N \frac{x - x_j}{x_i - x_j} \right] f(x_i) \qquad (4.6.12)$$

The Gauss quadrature is derived by introducing Eq. (4.6.12) into Eq. (4.6.10):

$$\int_{-1}^{1} f(x)\,dx = \sum_{i=1}^{N} f(x_i) \int_{-1}^{1} \left[\prod_{\substack{j=1 \\ j \neq i}}^{N} \frac{x - x_j}{x_i - x_j} \right] dx \tag{4.6.13}$$

Equation (4.6.13) becomes Eq. (4.6.1) if we define

$$w_i = \int_{-1}^{1} \left[\prod_{\substack{j=1 \\ j \neq i}}^{N} \frac{x - x_j}{x_i - x_j} \right] dx \tag{4.6.14}$$

As seen from the foregoing derivation, the Gauss quadrature is exact if the integrand $f(x)$ is a polynomial of order $2N - 1$ or less.

Table 4.4 indicates that all the weights are positive. An advantage of Gauss quadrature is that there is no round-off error problem unless the integrand changes sign in the middle of the domain because no subtractions of large numbers as in the Newton-Cotes formulas occur. Another advantage of the Gauss quadratures is that the function with a singularity at one or both integration limit(s) can be integrated because the functional values at the limits are not used.

For a large interval of integration, the domain can be divided into a number of small subintervals, and a Gauss quadrature can be repeatedly applied to each subinterval. The idea is the same as the extended trapezoidal or Simpson's rule.

4.6.2 Other Gauss Quadratures

Gauss quadratures discussed in the previous section are named Gauss-Legendre quadratures because they are based on the orthogonality of the Legendre polynomials. Similar quadratures based on Hermite polynomials, Laguerre polynomials [Froeberg], and Chebyshev polynomials are called respectively Gauss-Hermite, Gauss-Laguerre, and Gauss-Chebyshev quadratures.

Gauss-Hermite quadratures are suitable for

$$\int_{-\infty}^{\infty} \exp{(-x^2)} f(x)\,dx \tag{4.6.15}$$

and given by

$$\int_{-\infty}^{\infty} \exp{(-x^2)} f(x)\,dx \simeq \sum_{k=1}^{N} w_k f(x_k) \tag{4.6.16}$$

In Eq. (4.6.16) x_k are roots of the Hermite polynomial of order N and w_k are weights, some of which are shown in Table 4.5.

Gauss-Laguerre quadratures are suitable for

$$\int_{0}^{\infty} \exp{(-x)} f(x)\,dx \tag{4.6.17}$$

Table 4.5 Hermite points and weights

	$\pm x_i$	w_i
$N = 2$	0.70710678	0.88622692
$N = 3$	0.00000000	1.18163590
	1.22474487	0.29540897
$N = 4$	0.52464762	0.80491409
	1.65068012	0.08131283
$N = 5$	0.00000000	0.94530872
	0.95857246	0.39361932
	2.02018287	0.01995324

and are given by

$$\int_0^\infty \exp{(-x)}f(x)\,dx \simeq \sum_{k=1}^{N} w_k f(x_k) \tag{4.6.18}$$

where x_k are roots of the Laguerre polynomial of order N and w_k are weights, some of which are shown in Table 4.6.

Table 4.6 Laguerre points and weights

	x_i	w_i
$N = 2$	0.58578643	0.85355339
	3.41421356	0.14644660
$N = 3$	0.41577455	0.71109300
	2.29428036	0.27851773
	6.28994508	$0.10389256E - 1$
$N = 4$	0.32254768	0.60315410
	1.74576110	0.35741869
	4.53662029	$0.38887908E - 1$
	9.39507091	$0.53929470E - 3$

Gauss-Chebyshev quadratures are suitable for

$$\int_{-1}^{1} \frac{1}{\sqrt{1 - x^2}} f(x)\,dx \tag{4.6.19}$$

and given by

$$\int_{-1}^{1} \frac{1}{\sqrt{1 - x^2}} f(x)\,dx \simeq \sum_{k=1}^{N} w_k f(x_k) \tag{4.6.20}$$

In Eq. (4.6.20), x_k are roots of the Chebyshev polynomial of order N and w_k are weights. The roots of the Chebyshev polynomial of order N are

$$x_k = \cos \frac{k - 1/2}{N} \pi, \quad k = 1, 2, \ldots, N \tag{4.6.21}$$

The weights are given by

$$w_k = \frac{\pi}{N} \quad \text{for all } k \tag{4.6.22}$$

Therefore, Eq. (4.6.19) reduces to

$$\int_{-1}^{1} \frac{1}{\sqrt{1 - x^2}} f(x) \, dx \simeq \frac{\pi}{N} \sum_{k=1}^{N} f(x_k) \tag{4.6.23}$$

The integrating limits of $[-1, 1]$ can be changed to an arbitrary domain $[a, b]$ by Eq. (4.6.4).

All three Gauss quadratures explained in this subsection are exact if $f(x)$ is a polynomial of order $2N - 1$ or less. For more details of the Gauss quadratures see Carnahan/Luther/Wilkes and King.

SUMMARY OF THIS SECTION

(a) The Gauss (-Legendre) quadratures are based on integrating a polynomial fitted to the data points given at the roots of a Legendre polynomial.

(b) The order of accuracy of a Gauss quadrature is approximately twice as high as that of the Newton-Cotes closed formula using the same number of data points.

(c) Because all the coefficients are positive, no serious round-off errors occur provided that the integrand does not change sign within the limits of integration.

(d) Other Gauss quadratures are suitable for special integrations.

4.7 NUMERICAL INTEGRATION WITH INFINITE LIMITS OR SINGULARITIES

In this section, we study the following types of integrals that need special attention:

$$I = \int_{-\infty}^{\infty} \exp(-x^2) \, dx \tag{4.7.1}$$

$$I = \int_{0}^{1} \frac{1}{\sqrt{x}(e^x + 1)} \, dx \tag{4.7.2a}$$

$$I = \int_{0}^{1} x^{0.7} \cos(x) \, dx \tag{4.7.2b}$$

In Eq. (4.7.1), the integration is extended over an infinite domain. Equation (4.7.2a) involves a singularity of the integrand at $x = 0$ (the function approaches infinity as x approaches 0). Equation (4.7.2b) has no apparent singularity, yet it is not a trivial problem for any of the numerical integration methods described in the earlier sections. Indeed, if the extended trapezoidal rule or Simpson's rule is applied, the result keeps changing as the number of intervals is doubled. The reason is that the function is not analytic at $x = 0$.

This section discusses the methods based on the trapezoidal rule and the double exponential transformation that are robust and work with little care to a wide range of problems.

A function that is integrable in an infinite or semi-infinite domain is nearly zero except in a certain part of the domain. The major contribution to the integral comes from a relatively small domain where the function is significantly different from zero.

It has been shown [Takahashi/Mori; Mori/Piessens] that if $f(x)$ is analytical over $[-\infty, \infty]$, the most efficient method for numerical integration for

$$ I = \int_{-\infty}^{\infty} f(x)\,dx \tag{4.7.3} $$

is the extended trapezoidal rule:

$$ I = h \sum_{i=-M}^{M} f(x_i) \tag{4.7.4} $$

where $x_i = ih$, and M is an integer, and $I\infty$ in Eq. (4.7.3) are replaced by IMh. The extended trapezoidal rule is preferred to the extended Simpson's rule or any higher-order quadratures for improper integrals or in integrating singular functions.

Example 4.6 shows that the extended trapezoidal rule gives highly accurate results with a relatively small number of points.

Example 4.6

Numerically evaluate

$$ I = \frac{1}{\sqrt{\pi}} \int_{-\infty}^{\infty} \exp(-x^2)\,dx $$

by the extended trapezoidal rule with a total of 20, 40, and 80 intervals.

⟨Solution⟩

We replace the limits of integration by -10 and 10 as sufficiently large

$$ I = \frac{1}{\sqrt{\pi}} \int_{-10}^{10} \exp(-x^2)\,dx $$

Using the trapezoidal rule in PROGRAM 4-1, we obtain the following results:

N	I
20	1.000104
40	1.000001
80	1.000000

where N is the number of intervals. The exact value is 1.0, so the agreement is very good even with N = 20.

Next, we consider integrating a function in a finite domain, but the function is singular at one or both of the limits. See Eq. (4.7.2) for example. The finite domain of integration, say $[a, b]$, can be transformed to $[-\infty, \infty]$ by a coordinate transformation. Once it is reduced to the problem of Eq. (4.7.3), the extended trapezoidal rule is applied.

Consider

$$I = \int_b^a f(x)\,dx \tag{4.7.5}$$

where a and b are finite. The mapping may be written by

$$z = z(x)$$

or equivalently

$$x = x(z) \tag{4.7.6}$$

where z is a mapping function satisfying

$$z(a) = -\infty$$
$$z(b) = \infty \tag{4.7.7}$$

Then, Eq. (4.7.5) can be written as

$$I = \int_{-\infty}^{\infty} f[x(z)]\left(\frac{dx}{dz}\right)dz \tag{4.7.8}$$

An example for such transformation is the exponential transformation given by

$$x = \frac{1}{2}[a + b + (b - a)\tanh(z)] \tag{4.7.9}$$

or equivalently

$$z = \tanh^{-1}\left(\frac{2x - a - b}{b - a}\right) \tag{4.7.10}$$

The accuracy of the numerical integration, however, is affected by the choice of the transformation. The double exponential transformation given by

$$x = \frac{1}{2}\left[a + b + (b - a)\tanh\left(\frac{\pi}{2}\sinh(z)\right)\right] \tag{4.7.11}$$

has been proposed as an optimal choice. With this choice dx/dz becomes

$$\frac{dx}{dz} = \frac{(b - a)\dfrac{\pi}{4}\cosh(z)}{\cosh^2\left[\dfrac{\pi}{2}\sinh(z)\right]} \tag{4.7.12}$$

Introducing Eq. (4.7.11) and (4.7.12) into Eq. (4.7.8) and applying the extended trapezoidal rule give

$$I \simeq h \sum_{k=-N}^{N} f(x_k)\left(\frac{dx}{dz}\right)_k \tag{4.7.13}$$

where N is an integer, and $I\infty$ in Eq. (4.7.8) are replaced by INh, and

$$z_k = kh \tag{4.7.14}$$

$$x_k = \frac{1}{2}\left[a + b + (b - a)\tanh\left(\frac{\pi}{2}\sinh(z_k)\right)\right] \tag{4.7.15}$$

$$\left(\frac{dx}{dz}\right)_k = \frac{(b - a)\dfrac{\pi}{4}\cosh(z_k)}{\cosh^2\left[\dfrac{\pi}{2}\sinh(z_k)\right]} \tag{4.7.16}$$

In Eq. (4.7.13), a question is how large Nh is large enough, which may be answered by examining the denominator of Eq. (4.7.16). When z_k increases, it approaches

$$\cosh^2\left[\frac{\pi}{2}\sinh(z_k)\right] \longrightarrow \frac{1}{4}\exp\left[\frac{\pi}{2}\exp(z_k)\right] \tag{4.7.17}$$

which increases double-exponentially and causes an overflow. In single precision on IBM PC and a VAX computer the overflow occurs when

$$\frac{1}{4} \exp \left[\frac{\pi}{2} \exp (z_k) \right] \cong 2 \times 10^{38} \tag{4.7.18}$$

or equivalently if z_k is greater than approximately 4.0. This criterion determines the maximum Nh that can be used. On IBM PC the criterion is

$$Nh < 4 \quad \text{(IBM PC)} \tag{4.7.19}$$

On an IBM mainframe, the highest floating point value in FORTRAN-77 is 7.5×10^{75} so that the criterion for Nh is

$$Nh < 4.7 \quad \text{(IBM mainframe)} \tag{4.7.20}$$

Another problem is the round-off error in Eq. (4.7.15) that occurs when the hyperbolic tangent term becomes very close to -1 or 1. To avoid this, we first recognize that the hyperbolic tangent term can be written as

$$\tanh (p) = \left(s - \frac{1}{s} \right) \bigg/ \left(s + \frac{1}{s} \right) \tag{4.7.21}$$

where $s = \exp (p)$. Using Eq. (4.7.21), Eq. (4.7.15) is written in the form of

$$x_k = \left(bs + \frac{a}{s} \right) \bigg/ \left(s + \frac{1}{s} \right) \tag{4.7.22}$$

where

$$p = \frac{\pi}{2} \sinh (z_k)$$

or

$$s = \exp \left[\frac{\pi}{2} \sinh (z_k) \right]$$

Because Eq. (4.7.22) has no subtraction operation, no serious round-off occurs.

PROGRAM 4–5 performs integration according to the double exponential transformation. For the infinite limits $[-\infty, \infty]$, use the straightforward trapezoidal rule in PROGRAM 4–1 rather than that in PROGRAM 4–5.

Example 4.7

The length of the curve given by $y = g(x)$, $a < x < b$, is

$$I = \int_a^b \sqrt{1 + (g'(x))^2}\,dx$$

Calculate the length of the parabolic arc, $y^2 = 4x$, in $0 < x < 2$.

⟨Solution⟩

Since $g(x) = 2\sqrt{x}$, its derivative is $g'(x) = 1/\sqrt{x}$. The integral becomes

$$I = \int_0^2 \sqrt{1 + \frac{1}{x}}\,dx$$

The integrand in the foregoing equation is singular at $x = 0$.
   The computation is performed by PROGRAM 4–5 on IBM PC. In the program the limits of integration on the transformed coordinate are set to $z = -4$ and $z = 4$. The computed results are:

N	I
10	3.600710
20	3.595706
30	3.595706

where N is the number of intervals used in the extended trapezoidal rule.

SUMMARY OF THE SECTION

(a) The extended trapezoidal rule is optimal to integrate a function analytical in $[-\infty, \infty]$.

(b) Integrating a function with singularity may be performed by the extended trapezoidal rule with the double exponential transformation.

4.8 NUMERICAL INTEGRATION IN A TWO-DIMENSIONAL DOMAIN

Consider a domain as illustrated in Figure 4.5, where the left and right boundaries are vertical lines, and the top and bottom boundaries are given by curves $y = d(x)$ and $y = c(x)$, respectively. A double integration in the domain is written as

$$I = \int_a^b \left[\int_{c(x)}^{d(x)} f(x, y)\,dy \right] dx \tag{4.8.1}$$

Problems of double integrals are, however, not always written in the form of Eq. (4.8.1). They are often given in different forms such as

$$I = \int_a^b \int_{c(x)}^{d(x)} f(x, y)\, dy\, dx$$

$$I = \int_a^b dx \int_{c(x)}^{d(x)} dy\, f(x, y)$$

or even

$$I = \iint_A f(x, y)\, dx\, dy$$

where A is meant by the domain. In any case, the problem should be rewritten in the form of Eq. (4.8.1) before proceeding with numerical integrations. Exchange x and y if necessary.

Numerical double integration is essentially a combination of one-dimensional problems. If we define

$$G(x) \equiv \int_{c(x)}^{d(x)} f(x, y)\, dy \tag{4.8.2}$$

Eq. (4.8.1) becomes

$$I = \int_a^b G(x)\, dx \tag{4.8.3}$$

Any of the numerical quadrature formulas described earlier is applicable to Eq. (4.8.3). A numerical integration for Eq. (4.8.3) may be written as

$$I \simeq \sum_{i=0}^{N} w_i G(x_i) \tag{4.8.4}$$

where w_i are weights and x_i are points of the particular quadrature. The values of $G(x_i)$ are evaluated numerically also. Setting $x = x_i$, Eq. (4.8.2) becomes

$$G(x_i) \equiv \int_{c(x_i)}^{d(x_i)} f(x_i, y)\, dy \tag{4.8.5}$$

which is a one-dimensional problem because the only variable of the integrand is y. Equation (4.8.5) may be evaluated by using any one of the numerical quadratures.

As an illustration, we apply the extended trapezoidal rule to the double integration problem:

$$I = \int_a^b \left[\int_{c(x)}^{d(x)} f(x, y)\, dy \right] dx \tag{4.8.6}$$

The range of integration $[a, b]$ is divided into N equispaced intervals with the interval size $h_x = (b - a)/N$ (see Figure 4.5 where $N = 4$ is assumed.) The grid points will be denoted by $x_0, x_1, x_2, \ldots, x_N$. By applying the trapezoidal rule on the x axis, we get

$$I \simeq \frac{h_x}{2}\left[\int_{c(x_0)}^{d(x_0)} f(x_0, y)\, dy + 2 \int_{c(x_1)}^{d(x_1)} f(x_1, y)\, dy \right.$$

$$\left. + 2 \int_{c(x_2)}^{d(x_2)} f(x_2, y)\, dy + \cdots + \int_{c(x_N)}^{d(x_N)} f(x_N, y)\, dy \right] \tag{4.8.7}$$

Equation (4.8.7) can be written more compactly as

$$I \simeq (h_x/2)[G(x_0) + 2G(x_1) + 2G(x_2) + \cdots + G(x_N)] \tag{4.8.8}$$

where

$$G(x_i) = \int_{c(x_i)}^{d(x_i)} f(x_i, y)\, dy \tag{4.8.9}$$

In evaluating Eq. (4.8.9), the domain of integration, $[c(x_i), d(x_i)]$, is divided into N intervals with the interval size

$$h_y = \frac{1}{N}[d(x_i) - c(x_i)]$$

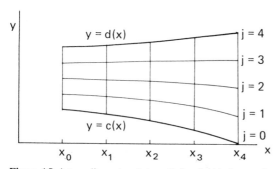

Figure 4.5 A two-dimensional domain for double integration

The y values of grid points are denoted by $y_{i,0}, y_{i,1}, y_{i,2}, \ldots, y_{i,N}$. Then, the extended trapezoidal integration becomes

$$G(x_i) = \int_{c(x_i)}^{d(x_i)} f(x_i, y)\, dy$$

$$= \frac{h_y}{2}[f(x_i, y_{i,0}) + 2f(x_i, y_{i,1}) + 2f(x_i, y_{i,2}) + \cdots + f(x_i, y_{i,N})] \tag{4.8.10}$$

The trapezoidal rule used in the foregoing integrals may be replaced by any other numerical integration method including Gauss quadratures and Simpson's rules [Press/Flannery/Teukolsky/Vetterling, p. 126].

Example 4.8

Evaluate the following double integral

$$I = \int_a^b \left[\int_{c(x)}^{d(x)} \sin (x + y)\, dy \right] dx$$

by the Simpson's 1/3 rule. The limits of integrations are

$$a = 1, \quad b = 3$$
$$c(x) = \ln (x)$$
$$d(x) = 3 + \exp (x/5)$$

⟨Solution⟩

For the Simpson's 1/3 rule, the grid points on the x axis are

$$x_0 = 1, \quad x_1 = 2, \quad x_2 = 3$$

See Figure E4.8 for the domain of integration and grid points. Applying Simpson's 1/3 rule to the first integral yields

$$I \simeq \frac{h_x}{3} [G(x_0) + 4G(x_1) + G(x_2)]$$

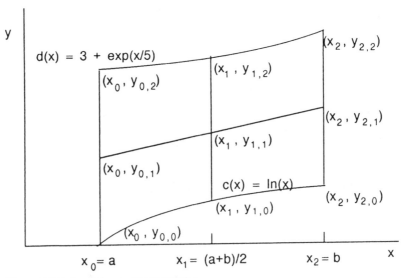

Figure E4.8 A grid for double integration

where $h_x = (b - a)/2 = 1$ and

$$G(x_i) = \int_{\ln(x_i)}^{3 + \exp(x_i/5)} \sin(x_i + y) \, dy$$

or more explicitly

$$I = \int_a^b \left[\int_{c(x)}^{d(x)} \sin(x + y) \, dy \right] dx$$

$$\simeq \frac{h_x}{3} \left[\int_{\ln(1)}^{3 + \exp(1/5)} \sin(1 + y) \, dy + 4 \int_{\ln(2)}^{3 + \exp(2/5)} \sin(2 + y) \, dy \right.$$

$$\left. + \int_{\ln(3)}^{3 + \exp(3/5)} \sin(3 + y) \, dy \right]$$

$$= \frac{h_x}{3} \left[\int_0^{4.2214} \sin(1 + y) \, dy + 4 \int_{0.6931}^{4.4918} \sin(2 + y) \, dy + \int_{1.0986}^{4.8221} \sin(3 + y) \, dy \right]$$

With Simpson's 1/3 rule, the first integral in the previous line becomes

$$\int_0^{4.2214} \sin(1 + y) \, dy$$

$$\simeq \frac{2.11070}{3} [\sin(1 + 0) + 4\sin(1 + 2.11070) + \sin(1 + 4.2214)]$$

$$= \frac{2.11070}{3} [0.84147 + (4)(0.03088) + (-0.87322)]$$

$$= 0.064581$$

Similar computations yield

$$\int_{0.6931}^{4.4918} \sin(2 + y) \, dx \simeq -2.1086$$

$$\int_{1.0986}^{4.8221} \sin(3 + y) \, dx \simeq -0.67454$$

Thus, the final value of the double integral becomes

$$I \simeq \frac{1}{3} [0.064581 + (4)(-2.1086) - 0.67454]$$

$$= -3.0148$$

SUMMARY OF THIS SECTION

(a) Double numerical integration is the application twice of a numerical integration method for single integration, once for the y direction and another for the x direction.

(b) Any numerical integration method for single integration can be applied to double integration.

PROGRAMS

PROGRAM 4–1 Extended Trapezoidal and Simpson's Rules

(A) Explanations

PROGRAM 4–1 integrates an analytical function by either the extended trapezoidal rule or extended Simpson's rule at the user's discretion. Before running the program, the user must define the integrand by changing one line in func(). The user can choose an integrating method, limits of integration, and the number of intervals by input from the keyboard interactively.

(B) List

```
/* CSL/c4-1.c   Trapezoidal/Simpson's Rules */
#include <stdio.h>
#include <stdlib.h>
#include <math.h>
#define TRUE 1
struct t_bc{float a, b, h;}   bc;
/*             isimp: 0, trapezoidal rule; 1, Simpson's rule
               bc.a, bc.b: lower and upper limits of integral
               n: number of intervals
               bc.t(h: interval size
               w: weights
               s, ss: integral                              */
main()
{
char simp[7];
int isimp, k, n;
float s;
void simps(), trapz();
    printf( "\nCSL/C4-1    Trapezoidal/Simpson's Rule \n\n" );
    printf( "The function to be integrated is hard-coded in func(). \n"
    while( TRUE ){
       printf( "Type 0 for trapezoidal, or 1 for Simpson's\n  " );
       scanf( "%d", &isimp );
       printf( "Number of intervals ?   " ); scanf( "%d", &n );
       if( !(n > 0 && isimp == 0) ) {
          if( !(isimp == 1 && n > 1) ){
             printf( "Input is invalid. \n" );
             exit(0);
          }
       }
       printf( "Lower limit of integration? " );   scanf( "%f", &bc.a );
       printf( "Upper limit of integration? " );   scanf( "%f", &bc.b );
       bc.h = (bc.b - bc.a)/n;
       if( isimp == 0 ){
          trapz( &s, n );     /*-- Trapezoidal rule */
       }
       else{
          simps( &s, n );     /*-- Simpson's rule */
       }
       printf( "----------------------------------------\n" );
```

```
            printf( " Result = %g \n", s );
            printf( "-------------------------------------------\n" );
            printf( "\nType 1 to continue, or 0 to stop.\n");
            scanf( "%d", &k );
            if( k != 1 ) exit(0);
        }
}

void trapz(ss, n)    /* Trapezoidal rule */
float *ss;       int n;
{
int i, i_;
float sum, w, x;
double func();
    sum = 0;
    for( i = 0; i <= n; i++ ){
        x = bc.a + i*bc.h;
        w = 2;
        if( i == 0 || i == n )    w = 1;
        sum = sum + w*func( x );
    }
    *ss = sum*bc.h/2;
    return;
}

void simps(ss, n)    /* Simpson's rule*/
float *ss;     int n;
{
int i, ls;      float sum, s, w, x;
double func();
    s = 0;   sum = 0;
    if( n/2*2 == n )   ls = 0;
    else {
        ls = 3;
        for( i = 0; i <= 3; i++ ) { /*  Simpson's 3/8 rule if N is odd */
            x = bc.a + bc.h*i;        w = 3;
            if( i == 0 || i == 3 )    w = 1;
            sum = sum + w*func( x );
        }
        sum = sum*bc.h*3/8L;
        if( n == 3 )   return;
    }
    for( i = 0; i <= (n - ls); i++ ){    /*  Simpson's 1/3 rule */
        x = bc.a + bc.h*(i + ls);         w = 2;
        if( (int)( i/2 )*2 + 1 == i )     w = 4;
        if( i == 0 || i == n - ls )       w = 1;
        s = s + w*func( x );
    }
    *ss = sum + s*bc.h/3;
    return;
}

double func(x)           /* Evaluates the function to be integrated. */
float x;
{
float func_v;
    func_v = pow(1  +  pow(x/2,  2),  2)*3.14159;
    return( func_v );
}
```

(C) Sample Output

```
CSL/C4-1    Trapezoidal/Simpson's Rule

The function to be integrated is hard-coded in func().
Type 0 for trapezoidal, or 1 for Simpson's
  0
Number of intervals ?   10
Lower limit of integration? 0
Upper limit of integration? 2
-------------------------------------------
 Result = 11.7705
-------------------------------------------


Type 0 for trapezoidal, or 1 for Simpson's
  1
Number of intervals ?   5
Lower limit of integration? 0
Upper limit of integration? 2
-----------------------------------------
 Result = 11.7309
-----------------------------------------
```

(D) Discussions

In the foregoing output, the integral

$$I = \int_a^b \left(1 + \left[\frac{x}{2} \right]^2 \right)^2 \pi \, dx$$

is evaluated. In the first calculation, the trapezoidal rule is selected with ten intervals, and $a = 0$ and $b = 2$ are specified by the input. The computed result is $I = 11.77047$. In the second calculation, the same integration is evaluated with Simpson's rule using five intervals, which yields $I = 11.73095$. In the latter case, Simpson's 1/3 rule and Simpson's 3/8 rule are mixed because the 1/3 rule alone does not work for $N = 5$.

Quiz: (1) What is the volume of the body of rotation made by rotating $y = \sin(x)$, $0 < x < \pi$, about the x axis? (2) If this is too easy, how about the volume of the body made by rotating e^x, $0 < x < 1$ about the y axis? (3) We tend to believe that the trapezoidal rule is inferior to Simpson's rules or more sophisticated numerical integration schemes. If you believe this also, read Section 4.7 and look at PROGRAM 4-5.

PROGRAM 4-2 Newton-Cotes Closed Formulas

(A) Explanations

PROGRAM 4-2 performs numerical integration using the Newton-Cotes closed formula. The constants for up to $N = 10$ are built into the program.

Before running the program, the user must define the function to be integrated in fun(), where F = SIN (X) is written as an example. The order of the formula and integrating limits are specified interactively as input.

Programs

(B) List

```
/*  CSL/c4-2.c      Newton-Cotes Closed Formula */
#include <stdio.h>
#include <math.h>
#define TRUE 1
/*              n : order of the Newton-Cotes formula (2 thru 10)
                a, b : lower and upper bounds of integration
                w[j][n] : j-th weight for the n-th order formula
                alpha : Q/R
                q, r : Q, R
                answer : result of integraion                       */
main()
{
int i, j, k, n, _i, _r;
float a, alpha, answer, b, f, h, q, r, x;
static float w[21][10];
double fun();
static int _aini = 1;
static float _itmp0[] = {1,1,1,2};
static float _itmp1[] = {1,4,1,1,3};
static float _itmp2[] = {1,3,3,1,3,8};
static float _itmp3[] = {7,32,12,32,7,2,45};
static float _itmp4[] = {19,75,50,50,75,19,5,288};
static float _itmp5[] = {41,216,27,272,27,216,41,1,140};
static float _itmp6[] = {751,3577,1323,2989,2989,1323,3577,751};
static float _itmp7[] = {7,17280};
static float _itmp8[] = {989,5888,-928,10496,-4540,10496,-928,5888,989};
static float _itmp9[] = {4,14175};
static float _itmp10[] = {2857,15741,1080,19344,5788,5788,19344,1080};
static float _itmp11[] = {15741,2857,9,89600};
static float _itmp12[] = {16067,106300,-48925,272400,-260550,42736};
static float _itmp13[] = {-260550,272400,-48525,106300,16067};
static float _itmp14[] = {5,299376};
    for( i = 0, _r = 0; i <= 3; i++ )   w[i][1] = _itmp0[_r++];
    for( i = 0, _r = 0; i <= 4; i++ )   w[i][2] = _itmp1[_r++];
    for( i = 0, _r = 0; i <= 5; i++ )   w[i][3] = _itmp2[_r++];
    for( i = 0, _r = 0; i <= 6; i++ )   w[i][4] = _itmp3[_r++];
    for( i = 0, _r = 0; i <= 7; i++ )   w[i][5] = _itmp4[_r++];
    for( i = 0, _r = 0; i <= 8; i++ )   w[i][6] = _itmp5[_r++];
    for( i = 0, _r = 0; i <= 7; i++ )   w[i][7] = _itmp6[_r++];
    for( i = 8, _r = 0; i <= 9; i++ )   w[i][7] = _itmp7[_r++];
    for( i = 0, _r = 0; i <= 8; i++ )   w[i][8] = _itmp8[_r++];
    for( i = 9, _r = 0; i <= 10; i++ )  w[i][9] = _itmp9[_r++];
    for( i = 0, _r = 0; i <= 7; i++ )   w[i][9] = _itmp10[_r++];
    for( i = 8, _r = 0; i <= 11; i++ )  w[i][9] = _itmp11[_r++];
    for( i = 0, _r = 0; i <= 5; i++ )   w[i][10] = _itmp12[_r++];
    for( i = 6, _r = 0; i <= 10; i++ )  w[i][10] = _itmp13[_r++];
    for( i = 11, _r = 0; i <= 12; i++ ) w[i][10] = _itmp14[_r++];
    printf( "\nCSL/C4-2      Newton-Cotes Closed Formula \n\n" );
    while( TRUE ) {
        printf( "Number of data points (2-10)   ? " ); scanf( "%d", &k );
        printf( "Lower limit of integration, a ? " ); scanf( "%f", &a );
        printf( "Upper limit of integration, b ? " ); scanf( "%f", &b );
        n = k - 1;
        q = w[n + 1][n];
        r = w[n + 2][n];
        printf(  "Q = %g       R = %g \n", q, r );
        alpha = q/r;
```

```
    h = (b - a)/n;            /* grid interval */
    printf( "-------------------------------------------------- \n" );
    printf( " n              x                f(x)                 w            \n" );
    printf( "-------------------------------------------------- \n" );
    answer = 0;                    /* Newton-Cotes formula initialization */
    for( j = 0; j <= n; j++ ) {
        x = a + j*h;
        f = fun( x );
        printf(   "%2d     %12.5e     %12.5e     %12.5e \n",
                                                 j, x, f, w[j][n] );
        answer = answer + f*w[j][n];    /*-- Newton-Cotes formula */
    }
    answer = answer*h*alpha;
    printf( "-------------------------------------------------- \n" );
    printf( "      Result          I =%14.6e \n", answer );
    printf( "-------------------------------------------------- \n" );
    printf( "\n Type 1 to continue, or 0 to stop.\n" );
    scanf( "%d", &k );
    if( k != 1 ) exit(0);
    }
}

double fun(x)     /* Defines the function to be integrated.*/
float x;
{
float fun_v;
    fun_v = sin( x );
    return( fun_v );
}
```

(C) Sample Output

```
CSL/C4-2     Newton-Cotes Closed Formula

Number of data points (2-10)   ? 6
Lower limit of integration, a ? 0
Upper limit of integration, b ? 2
Q = 5        R = 288
-----------------------------------------------------
 n           x               f(x)                w
-----------------------------------------------------
 0      0.00000e+00      0.00000e+00        1.90000e+01
 1      4.00000e-01      3.89418e-01        7.50000e+01
 2      8.00000e-01      7.17356e-01        5.00000e+01
 3      1.20000e+00      9.32039e-01        5.00000e+01
 4      1.60000e+00      9.99574e-01        7.50000e+01
 5      2.00000e+00      9.09297e-01        1.90000e+01
-----------------------------------------------------
     Result      I =  1.416117e+00
-----------------------------------------------------
```

(D) Discussions

The foregoing output shows the computation of

$$I = \int_0^2 \sin(x)\, dx$$

by the Newton-Cotes closed formula of order 5. Notice in the output that the Newton-Cotes formula uses equispaced abscissas of six data points and includes the functional values at two end points. Compare the result to the exact value, $1 - \cos(2) = 1.4161468$. The error of the computation is, therefore, 0.002%.

PROGRAM 4–3 Newton-Cotes Open Formulas

(A) Explanations

PROGRAM 4–3 performs numerical integration using the Newton-Cotes open formulas.

The structure of PROGRAM 4–3 is very similar to PROGRAM 4–2. The user must define the integrand in fun(). After starting, the program asks for the order of the formula, and the lower and upper limits of the integral. The program is reexecuted after the results are printed out.

(B) List

```
/* CSL/c4-3.c   Newton-Cotes Open Formula */
#include <stdio.h>
#include <stdlib.h>
#include <math.h>
#define TRUE 1
/*              n : order of the Newton-Cotes open formula
                a, b : lower and upper limits of integration
                w[j][n] : j-th weight for the n-th order formula
                al : Q/R
                q, r : Q, R
                ans : result of integration                      */
main()
{
int i, j, k, m, n, _i, _r;
float a, al, ans, b, h, q, r, x, y;
static float w[11][9];
double fun();
static float _itmp0[] = {1,1,3,2};
static float _itmp1[] = {2,-1,2,4,3};
static float _itmp2[] = {11,1,1,11,5,24};
static float _itmp3[] = {11,-14,26,-14,11,6,20};
static float _itmp4[] = {611,-453,562,562,-453,611,7,1440};
static float _itmp5[] = {460,-954,2196,-2459,2196,-954,460,8,945};
    for( i = 1, _r = 0; i <= 4; i++ )   w[i][1] = _itmp0[_r++];
    for( i = 1, _r = 0; i <= 5; i++ )   w[i][2] = _itmp1[_r++];
    for( i = 1, _r = 0; i <= 6; i++ )   w[i][3] = _itmp2[_r++];
    for( i = 1, _r = 0; i <= 7; i++ )   w[i][4] = _itmp3[_r++];
    for( i = 1, _r = 0; i <= 8; i++ )   w[i][5] = _itmp4[_r++];
    for( i = 1, _r = 0; i <= 9; i++ )   w[i][6] = _itmp5[_r++];
    printf( "\nCSL/C4-3   Newton-Cotes Open Formula \n" );
    while( TRUE ){
        printf( "Number of grid points? (Select from 2,3,4,5,6,7)\n" );
```

```
        scanf( "%d", &m );
        n = m - 1;
        printf( "Lower bound of integration ?" );   scanf( "%f", &a );
        printf( "Upper bound of integration?" );    scanf( "%f", &b );
        q = w[m + 1][n];
        r = w[m + 2][n];
        printf(  "Q = %12.5e   R = %12.5e   \n", q, r );
        al = q/r;                      /* alpha=Q/R */
        h = (b - a)/(n + 2);           /* grid interval */
        printf( "---------------------------------------------- \n" );
        printf( "  i      x              f(x)            w          \n" );
        printf( "---------------------------------------------- \n" );
        ans = 0;
        for( j = 1; j <= m; j++ ){
           x = a + j*h;
           y = fun( x );
           ans = ans + y*w[j][n];  /* Newton-Cotes open formula */
           printf("  %d %12.5e %12.5e %12.5e \n",
                                        j, x, y, w[j][n] );
        }
        ans = ans*h*al;
        printf( "---------------------------------------------- \n" );
        printf( "       Final result     I =%13.6e \n", ans );
        printf( "---------------------------------------------- \n" );
        printf( "Type 1 to continue, or 0 to stop.\n" );
        scanf( "%d", &k );   if( k != 1 ) exit(0);
   }
}

double fun(x)    /* Defines the function to be integrated */
float x;
{
float f_v;
   f_v = sin( x );
   return( f_v );
}
```


(C) Sample Output


```
CSL/C4-3    Newton-Cotes Open Formula

Number of grid points? (Select from 2,3,4,5,6,7)   6
Lower bound of integration ?  0
Upper bound of integration?   2
Q =  7.00000e+00   R =  1.44000e+03
----------------------------------------------
   i      x         f(x)           w
----------------------------------------------
   1  2.85714e-01  2.81843e-01   6.11000e+02
   2  5.71429e-01  5.40834e-01  -4.53000e+02
   3  8.57143e-01  7.55975e-01   5.62000e+02
   4  1.14286e+00  9.09823e-01   5.62000e+02
   5  1.42857e+00  9.89903e-01  -4.53000e+02
   6  1.71429e+00  9.89723e-01   6.11000e+02
----------------------------------------------
       Final result     I = 1.416224e+00
----------------------------------------------
```

(D) Discussions

The same equation as in PROGRAM 4–2 is evaluated in the foregoing output. Notice that the functional values at the limits at $x = 0$ and $x = 2$ are not used and the values of w take large values with plus and minus signs, typical for a high-order Newton-Cotes formula. Large weights with changing signs cause truncation errors as discussed in Section 4.5. The exact value of integration is $1 - \cos(2) = 1.4161468$. Therefore, the error of the computation is 0.0054%.

PROGRAM 4–4 Gauss Quadrature

(A) Explanations

PROGRAM 4–4 integrates a function by a Gauss quadrature. The Gauss quadrature table is contained in the program.

Before executing the program, the user must define the integrand in fun(). When executed, the program asks for the number of Gauss points, and integration limits.

(B) List

```
/* CSL/c4-4.c    Gauss Quadrature */
#include <stdio.h>
#include <stdlib.h>
#include <math.h>
#define TRUE 1
/*             n : number of Gauss points to be used
               a, b : lower and upper limits of integration
               xa[j] : x-coordinate of the j-th point
               w[j] :  weight of the j-th point
               f : functional value at x
               xi : result of integraion                              */
main()
{
int j, k, n;
float a, b, f, pi = 3.1415927, w[11], x, xa[11], xi;
void fun();
    printf( "\n\nCSL/C4-4       Gauss Quadrature \n\n" );
    while (TRUE){
        while (TRUE){
            printf( "Select the number of Gauss points from:") ;
            printf( " n=2,3,4,5,6,8,10\n" );
            printf( "n ?  " );   scanf( "%d", &n );
            if( (n<0) || (n == 7) || (n == 9) ){
                if (n<0) printf( "n<0 is invalid;  repeat input.\n");
                if ( (n==7) || (n==9))
                    printf( "n=7 and 9 are not available: repeat input.\n" );
            }
            else     break;
        }
        printf( "Lower limit, a ?  " ); scanf( "%f", &a );
        printf( "Upper limit, b ?  " ); scanf( "%f", &b );
        if( n == 2 ){
            xa[2] = 0.5773502691;   w[2] = 1;
        }
```

```
    if( n == 3 ){
       xa[2] = 0;              xa[3] = 0.7745966692;
       w[2] = 0.8888888888; w[3] = 0.5555555555;
    }
    if( n == 4 ){
       xa[3] = 0.3399810435; xa[4] = 0.8611363115;
       w[3] = 0.6521451548;  w[4] = 0.3478548451;
    }
    if( n == 5 ){
       xa[3] = 0;              xa[4] = 0.5384693101; xa[5] = 0.9061798459;
       w[3] = 0.5688888888;  w[4] = 0.4786286704;  w[5] = 0.2369268850;
    }
    if( n == 6 ){
       xa[4] = 0.2386191860; xa[5] = 0.6612093864; xa[6] = 0.9324695142;
       w[4] = 0.4679139345;  w[5] = 0.3607615730;  w[6] = 0.1713244923;
    }
    if( n == 8 ){
       xa[5] = 0.1834346424; xa[6] = 0.5255324099; xa[7] = 0.7966664774;
       xa[8] = 0.9602898564;
       w[5] = 0.3626837833;  w[6] = 0.3137066458;  w[7] = 0.2223810344;
       w[8] = 0.1012285362;
    }
    if( n == 10 ){
       xa[6] = 0.1488743389; xa[7] = 0.4333953941; xa[8] = 0.6794095682;
       xa[9] = 0.8650633666; xa[10] = 0.9739065285;
       w[6] = 0.2955242247;  w[7] = 0.2692667193;  w[8] = 0.2190863625;
       w[9] = 0.1494513491;  w[10] = 0.0666713443;
    }
    for( j = 1; j <= ( n/2 ); j++ ){
       w[j] = w[n + 1 - j];
       xa[j] = -xa[n + 1 - j];
    }
    printf( "---------------------------------------------------------- \n" );
    printf( "    k        x                f(x)                w         \n" );
    printf( "---------------------------------------------------------- \n" );
    for( j = 1; j <= n; j++ ){
       xa[j] = (xa[j]*(b - a) + a + b)/2;   /* Gauss points */
    }
    xi = 0;                 /* Initializing the Gauss quadrature formula. */
    for( j = 1; j <= n; j++ ){
       x = xa[j];
       fun( &f, x );                 /* Finding functional values */
       xi = xi + f*w[j];             /* Gauss quadrature formula */
       printf( "   %3d  %12.5e    %12.5e    %12.5e \n", j, x, f, w[j] );
    }
    xi = xi*(b - a)/2;
    printf( "---------------------------------------------------------- \n" );
    printf( "    Final result         I =%13.6e \n", xi );
    printf( "---------------------------------------------------------- \n" );
    printf( "\nType 1 to continue, or 0 to stop.\n" ); scanf("%d",&k );
    if( k != 1 ) exit(0);
  }
}

void fun(f, x)   /* Defines the function to be integrated. */
float *f, x;
{   *f = sin( x );            return;
}
```

(C) Sample Output

```
CSL/C4-4     Gauss Quadrature

Orders of the quadratures available are: n=2,3,4,5,6,8,10
n ?  4
Lower limit, a ?  0
Upper limit, b ?  2
-----------------------------------------------------------
     k          x              f(x)              w
-----------------------------------------------------------
     1    1.38864e-01    1.38418e-01       3.47855e-01
     2    6.60019e-01    6.13132e-01       6.52145e-01
     3    1.33998e+00    9.73480e-01       6.52145e-01
     4    1.86114e+00    9.58147e-01       3.47855e-01
-----------------------------------------------------------
     Final result       I = 1.416147e+00
-----------------------------------------------------------
```

(D) Discussions

In the foregoing output, the same integral as in PROGRAMS 4–2 and 4–3 is again evaluated. In the calculation, four Gauss points are used. The result is exact within a round-off error in the last digit.

Weights of the Gauss quadratures are all positive and significantly smaller in magnitude compared to those in Newton-Cotes formulas.

PROGRAM 4–5 Integration of Singular Function

(A) Explanations

This program integrates a function by the double exponential method and can integrate improper integrals. The integrand is defined fun(). Double precision is used throughout the program.

(B) List

```
/* CSL/c4-5.c    Integration of a Singular Function    */
#include <stdio.h>
#include <stdlib.h>
#include <math.h>
#define TRUE 1
/*            a : lower limit of integration
              b : upper limit of integration
              h : grid interval
              n : number of intervals
              dxdz : dx/dz
              hcos : hyperbolic cosine
              hsin : hyperbolic sine
              ss : result of integration                */
main()
```

```
{
int k, n;       float a, b, h;
double dxdz, exz, fun(), hcos, hsin, p, pai = 3.1415927, s, ss, w, x, z;
   printf( "\n\nCSL/C4-5.C  Integration of a Singular Function ");
   while (TRUE) {
      printf( "\n\nTotal number of intervals, n?" ); scanf( "%d", &n );
      printf( "Lower limit of integration, a?   " ); scanf( "%f", &a );
      printf( "Upper limit of integration, b?   " ); scanf( "%f", &b );
      n = n/2;          /* Half of the total number of grid points */
      h = 4.0/n;        /* Determines h; numerator 4.0 can be increased */
                        /* if 64 bits are used for floating values */

      ss = 0;
      for( k = -n; k <= n; k++ ){
         z = h*k;                       exz = exp( z );
         hcos = (exz + 1.0/exz)/2.;     hsin = (exz - 1.0/exz)/2.;
         s = exp( pai*0.5*hsin );
         x = (b*s + a/s)/(s + 1.0/s);
         if( x != a && x != b ){
            p = pai/2.0*hsin;           w = exp( p );
            dxdz = (b - a)*pai/2.0*hcos/pow((w + 1.0/w)/2.0, 2.0)/2.0;
            ss = ss + h*fun( x )*dxdz;
         }
      }
      printf( "-----------------------------------------------\n" );
      printf( "    Final result:     I= %13.6e \n", ss );
      printf( "-----------------------------------------------\n" );
      printf( "\nType 1 to continue, or 0 to stop.\n"); scanf("%d", &k);
      if( k != 1 ) exit(0);
   }
}

double fun(x)      /* Evaluates the function to be integrated. */
double x;
{
double fun_v;
   fun_v = sqrt( 1.0 + 1/ x );
   return( fun_v );
}
```

(C) Sample Output

```
CSL/C4-5  Integration of a Singular Function

Total number of points, n?     10
Lower limit of integration, a?   0
Upper limit of integration, b?   1
-----------------------------------
    Final result:     I=  2.297836e+00
-----------------------------------


Total number of points, n?     20
Lower limit of integration, a?   0
Upper limit of integration, b?   1
-----------------------------------
    Final result:     I=  2.295587e+00
-----------------------------------
```

```
Total number of points, n?    40
Lower limit of integration, a?   0
Upper limit of integration, b?   1
-------------------------------------------
    Final result:      I=  2.295587e+00
-------------------------------------------
```

(D) Discussions

In the foregoing output, the integral is evaluated three times for

$$I = \int_0^1 \sqrt{1 + \frac{1}{x}}\, dx$$

where the integrand is singular at $x = 0$. The first calculation is done with 10 points, the second with 20, and the third with 40 points. The results with 20 and 40 points agree exactly.

One problem with the double exponential transformation is that the abscissas of the data points in terms of x becomes very close to $x = a$ or b, at which the function may be singular. This may cause overflow or zero division. If this is the case, one remedy is to reduce the limiting value of z, which is currently set to 4, to a slightly smaller value.

PROGRAM 4–6 Double Integration

(A) Explanations

PROGRAM 4–6 performs double integration using the extended Simpson's 1/3 rule for both x and y directions. When executed, the program asks for the number of intervals to be used in the x and y directions. It also asks for A and B.

The user must define three functions in the program before running: the integrand in fun(), the lower bound curve in lower_curve(), and the upper bound curve in upper_curve(). The number of intervals and integrating limits in x are given interactively.

(B) List

```
/* CSL/c4-6.c   Double Integration by Simpson's Rule */
#include <stdio.h>
#include <stdlib.h>
#include <math.h>
#define TRUE 1
/*          a : lower limit of integraion of x
            b : upper limit of integration of x
            c : lower bound curve (function of x)
            d : upper bound curve (function of x)
            f : integrand
            hx, hy : interval sizes
            t : result of integraion                    */
```

```
main()
{
int i, j, k, m, n;
double a, b, c, d, f, hx, hy, s, t, w, x, y;
float a_inp, b_inp;
void lower_curve(), upper_curve(), fun();
   printf( "\n\nCSL/C4-6   Double Integration by the Simpson's Rule\n");
   while( TRUE ) {
       printf("\nNumber of intervals in x-direction?"); scanf( "%d", &m );
       printf( "Number of intervals in y-direction?"); scanf( "%d", &n );
       if( m/2*2 != m || n/2*2 != n ){
           printf( "Number of intervals must be even. Repeat.\n" );
       }
       else{
           printf( "Lower limit of x, a? " );     scanf( "%f", &a_inp );
           printf( "Upper limit of x, b? " );     scanf( "%f", &b_inp );
           printf( "---------------------------------------------\n" );
           printf( " Point on     Integral in y          Weight\n" );
           printf( " x axis       direction\n" );
           printf( "   i              Iy                  w\n" );
           printf( "---------------------------------------------\n" );
           a = a_inp; b = b_inp;
           hx = (b - a)/m;
           t = 0;
           for( i = 0; i <= m; i++ ){
               x = a + i*hx;
               lower_curve( &c, x );  /* Finds the lower limit of y value. */
               upper_curve( &d, x );  /* Finds the upper limit of y value. */
               hy = (d - c)/n;        /* Interval size in y-direction */
               s = 0;
               for( j = 0; j <= n; j++ ){
                   y = c + j*hy;      /* y value of grid points */
                   fun( &f, x, y );       /* Finds f value. */
                                           w = 4;
                   if( ( j/2 )*2 == j )    w = 2;
                   if( (j == 0) || (j == n) )   w = 1;
                   s = s + w*f;        /* Integration in y-direction */
               }
               s = s*hy/3;
                                       w = 4;
               if( ( i/2 )*2 == i )    w = 2;
               if( (i == 0) || (i == m) ) w = 1;
               printf("%4d       %12.5e       %12.5e   \n", i, s, w );
               t = t + w*s;       /*Integration in x-direction */
           }
           t = t*hx/3;            /*Result of integration */
           printf( "---------------------------------------------\n" );
           printf( "Final result      I = %13.6e \n", t );
           printf( "---------------------------------------------\n" );
           printf( "\nType 1 to continue, or 0 to stop. \n" );
           scanf( "%d", &k );     if( k != 1 ) exit(0);
       }
   }
}

void lower_curve(c, x)             /* Lower bound curve */
double *c, x;
{
```

```
    *c = log( x );
    return;
}

void upper_curve(d, x)              /* Upper bound curve */
double *d, x;
{
    *d = 3 + exp( x/5 );
    return;
}

void fun(f, x, y)         /* Defines and evaluates integrand. */
double *f, x, y;
{
    *f = sin( x + y );
    return;
}
```

(C) Sample Output

```
CSL/C4-6    Double Integration by the Simpson's Rule

Number of intervals in x-direction?   6
Number of intervals in y-direction?   6
Lower limit of x, a? 1
Upper limit of x, b? 3
------------------------------------------------
Point on       Integral in y           Weight
x axis         direction
   i                Iy                     w
------------------------------------------------
   0            5.30626e-02            1.00000e+00
   1           -8.50753e-01            4.00000e+00
   2           -1.54744e+00            2.00000e+00
   3           -1.88120e+00            4.00000e+00
   4           -1.80004e+00            2.00000e+00
   5           -1.34007e+00            4.00000e+00
   6           -6.08341e-01            1.00000e+00
------------------------------------------------
Final result       I = -2.615372e+00
------------------------------------------------
```

(D) Discussions

In the foregoing output, the equation

$$I = \int_{a=1}^{b=3} \int_{\log(x)}^{3+\exp(x/5)} \sin(x+y)\,dy\,dx$$

is evaluated. The upper and lower limits of x are specified by interactive input. The integrand and the lower and upper limit curves are defined within the program.

The extended Simpson's 1/3 rule is first used to evaluate the inner integral

$$I_y(x) = \int_{\log(x)}^{3+\exp(x/5)} \sin(x+y)\,dy$$

for seven values of x. The results are printed out in the table. The outer integration is also performed by the extended Simpson's rule. To examine the accuracy of the result, one should execute the program with twice as many intervals in both x and y directions and check if the results change significantly.

QUIZ: A man had a perfectly spherical cheese ball of 3-cm radius. Assuming that the origin of the three-dimensional coordinates is at the center of the cheese, he cut it to two equal parts along the horizontal center plane, $z = 0$. He saved one half in his refrigerator. He then cut the remaining cheese into two pieces again, but along the curve of $y = 2(x^2/9 - 1)$ and ate the smaller piece near the edge. What is the volume of the cheese remaining? Assume the cheese had no holes like Swiss cheese.

PROBLEMS

(**4.1**) Integrate the following functions over the given interval by using the extended trapezoidal rule with intervals of $N = 2, 4$, and 8 (16 and 32 also if a computer program is used.)

$3x^3 + 5x - 1$	$[0, 1]$
$x^3 - 2x^2 + x + 2$	$[0, 3]$
$x^4 + x^3 - x^2 + x + 3$	$[0, 1]$
$\tan(x)$	$[0, \pi/4]$
e^x	$[0, 1]$
$1/(2 + x)$	$[0, 1]$

(**4.2**) Calculate the integral

$$I = \int_0^{\pi/2} \sin(x)\,dx$$

by the extended trapezoidal rule with $N = 2, 4, 8, 25$, and 100 intervals. Then, evaluate the error of the numerical results by comparison to the exact value.

(**4.3**) A function table is given as follows:

x	$f(x)$
0.0	0
0.1	2.1220
0.2	3.0244
0.3	3.2568
0.4	3.1399
0.5	2.8579
0.6	2.5140
0.7	2.1639
0.8	1.8358

Evaluate the integral

$$\int_0^{0.8} f(x)\,dx$$

by the extended trapezoidal rule with $h = 0.4$, $h = 0.2$ and $h = 0.1$.

(4.4) By applying the Romberg integration to the results of the trapezoidal rule with $h = 0.1$ and $h = 0.2$ for Problem 4.3, estimate a more accurate integral.

(4.5) A function table follows:

i	x_i	$f(x_i)$
1	0	0.9162
2	0.25	0.8109
3	0.5	0.6931
4	0.75	0.5596
5	1.0	0.4055

(a) Calculate

$$I = \int_0^1 f(x)\,dx$$

by using the extended trapezoidal rule with $h = 0.25$ and $h = 0.5$.

(b) By applying the Romberg integration to the results of the question (a), estimate a more accurate value of I.

(4.6) Repeat the problem in Example 4.1 by using the extended Simpson's 1/3 rule with $N = 2, 4, 8,$ and 16.

(4.7) Repeat Problem 4.1 by using Simpson's rule with $N = 4, 8,$ and 16.

(4.8) Derive Simpson's 1/3 rule by integrating the Newton forward interpolation polynomial fitted at x_0, $x_0 + h$ and $x_0 + 2h$.

(4.9) Prove that the Romberg integration based on I_{2h} and I_h of the extended trapezoidal rule is identical with the result of Simpson's rule using h as the interval size.

(4.10) Evaluate the integral of the following functions extended over the interval indicated for each by using Simpson's rule with $N = 2, 4, 8, 16,$ and 32:

(a) $y = \dfrac{1}{2 + \cos(x)}$ $\qquad\qquad$ $[0, \pi]$

(b) $y = \dfrac{\log(1 + x)}{x}$ $\qquad\qquad$ $[1, 2]$

(c) $y = \dfrac{1}{1 + \sin^2(x)}$ $\qquad\qquad$ $[0, \pi/2]$

(4.11) Evaluate the integral of the following functions extended over the interval indicated for each by using Simpson's rule with $N = 2, 4, 8, 16,$ and 32:

(a) $y = x \exp(2x)$ $\qquad\qquad$ $[0, 1]$

(b) $y = x^{-x}$ $\qquad\qquad$ $[0, 1]$

(c) $y = \exp(2x) \sin^2(x)$ $\qquad\qquad$ $[0, 2\pi]$

(4.12) Repeat the problem in Example 4.1 by using the extended Simpson's rule with $N = 3, 7,$ and 11 intervals.

(4.13) Suppose you are an architect and planning to use a large arch of the parabolic shape given by

$$y = 0.1x(30 - x) \text{ meters}$$

where y is the height above the ground and x is in meters. Calculate the total length of the arch by using the extended Simpson's rule. (Divide the domain from $x = 0$ to $x = 30$ m into 10 equally spaced intervals.) The total length of the arc is given by

$$L = \int_0^{30} \sqrt{1 + (dy/dx)^2} \, dx$$

(4.14) An automobile of mass M = 5400 kg is moving at a speed of 30 m/s. The engine is disengaged suddenly at $t = 0$ sec. Assume that the equation of motion after $t = 0$ is given by

$$5400v \frac{dv}{dx} = -8.276v^2 - 2000$$

where $v = v(t)$ is the speed (m/sec) of the car at t. The left side represents $Mv(dv/dx)$. The first term on the right side is the aerodynamic drag, and the second term is the rolling resistance of the tires. Calculate how far the car travels until the speed reduces to 15 m/sec. (*Hint:* The equation of motion may be integrated as

$$\int_{15}^{30} \frac{5400v \, dv}{8.276v^2 + 2000} = \int dx = x$$

Evaluate the preceding equation using Simpson's rule.)

(4.15) If $f(x)$ is a polynomial of order N or less, the Newton-Cotes closed formula of order N (using $N + 1$ points) becomes exact. Explain the reason.

(4.16) The length of a curve defined by $x = \theta(t)$, $y = \psi(t)$, $a < t < b$, is given by

$$s = \int_a^b ([\theta'(t)]^2 + [\psi'(t)]^2)^{1/2} \, dt$$

Using the Gauss quadratures with $N = 2, 4$, and 6, find the curve length of the cycloid defined by

$$x = 3[t - \sin(t)], \quad y = 2 - 2\cos(t), \quad 0 < t < 2\pi$$

(4.17) If $f(x)$ is a polynomial of order $2N - 1$ or less, the Gauss quadrature using N Legendre points becomes exact. Explain why.

(4.18) Evaluate the following integral by the Gauss quadrature of $N = 4$ and $N = 6$:

$$I = \int_0^1 \frac{\ln(1 + x)}{x} \, dx$$

(4.19) Evaluate the following improper integral as accurately as possible by using the extended trapezoidal rule.

$$\int_{-\infty}^{\infty} \frac{\exp(-x^2)}{(1 + x^2)} \, dx$$

(4.20) Evaluate the following improper integrals as accurately as possible by using the extended trapezoidal rule with the double exponential transformation.

(a) $\int_0^1 \frac{\tan(x)}{x^{0.7}} \, dx$

(b) $\int_0^1 \frac{\exp(x)}{\sqrt{1 - x^2}} \, dx$

(4.21) Repeat (4.18) with PROGRAM 4–5.

(4.22) Calculate the following integrals by the Gauss quadrature of $N = 6$:

(a) $\int_0^\pi \dfrac{1}{2 + \cos(x)}\, dx$

(b) $\int_1^2 \dfrac{\ln(1 + x)}{x}\, dx$

(c) $\int_0^1 x \exp(2x)\, dx$

(d) $\int_0^1 x^{-x}\, dx$

(4.23) Calculate the following integral by using the extended trapezoidal rule in each direction:

$$\int_1^2 dx \int_0^1 dy \sin(x + y)$$

(Use only two intervals in each direction; the sine function is in radian.)

(4.24) Evaluate the following integral by the Simpson 1/3 rule:

$$I = \int_0^1 \int_0^x \sqrt{x + y}\, dy\, dx$$

(4.25) The area of a unit circle is π. Accuracy of a numerical scheme for double integration may be tested for the following problem:

$$I = \iint_D dy\, dx$$

where D means that the integration is extended over the interior of

$$x^2 + y^2 \leqslant 2x$$

which is a unit circle. Perform the numerical evaluation of the preceding double integration by the extended Simpson rule in both directions with 2×2, 4×4, 8×8, 16×16, 32×32, and 64×64 intervals.

(4.26) By using the extended Simpson's rule with 10 intervals in each direction, evaluate the double integral

$$I = \int_0^\pi \int_0^{\sin(x)} \exp(-x^2 - y^2)\, dy\, dx$$

(4.27) Evaluate the following double integral by the Simpson's 1/3 rule:

$$I = \int_1^2 \int_0^{2 - 0.5x} \sqrt{x + y}\, dy\, dx$$

(4.28) Repeat the problem in Example 4.8 by using the Gauss quadrature of $N = 3$.

REFERENCES

Abramovitz, M., and I. A. Stegun, eds., *Handbook of Mathematical Functions*, National Bureau of Standards, 1970.

Carnahan, B., H. A. Luther, and J. O. Wilkes, *Applied Numerical Methods*, Wiley, 1969.

Ferziger, J. H., *Numerical Methods for Engineering Application*, Wiley-Interscience, 1981.

Froeberg, C. E., *Numerical Mathematics—Theory and Computer Applications*, Benjamine/Cummings, 1985.

Gerald, C. F., and P. O. Wheatley, *Applied Numerical Analysis*, 4th ed., Addison-Wesley, 1989.

James, M. L., G. M. Smith, and J. C. Wolford, *Applied Numerical Methods for Digital Computation*, 3rd ed., Harper & Row, 1985.

King, J. T., *Introduction to Numerical Computation*, McGraw-Hill, 1984.

Mori, M., Quadrature Formulas Obtained by Variable Transformation and DE-rule, *J. Comp. Appl. Math.*, Vol 12–13, p. 119-130, 1980.

Mori, M., and R. Piessens, ed., *Numerical Quadrature*, North-Holland, 1987.

Press, W. P., B. P. Flannery, S. A. Teukolsky, and W. T. Vetterling, *Numerical Recipes*, Cambridge University Press, 1986.

Stoer, J., and R. Burlish, *Introduction to Numerical Analysis*, Springer-Verlag, 1980.

Takahashi, H., and M. Mori, "Double Exponential Formulas for Numerical Integration," *Publ. RIMS*, Kyoto University, Vol. 9, No. 3, 1974.

5

Numerical Differentiation

5.1 INTRODUCTION

Numerical differentiation, or difference approximation, is used to evaluate derivatives of a function using the functional values given at grid points. Difference approximations are important in the numerical solution of both ordinary and partial differential equations.

To illustrate numerical differentiation, consider a function $f(x)$ as depicted in Figure 5.1. Suppose the first derivative of $f(x)$ at $x = x_0$ is to be evaluated. If the values of f at $x_0 - h$, x_0, and $x_0 + h$ are given where h is an interval between two consecutive points on the x axis, then $f'(x_0)$ may be approximated by the gradient of linear interpolation A, B, or C depicted in Figure 5.1. These three approximations using the gradients of A, B, and C are called respectively *forward*, *backward*, and *central* difference approximations. The mathematical formulas of them are as follows:

(a) Approximation using A (forward difference approximation)

$$f'(x_0) \simeq \frac{f(x_0 + h) - f(x_0)}{h} \tag{5.1.1}$$

(b) Approximation using B (backward difference approximation)

$$f'(x_0) \simeq \frac{f(x_0) - f(x_0 - h)}{h} \tag{5.1.2}$$

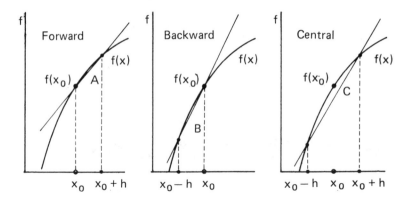

The $f'(x_0)$ is approximated by the gradient
of a line passing $(f(x_0), x_0)$

Figure 5.1 Graphical explanation of difference approximations for $f'(x_0)$

(c) Approximation using C (central difference approximation)

$$f'(x_0) \simeq \frac{f(x_0 + h) - f(x_0 - h)}{2h} \tag{5.1.3}$$

There are three different approaches for deriving difference approximations. The first is based on Taylor expansions of the function about a grid point, the second is to use difference operators, and the third is to differentiate the interpolating polynomials. Advantages and disadvantages of each approach are summarized in Table 5.1. A program to generate difference approximation formulas is given as PROGRAM 5–1 (see Section 5.3 for more details).

Table 5.1 Brief summary of the three methods of deriving numerical differentiation formulas

Method of derivation	Advantages	Disadvantages
Taylor expansion	Error terms are explicitly given. Applicable to nonuniform grids.	Only one formula can be derived at a time.
Difference operator	Close similarity between derivatives and difference approximations.	Needs Taylor expansion to analyze error.
Differentiating interpolation polynomials	Numerous difference approximation formulas are obtained systematically.	Difficult to apply on nonuniformly spaced grid points.

5.2 USING TAYLOR EXPANSION

When a function is numerically represented on discrete points, the function is approximated by means of interpolation. Numerical integration of the function is derived by integrating the interpolation formula as described in Chapter 4. Likewise, numerical differentiation formulas may be derived by differentiating the interpolation formula.

We start the derivations using Taylor expansion, which is equivalent to differentiation of an interpolation and yields exactly the same results. In this section, derivation of difference approximation using Taylor expansion is explained in an ad hoc manner. A generic approach is explained in the next section.

For a derivative of order p, the smallest number of data points necessary to derive a difference approximation is $p + 1$. For example, a difference approximation for the first derivative of a function needs at least two points.

We start derivation of difference approximation for $f'_i = f'(x_i)$ using $f_i = f(x_i)$ and $f_{i+1} = f(x_{i+1})$. The Taylor expansion of f_{i+1} about x_i is

$$f_{i+1} = f_i + hf'_i + \frac{h^2}{2} f''_i + \frac{h^3}{6} f'''_i + \frac{h^4}{24} f''''_i + \cdots \qquad (5.2.1)$$

Solving Eq. (5.2.1) for f'_i yields

$$f'_i = \frac{f_{i+1} - f_i}{h} - \frac{1}{2} hf''_i - \frac{1}{6} h^2 f'''_i - \cdots \qquad (5.2.2)$$

If we truncate after the first term, Eq. (5.2.2) becomes the forward difference approximation, which is already shown as Eq. (5.1.1). The terms truncated constitute the truncation error. The truncation error is represented by the leading term $(-(h/2)f''_i$ in this case) because other terms vanish more rapidly than this when h is decreased. The forward difference approximation is expressed, including the truncation error effect, as

$$f'_i = \frac{f_{i+1} - f_i}{h} + 0(h) \qquad (5.2.3)$$

where

$$0(h) = -\frac{1}{2} hf''_i$$

The term $0(h)$ indicates that error is approximately proportional to the grid interval h.

The backward difference approximation for the first derivative using f_{i-1} and f_i is obtained in a similar manner. The Taylor expansion of f_{i-1} is

$$f_{i-1} = f_i - hf'_i + \frac{h^2}{2} f''_i - \frac{h^3}{6} f'''_i + \frac{h^4}{24} f''''_i - \cdots \qquad (5.2.4)$$

Solving for f'_i, the backward difference approximation is obtained as

$$f'_i = \frac{f_i - f_{i-1}}{h} + 0(h) \tag{5.2.5}$$

where

$$0(h) = \frac{1}{2}hf''_i$$

The central difference approximation using f_{i+1} and f_{i-1} may be derived by the Taylor expansions of f_{i+1} and f_{i-1} already given as Eq. (5.2.1) and (5.2.4) respectively. Subtracting Eq. (5.2.4) from Eq. (5.2.1) yields

$$f_{i+1} - f_{i-1} = 2hf'_i + \frac{1}{3}h^3f'''_i + \cdots \tag{5.2.6}$$

where the f''_i term has been automatically eliminated. Then by solving for f'_i we get

$$f'_i = \frac{f_{i+1} - f_{i-1}}{2h} - \frac{1}{6}h^2f'''_i + \cdots \tag{5.2.7}$$

The central difference approximation is expressed as

$$f'_i = \frac{f_{i+1} - f_{i-1}}{2h} + 0(h^2) \tag{5.2.8}$$

where

$$0(h^2) = -\frac{1}{6}h^2f'''_i$$

It is remarkable that, because of cancellation of the f'' term, the error of the central difference approximation is proportional to h^2 rather than h. When h is decreased, the error decreases more quickly than in the other two approximations.

As explained before, a difference approximation for $f_i^{(p)}$ needs at least $p + 1$ data points. If more data points are used, a more accurate difference approximation may be derived. With given data points, a difference equation of the highest accuracy is such that the error term is of the highest order possible.

To illustrate the meaning of this, we derive a difference approximation for f'_i using f_i f_{i+1} and f_{i+2}. Since the minimum number of data points for f' is two, we have one more data point than the minimum. Expansions of f_{i+1} and f_{i+2} are written as

$$f_{i+1} = f_i + hf'_i + \frac{h^2}{2} f''_i + \frac{h^3}{6} f'''_i + \frac{h^4}{24} f''''_i + \cdots \tag{5.2.9}$$

$$f_{i+2} = f_i + 2hf'_i + 4\frac{h^2}{2} f''_i + 8\frac{h^3}{6} f'''_i + 16\frac{h^4}{24} f''''_i + \cdots \tag{5.2.10}$$

To eliminate the second derivative term, we subtract Eq. (5.2.11) from 4 times Eq. (5.2.9):

$$4f_{i+1} - f_{i+2} = 3f_i + 2hf'_i - \frac{2}{3}h^3 f'''_i + \cdots \tag{5.2.11}$$

so the leading term of the truncation errors is the third-order derivative term. Solving Eq. (5.2.11) for f'_i yields

$$f'_i = \frac{-f_{i+2} + 4f_{i+1} - 3f_i}{2h} + O(h^2) \tag{5.2.12}$$

where the error term is given by

$$O(h^2) = \frac{1}{3}h^2 f'''_i$$

Equation (5.2.12) is called *three-point forward difference approximation* for f'_i, and the error is of the same order as the central difference approximation.

Similarly, *three-point backward difference approximation* may be derived using f_i, f_{i-1}, and f_{i-2} as follows:

$$f'_i = \frac{3f_i - 4f_{i-1} + f_{i-2}}{2h} + O(h^2) \tag{5.2.13}$$

where

$$O(h^2) = \frac{1}{3}h^2 f'''_i$$

Example 5.1

Calculate the first derivative of tan (x) at $x = 1$ by the five difference approximations derived in this section using $h = 0.1, 0.05$, and 0.02. Then evaluate the percentage of error of each approximation by comparison to the exact value.

〈**Solution**〉

By introducing $f_i = f(1 + ih) = \tan(1 + ih)$ into Eqs. (5.2.5), (5.2.3), (5.2.8), (5.2.13), and (5.2.12), the following results are obtained:

	h=0.1	h=0.05	h=0.02
`[tan(1)-tan(1-h)]/h`	2.9724	3.1805	3.3224
	(13.2)[a]	(7.1)	(3.0)
`[tan(1+h)-tan(1)]/h`	4.0735	3.7181	3.5361
	(-18.9)	(-8.5)	(-3.2)
`[tan(1+h)-tan(1-h)]/2h`	3.5230	3.4493	3.4293
	(-2.8)	(-0.69)	(-0.11)
`[3tan(1)-4tan(1-h)+tan(1-2h)]/2h`	3.3061	3.3885	3.4186
	(3.5)	(1.08)	(0.20)
`[-tan(1+2h)+4tan(1+h)-3tan(1)]/2h`	3.0733	3.3627	3.4170
	(10.3)	(1.83)	(0.25)

[a] Percentage of error.

Notice that errors of the first two approximations decrease in proportion to h, whereas errors of the last three approximations decrease in proportion to h^2. It is clear that the rate of reduction of error becomes faster as the order of accuracy becomes higher.

Next we derive difference approximations for the second derivative. The basic principle in deriving a second-order difference approximation is to eliminate the first derivative from the Taylor expansions and, if possible, as many terms of order higher than 2 as possible.

As an example, we derive a difference approximation for f_i'' using f_{i+1}, f_i, and f_{i-1}. Taylor expansions of f_{i+1} and f_{i-1} are given by Eq. (5.2.1) and Eq. (5.2.4). By adding the two expansions we get

$$f_{i+1} + f_{i-1} = 2f_i + h^2 f_i'' + \frac{1}{12} h^4 f_i'''' + \cdots$$

Subtracting $2f_i$ from the both sides gives

$$f_{i+1} - 2f_i + f_{i-1} = h^2 f_i'' + \frac{1}{12} h^4 f_i'''' + \cdots$$

Then, truncating after the f'' term and rewriting yield

$$f_i'' = \frac{f_{i+1} - 2f_i + f_{i-1}}{h^2} + 0(h^2) \qquad\qquad (5.2.14)$$

Equation (5.2.14) is called a *central difference approximation* for f'', and the error is represented by

$$O(h^2) = -\frac{1}{12}h^2 f_i''''$$

Another difference approximation for f_i'' may be derived using f_i, f_{i-1}, and f_{i-2} (because $p = 2$, the minimum number of data points necessary is 3). Subtracting 2 times the Taylor expansion of f_{i-1} from that of f_{i-2} results in

$$f_{i-2} - 2f_{i-1} = -f_i + h^2 f_i'' - h^3 f_i''' + \cdots$$

Solving the preceding equation for f_i'' yields

$$f_i'' = \frac{f_{i-2} - 2f_{i-1} + f_i}{h^2} + O(h) \qquad O(h) = hf_i''' \qquad (5.2.15)$$

Equation (5.2.15) is called the *backward difference approximation* for f_i''.

Difference approximations for higher derivatives can be obtained by appropriate linear combinations of the Taylor expansions. The derivation becomes increasingly more cumbersome as the number of points or the order of derivative increases. Indeed, the difference approximations printed in the literature have frequent mistakes, particularly in error terms for higher-order difference approximations. For this reason, a more systematic algorithm and a computer program (PROGRAM 5–1) based on it are described in the next section. Using the program, verifying the difference approximations is now very easy.

So far in this section we have assumed that grid points are equispaced. However, difference approximations on nonuniformly spaced grids may be derived by using the Taylor expansions.

The difference approximations that are frequently used are listed in Table 5.2.

Table 5.2 Difference approximations[a]

First Derivative

(a) Forward difference approximations:

$$f_i' = \frac{f_{i+1} - f_i}{h} + O(h), \qquad\qquad O(h) = -\frac{1}{2}hf_i''$$

$$f_i' = \frac{-f_{i+2} + 4f_{i+1} - 3f_i}{2h} + O(h^2), \qquad O(h^2) = \frac{1}{3}h^2 f_i'''$$

$$f_i' = \frac{2f_{i+3} - 9f_{i+2} + 18f_{i+1} - 11f_i}{6h} + O(h^3), \quad O(h^3) = -\frac{1}{4}h^3 f_i''''$$

(*Continued*)

Table 5.2 (Continued)

(b) Backward difference approximations:

$$f'_i = \frac{f_i - f_{i-1}}{h} + O(h), \qquad\qquad O(h) = \frac{1}{2}hf''_i$$

$$f'_i = \frac{3f_i - 4f_{i-1} + f_{i-2}}{2h} + O(h^2), \qquad\qquad O(h^2) = \frac{1}{3}h^2 f'''_i$$

$$f'_i = \frac{11f_i - 18f_{i-1} + 9f_{i-2} - 2f_{i-3}}{6h} + O(h^3), \quad O(h^3) = \frac{1}{4}h^3 f''''_i$$

(c) Central difference approximation:

$$f'_i = \frac{f_{i+1} - f_{i-1}}{2h} + O(h^2), \qquad\qquad O(h^2) = -\frac{1}{6}h^2 f'''_i$$

$$f'_i = \frac{-f_{i+2} + 8f_{i+1} - 8f_{i-1} + f_{i-2}}{12h} + O(h^4), \quad O(h^4) = \frac{1}{30}h^4 f_i^{(v)}$$

Second Derivative

(d) Forward difference approximations:

$$f''_i = \frac{f_{i+2} - 2f_{i+1} + f_i}{h^2} + O(h), \qquad\qquad O(h) = -hf'''_i$$

$$f''_i = \frac{-f_{i+3} + 4f_{i+2} - 5f_{i+1} + 2f_i}{h^2} + O(h^2), \quad O(h^2) = \frac{11}{12}h^2 f''''_i$$

(e) Backward difference approximations:

$$f''_i = \frac{f_i - 2f_{i-1} + f_{i-2}}{h^2} + O(h), \qquad\qquad O(h) = hf'''_i$$

$$f''_i = \frac{2f_i - 5f_{i-1} + 4f_{i-2} - f_{i-3}}{h^2} + O(h^2), \quad O(h^2) = \frac{11}{12}h^2 f''''_i$$

(f) Central difference approximation:

$$f''_i = \frac{f_{i+1} - 2f_i + f_{i-1}}{h^2} + O(h^2), \qquad\qquad O(h^2) = -\frac{1}{12}h^2 f''''_i$$

$$f''_i = \frac{-f_{i+2} + 16f_{i+1} - 30f_i + 16f_{i-1} - f_{i-2}}{12h^2} + O(h^4), \quad O(h^4) = \frac{1}{90}h^4 f_i^{(vi)}$$

Third Derivative

(g) Forward difference approximation:

$$f'''_i = \frac{f_{i+3} - 3f_{i+2} + 3f_{i+1} - f_i}{h^3} + O(h), \quad O(h) = -\frac{3}{2}hf''''_i$$

(h) Backward difference approximation:

$$f'''_i = \frac{f_i - 3f_{i-1} + 3f_{i-2} - f_{i-3}}{h^3} + O(h), \quad O(h) = \frac{3}{2}hf''''_i$$

(i) Central difference approximation:

$$f'''_i = \frac{f_{i+2} - 2f_{i+1} + 2f_{i-1} - f_{i-2}}{2h^3} + O(h^2), \quad O(h^2) = -\frac{1}{4}h^2 f_i^{(v)}$$

---

[a] The difference approximations in this table are generated by using PROGRAM 5–1.

SUMMARY OF THIS SECTION

(a) A difference approximation for $f_i^{(p)}$ needs at least $p + 1$ data points.

(b) The difference approximation is derived by expanding f_j into a Taylor series about x_i.

(c) The derivatives of order less than p must be eliminated. This is possible with a minimum of $p + 1$ data points.

(d) The error term is the lowest-order term truncated.

5.3 A GENERIC ALGORITHM TO DERIVE A DIFFERENCE APPROXIMATION

The objective of this section is to describe a generic algorithm to derive a difference approximation using a specified set of grid points for a given order of derivative. The algorithm is then implemented as PROGRAM 5–1.

Suppose that the total number of the grid points is L and the grid points are numbered as $i = \alpha, \beta, \ldots, \lambda$. We assume $L \geqslant p + 1$ where p is the order of the derivative to be approximated. The abscissas of the grid points are $x_i = \alpha h, \beta h, \ldots, \lambda h$ with $i = \alpha, \beta, \ldots, \lambda$.

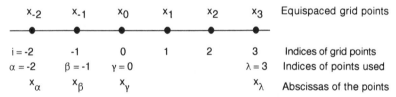

Figure 5.2 Illustration of grid points used for difference approximation

The difference approximation for the pth derivative of $f(x)$ using these grid points may be written in the form:

$$f_0^{(p)} = \frac{a_\alpha f_\alpha + a_\beta f_\beta + \cdots + a_\lambda f_\lambda}{h^p} + E \qquad (5.3.1)$$

where a_α through a_λ are L undetermined coefficients; $f_\alpha = f(x_\alpha), f_\beta = f(x_\beta), \ldots$ are ordinates to be used; and E is the error written by

$$E \simeq c_1 h^{L-p} f^{(L)} + c_2 h^{L-p+1} f^{(L+1)} \qquad (5.3.2)$$

The essence of the algorithm is to introduce the Taylor expansions of f_i into Eq. (5.3.1) and to determine the undetermined coefficients so that the error term is minimized, or equivalently, the order of E becomes the highest order possible.

For simplicity of further explanation, let us assume $p = 1$, $L = 3$, $\alpha = 0$, $\beta = 1$ and $\gamma = 2$. Then, Eq. (5.3.1) becomes

$$f'_0 = \frac{a_0 f_0 + a_1 f_1 + a_2 f_2}{h} + E \qquad (5.3.3)$$

where a_0, a_1, and a_2 are three undetermined coefficients, and $x_0 = 0$, $x_1 = h$ and $x_2 = 2h$ are grid points to be used. Introducing the Taylor expansions of f_1 and f_2 about $x = 0$ into Eq. (5.3.3) yields

$$f'_0 = \frac{a_0 f_0}{h} + \frac{a_1}{h} \left[f_0 + h f'_0 + \frac{h^2}{2} f''_0 + \frac{h^3}{6} f'''_0 + \cdots \right]$$

$$+ \frac{a_2}{h} \left[f_0 + 2h f'_0 + \frac{4h^2}{2} f''_0 + \frac{8h^3}{6} f'''_0 + \cdots \right] + E$$

or after reorganizing terms,

$$f'_0 = f_0 [a_0 + a_1 + a_2] \frac{1}{h} + f'_0 [0 + a_1 + 2a_2] + f''_0 [0 + a_1 + 4a_2] \frac{h}{2}$$

$$+ f'''_0 [0 + a_1 + 8a_2] \frac{h^2}{6} + f''''_0 [0 + a_1 + 16a_2] \frac{h^3}{24} + \cdots + E \qquad (5.3.4)$$

Equation (5.3.4) has three undetermined coefficients, which can be determined by imposing three conditions. To minimize error of Eq. (5.3.4), we set the coefficients of f_0, f'_0 and f''_0 to 0, 1, and 0, respectively:

$$a_0 + a_1 + a_2 = 0$$

$$0 + a_1 + 2a_2 = 1 \qquad (5.3.5)$$

$$0 + a_1 + 4a_2 = 0$$

By solving the foregoing equations, the three undetermined coefficients become $a_0 = -\frac{3}{2}$, $a_1 = 2$, and $a_2 = -\frac{1}{2}$.

The higher-order terms in Eq. (5.3.4) that do not vanish constitute the error, namely

$$E = -f'''_0 [0 + a_1 + 8a_2] \frac{h^2}{6} - f''''_0 [0 + a_1 + 16a_2] \frac{h^3}{24} + \cdots \qquad (5.3.6)$$

By comparing Eq. (5.3.6) to Eq. (5.3.2), c_1 and c_2 in Eq. (5.3.2) are found to be

$$c_1 = -\frac{1}{6}(a_1 + 8a_2)$$

$$c_2 = -\frac{1}{24}(a_1 + 16a_2)$$

which become, by introducing $a_1 = 2$ and $a_2 = -\frac{1}{2}$,

$$c_1 = -\frac{1}{6}\left(2 - \frac{8}{2}\right) = \frac{1}{3}$$

$$c_2 = -\frac{1}{24}\left(2 - \frac{16}{2}\right) = \frac{1}{4}$$

Because the first term of Eq. (5.3.2) is not zero, we ignore the second term and write the error term as

$$E = \frac{1}{3}h^2 f_0'''$$ (5.3.7)

If on the other hand the first term of Eq. (5.3.6) were zero, the second term would represent the error.

The final result of the present derivation is

$$f_0' = \frac{1}{h}\left[-\frac{3}{2}f_0 + 2f_1 - \frac{1}{2}f_2\right] + E$$

or equivalently

$$f_0' = \frac{-3f_0 + 4f_1 - f_2}{2h} + E$$ (5.3.8)

where

$$E = \frac{1}{3}h^2 f_0'''$$

In more general terms, using L data points, we can determine L undetermined coefficients of Eq. (5.3.1) by correctly fixing the first L terms of the Taylor expansion of Eq. (5.3.1). Thus, the error term becomes proportional to the $(L + 1)$th term or equivalently the Lth derivative if its coefficient is not zero. If it is zero, the error term becomes one order higher.

The present algorithm works even when the point indices, $\alpha, \beta, \ldots$, are not integers. This means that difference approximation on a nonequispaced grid may be derived by the same algorithm.

As described earlier, a program to perform the present algorithm is available as PROGRAM 5–1.

SUMMARY OF THIS SECTION.

The generic algorithm described in this section is essentially the same as the algorithms discussed in the previous section. However, its general formulation makes it possible to develop a computer program.

5.4 USING DIFFERENCE OPERATORS

We now define three difference operators [Ralston; Isaacson/Keller]:

(a) Forward difference operator: Δ

$$\Delta f_i = f_{i+1} - f_i \qquad (5.4.1)$$

(b) Backward difference operator: ∇

$$\nabla f_i = f_i - f_{i-1} \qquad (5.4.2)$$

(c) Central difference operator: δ

$$\delta f_i = f_{i+\frac{1}{2}} - f_{i-\frac{1}{2}}$$

or

$$\delta f_{i+\frac{1}{2}} = f_{i+1} - f_i \qquad (5.4.3)$$

where

$$f_{i+\frac{1}{2}} = f\left(x_i + \frac{h}{2}\right).$$

Higher-order difference operators may be written as powers of the above difference operators: for example, Δ^n, ∇^n and δ^n are nth-order difference operators. Additional nth-order difference operators may be obtained by applying both ∇ and Δ in the form $\nabla^{n-m}\Delta^m$ where $1 \leqslant m \leqslant n$. In the case of $n = 2$, the difference operators produce

$$\Delta^2 f_i = \Delta(f_{i+1} - f_i) = f_{i+2} - 2f_{i+1} + f_i \qquad (5.4.4a)$$
$$\nabla^2 f_i = \nabla(f_i - f_{i-1}) = f_i - 2f_{i-1} + f_{i-2} \qquad (5.4.4b)$$
$$\delta^2 f_i = \delta(f_{i+\frac{1}{2}} - f_{i-\frac{1}{2}}) = f_{i+1} - 2f_i + f_{i-1} \qquad (5.4.4c)$$
$$\Delta\nabla f_i = \Delta(f_i - f_{i-1}) = f_{i+1} - 2f_i + f_{i-1} \qquad (5.4.4d)$$
$$\nabla\Delta f_i = \nabla(f_{i+1} - f_i) = f_{i+1} - 2f_i + f_{i-1} \qquad (5.4.4e)$$

Notice that in the foregoing equations the last three differences are identical. Indeed, we can write the identity relations:

$$\delta^2 = \Delta\nabla = \nabla\Delta \qquad (5.4.5)$$

If n is even and $m = n/2$, $\nabla^{n-m}\Delta^m$ becomes $\nabla^m\Delta^m$, which is the nth-order central difference operator.

Difference approximations are derived by approximating differential operators by difference operators. For example, the first-order ordinary differential operator may be approximated in three different ways:

$$\frac{d}{dx} \simeq \frac{\Delta}{\Delta x}$$

$$\frac{d}{dx} \simeq \frac{\nabla}{\nabla x} \qquad (5.4.6)$$

$$\frac{d}{dx} \simeq \frac{\delta}{\delta x}$$

By applying the approximations of Eq. (5.4.6) to a function $f(x)$, difference approximations for f_i' are obtained. Using $\dfrac{\Delta}{\Delta x}$ yields the forward difference approximation:

$$\left[\frac{d}{dx} f(x)\right]_{x_i} \simeq \frac{\Delta}{\Delta x} f_i = \frac{f_{i+1} - f_i}{h} \qquad (5.4.6)$$

where Δx in the denominator is interpreted as $\Delta x_i = x_{i+1} - x_i = h$. Similarly, using $\dfrac{\nabla}{\nabla x}$ yields

$$\left[\frac{d}{dx} f(x)\right]_{x_i} \simeq \frac{\nabla}{\nabla x} f_i = \frac{f_i - f_{i-1}}{h} \qquad (5.4.7)$$

The central difference approximation for f_i' is derived by applying $\dfrac{\delta}{\delta x}$ based on the grid spacing of $2h$, namely

$$\left[\frac{d}{dx} f(x)\right]_{x_i} \simeq \frac{\delta}{\delta x} f_i = \frac{f_{i+1} - f_{i-1}}{2h} \qquad (5.4.8)$$

where δx in the denominator is interpreted as $\delta x_i = x_{i+1} - x_{i-1} = 2h$. The central difference approximation may also be derived by taking the arithmetic average of the forward and backward difference approximations as

$$\left[\frac{d}{dx} f(x)\right]_{x_i} \simeq \frac{1}{2}\left[\frac{\Delta}{\Delta x} + \frac{\nabla}{\nabla x}\right] f_i = \frac{f_{i+1} - f_{i-1}}{2h} \qquad (5.4.9)$$

The approximations for the second-order differential operator may be written by applying the first-order approximation twice:

$$\frac{d^2}{dx^2} \simeq \frac{\Delta^2}{\Delta x^2}$$

$$\frac{d^2}{dx^2} \simeq \frac{\nabla^2}{\nabla x^2}$$

$$\frac{d^2}{dx^2} \simeq \frac{\nabla}{\nabla x}\left(\frac{\Delta}{\Delta x}\right) = \frac{\Delta}{\Delta x}\left(\frac{\nabla}{\nabla x}\right) \tag{5.4.10}$$

$$\frac{d^2}{dx^2} \simeq \frac{\delta^2}{\delta x^2}$$

Summary of This Section

(a) Three basic difference operators are introduced.

(b) Difference approximations are derived by approximating differential operators by difference operators.

(c) By combining the three difference operators, various difference approximations may be derived.

5.5 USING DIFFERENTIATION OF NEWTON INTERPOLATION POLYNOMIALS

Newton interpolation polynomials of both forward and backward types are useful in deriving difference approximations, although we consider only the forward type here. By using Newton interpolation polynomials, numerous difference approximations are systematically derived [Carnahan/Luther/Wilkes; Cheney/Kincaid].

The Newton forward interpolation formula fitted to $N + 1$ data points is written as

$$g(x) = g(x_k + sh) = \sum_{n=0}^{N}\binom{s}{n}\Delta^n f_k$$

$$= f_k + s\Delta f_k + \frac{1}{2}s(s-1)\Delta^2 f_k + \frac{1}{6}s(s-1)(s-2)\Delta^3 f_k$$

$$+ \frac{1}{24}s(s-1)(s-2)(s-3)\Delta^4 f_k + \cdots + \binom{s}{N}\Delta^N f_k \tag{5.5.1}$$

where

$$s = \frac{x - x_k}{h}$$

and $f_k, f_{k+1}, \ldots, f_{k+N}$ are functional values at $x_k, x_{k+1}, \ldots,$ and x_{k+N}, respectively.

As described in Chapter 2, the Newton forward interpolation formula fitted to $N + 1$ data points is a Nth-order polynomial. Its derivatives approximate derivatives of $f(x)$. The accuracy of the approximations varies depending on N as well as which point in the interpolating range the derivative is derived for. Since accuracy of a Newton interpolation is best at the center of the domain of interpolation, accuracy of the difference approximation is best at the center, too.

To explain derivation of difference approximations, we consider $N = 2$, for which Eq. (5.5.1) becomes

$$g(x) = f_k + s\Delta f_k + \frac{1}{2}s(s-1)\Delta^2 f_k \tag{5.5.2}$$

Differentiating once yields

$$g'(x) = \frac{1}{h}\left[\Delta f_k + \frac{1}{2}(2s - 1)\Delta^2 f_k\right] \tag{5.5.3}$$

For $s = 0$, 1, and 2 it becomes respectively

$$g'(x_k) = \frac{1}{2h}[2\Delta f_k - \Delta^2 f_k] = \frac{1}{2h}[-f_{k+2} + 4f_{k+1} - 3f_k]$$

$$g'(x_{k+1}) = \frac{1}{2h}[2\Delta f_k + \Delta^2 f_k] = \frac{1}{2h}[f_{k+2} - f_k]$$

$$g'(x_{k+2}) = \frac{1}{2h}[2\Delta f_k + 3\Delta^2 f_k] = \frac{1}{2h}[3f_{k+2} - 4f_{k+1} + f_k]$$

The foregoing equations are respectively the forward difference approximation at grid point k, the central difference approximation at grid point $k + 1$, and the backward difference approximation at grid point $k + 2$. Replacing k in the first, second, and third equations by i, $i - 1$, and $i - 2$, respectively, yields

$$g'(x_i) = \frac{1}{2h}[2\Delta f_i - \Delta^2 f_i] = \frac{1}{2h}[-f_{i+2} + 4f_{i+1} - 3f_i] \tag{5.5.4}$$

$$g'(x_i) = \frac{1}{2h}[2\Delta f_{i-1} + \Delta^2 f_{i-1}] = \frac{1}{2h}[f_{i+1} - f_{i-1}] \tag{5.5.5}$$

$$g'(x_i) = \frac{1}{2h}[2\Delta f_{i-2} + 3\Delta^2 f_{i-2}] = \frac{1}{2h}[3f_i - 4f_{i-1} + f_{i-2}] \tag{5.5.6}$$

These are the equations already given by Eqs. (5.2.12), (5.2.8), and (5.2.13).

Error of a Newton forward interpolation polynomial is represented by the term that would be added if the interpolation is fitted at one more grid point (see Subsection 2.4.2). Using this rule, we evaluate the error of the difference approximations in

Eqs. (5.5.4), (5.5.5), and (5.5.6). If N for Eq. (5.5.1) is increased from $N = 2$ to $N = 3$, the additional term added to Eq. (5.5.2) is

$$\frac{1}{6}s(s-1)(s-2)\Delta^3 f_k \qquad (5.5.7)$$

Its first derivative with respect to x is

$$\frac{1}{6h}[3s^2 - 6s + 2]\Delta^3 f_k \qquad (5.5.8)$$

which becomes

$$\frac{1}{3h}\Delta^3 f_k, \quad \text{for } s = 0$$

$$-\frac{1}{6h}\Delta^3 f_k, \quad \text{for } s = 1 \qquad (5.5.9)$$

$$\frac{1}{3h}\Delta^3 f_k, \quad \text{for } s = 2$$

We note that the Nth derivative of the Newton forward interpolation of order N becomes

$$\frac{d^N}{dx^N}g(x) = \frac{1}{h^N}\Delta^N f_i \qquad (5.5.10)$$

The equation is an approximation for the Nth derivative of $f(x)$ in the whole range of interpolation. Therefore, we can write

$$\Delta^N f_i \simeq h^N f^{(N)}(x) \qquad (5.5.11)$$

where $f^{(N)}$ denotes the Nth derivative of $f(x)$. By using Eq. (5.5.11), Eq. (5.5.9) becomes approximately

$$\frac{1}{3}h^2 f_k''', \quad \text{for } s = 0 \qquad (5.5.12)$$

$$-\frac{1}{6}h^2 f_k''', \quad \text{for } s = 1 \qquad (5.5.13)$$

$$\frac{1}{3}h^2 f_k''', \quad \text{for } s = 2 \qquad (5.5.14)$$

These represent errors of Eqs. (5.5.4), (5.5.5), and (5.5.6), respectively, and agree with the errors derived in Section 5.2 by using the Taylor expansions.

In general, a difference approximation of order p is derived by differentiating a Newton forward interpolation polynomial of order p or greater. The difference approximation of a higher accuracy is obtained by increasing the order of the Newton forward interpolation polynomial. Therefore, a more accurate difference approximation requires more grid points. The accuracy of a difference approximation is best at the center of the range of interpolation. The central difference approximation may be regarded as the derivative of the Newton interpolation at the center of the range of interpolation. On the other hand, forward and backward difference approximations are the derivatives of the Newton interpolation polynomial at the edges of the interpolation range. Therefore, the central difference approximation is always more accurate than forward or backward or other one-sided difference approximations using the same Newton interpolation polynomial.

SUMMARY OF THIS SECTION

(a) Difference approximations may be derived by differentiating an interpolation polynomial, for example, the Newton forward interpolation polynomial.

(b) The error term of the difference approximation is derived by using the additional term that comes when an additional data point is used.

(c) The difference approximations obtained by using the interpolation formula are consistent with those obtained by using the Taylor expansions.

(d) Accuracy of an interpolation formula based on equispaced grid points is highest at the center of the interpolation domain. Consequently the difference approximation is most accurate if the derivative of an interpolation formula at the center of the domain is used. This explains why central difference approximation is always more accurate than the forward or backward difference approximations using the same number of data points.

5.6   DIFFERENCE APPROXIMATIONS OF PARTIAL DERIVATIVES

Difference approximations for partial derivatives of multidimensional functions are essentially the same as the numerical differentiation of one-dimensional functions.

Consider a two-dimensional function $f(x, y)$. The difference approximation for the partial derivative

$$f_x = \frac{\partial}{\partial x} f(x, y) \quad \text{at } x = x_0 \text{ and } y = y_0 \tag{5.6.1}$$

can be derived by fixing y to y_0 and considering $f(x, y_0)$ as a one-dimensional function. Therefore, the forward, central, and backward difference approximations for the preceding partial derivatives may be written respectively as

$$f_x \simeq \frac{f(x_0 + \Delta x, y_0) - f(x_0, y_0)}{\Delta x} \tag{5.6.2a}$$

$$f_x \simeq \frac{f(x_0 + \Delta x, y_0) - f(x_0 - \Delta x, y_0)}{2\Delta x} \tag{5.6.2b}$$

$$f_x \simeq \frac{f(x_0, y_0) - f(x_0 - \Delta x, y_0)}{\Delta x} \tag{5.6.2c}$$

The central difference approximations for the second partial derivatives of $f(x, y)$ at x_0 and y_0 are

$$f_{xx} = \frac{\partial^2}{\partial x^2} f \simeq \frac{f(x_0 + \Delta x, y_0) - 2f(x_0, y_0) + f(x_0 - \Delta x, y_0)}{\Delta x^2} \tag{5.6.3a}$$

$$f_{yy} = \frac{\partial^2}{\partial y^2} f \simeq \frac{f(x_0, y_0 + \Delta y) - 2f(x_0, y_0) + f(x_0, y_0 - \Delta y)}{\Delta y^2} \tag{5.6.3b}$$

$$f_{xy} = \frac{\partial^2}{\partial x \partial y} f \simeq \frac{f(x_0 + \Delta x, y_0 + \Delta y) - f(x_0 + \Delta x, y_0 - \Delta y)}{4\Delta x \Delta y}$$

$$+ \frac{-f(x_0 - \Delta x, y_0 + \Delta y) + f(x_0 - \Delta x, y_0 - \Delta y)}{4\Delta x \Delta y} \tag{5.6.3c}$$

Example 5.2

The table of a two-dimensional function $f(x, y)$ is given in the following table:

y	x = 1.0	1.5	2.0	2.5	3.0
1.0	1.63	2.05	2.50	2.98	3.49
1.5	1.98	2.51	3.08	3.69	4.33
2.0	2.28	2.91	3.61	4.37	5.17
2.5	2.64	3.25	4.08	5.00	5.98
3.0	2.65	3.50	4.48	5.57	6.76

(a) Using the central difference approximations, evaluate the following partial derivatives:

$$f_x(2, 2),\ f_y(2, 2),\ f_{yy}(2, 2) \text{ and } f_{xy}(2, 2)$$

(b) Using the three-point forward difference approximation, evaluate the following partial derivatives:

$$f_x(2, 2),\ f_y(2, 2)$$

⟨**Solution**⟩

Using the definition $\Delta x = \Delta y = h = 0.5$, the calculations are performed as follows:

(a) $f_x(2, 2) \simeq \dfrac{f(2 + h, 2) - f(2 - h, 2)}{2h} = \dfrac{4.37 - 2.91}{(2)(0.5)} = 1.46$

$f_y(2, 2) \simeq \dfrac{f(2, 2 + h) - f(2, 2 - h)}{2h} = \dfrac{4.08 - 3.08}{(2)(0.5)} = 1.00$

$f_{yy}(2, 2) \simeq \dfrac{f(2, 2 + h) - 2f(2, 2) + f(2, 2 - h)}{h^2}$

$= \dfrac{4.08 - 2(3.61) + 3.08}{(0.5)^2} = -0.24$

$f_{xy}(2, 2) \simeq \dfrac{\begin{array}{c} f(2 + h, 2 + h) - f(2 + h, 2 - h) \\ - f(2 - h, 2 + h) + f(2 - h, 2 - h) \end{array}}{(2h)^2}$

$= \dfrac{5.0 - 3.69 - 3.25 + 2.51}{[2(0.5)]^2} = 0.57$

(b) $f_x(2, 2) \simeq \dfrac{-f(2 + 2h, 2) + 4f(2 + h, 2) - 3f(2, 2)}{2h}$

$= \dfrac{-(5.17) + 4(4.37) - 3(3.61)}{(2)(0.5)} = 1.48$

$f_y(2, 2) \simeq \dfrac{-f(2, 2 + 2h) + 4f(2, 2 + h) - 3f(2, 2)}{2h}$

$= \dfrac{-(4.48) + 4(4.08) - 3(3.61)}{(2)(0.5)} = 1.01$

SUMMARY OF THIS SECTION. Difference approximations for partial derivatives are essentially the same as for ordinary derivatives. Therefore, all the difference approximations developed for ordinary derivatives are applicable to partial derivatives.

PROGRAM

PROGRAM 5–1 Difference Approximation Finder

(A) Explanations

The program finds the difference approximation for the derivative of a desired order using the grid points specified by the user. The difference approximation for the derivative of a function is given by Eq. (5.3.1). The error can be expressed in the form of Eq. (5.3.2).

The program asks for (1) how many grid points, L, are to be used in the difference approximation formula (a maximum of 10), (2) the grid point indices, $\alpha, \beta, \ldots \lambda$, and (3) the order of the derivative to be approximated. The grid interval is denoted by h but no numerical value is attached. The simultaneous linear equations are solved in function gauss(). See Section 6.2 for more details of the Gauss elimination method. All input data are given interactively. No change of the program is necessary.

(B) List

```
/* CSL/c5-1.c     Difference Approximation Finder*/
#include <stdio.h>
#include <stdlib.h>
#include <math.h>
#define TRUE 1
/*             km : number of grid points
                    in the finite difference approximation (L)
               el[k] : grid point index for the k-th point counted
                       from the left
               kdr : order of polynomial to be approximated (p)
               a[k][l] : coefficients of the linear equation
               c[k] : coefficient of the k-th functional valud in the
                    numerator of the difference approximation for k=km,
                    or coefficient of the error term for k>km
               f : reciprocal of the denominator
                   of the difference approximation
               n : order of matrix (equals the number of points)      */
main ()
{
int i, k, kdr, kk, km, km1, kmp2, L, mt, nh, n;
double cm, cpdd, dd, f, ff, finv, u, z;
double a[10][11], el[10], b[10][11], c[10], cf[11];
float x_inp;
void gauss();
    mt = 6;
    printf( "\nCSL/C5-1    Difference Approximation Finder  \n" );
    while( TRUE ){
        printf( "\nNumber of points ?  " );   scanf( "%d", &km );
        if( km >=2 && km <= 10 ){
            for( k = 1; k <= km; k++ ){
                printf( "\nGrid point index of point (%d) ?   ",k );
                scanf( "%f", &x_inp ); el[k] = x_inp;
            }
            printf( "\nOrder of difference scheme to be derived ? " );
            scanf( "%d", &kdr );
            z = 1.0;
            for( i = 1; i <= kdr; i++ ) z = z*(float)( i );
            for( k = 1; k <= (km + 2); k++ ){
                for( L = 1; L <= km; L++ ){
                    a[k][L] = 1.0;
                    if( k > 1 ){
                        for (i = 1; i<=k-1; i++)
                            a[k][L] = el[L]*a[k][L];
                    }
                    b[k][L] = a[k][L];
                }
```

```
         }
         ff = 1;
         for( k = 1; k <= km; k++ ){
            a[k][km+1] = 0;
            if( k - 1 == kdr )    a[k][km+1] = z;
         }
         n = km;
         kmp2 = km + 2;
         gauss( n,a );
         for( k = 1; k <= (km + 2); k++ ){
            c[k] = 0.0;
            for( L = 1; L <= km; L++ ) c[k] = c[k] + b[k][L]*a[L][km+1];
         }
         f = 1000.0;
         for( k = 1; k <= km; k++ ){
            if( a[k][km+1] != 0 ){
               if( fabs( a[k][km+1] ) >= 0.0001 ){
                  u = fabs( a[k][km+1] );
                  if( u < f )   f = u;
               }
            }
         }
         for( k = 1; k <= km; k++ ) cf[k] = a[k][km+1]/f;
         printf( "Difference scheme:\n" );
         for( k = 1; k <= km; k++ ){
            finv = 1.0/f;
            printf(   " +[ %10.5f/( %8.5f h^%d)] f( %gh ) \n",
                                 cf[k],finv, kdr, el[k] );
         }
         printf("\nError term\n");
         for( k = 1; k <= (km + 2); k++ ){
            if( fabs( c[k] ) < 0.00000001 ) c[k] = 0;
         }
         dd = 1.0;
         for( k = 1; k <= km; k++ )  dd = dd*(float)( k );
         for( k = km + 1; k <= (km + 2); k++ ){
            cm = -c[k];     cpdd = -c[k]/dd;
            km1 = k - 1;    nh = km1 - kdr;
            if( k == km + 1 && cm != 0 ) { /* Prints error terms */
               printf(   "      (%7.3f/%7.3f)h^%d f", cm, dd, nh );
               for (i=1; i<=km1; i++)    printf( "'" );
            }
            if( k == km + 2 ){
               printf(   "\n    +(%7.3f/%7.3f)h^%d f", cm, dd, nh );
               for (i=1; i<=km1; i++)     printf( "'" );
            }
            dd = dd*(float)( k );
         }
         printf( "\n\nType 1 to continue, or 0 to stop.\n" );
         scanf( "%d", &kk );  if( kk != 1 )  exit(0);
      }
      else {
         printf( " Invalid input.  Please repeat input.\n" );
         exit(0);
      }
   }
}
}
```

```
void gauss(n, a)
int n;    double a[][11];
{
int i, j, jc, jr, k, kc, nv, pv;
double det, eps, ep1, eps2, r, temp, tm, va;
   eps = 1.0; ep1 = 1.0 ;              /* eps = Machine epsilon */
   while( ep1  > 0 ){
      eps = eps/2.0; ep1 = eps*0.98 + 1; ep1 = ep1 - 1;
   }
   eps = eps*2;        eps2 = eps*2;
   printf( "                 Machine epsilon=%g \n", eps );
   det = 1;                            /* Initialization of determinant */
   for( i = 1; i <= (n - 1); i++ ){
      pv = i;
      for( j = i + 1; j <= n; j++ ){
         if( fabs( a[pv][i] ) < fabs( a[j][i] ) )    pv = j;
      }
      if( pv != i ){
         for( jc = 1; jc <= (n + 1); jc++ ){
            tm = a[i][jc];  a[i][jc] = a[pv][jc];      a[pv][jc] = tm;
         }
         det = -det;
      }
      if( a[i][i] == 0 ){
         printf( "Matrix is singular.\n" );
         exit(0);
      }
      for( jr = i + 1; jr <= n; jr++){/* Elimination of below-diagonal.*/
         if( a[jr][i] != 0 ){
            r = a[jr][i]/a[i][i];
            for( kc = i + 1; kc <= (n + 1); kc++ ){
               temp = a[jr][kc];
               a[jr][kc] = a[jr][kc] - r*a[i][kc];
               if( fabs( a[jr][kc] ) < eps2*temp ) a[jr][kc] = 0.0;
            }
         }
      }
   }
   for( i = 1; i <= n; i++ ) {
      det = det*a[i][i];               /*  Determinant is calculated. */
   }
   if( det == 0 ){
      printf( "Matrix is singular.\n" );
      exit(0);
   }
   else{                              /* Backward substitution starts. */
      a[n][n+1] = a[n][n+1]/a[n][n];
      for( nv = n - 1; nv >= 1; nv-- ){
         va = a[nv][n+1];
         for( k = nv + 1; k <= n; k++ ) {va = va - a[nv][k]*a[k][n+1];}
         a[nv][n+1] = va/a[nv][nv];
      }
      printf( "                Determinant = %g \n", det );
      return;
   }
}
```

(C) Sample Output

```
CSL/C5-1   Difference Approximation Finder

Number of points ?  3
Grid point index of point (1) ?    2
Grid point index of point (2) ?    1
Grid point index of point (3) ?    0
Order of difference scheme to be derived ? 1
       Machine epsilon=2.775558e-17
       Determinant = -2
Difference scheme:
 +[   -1.00000/(  2.00000 h^1)] f( 2h )
 +[    4.00000/(  2.00000 h^1)] f( 1h )
 +[   -3.00000/(  2.00000 h^1)] f( 0h )

Error term
    (  2.000/  6.000)h^2 f'''
  +(  6.000/ 24.000)h^3 f''''


Number of points ?  5
Grid point index of point (1) ?   -2
Grid point index of point (2) ?   -1
Grid point index of point (3) ?    0
Grid point index of point (4) ?    1
Grid point index of point (5) ?    2
Order of difference scheme to be derived ? 2
       Machine epsilon=2.775558e-17
       Determinant = 288
Difference scheme:
 +[   -1.00000/( 12.00000 h^2)] f( -2h )
 +[   16.00000/( 12.00000 h^2)] f( -1h )
 +[  -30.00000/( 12.00000 h^2)] f( 0h )
 +[   16.00000/( 12.00000 h^2)] f( 1h )
 +[   -1.00000/( 12.00000 h^2)] f( 2h )

Error term
  +(  8.000/720.000)h^4 f''''''
```

(D) Discussions

Case 1 derives the difference approximation for f'_0 using three points, $i = 0$, 1 and 2. The result is

$$f'_0 = \frac{-3f_0 + 4f_1 - f_2}{2h} + E$$

with

$$E = \frac{1}{3}h^2 f'''_0$$

The result of Case 2 is

$$f_0'' = \frac{-f_{-2} + 16f_{-1} - 30f_0 + 16f_1 - f_2}{12h^2} + E$$

with

$$E = \frac{1}{90}h^4 f_0^{(vi)}$$

PROBLEMS

(5.1) Evaluate the first derivative of $y(x) = \sin(x)$ for $x = 1$ and $h = 0.001$, 0.005, 0.01, 0.05, 0.1, and 0.5 by the three different schemes:

(a) $y'(1) \simeq [y(1+h) - y(1)]/h$

(b) $y'(1) \simeq [y(1) - y(1-h)]/h$

(c) $y'(1) \simeq [y(1 + h/2) - y(1 - h/2)]/h$

Evaluate the errors by comparison to the exact values.

(5.2) Calculate $d\sqrt{x}/dx$ at $x = 1$ by using the forward, backward, and central difference approximations with $h = 0.1, 0.05$, and 0.025. Evaluate the error of each result by **(a)** comparison to the exact value, and **(b)** using the error terms shown in Table 5.2, namely, $\frac{1}{2}hf''$, $\frac{1}{2}hf''$ and $-\frac{1}{6}h^2 f'''$, respectively.

(5.3) (a) Derive a difference approximation and the error term for f_i' using **(i)** f_{i-1} and f_{i+2}, **(ii)** f_{i-1}, f_i and f_{i+2}, and **(iii)** f_{i-2} and f_{i+2}. Assume grid points are equispaced.

(5.4) Derive a difference approximation and the error term for f_i'' using f_i, f_{i-1} and f_{i-2} (three-point backward difference approximation for f_i'').

(5.5) Repeat Problem 5.2 with the second-order-accurate forward and backward difference approximations:

(a) $y'(1) \simeq [-y(1 + 2h) + 4y(1 + h) - 3y(1)]/2h$

(b) $y'(1) \simeq [3y(1) - 4y(1 - h) + y(1 - 2h)]/2h$

and evaluate the errors by comparison to the exact value of $y'(1)$.

(5.6) Calculate the first derivative $y'(1)$, where $y(x) = \sin(x)$, by the second-order-accurate forward and backward difference approximations used in Problem 5.5 for $h = 0.001, 0.005, 0.01, 0.1$, and 0.5. Then, evaluate the error of each numerical approximation by comparison to the exact value. If an increase of error with reduction of h is observed, explain the reason.

(5.7) Considering a uniform beam of 1 m long simply supported at both ends, the bending moment is given by the following table:

$$y'' = M(x)/EI$$

where $y(x)$ is the deflection, $M(x)$ is the bending moment, and EI is the flexural rigidity. Calculate the bending moment at each grid point including the two end points, assuming the deflection distribution is among the following:

i	x_i	y_i
0	0.0 (m)	0.0 (cm)
1	0.2	7.78
2	0.4	10.68
3	0.6	8.37
4	0.8	3.97
5	1.0	0.0

Assume $EI = 1.2 \times 10^7$ Nm^2. Use central difference approximation for the grid points except at the boundaries. For the grid points at the ends, use the forward or backward difference approximation using four data points.

(5.8) Evaluate the second derivative of tan (x) at $x = 1$ by the central difference formula using $h = 0.1, 0.05$, and 0.02. Evaluate the error by comparison to the exact value and show that the error is proportional to h^2.

(5.9) The velocity distribution of a fluid near a flat surface is given by the following table

i	y_i (m)	u_i (m/s)
0	0.0	0.0
1	0.002	0.006180
2	0.004	0.011756
3	0.006	0.016180
4	0.008	0.019021

Newton's law for shear stress is given by

$$\tau = \mu \frac{d}{dy} u$$

where μ is the viscosity and assumed here to be 0.001 Ns/m^2. Calculate the shear stress at $y = 0$ by the difference approximations using the following grid points: (i) $i = 0$ and 1; and (ii) $i = 0, 1$, and 2.

(5.10) (a) Knowing the error term for

$$f'_i \simeq \frac{f_i - f_{i-1}}{h}$$

estimate the error term for

$$f'_i \simeq \frac{f_i - f_{i-2}}{2h}$$

(b) Accuracy of a difference approximation can be improved by a linear combination of two difference approximations so that the lowest-order truncation errors of the two approximations are cancelled. Determine α of the following approximation so that the accuracy is optimized:

$$f'_i \simeq \alpha \frac{f_i - f_{i-1}}{h} + (1 - \alpha) \frac{f_i - f_{i-2}}{2h}$$

(5.11) Determine α of

$$f''_i \simeq \alpha \frac{f_{i+1} - 2f_i + f_{i-1}}{h^2} + (1 - \alpha) \frac{f_{i+2} - 2f_i + f_{i-2}}{(2h)^2}$$

so the accuracy is maximized. Hint: eliminate the leading error for

$$f''_i \simeq \frac{f_{i+1} - 2f_i + f_{i-1}}{h^2}$$

(5.12) Derive the most accurate difference approximations for f'_i and f''_i using $f_{i-2}, f_{i-1}, f_i, f_{i+1}$, and f_{i+2}. Assume that the grid spacing is constant.

(5.13) By applying the Taylor expansions, derive the difference approximations for f'_i and f''_i using f_i, f_{i+1}, f_{i+2} and f_{i+3} with the highest possible accuracy for each. Assume that the grid spacing is constant.

(5.14) A function table is given as follows:

x	f
-0.1	4.157
0	4.020
0.2	4.441

(a) Derive the best difference approximation to calculate $f'(0)$ with the data given.

(b) What is the error term for the difference approximation?

(c) Calculate $f'(0)$ by the formula you derived.

(5.15) Evaluate the truncation error of the following difference formula:

$$f'_i \simeq \frac{-f_{i+3} + 9f_{i+1} - 8f_i}{6h}$$

(5.16) Two difference approximations for the fourth derivative are given by

$$f''''_i \simeq \frac{f_{i+4} - 4f_{i+3} + 6f_{i+2} - 4f_{i+1} + f_i}{h^4} + O(h)$$

$$f''''_i \simeq \frac{f_{i+2} - 4f_{i+1} + 6f_i - 4f_{i-1} + f_{i-2}}{h^4} + O(h^2)$$

By using the Taylor expansion, find the error terms.

(5.17) Prove the following equations:

$$\nabla^3 f_i = \Delta^3 f_{i-3}$$
$$\Delta^4 f_i = \nabla^4 f_{i+4}$$
$$\Delta^3 \nabla f_i = \Delta^4 f_{i-1}$$
$$\delta^2 f_i = \Delta^2 f_{i-1}$$
$$\Delta^n \nabla^m f_i = \Delta^{n+m} f_{i-m}$$

(5.18) Find m if $\nabla^5 f_i = \Delta^4 \nabla f_{i+m}$.

(5.19) Write the following difference approximations explicitly in terms of f_i's and estimate the order of error for each:

(a) $f_i'' \simeq \Delta^2 f_i/h^2$

(b) $f_i'' \simeq \nabla^2 f_i/h^2$

(c) $f_i'' \simeq \nabla\Delta \, f_i/h^2$

(5.20) Write the following difference approximations explicitly in terms of f_i's and derive the error terms:

(a) $f_i'' \simeq \Delta^3 f_i/h^3$

(b) $f_i''' \simeq \Delta^2 \nabla f_i/h^3$

(c) $f_i''' \simeq \Delta \nabla^2 f_i/h^3$

(d) $f_i''' \simeq \nabla^3 f_i/h^3$

(e) $f_i''' \simeq \dfrac{1}{2}[\Delta^2 \nabla f_i/h^3 + \Delta \nabla^2 f_i/h^3]$

(5.21) Derive the central difference approximation for f_i'' using the Newton forward interpolation of order 4 (fitted to five data points), and evaluate the error term by using the Newton interpolation formula of one order higher.

(5.22) Show that the first derivative of Eq. (5.4.1) is given by

$$g'(x) = \frac{1}{h}\left[\Delta f_k + \frac{1}{2}(2s-1)\Delta^2 f_k + \frac{1}{6}(3s^2 - 6s + 2)\Delta^3 f_k \right.$$
$$\left. + \frac{1}{24}(4s^3 - 18s^2 + 22s - 6)\Delta^4 f_k + \cdots + \frac{d}{ds}\binom{s}{N}\Delta^N f_k \right]$$

where $x = x_i + sh$.

(5.23) Express the following formulas by using the shift operator:

(a) $\dfrac{1}{2}(\nabla^2\Delta + \Delta\nabla^2)f_i$

(b) $\nabla^2\Delta^2 f_i$

(5.24) Derive difference approximations for f_i''' and f_i'''' using the differences in the previous problem, and evaluate the order of error by using the Taylor expansion.

(5.25) The following are difference approximations for $y^{(iv)}$:

$$\nabla^4 y_i/h^4, \quad \nabla^3 \Delta y_i/h^4, \quad \nabla^2 \Delta^2 y_i/h^4, \quad \nabla\Delta^3 y_i/h^4, \quad \Delta^4 y_i/h^4$$

(a) Write the difference approximations explicitly in terms of f_i.

(b) Evaluate the error term for each approximation by using the Newton forward interpolation polynomials. (See Section 5.4 for evaluating errors of backward formulas using the forward Newton interpolation polynomials.)

(5.26) The velocity distribution of a fluid near a flat surface is given by

i	$y_i\ (m)$	$u_i\ (m/s)$
0	0.0	0.0
1	0.001	0.4171
2	0.003	0.9080
3	0.006	1.6180

where y is the distance from the surface and u is the velocity. Assuming that the flow is laminar and $\mu = 0.001\ Ns/m^2$, calculate the shear stress at $y = 0$ by using the following data points:

(a) $i = 0$ and 1

(b) $i = 0, 1$, and 2

(*Hint:* see Problem 5.9 for Newton's viscosity law.)

(5.27) Derive difference approximations for f_0' by differentiating the Newton forward interpolation formulas fitted to the following data and evaluate the error terms for each approximation formula:

(a) f_0, f_1

(b) f_0, f_1, f_2

(c) f_0, f_1, f_2, f_3

(5.28) Derive the difference approximations for f_i' and f_i'' using f_i, f_{i+1}, f_{i+2} and f_{i+3} that has the highest possible accuracy. Assume that the grid spacing is constant.

(5.29) Repeat Problem (5.27) using the following data

(a) f_{-1}, f_0, f_1

(b) f_{-2}, f_{-1}, f_0

(c) $f_{-2}, f_{-1}, f_0, f_1, f_2$

(5.30) Prove by induction the following relations:

(a) $\Delta^n y(x) = \displaystyle\sum_{k=0}^{n} (-1)^k \frac{n!}{k!(n-k)!}\, y(x + nh - kh)$

(b) $\nabla^n y(x) = \displaystyle\sum_{k=0}^{n} (-1)^k \frac{n!}{k!(n-k)!}\, y(x - kh)$

(c) $\delta^{2n} y(x) = \displaystyle\sum_{k=0}^{2n} (-1)^k \frac{2n!}{k!(2n-k)!}\, y(x + 2nh - kh)$

(5.31) Derive the forward difference approximation for the first derivative that is third-order accurate (error is proportional to f'''') for an equispaced grid using the Newton forward interpolation polynomial.

(5.32) Derive the difference approximation for $f''(x_i)$ using the following three grid points:

Figure P5.32

(5.33) Derive the difference approximation for $f'''(x_i)$ using the following four grid points:

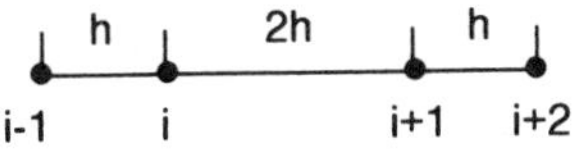

Figure P5.33

(5.34) The function table for $f(x, y)$ follows:

y	$x = 0.0$	0.5	1.0	1.5	2.0
0.0	0.0775	0.1573	0.2412	0.3309	0.4274
0.5	0.1528	0.3104	0.4767	0.6552	0.8478
1.0	0.2235	0.4547	0.7002	0.9653	1.2533
1.5	0.2866	0.5846	0.9040	1.2525	1.6348

(a) Evaluate $(\partial/\partial y)f$ at $x = 1.0$ and $y = 0$ using the forward difference approximation with an error of order h^2 where $h = 0.5$.

(b) Evaluate $(\partial^2/\partial x^2)f$ at $x = 1.0$ and $y = 1.0$ using the central difference approximation with an error of order h^2 where $h = 0.5$.

(c) Evaluate $(\partial^2/\partial x \partial y)f$ at $x = 0$ and $y = 0$ using the forward difference approximation with an error of order h^2 where $h = 0.5$.

REFERENCES

Carnahan, B., H. A. Luther, J. O. Wilkes, *Applied Numerical Methods*, Wiley, 1969.

Cheney, W., and D. Kincaid, *Numerical Mathematics and Computing*, Brooks/Cole, 1985.

Hornbeck, R. W., *Numerical Methods*, Quantum, 1975.

Isaacson, E., and H. B. Keller, *Analysis of Numerical Methods*, Wiley, 1966.

James, M. L., G. M. Smith, and J. C. Wolford, *Applied Numerical Methods for Digital Computation*, 3rd ed., Harper & Row, 1985.

Ralston, A., *A First Course in Numerical Analysis*, McGraw-Hill, 1965.

6

Numerical Linear Algebra

6.1 INTRODUCTION

The primary objective of this chapter is to study basic computational methods for solving inhomogeneous sets of linear equations. Linear algebra is so fundamental in both scientific analyses and numerical methods that we cannot do much without basic knowledge of it.

The first subjects of this chapter are the Gauss and Gauss-Jordan eliminations to solve linear equations. They will be described without matrix or vector notations. After this, however, a minimal amount of matrix/vector notations and basic rules are introduced. Then, three related subjects, namely, matrix inversion, LU decomposition, and the determinant are discussed. Finally, solution of m equations with n unknowns is described.

Table 6.1 Comparison of three methods for linear equations

Method	Advantages	Disadvantages
Gauss elimination	The most fundamental solution algorithm.	Solution of one set of linear equations at a time.
Gauss-Jordan elimination	Basis for computing inverse; can solve multiple sets of equations.	Less efficient for a single set of equations.
LU decomposition	Efficient if one set of linear equations is repeatedly solved with different inhomogeneous terms (for example, in the inverse power method).	Less efficient and more cumbersome than Gauss elimination if used only once.

The subjects of homogeneous sets of linear equations are left for the next chapter.

Advantages and disadvantages of using Gauss and Gauss-Jordan eliminations and *LU* decomposition are summarized in Table 6.1.

6.2 GAUSS AND GAUSS-JORDAN ELIMINATIONS FOR SIMPLE IDEAL PROBLEMS

Gauss elimination is most widely used to solve a set of linear equations. In this section we study Gauss elimination and its variant, Gauss-Jordan elimination without pivoting. Pivoting in Gauss elimination is discussed in the next section.

A set of N equations is presented by

$$
\begin{aligned}
a_{1,1}x_1 + a_{1,2}x_2 + a_{1,3}x_3 + \cdots + a_{1,N}x_N &= y_1 \\
a_{2,1}x_1 + a_{2,2}x_2 + a_{2,3}x_3 + \cdots + a_{2,N}x_N &= y_2 \\
&\vdots \\
a_{N,1}x_1 + a_{N,2}x_2 + a_{N,3}x_3 + \cdots + a_{N,N}x_N &= y_N
\end{aligned}
\tag{6.2.1}
$$

where $a_{i,j}$ are coefficients, x_i are unknowns, and y_i are known terms called *inhomogeneous terms*. Here, the number of unknowns equals the number of equations as the most usual form of a set of linear equations. If the number of equations and that of unknowns are different, solutions can exist but should be approached very carefully. The problems in which the number of equations is not equal to that of unknowns are reserved for Sections 6.10.

When at least one of the inhomogeneous terms is not zero in Eq. (6.2.1), the set is said to be inhomogeneous. Gauss elimination applies only to inhomogeneous sets of equations. Solution of a set of linear equations is not always easy because it may not have a unique solution. Even if it has a unique solution, the computed solution can be inaccurate if it is an ill-conditioned problem.

However, for the sake of simplicity, we consider an ideal problem in that the set of equations has a unique solution and nothing difficult happens in the solution process. Gauss elimination consists of (a) forward elimination, and (b) backward substitution. The forward elimination proceeds as follows.

The first equation times $a_{2,1}/a_{1,1}$ is subtracted from the second equation to eliminate the first term of the second equation; likewise the first term of every equation thereafter, $i > 2$, is eliminated by subtracting the first equation times $a_{i,1}/a_{1,1}$. Then, the equations should look like

$$
\begin{aligned}
a_{1,1}x_1 + a_{1,2}x_2 + a_{1,3}x_3 + \cdots + a_{1,N}x_N &= y_1 \\
a'_{2,2}x_2 + a'_{2,3}x_3 + \cdots + a'_{2,N}x_N &= y'_2 \\
&\vdots \\
a'_{N,2}x_2 + a'_{N,3}x_3 + \cdots + a'_{N,N}x_N &= y'_N
\end{aligned}
\tag{6.2.2}
$$

where

$$a'_{i,j} = a_{i,j} - (a_{i,1}/a_{1,1})a_{1,j}$$

Notice that the first equation is unchanged.

Next, the second term of every equation in the third through the last equation, $i > 2$, is eliminated by subtracting the second equation times $a'_{i,2}/a'_{2,2}$. After this step is completed, the third terms of the fourth through the last equations are eliminated. When the forward elimination process is finished, the set of the equations will be in the following form:

$$a_{1,1}x_1 + a_{1,2}x_2 + a_{1,3}x_3 + \cdots + a_{1,N}x_N = y_1$$
$$a'_{2,2}x_2 + a'_{2,3}x_3 + \cdots + a'_{2,N}x_N = y'_2$$
$$a''_{3,3}x_3 + \cdots + a''_{3,N}x_N = y''_3 \qquad (6.2.3)$$
$$\vdots$$
$$a^{(N-1)}_{N,N}x_N = y^{(N-1)}_N$$

The leading terms in each of the foregoing equations are called *pivots*. Each equation could have been normalized by dividing through by the leading coefficient, but no normalization is used in Gauss elimination. The primary reason is that normalization of the equations increases the overall computing time.

The backward substitution procedure starts with the last equation. The solution for x_N is obtained from the last equation:

$$x_N = y^{(N-1)}_N / a^{(N-1)}_{N,N}$$

Subsequently,

$$x_{N-1} = \left[y^{(N-2)}_{N-1} - a^{(N-2)}_{N-1,N} x_N \right] / a^{(N-2)}_{N-1,N-1}$$
$$\vdots \qquad\qquad (6.2.4)$$
$$x_1 = \left[y_1 - \sum_{j=2}^{N} a_{1,j}x_j \right] \Big/ a_{1,1}$$

Thus, Gauss elimination is completed.

Gauss elimination can be carried out by writing only the coefficients and the right-side terms in an array form. Indeed, this is exactly what a computer program for Gauss elimination does. Even for hand calculation, the array form is more convenient than writing all equations. The array expression of Eq. (6.2.1) is

$$
\begin{array}{ccccccc}
a_{1,1} & a_{1,2} & a_{1,3} & \cdots & a_{1,N-1} & a_{1,N} & y_1 \\
a_{2,1} & a_{2,2} & a_{2,3} & \cdots & a_{2,N-1} & a_{2,N} & y_2 \\
\vdots & & & & & & \\
a_{N,1} & a_{N,2} & a_{N,3} & \cdots & a_{N,N-1} & a_{N,N} & y_N
\end{array}
\qquad (6.2.5)
$$

All the intermediate stages of forward elimination are written in the array form. The array after the forward elimination becomes

$$
\begin{array}{ccccccc}
a_{1,1} & a_{1,2} & a_{1,3} & \cdots & a_{1,N-1} & a_{1,N} & y_1 \\
0 & a'_{2,2} & a'_{2,3} & \cdots & a'_{2,N-1} & a'_{2,N} & y'_2 \\
0 & 0 & a''_{3,3} & \cdots & a''_{3,N-1} & a''_{3,N} & y''_3 \\
\vdots & & & & & & \\
0 & 0 & 0 & \cdots & a^{(N-2)}_{N-1,N-1} & a^{(N-2)}_{N-1,N} & y^{(N-2)}_{N-1} \\
0 & 0 & 0 & \cdots & 0 & a^{(N-1)}_{N,N} & y^{(N-1)}_{N}
\end{array}
\qquad (6.2.6)
$$

Example 6.1

Solve the following linear equations in the array form:

$$
\begin{aligned}
2x_1 + x_2 - 3x_3 &= -1 \\
-x_1 + 3x_2 + 2x_3 &= 12 \\
3x_1 + x_2 - 3x_3 &= 0
\end{aligned}
\qquad (A)
$$

⟨**Solution**⟩

An array expression of the equations is

$$
\begin{array}{cccc}
2 & 1 & -3 & -1 \\
-1 & 3 & 2 & 12 \\
3 & 1 & -3 & 0
\end{array}
\qquad (B)
$$

In the array B, the first three columns are the coefficients of Eq. (A), and the last column represents the inhomogeneous term.

To start forward elimination, the first row times $-1/2$ is subtracted from the second row. The first row times $3/2$ is subtracted from the third row. The array is now

$$
\begin{array}{cccc}
2 & 1 & -3 & -1 \\
0 & 7/2 & 1/2 & 23/2 \\
0 & -1/2 & 3/2 & 3/2
\end{array}
\qquad (C)
$$

Continuing the forward elimination, the second row times $-1/7$ is subtracted from the third row:

$$
\begin{array}{cccc}
2 & 1 & -3 & -1 \\
0 & 7/2 & 1/2 & 23/2 \\
0 & 0 & 11/7 & 22/7
\end{array}
\qquad (D)
$$

This is the end of forward elimination.

The backward substitution starts with the last row. Interpreting the last row as

$$(11/7)x_3 = 22/7$$

we get

$$x_3 = 2$$

Similarly

$$x_2 = 3$$

and

$$x_1 = 1$$

Gauss-Jordan elimination is a variant of Gauss elimination and shares with Gauss elimination the same forward elimination procedure, but it is different in the backward process. The backward process of Gauss-Jordan elimination is called *backward elimination*.

Starting from Eq. (6.2.6), the backward elimination makes the coefficients at the pivoting position unity and eliminates all other coefficients. First, dividing the last row by $a_{N,N}^{(N-1)}$ yields

$$0 \quad 0 \quad 0 \quad \cdots \quad 1 \quad \bar{y}_N \tag{6.2.7}$$

where

$$\bar{y}_N = y_N^{(N-1)}/a_{N,N}^{(N-1)}$$

The Nth coefficients of each row except for the last row is eliminated by subtracting the last row, Eq. (6.2.7), times the Nth coefficient in the ith row:

$$
\begin{array}{ccccccc}
a_{1,1} & a_{1,2} & a_{1,3} & \cdots & a_{1,N-1} & 0 & \bar{y}_1 \\
0 & a'_{2,2} & a'_{2,3} & \cdots & a'_{2,N-1} & 0 & \bar{y}_2 \\
0 & 0 & a''_{3,3} & \cdots & a''_{3,N-1} & 0 & \bar{y}_3 \\
\vdots & & & & & & \\
0 & 0 & 0 & \cdots & a_{N-1,N-1}^{(N-2)} & 0 & \bar{y}_{N-1} \\
0 & 0 & 0 & \cdots & 0 & 1 & \bar{y}_N
\end{array}
\tag{6.2.8}
$$

where

$$\bar{y}_i = y_i^{(i-1)} - a_{i,N}^{(i-1)}\bar{y}_N$$

Now, Eq. (6.2.8) has the same configuration as Eq. (6.2.6) except for the last row and the Nth column. Therefore, the $(N-1)$th row may be normalized, and the $(N-1)$th column may be eliminated in a similar way. We divide the $(N-1)$th row by $a_{N-1,N-1}^{(N-2)}$. Then, the $(N-1)$th coefficients in all the rows above the $(N-1)$th row are eliminated by subtracting the $(N - 1)$th row times the $(N - 1)$th coefficient in the row to be eliminated:

$$
\begin{array}{cccccccc}
a_{1,1} & a_{1,2} & a_{1,3} & \cdots & 0 & 0 & \bar{y}'_1 \\
0 & a'_{2,2} & a'_{2,3} & \cdots & 0 & 0 & \bar{y}'_2 \\
0 & 0 & a''_{3,3} & \cdots & 0 & 0 & \bar{y}'_3 \\
\vdots & & & & & & \\
0 & 0 & 0 & \cdots & 1 & 0 & \bar{y}'_{N-1} \\
0 & 0 & 0 & \cdots & 0 & 1 & \bar{y}_N
\end{array}
\tag{6.2.9}
$$

By repeating the elimination process, finally the array becomes

$$
\begin{array}{cccccccc}
1 & 0 & 0 & \cdots & 0 & 0 & \bar{y}_1^{(N-1)} \\
0 & 1 & 0 & \cdots & 0 & 0 & \bar{y}_2^{(N-2)} \\
0 & 0 & 1 & \cdots & 0 & 0 & \bar{y}_3^{(N-3)} \\
\vdots & & & & & & \\
0 & 0 & 0 & \cdots & 1 & 0 & \bar{y}'_{N-1} \\
0 & 0 & 0 & \cdots & 0 & 1 & \bar{y}_N
\end{array}
\tag{6.2.10}
$$

This is the end of backward elimination. Two aspects should be recognized in Eq. (6.2.10). First, all the coefficients are zero except the pivots are unity. Second, since Eq. (6.2.10) is an array form of a set of equations, the each row is interpreted as

$$ x_i = \bar{y}_i^{(N-i)} $$

that is, the rightmost column is the final solution.

Example 6.2

Solve the same problem of Example 6.1 by Gauss-Jordan elimination.

⟨**Solution**⟩

The forward elimination of Gauss-Jordan elimination is the same as in Gauss elimination, so we consider backward elimination starting from Eq. (D) in Example 6.1.

The third row of Eq. (D) in Example 6.1 is divided by 11/7. The third row times 1/2 is subtracted from the second row, and the third row times -3 is subtracted from the first row:

$$
\begin{array}{cccc}
2 & 1 & 0 & 5 \\
0 & 7 & 0 & 21 \\
0 & 0 & 1 & 2
\end{array}
\tag{E}
$$

The second row is divided by 7:

$$
\begin{array}{cccc}
2 & 1 & 0 & 5 \\
0 & 1 & 0 & 3 \\
0 & 0 & 1 & 2
\end{array}
\qquad \text{(F)}
$$

The second row times 1 is subtracted from the first row:

$$
\begin{array}{cccc}
2 & 0 & 0 & 2 \\
0 & 1 & 0 & 3 \\
0 & 0 & 1 & 2
\end{array}
\qquad \text{(G)}
$$

Finally, the first row is divided by 2 to complete the solution:

$$
\begin{array}{cccc}
1 & 0 & 0 & 1 \\
0 & 1 & 0 & 3 \\
0 & 0 & 1 & 2
\end{array}
\qquad \text{(H)}
$$

We can see that the last column is the solution [compare to Eq. (6.3.4)], and that the first three columns are all zero except for the unity in each diagonal position. The procedure explained by this array may be extended to a linear equation set of any size.

When Gauss elimination and Gauss-Jordan elimination are used with hand computations, it is useful to write down the products of a row and a constant as shown in the next example.

Example 6.3

Solve the following equations by Gauss elimination with hand calculation:

$$-0.04x_1 + 0.04x_2 + 0.12x_3 = 3$$
$$0.56x_1 - 1.56x_2 + 0.32x_3 = 1$$
$$-0.24x_1 + 1.24x_2 - 0.28x_3 = 0$$

⟨**Solution**⟩

The array expression of the problem is

$$
\begin{array}{cccc}
-0.04 & 0.04 & 0.12 & 3 \\
0.56 & -1.56 & 0.32 & 1 \\
-0.24 & 1.24 & -0.28 & 0
\end{array}
$$

The forward elimination proceeds as:

row 1	-0.04	0.04	0.12	3
row 2	0.56	-1.56	0.32	1
row 3	-0.24	1.24	-0.28	0
row 1 times $0.56/(-0.04) = 14$:	-0.56	-0.56	-1.68	-42 (A)
row 1 times $-0.24/(0.04) = 6$:	-0.24	0.24	0.72	18 (B)

Subtracting (A) from row, 2, and subtracting (B) from row 3 yield

row 1	−0.04	0.04	0.12	3
row 2	0	−1	2	43
row 3	0	+1	−1	−18

The second coefficient of the third row is eliminated by subtracting row 2 times −1 from row 3:

row 1	−0.04	0.04	0.12	3
row 2	0	−1	2	43
row 3	0	0	1	25

The backward substitutions of Gauss elimination is straightforward:

$$x_3 = 25/(1) = 25$$
$$x_2 = [43 − (2)(25)]/(−1) = 7$$
$$x_1 = [3 − (0.12)(25) − (0.04)(7)]/(−0.04) = 7$$

Comment: Whenever multiples of a row become necessary in hand calculations, do not hesitate to write them down as shown in (A) and (B).

In the Gauss-Jordan elimination, the forward and backward eliminations need not be separated. This is possible because the coefficients not only below but also above a pivot can be eliminated at the same time. If this approach is taken, the form of the coefficients become diagonal when eliminations by the last pivot are completed.

SUMMARY OF THIS SECTION

(a) Gauss elimination consists of forward elimination and backward substitution. The forward elimination is performed using an array consisting of coefficients and the inhomogeneous terms.

(b) The forward elimination of Gauss-Jordan elimination is identical to that of Gauss elimination. However, Gauss-Jordan elimination uses backward elimination rather than backward substitution.

6.3 PIVOTING AND STANDARD GAUSS ELIMINATION

In Section 6.2, Gauss elimination is applied to an ideally simple problem with no null coefficients. However, the method does not work if the first coefficient of the first row is zero or if a diagonal coefficient becomes zero in the process of solution, because they are used as denominators in forward elimination.*

---

* In Eq. (6.2.1) the diagonal coefficients, or pivots, are the coefficient of x_1 in the first equation, that of x_2 in the second equation, and that of x_3 in the third equation. In general, the coefficient of x_n in the nth equation is a diagonal coefficient.

Pivoting is used to change sequential order of the equations for two purposes—first, to prevent diagonal coefficients from becoming zero, and second, to make each diagonal coefficient larger in magnitude than any other coefficients below it. The equations are not mathematically affected by changes of the sequential order, but changing the order makes the computation possible whenever the diagonal coefficient becomes zero. Even when all diagonal coefficients are nonzero, the changes of order increases accuracy of the computations.

Pivoting explained in the remainder of this section is more suitable in computer programs than for hand calculations, because pivoting tends to increase the amount of effort substantially. Therefore, in quick hand calculation for presumably well-behaved problems encountered in such situations as tests, students may avoid it except when the diagonal coefficient becomes zero (unless otherwise specified by the instructor).

To explain pivoting, consider the array

$$
\begin{array}{cccc}
0 & 10 & 1 & 2 \\
1 & 3 & -1 & 6 \\
2 & 4 & 1 & 5
\end{array}
\tag{6.3.1}
$$

Elimination of the first number in the second and third rows is impossible because the first number in the first row is zero. In our first pivoting, the first row and the last row are exchanged. Although the first row and the second row could be exchanged instead, the third row is brought to the top because the first number in the third row has the largest modulus (absolute number) in the first column. Bringing the largest number in the column to the diagonal position has the advantage that round-off error is reduced. After this pivoting, the array becomes

$$
\begin{array}{cccc}
2 & 4 & 1 & 5 \\
1 & 3 & -1 & 6 \\
0 & 10 & 1 & 2
\end{array}
\tag{6.3.2}
$$

Next we eliminate the first number in the second row by subtracting the first row times 1/2 from the second:

$$
\begin{array}{cccc}
2 & 4 & 1 & 5 \\
0 & 1 & -3/2 & 7/2 \\
0 & 10 & 1 & 2
\end{array}
\tag{6.3.3}
$$

The first number of the third row is already zero, so we proceed to the elimination of the second number, 10, in the third row. However, this number, 10, is greater than the second number (diagonal coefficient) in the second row. In general, eliminating a number larger in magnitude than the diagonal term is undesirable as already mentioned. Therefore, we exchange the order of the second and the third rows:

$$\begin{array}{cccc} 2 & 4 & 1 & 5 \\ 0 & 10 & 1 & 2 \\ 0 & 1 & -3/2 & 7/2 \end{array} \qquad (6.3.4)$$

After we eliminate the second number in the third row, Eq. (6.3.4) becomes

$$\begin{array}{cccc} 2 & 4 & 1 & 5 \\ 0 & 10 & 1 & 2 \\ 0 & 0 & -16/5 & 33/5 \end{array} \qquad (6.3.5)$$

The entries of 2, 10 and $-16/5$ are called *diagonal coefficients*, or *pivots*.
    The backward substitutions give

$$x_3 = -2.0625$$
$$x_2 = (2 - x_3)/10 = 0.4062$$
$$x_1 = (5 - 4x_2 - x_3) = 2.7187$$

The Gauss-Jordan elimination also gives

$$\begin{array}{cccc} 1 & 0 & 0 & 2.7187 \\ 0 & 1 & 0 & 0.4062 \\ 0 & 0 & 1 & -2.0625 \end{array}$$

(The example problem being considered here, for simplicity of explanation, is a kind of well-conditioned, so accuracy is not affected by pivoting.)
    If a zero diagonal element is unavoidable in spite of pivoting, it indicates that the problem is one of the unsolvable problems. If this occurs, we stop the effort of computation.
    Pivoting cannot cure all difficulties involved in the solution of linear equations, however. If the results are not accurate even with pivoting, double precision should be used. In general, the linear equations associated with boundary value problems of differential equations are well conditioned and seldom have accuracy problems with single precision. On the other hand, the linear equations associated with curve fitting based on the least square are very often ill conditioned and thus require a high precision. Ill-conditioned problems will be discussed in more detail in Section 6.9.
    PROGRAM 6–1 performs Gauss elimination with pivoting.
    Example 6.4 shows the effect of pivoting in a typical set of linear equations.

Example 6.4

The exact solution of the following problem in an array form is that all the unknowns become unity, because the inhomogeneous terms are summations of the coefficients on the same line:

1.334E−4	4.123E+1	7.912E+2	−1.544E+3	−711.5698662
1.777	2.367E−5	2.070E+1	−9.035E+1	−67.87297633
9.188	0	−1.015E+1	1.988E−4	−0.961801200
1.002E+2	1.442E+4	−7.014E+2	5.321	13824.12100

(a) Solve the equations without pivoting and then with pivoting using single precision.

(b) Repeat by double precision.

⟨**Solution**⟩

The solution with pivoting may be obtained by running PROGRAM 6–1. The solution without pivoting is also done by the same program by deleting a few lines used for pivoting. The results are as follows:

(a) Single precision:

i	Without pivoting[a] x_i	With pivoting[a] x_i
1	0.95506	0.99998
2	1.00816	1
3	0.96741	1
4	0.98352	1

The results in single precision without pivoting are very poor, but pivoting significantly improves accuracy.

(b) Double precision:

i	Without Pivoting[a] x_i	With Pivoting[a] x_i
1	0.9999 9999 9861 473	1.0000 0000 0000 002
2	1.0000 0000 0000 784	1.0000 0000 0000 000
3	0.9999 9999 9984 678	1.0000 0000 0000 000
4	0.9999 9999 9921 696	1.0000 0000 0000 000

[a] The present computations were performed by VAX, the accuracy of which is almost the same as IBM PC and IBM mainframe computers. The single precision of Cray is approximately equal to the double precision of VAX, IBM PC, and IBM mainframe. Therefore, if Cray is used for the present problem in single precision, the results will be equivalent to those of double precision that we have shown. See Chapter 1 for comparisons of various computers.

Double precision increases accuracy even without pivoting. However, with pivoting, accuracy is further increased.

(a) Pivoting has two purposes: one, to overcome the difficulty with null coefficients at diagonal positions; and second, to decrease the round-off errors.

(b) For quick hand calculations, pivoting may be avoided except when diagonal coefficients become zero (provided that the problem is a well-designed exercise problem).

(c) For the best accuracy, the use of both double precision and pivoting is desirable.

6.4 UNSOLVABLE PROBLEMS

A set of linear equations is not always numerically solvable. The following three sets of equations are simple but important examples:

(a) $\begin{aligned} -x + y &= 1 \\ -2x + 2y &= 2 \end{aligned}$

(b) $\begin{aligned} -x + y &= 1 \\ -x + y &= 0 \end{aligned}$

(c) $\begin{aligned} -x + y &= 1 \\ x + 2y &= -2 \\ 2x - y &= 0 \end{aligned}$

The equations in each set are plotted in Figure 6.1.

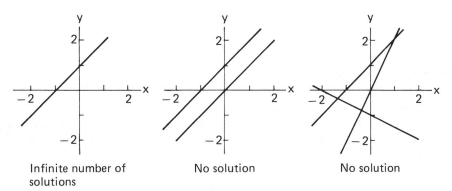

Infinite number of solutions   No solution   No solution

Figure 6.1 Plotting of three sets of linear equations

In the set (a), the second equation is 2 times the first equation, so they are mathematically identical. Any point (x, y) satisfying one equation solves the other also. Therefore, the number of solutions is infinite. In other words, there is no unique solution. If one equation is a multiple of another or can be obtained by adding or subtracting other equations, that equation is said to be *linearly dependent* on others.

If none of the equations is linearly dependent, all the equations are said to be *linearly independent.*

In the set (b), the two equations are parallel lines that never intersect, so there is no solution. Such a system is called an *inconsistent system.* A set of equations is inconsistent if the left side of at least one equation can be completely eliminated by adding or subtracting other equations, while the right side remains nonzero.

In the third set, there are three independent equations for two unknowns. As seen in Figure 6.1(c), these three equations can never be simultaneously satisfied.

A case such as (c) cannot happen if the number of equations equals the number of unknowns. Yet, lack of linear independence, such as in (a), or inconsistency, as in (b), may occur. If the number of equations is more than two, lack of linear independence and inconsistency are less obvious. However, a program of Gauss elimination attempted on such a set stops in the middle of computation because of an arithmeric error such as overflow or zero division. Indeed, if a set of equations is inconsistent or linearly dependent, a row of coefficients (not including the last number corresponding to the right side term) in the array becomes zero during forward elimination. In PROGRAM 6–1, such cases are detected and the program is stopped after printing *"matrix is singular."*

SUMMARY OF THIS SECTION. The conditions necessary for a unique solution are stated as follows:

(a) The number of equations must equal that of the unknowns.
(b) Each equation is linearly independent; equivalently, no equation can be eliminated by adding or subtracting other equations.

6.5 MATRICES AND VECTORS

This section introduces the operations of matrices and vectors.

A *matrix* is a rectangular array of numbers, such as those already seen in the previous section. When the array is square, it is called a *square matrix.* The following matrices are square matrices:

$$A = \begin{bmatrix} 2 & 1 & 0 \\ -1 & 3 & 1 \\ 1 & 1 & 1 \end{bmatrix}$$

$$B = \begin{bmatrix} 3 & -1 & 1 \\ 2 & 1 & 3 \\ -2 & 4 & 5 \end{bmatrix}$$

Matrices are often written in the symbolic form as follows:

$$A = \begin{bmatrix} a_{11} & a_{12} & a_{13} \\ a_{21} & a_{22} & a_{23} \\ a_{31} & a_{32} & a_{33} \end{bmatrix} \qquad (6.5.1)$$

$$B = \begin{bmatrix} b_{11} & b_{12} & b_{13} \\ b_{21} & b_{22} & b_{23} \\ b_{31} & b_{32} & b_{33} \end{bmatrix} \qquad (6.5.2)$$

Notice that in Eqs. (6.5.1) and (6.5.2) the first subscript in a matrix changes in the vertical direction and the second subscript changes in the horizontal direction. Equations (6.5.1) and (6.5.2) are often abbreviated by

$$A = [a_{ij}] \quad \text{and} \quad B = [b_{ij}]$$

respectively.

If a matrix is rectangular with m rows and n columns, we call it a $m \times n$ *matrix*. For example,

$$A = \begin{bmatrix} 5 & 9 & 2 \\ 3 & 0 & 4 \end{bmatrix}$$

is a 2×3 matrix.

Matrices have four basic arithmetic operations similar to those for numbers: addition, subtraction, multiplication, and division. Among these, the first two are straightforward, but the last two are somewhat more involved. Addition, subtraction, and multiplication are defined as follows:

Addition

$$A + B = C$$

where $C = [c_{ij}]$ is a matrix with each element given by

$$c_{ij} = a_{ij} + b_{ij}$$

Subtraction

$$A - B = C$$

where $C = [c_{ij}]$ with

$$c_{ij} = a_{ij} - b_{ij}$$

Multiplication

$$AB = C$$

where $C = [c_{ij}]$ with

$$c_{ij} = \sum_{k=1}^{N} a_{ik} b_{kj}$$

As can be seen easily, the product AB is not equal to BA in general. If $AB = BA$, then matrices A and B are said *to commute*. If A and B are rectangular matrices, the product exists only if A is an $m \times n$ matrix and B is an $n \times k$ matrix (the number of columns of A and the number of rows of B are the same).

Division of one matrix by another matrix is defined as follows:

Division

$$B^{-1}A = C$$

where A is divided by B, and B^{-1} is called the inverse of B. The division is equivalent to

$$A = BC$$

Division is much more restrictive than other operations because B^{-1} can exist only if B is a square matrix. More details of inverse matrices are described in Section 6.6.

A column vector is a column array of numbers or variables, for example:

$$x = \begin{bmatrix} x_1 \\ x_2 \\ x_3 \end{bmatrix}, \quad y = \begin{bmatrix} y_1 \\ y_2 \\ y_3 \end{bmatrix}$$

If a column vector consists of N numbers (or elements), the order of the vector is said to be N. A column vector is considered also as a $N \times 1$ matrix. A row vector is a row array of numbers, for example:

$$a = [a_1, a_2, a_3, a_4]$$

A row vector is considered as a $1 \times N$ matrix. When "vector" is used without "column" or "row", it usually means a column vector. Because vectors are special cases of matrices, all the rules for matrices apply to vectors.

Addition of vectors is defined by

$$x + y = z$$

where x, y, and z are vectors of the same order, and the ith elements of the vectors have the relation

$$x_i + y_i = z_i$$

Subtraction of one vector from another vector is

$$x - y = z$$

where

$$x_i - y_i = z_i$$

Multiplication of a matrix and a vector is defined by

$$Ax = y$$

where A is a matrix and y and x are vectors. In the foregoing equation, the y_i's may be explicitly written as

$$y_i = \sum_{k=1}^{N} a_{ik} x_k$$

Example 6.5

Square matrices and vectors are defined by

$$A = \begin{bmatrix} 1 & 2 & 4 \\ 3 & 1 & 2 \\ 4 & 1 & 3 \end{bmatrix}, \quad B = \begin{bmatrix} 7 & 3 & 1 \\ 2 & 3 & 5 \\ 8 & 1 & 6 \end{bmatrix}$$

$$x = \begin{bmatrix} 1 \\ 4 \\ 2 \end{bmatrix}, \quad y = \begin{bmatrix} 3 \\ 9 \\ 4 \end{bmatrix}$$

Calculate $A + B$, $B - A$, AB, BA, $x + y$, $x - y$, and Ax.

⟨**Solution**⟩

The calculations are shown next:

$$A + B = \begin{bmatrix} 1+7 & 2+3 & 4+1 \\ 3+2 & 1+3 & 2+5 \\ 4+8 & 1+1 & 3+6 \end{bmatrix} = \begin{bmatrix} 8 & 5 & 5 \\ 5 & 4 & 7 \\ 12 & 2 & 9 \end{bmatrix}$$

$$A - B = \begin{bmatrix} 1-7 & 2-3 & 4-1 \\ 3-2 & 1-3 & 2-5 \\ 4-8 & 1-1 & 3-6 \end{bmatrix} = \begin{bmatrix} -6 & -1 & 3 \\ 1 & -2 & -3 \\ -4 & 0 & -3 \end{bmatrix}$$

$$AB = \begin{bmatrix} 1 & 2 & 4 \\ 3 & 1 & 2 \\ 4 & 1 & 3 \end{bmatrix} \begin{bmatrix} 7 & 3 & 1 \\ 2 & 3 & 5 \\ 8 & 1 & 6 \end{bmatrix}$$

$$= \begin{bmatrix} 1 \times 7 + 2 \times 2 + 4 \times 8 & 1 \times 3 + 2 \times 3 + 4 \times 1 & 1 \times 1 + 2 \times 5 + 4 \times 6 \\ 3 \times 7 + 1 \times 2 + 2 \times 8 & 3 \times 3 + 1 \times 3 + 2 \times 1 & 3 \times 1 + 1 \times 5 + 2 \times 6 \\ 4 \times 7 + 1 \times 2 + 3 \times 8 & 4 \times 3 + 1 \times 3 + 3 \times 1 & 4 \times 1 + 1 \times 5 + 3 \times 6 \end{bmatrix}$$

$$= \begin{bmatrix} 43 & 13 & 35 \\ 39 & 14 & 20 \\ 54 & 18 & 27 \end{bmatrix}$$

$$BA = \begin{bmatrix} 7 & 3 & 1 \\ 2 & 3 & 5 \\ 8 & 1 & 6 \end{bmatrix} \begin{bmatrix} 1 & 2 & 4 \\ 3 & 1 & 2 \\ 4 & 1 & 3 \end{bmatrix} = \begin{bmatrix} 20 & 18 & 37 \\ 31 & 12 & 29 \\ 35 & 23 & 52 \end{bmatrix}$$

$$x + y = \begin{bmatrix} 1 \\ 4 \\ 2 \end{bmatrix} + \begin{bmatrix} 3 \\ 9 \\ 4 \end{bmatrix} = \begin{bmatrix} 4 \\ 13 \\ 6 \end{bmatrix}$$

$$x + y = \begin{bmatrix} 1 \\ 4 \\ 2 \end{bmatrix} - \begin{bmatrix} 3 \\ 9 \\ 4 \end{bmatrix} = \begin{bmatrix} -2 \\ -5 \\ -2 \end{bmatrix}$$

$$AX = \begin{bmatrix} 1 & 2 & 4 \\ 3 & 1 & 2 \\ 4 & 1 & 3 \end{bmatrix} \begin{bmatrix} 1 \\ 4 \\ 2 \end{bmatrix} = \begin{bmatrix} 1 \times 1 + 2 \times 4 + 4 \times 2 \\ 3 \times 1 + 1 \times 4 + 2 \times 2 \\ 4 \times 1 + 1 \times 4 + 3 \times 2 \end{bmatrix} = \begin{bmatrix} 17 \\ 11 \\ 11 \end{bmatrix}$$

Comment: Notice $AB \neq BA$.

Example 6.6

Calculate the following products:

$$\begin{bmatrix} 1 & 2 \\ 4 & 3 \\ 0 & 2 \end{bmatrix} \begin{bmatrix} 5 \\ 1 \end{bmatrix}$$

$$\begin{bmatrix} 2 & 1 & 7 \end{bmatrix} \begin{bmatrix} 1 & 2 \\ 4 & 3 \\ 0 & 2 \end{bmatrix}$$

$$\begin{bmatrix} 8 & 1 & 3 \\ 1 & 5 & 2 \end{bmatrix} \begin{bmatrix} 1 & 2 \\ 4 & 3 \\ 0 & 2 \end{bmatrix}$$

$$\begin{bmatrix} 1 & 2 \\ 4 & 3 \\ 0 & 2 \end{bmatrix} \begin{bmatrix} 8 & 1 & 3 \\ 1 & 5 & 2 \end{bmatrix}$$

⟨**Solution**⟩

$$\begin{bmatrix} 1 & 2 \\ 4 & 3 \\ 0 & 2 \end{bmatrix} \begin{bmatrix} 5 \\ 1 \end{bmatrix} = \begin{bmatrix} 1 \times 5 + 2 \times 1 \\ 4 \times 5 + 3 \times 1 \\ 0 \times 5 + 2 \times 1 \end{bmatrix}$$

$$[2 \quad 1 \quad 7] \begin{bmatrix} 1 & 2 \\ 4 & 3 \\ 0 & 2 \end{bmatrix} = [2 \times 1 + 1 \times 4 + 7 \times 0 \quad 2 \times 2 + 1 \times 3 + 7 \times 2] = [6 \quad 21]$$

$$\begin{bmatrix} 8 & 1 & 3 \\ 1 & 5 & 2 \end{bmatrix} \begin{bmatrix} 1 & 2 \\ 4 & 3 \\ 0 & 2 \end{bmatrix} = \begin{bmatrix} 8 \times 1 + 1 \times 4 + 3 \times 0 & 8 \times 2 + 1 \times 3 + 3 \times 2 \\ 1 \times 1 + 5 \times 4 + 2 \times 0 & 1 \times 2 + 5 \times 3 + 2 \times 2 \end{bmatrix}$$

$$= \begin{bmatrix} 12 & 25 \\ 21 & 21 \end{bmatrix}$$

$$\begin{bmatrix} 1 & 2 \\ 4 & 3 \\ 0 & 2 \end{bmatrix} \begin{bmatrix} 8 & 1 & 3 \\ 1 & 5 & 2 \end{bmatrix} = \begin{bmatrix} 1 \times 8 + 2 \times 1 & 1 \times 1 + 2 \times 5 & 1 \times 3 + 2 \times 2 \\ 4 \times 8 + 3 \times 1 & 4 \times 1 + 3 \times 5 & 4 \times 3 + 3 \times 2 \\ 0 \times 8 + 2 \times 1 & 0 \times 1 + 2 \times 5 & 0 \times 3 + 2 \times 2 \end{bmatrix}$$

$$= \begin{bmatrix} 10 & 11 & 7 \\ 35 & 19 & 18 \\ 2 & 10 & 4 \end{bmatrix}$$

Special matrices and vectors are defined next:

Null Matrix. All the elements of the null matrix are zero:

$$A = \begin{bmatrix} 0 & 0 & 0 \\ 0 & 0 & 0 \\ 0 & 0 & 0 \end{bmatrix}$$

Identity Matrix. All the elements are zero except the diagonal elements, which are all unity. An identity matrix is denoted by I, namely

$$I = \begin{bmatrix} 1 & 0 & 0 \\ 0 & 1 & 0 \\ 0 & 0 & 1 \end{bmatrix}$$

Transposed Matrix. For a matrix defined by $A = [a_{ij}]$, its transpose is defined by $A^t = [a_{ji}]$ (i and j are exchanged). For example:

$$A = \begin{bmatrix} 2 & 3 \\ 0 & 5 \end{bmatrix} \quad \text{then} \quad A^t = \begin{bmatrix} 2 & 0 \\ 3 & 5 \end{bmatrix}$$

$$B = \begin{bmatrix} 5 & 0 \\ 2 & 7 \\ 1 & 2 \end{bmatrix} \quad \text{then} \quad B^t = \begin{bmatrix} 5 & 2 & 1 \\ 0 & 7 & 2 \end{bmatrix}$$

Inverse Matrix. The inverse of a square matrix A is written as A^{-1} and satisfies $AA^{-1} = A^{-1}A = I$. An explanation of calculations to determine A^{-1} is reserved for Section 6.6.

Orthogonal Matrix. A matrix that has orthonormal columns. It satisfies

$$Q^tQ = I, \quad QQ^t = I, \quad \text{and} \quad Q^t = Q^{-1}$$

Null Vector. All the elements of the null vector are zero:

$$x = \begin{bmatrix} 0 \\ 0 \\ 0 \end{bmatrix}$$

Unit Vectors. All the elements are zero except one element that is unity:

$$u = \begin{bmatrix} 1 \\ 0 \\ 0 \end{bmatrix}, \quad v = \begin{bmatrix} 0 \\ 1 \\ 0 \end{bmatrix}, \quad w = \begin{bmatrix} 0 \\ 0 \\ 1 \end{bmatrix}$$

Transposed Vector. If a vector is given by

$$v = \begin{bmatrix} x_1 \\ x_2 \\ x_3 \end{bmatrix}$$

then its transpose is written as v^t and defined by

$$v^t = \begin{bmatrix} x_1 & x_2 & x_3 \end{bmatrix}$$

Transpose of a column vector is a row vector.

SUMMARY OF THIS SECTION

(a) A column vector is a matrix with only one column. A row vector is a matrix with only one row and can be expressed as transpose of a column vector.

(b) Two matrices with the same number of columns and rows can be added or subtracted.

(c) A matrix B may be premultiplied by another matrix A if the number of columns of A and the number of rows of B are the same.

(d) If $BA = I$ or $AB = I$ where I is an identity matrix, then $B = A^{-1}$.

6.6 INVERSION OF A MATRIX

The inverse of a matrix may be calculated by applying Gauss-Jordan elimination. Consider a linear equation in matrix notation:

$$Ax = y \tag{6.6.1}$$

where A is a square matrix. Assuming that no pivoting is necessary, a premultiplication of Eq. (6.6.1) by a square matrix G yields

$$GAx = Gy \tag{6.6.2}$$

If G is chosen to be the inverse of A, namely A^{-1}, Eq. (6.6.2) reduces to

$$x = A^{-1}y \tag{6.6.3}$$

which is the solution. In other words, Gauss-Jordan elimination is equivalent to premultiplication by $G = A^{-1}$.

Therefore, if we apply the same operations performed in Gauss-Jordan elimination to the identity matrix (that is, multiplying rows by the same numbers as used in Gauss-Jordan elimination and subtracting rows in the same manner), then the identity matrix must be transformed to A^{-1}. This may be written symbolically as

$$GI = A^{-1} \tag{6.6.4}$$

To compute A^{-1}, we write A and I in an augmented array form

$$
\begin{array}{cccccc}
a_{1,1} & a_{1,2} & a_{1,3} & 1 & 0 & 0 \\
a_{2,1} & a_{2,2} & a_{2,3} & 0 & 1 & 0 \\
a_{3,1} & a_{3,2} & a_{3,3} & 0 & 0 & 1
\end{array}
\tag{6.6.5}
$$

Then we follow Gauss-Jordan elimination in exactly the same way as in solving a linear set of equations. When the left half of the augmented matrix is reduced to a unit matrix, the right half becomes A^{-1}.

Example 6.7

Calculate the inverse of the matrix:

$$A = \begin{bmatrix} 2 & 1 & -3 \\ -1 & 3 & 2 \\ 3 & 1 & -3 \end{bmatrix} \tag{A}$$

⟨**Solution**⟩

We write A and I in one array:

$$\begin{array}{rrrrrr} 2 & 1 & -3 & 1 & 0 & 0 \\ -1 & 3 & 2 & 0 & 1 & 0 \\ 3 & 1 & -3 & 0 & 0 & 1 \end{array}$$

The elimination procedure shown next is the Gauss-Jordan elimination described in Section 6.2.

Forward elimination proceeds as follows. The first row times $-1/2$ is subtracted from the second row, and the first row times $3/2$ is subtracted from the third row:

$$\begin{array}{rrrrrr} 2 & 1 & -3 & 1 & 0 & 0 \\ 0 & 3.5 & 0.5 & 0.5 & 1 & 0 \\ 0 & -0.5 & 1.5 & -1.5 & 0 & 1 \end{array}$$

The second row times $-0.5/3.5 = -1/7$ is subtracted from the third row:

$$\begin{array}{rrrrrr} 2 & 1 & -3 & 1 & 0 & 0 \\ 0 & 3.5 & 0.5 & 0.5 & 1 & 0 \\ 0 & 0 & 1.5714 & -1.4285 & 0.14285 & 1 \end{array} \tag{B}$$

Now, the backward elimination proceeds as follows. The last row is divided by 1.5714:

$$\begin{array}{rrrrrr} 2 & 1 & -3 & 1 & 0 & 0 \\ 0 & 3.5 & 0.5 & 0.5 & 1 & 0 \\ 0 & 0 & 1 & -0.90909 & 0.090909 & 0.63636 \end{array}$$

The second row is subtracted by 0.5 times the last row, the first row is added by 3 times the last row, and the second row is divided by 3.5:

$$\begin{array}{rrrrrr} 2 & 1 & 0 & -1.72727 & 0.27272 & 1.90908 \\ 0 & 1 & 0 & 0.27272 & 0.27272 & -0.09090 \\ 0 & 0 & 1 & -0.90909 & 0.09090 & 0.63636 \end{array}$$

The first row is subtracted by 1/2 times the second row and divided by 2:

$$\begin{array}{rrrrrr} 1 & 0 & 0 & -1 & 0 & 1 \\ 0 & 1 & 0 & 0.27272 & 0.27272 & -0.09090 \\ 0 & 0 & 1 & -0.90909 & 0.09090 & 0.63636 \end{array}$$

The last three columns in the foregoing augmented array constitute the inverse of the matrix A. This can be verified by premultiplying or postmultiplying A by A^{-1} as follows:

$$\begin{bmatrix} 2 & 1 & -3 \\ -1 & 3 & 2 \\ 3 & 1 & -3 \end{bmatrix} \begin{bmatrix} -1 & 0 & 1 \\ 0.27272 & 0.27272 & -0.09090 \\ -0.90909 & 0.09090 & 0.63636 \end{bmatrix} = \begin{bmatrix} 1 & 0 & 0 \\ 0 & 1 & 0 \\ 0 & 0 & 1 \end{bmatrix}$$

$$\begin{bmatrix} -1 & 0 & 1 \\ 0.27272 & 0.27272 & -0.09090 \\ -0.90909 & 0.09090 & 0.63636 \end{bmatrix} \begin{bmatrix} 2 & 1 & -3 \\ -1 & 3 & 2 \\ 3 & 1 & -3 \end{bmatrix} = \begin{bmatrix} 1 & 0 & 0 \\ 0 & 1 & 0 \\ 0 & 0 & 1 \end{bmatrix}$$

Although we have not used pivoting in the preceding explanation of matrix inversion, pivoting is necessary for a matrix inversion because the inversion scheme is essentially a Gauss elimination. Fortunately, however, the inverse matrix is not affected by a change in the sequential order of equations. The first column of A^{-1} is the solution of $Ax = \text{col}\,[1, 0, 0]$, and the second and third columns of A^{-1} are the solutions of $Ax = \text{col}\,[0, 1, 0]$ and $Ax = \text{col}\,[0, 0, 1]$, respectively. The sequential order of elements in x is not influenced by shuffling the order of equations. Thus, A^{-1} is not affected by pivoting.

When you calculate the inverse of a matrix by hand calculations, it is suggested that you write the values of a row times a constant as shown in the next example.

Example 6.8

Calculate the inverse of the following matrix by hand calculations with pivoting:

$$\begin{array}{rrr} -0.04 & 0.04 & 0.12 \\ 0.56 & -1.56 & 0.32 \\ -0.24 & 1.24 & -0.28 \end{array}$$

⟨**Solution**⟩

The augmented array is

$$\begin{array}{rrrrrr} -0.04 & 0.04 & 0.12 & 1 & 0 & 0 \\ 0.56 & -1.56 & 0.32 & 0 & 1 & 0 \\ -0.24 & 1.24 & -0.28 & 0 & 0 & 1 \end{array}$$

The first pivoting is done because the top leftmost element is smaller than the element just below:

row 1	0.56	−1.56	0.32	0	1	0
row 2	−0.04	0.04	0.12	1	0	0
row 3	−0.24	1.24	−0.28	0	0	1

Forward elimination proceeds as follows:

row 1 times 0.04/0.56   0.04   −0.114285   0.022857 0   0.071428 0   (A)

row 1 times 0.24/0.56   0.24   −0.668571   0.137142 0   0.428571 0   (B)

Adding (A) to row 1, and adding (B) to row 3:

row 1	0.56	−1.56	0.32	0	1	0
row 2	0	−0.071428	0.142857	1	0.071428	0
row 3	0	0.571428	−0.142857	0	0.428571	1

Row 2 and row 3 are now exchanged for pivoting:

row 1	0.56	−1.56	0.32	0	1	0
row 2	0	0.571428	−0.142857	0	0.428571	1
row 3	0	−0.071428	0.142857	1	0.071428	0

Row 2 times 0.071428/0.571428 = 0.125:

	0	0.071428	−0.017857	0	0.053571	0.125(C)

Adding (C) to row 3:

	0	0	0.124993	1	0.124993	0.125

This completes the forward elimination. The array is now:

row 1	0.56	−1.56	0.32	0	1	0
row 2	0	0.571428	−0.142857	0	0.428571	1
row 3	0	0	0.124993	1	0.124993	0.125

To start backward elimination, the last row is divided by 0.124993:

row 1	0.56	−1.56	0.32	0	1	0
row 2	0	0.571428	−0.142857	0	0.428571	1
row 3	0	0	1	8	1	1

Adding 0.142857 times row 3 to row 2, subtracting 0.32 times row 3, and dividing row 2 by 0.571428:

row 1	0.56	−1.56	0	−2.56	0.68	−0.32
row 2	0	1	0	2	1	2
row 3	0	0	1	8	1	1

Adding 1.56 times the second row to row 1 and dividing row 1 by 0.56:

row 1	1	0	0	1	4	5
row 2	0	1	0	2	1	2
row 3	0	0	1	8	1	1

Thus, the inverse is

$$\begin{bmatrix} 1 & 4 & 5 \\ 2 & 1 & 2 \\ 8 & 1 & 1 \end{bmatrix}$$

So far we used the Gauss-Jordan elimination to compute inverse of a matrix, but the same can be done by the Gauss elimination. The reason is as follows. The Gauss-Jordan elimination of the augmented matrix for a matrix of order N can be separated into solutions of N sets of linear equations with the same coefficient matrix A. The first set has the inhomogeneous term equal to the first column of the identity matrix, the second set has the inhomogeneous term equal to the second column of the identity matrix, and so on. The solution of the first set becomes the first column of the inverse matrix, and the solution for the second set the second column, and so on. In actual calculations, each set does not have to be computed separately, but can be computed simultaneously, because the computational procedure for all the sets are identical. The amount of computations with Gauss elimination is smaller than with Gauss-Jordan elimination, because the left half of the augmented matrix need not be reduced to an identity matrix in Gauss elimination. For this reason, PROGRAM 6–2 uses Gauss elimination rather than Gauss-Jordan elimination.

SUMMARY OF THIS SECTION

(a) The inverse of a matrix may be computed by applying Gauss-Jordan elimination to the augmented array consisting of the matrix to be inverted and the identity matrix.

(b) Once the augmented array is formed, pivoting applied thereafter does not affect the result of the Gauss-Jordan elimination.

(c) The inverse of a matrix can be computed by Gauss elimination also.

6.7 *LU* DECOMPOSITION

The LU decomposition scheme is a transformation of a matrix A to a product of two matrices,

$$A = LU$$

where L is a lower triangular matrix and U is a upper triangular matrix. When one has to solve a number of linear equation sets in which the coefficient matrices are all identical but the inhomogeneous (right side) terms are different, solving the equations using the LU decomposition is more efficient than Gauss elimination.

The LU decomposition for a 3×3 matrix is illustrated as

$$\begin{bmatrix} a_{1,1} & a_{1,2} & a_{1,3} \\ a_{2,1} & a_{2,2} & a_{2,3} \\ a_{3,1} & a_{3,2} & a_{3,3} \end{bmatrix} = \begin{bmatrix} 1 & 0 & 0 \\ l_{2,1} & 1 & 0 \\ l_{3,1} & l_{3,2} & 1 \end{bmatrix} \begin{bmatrix} u_{1,1} & u_{1,2} & u_{1,3} \\ 0 & u_{2,2} & u_{2,3} \\ 0 & 0 & u_{3,3} \end{bmatrix} \qquad (6.7.1)$$

Notice that all the diagonal elements of L are unity.

To evaluate $u_{i,j}$ and $l_{i,j}$ in Eq. (6.7.1) without pivoting, we first multiply the first row of L by each column of U and compare the result to the first row of A. It is found that the first row of U is identical to that of A:

$$u_{1,j} = a_{1,j}, \quad j = 1 \text{ to } 3 \tag{6.7.2}$$

Multiplying the second and third rows of L by the first column of U, respectively, and comparing to the left side yield

$$a_{2,1} = l_{2,1}u_{1,1}, \quad a_{3,1} = l_{3,1}u_{1,1}$$

or equivalently

$$l_{2,1} = a_{2,1}/u_{1,1}, \quad l_{3,1} = a_{3,1}/u_{1,1} \tag{6.7.3}$$

Multiplying the second row of L by the second and third columns of U and comparing to the left side yield

$$a_{2,2} = l_{2,1}u_{1,2} + u_{2,2}, \quad a_{2,3} = l_{2,1}u_{1,3} + u_{2,3}$$

or equivalently

$$u_{2,2} = a_{2,2} - l_{2,1}u_{1,2}, \quad u_{2,3} = a_{2,3} - l_{2,1}u_{1,3} \tag{6.7.4}$$

By multiplying the third row of L by the second column of U, we obtain

$$a_{3,2} = l_{3,1}u_{1,2} + l_{3,2}u_{2,2}$$

or equivalently

$$l_{3,2} = [a_{3,2} - l_{3,1}u_{1,2}]/u_{2,2} \tag{6.7.5}$$

Finally $l_{3,3}$ is obtained by multiplying the last column of U by the last row of L and equating to $a_{3,3}$ as

$$l_{3,1}u_{1,3} + l_{3,2}u_{2,3} + u_{3,3} = a_{3,3}$$

or equivalently

$$u_{3,3} = a_{3,3} - l_{3,1}u_{1,3} - l_{3,2}u_{2,3} \tag{6.7.6}$$

Example 6.9

Decompose the following matrix into L and U matrices:

$$A = \begin{bmatrix} 2 & 1 & -3 \\ -1 & 3 & 2 \\ 3 & 1 & -3 \end{bmatrix}$$

⟨**Solution**⟩

Following the procedure of Eqs. (6.7.2) through (6.7.6) we get:

$$u_{1,1} = 2, \quad u_{1,2} = 1, \quad u_{1,3} = -3$$
$$l_{2,1} = -0.5, \quad l_{3,1} = 1.5$$
$$u_{2,2} = 3 - (-0.5)(1) = 3.5$$
$$u_{2,3} = 2 - (-0.5)(-3) = 0.5$$
$$l_{3,2} = [1 - (1.5)(1)]/3.5 = -0.142857$$
$$u_{3,3} = -3 - (1.5)(-3) - (-0.142857)(-0.5) = 1.57142$$

Then,

$$L = \begin{bmatrix} 1 & 0 & 0 \\ -0.5 & 1 & 0 \\ 1.5 & -0.1428 & 1 \end{bmatrix} \quad U = \begin{bmatrix} 2 & 1 & -3 \\ 0 & 3.5 & 0.5 \\ 0 & 0 & 1.5714 \end{bmatrix}$$

The preceding results may be verified by substituting back into Eq. (6.7.1).

A general scheme of the *LU* decomposition for a matrix of order N is as follows:

(a) The first row of U, $u_{1,j}$ for $j = 1$ to N, is obtained by

$$u_{1,j} = a_{1,j}, \quad j = 1 \text{ to } N \tag{6.7.7}$$

(b) The first column of L, $l_{i,1}$ for $i = 2$ to N, is obtained by

$$l_{i,1} = a_{i,1}/u_{1,1}, \quad i = 2 \text{ to } N \tag{6.7.8}$$

(c) The second row U is obtained by

$$u_{2,j} = a_{2,j} - l_{2,1}u_{1,j}, \quad j = 2 \text{ to } N \tag{6.7.9}$$

(d) The second column of L is obtained by

$$l_{i,2} = [a_{i,2} - l_{i,1}u_{1,2}]/u_{2,2}, \quad i = 3 \text{ to } N \tag{6.7.10}$$

(e) The nth row of U is obtained by

$$u_{n,j} = a_{n,j} - \sum_{k=1}^{n-1} l_{n,k}u_{k,j}, \quad j = n \text{ to } N \tag{6.7.11}$$

(f) The nth column of L is obtained by

$$l_{i,n} = [a_{i,n} - \sum_{k=1}^{n-1} l_{i,k}u_{k,n}]/u_{n,n}, \quad i = n+1 \text{ to } N \tag{6.7.12}$$

In the foregoing process, the diagonal elements of L, namely $l_{i,i}$, are not calculated because they are all unity.

As we have noticed, the elements in the upper triangular part of L are all zero. Also, the elements in the lower triangular part of matrix U are all zero. Therefore, the elements of L and U can be stored in one array for the purpose of saving memory space. The L and U matrices in Eq. (6.7.1), for example, may be combined into one array as

$$
\begin{array}{ccc}
u_{1,1} & u_{1,2} & u_{1,3} \\
l_{2,1} & u_{2,2} & u_{2,3} \\
l_{3,1} & l_{3,2} & u_{3,3}
\end{array}
$$

In the array above, the diagonal elements of L are not stored because they are all unity. To reduce the use of memory space further, the results of the factorization can be overwritten on the memory space of A. This is possible because each element $a_{i,j}$ of A is used only once for calculation of $l_{i,j}$ or $u_{i,j}$ in the entire factorization. Therefore, as soon as $a_{i,j}$ is used, its memory space can be used to store $l_{i,j}$ or $u_{i,j}$.

We now study ways to solve a set of linear equations. The equation $Ax = y$ is equivalently written as,

$$LUx = y \tag{6.7.13}$$

where $LU = A$. Equation (6.7.13) is solved as follows. By setting

$$Ux = z \tag{6.7.14}$$

Eq. (6.7.13) becomes

$$Lz = y \tag{6.7.15}$$

Solution of Eq. (6.7.15) for z is easy because of the triangular form of L. Once z is obtained, Eq. (6.7.14) is solved for x.

In the case of a 3×3 matrix for illustration, we can write Eq. (6.7.15) as

$$\begin{bmatrix} 1 & 0 & 0 \\ l_{2,1} & 1 & 0 \\ l_{3,1} & l_{3,2} & 1 \end{bmatrix} \begin{bmatrix} z_1 \\ z_2 \\ z_3 \end{bmatrix} = \begin{bmatrix} y_1 \\ y_2 \\ y_3 \end{bmatrix} \tag{6.7.16}$$

The solution is calculated recursively as

$$\begin{aligned} z_1 &= y_1 \\ z_2 &= [y_2 - z_1 l_{2,1}] \\ z_3 &= [y_3 - z_1 l_{3,1} - z_2 l_{3,2}] \end{aligned} \tag{6.7.17}$$

By writing Eq. (6.7.14) more explicitly as

$$\begin{bmatrix} u_{1,1} & u_{1,2} & u_{1,3} \\ 0 & u_{2,2} & u_{2,3} \\ 0 & 0 & u_{3,3} \end{bmatrix} \begin{bmatrix} x_1 \\ x_2 \\ x_3 \end{bmatrix} = \begin{bmatrix} z_1 \\ z_2 \\ z_3 \end{bmatrix}$$

the solution becomes

$$x_3 = \frac{z_3}{u_{3,3}}$$

$$x_2 = \frac{z_2 - u_{2,3} x_3}{u_{2,2}}$$

$$x_1 = \frac{z_1 - u_{1,2} x_2 - u_{1,3} x_3}{u_{1,1}}$$

For a matrix of order N, forward elimination and backward substitution are summarized as follows.

(a) Forward elimination step:

$$z_1 = y_1$$

$$z_i = y_i - \left[\sum_{j=1}^{i-1} l_{i,j} z_j \right], \quad i = 2, 3, \ldots, N$$

(b) Backward substitution step:

$$x_N = \frac{z_N}{u_{N,N}}$$

$$x_i = \frac{\left[z_i - \sum\limits_{j=i+1}^{N} u_{i,j} x_j \right]}{u_{i,i}}, \quad i = N-1, N-2, \ldots, 3, 2, 1$$

Thus far in this section, we have not used pivoting for the sake of simplicity. However, pivoting is important for the same reason as in the Gauss elimination. We should remember that pivoting in Gauss elimination is equivalent to shuffling equations in the set. In matrix form, it means that rows of coefficients are shuffled together with the terms on the right side. This indicates that pivoting may be applied to the LU decomposition as long as the shuffling is applied to the left and right terms in the same way. When performing pivoting in the LU decomposition, the changes in the order of the rows are recorded. The same reordering is then applied to the right side terms before starting the solution in accordance with the steps (a) and (b) discussed above.

SUMMARY OF THIS SECTION

(a) Any nonsingular matrix can be decomposed to the LU form.

(b) If a set of linear equations has to be repeatedly solved with different inhomogeneous terms, the LU decomposition is recommended.

(c) The U matrix is identical to the coefficient array that appears in Gauss elimination when forward elimination is completed.

(d) LU decomposition is also useful in evaluating the determinant, as discussed in the next section.

6.8 DETERMINANT

We have been exposed to determinants, but an explanation in detail has been deferred until now.

The determinant is an important quantity associated with each square matrix. Indeed, an inhomogeneous set of linear equations cannot be uniquely solved if the determinant of the coefficient matrix is zero. On the other hand, a homogeneous set of linear equations has solutions only when the determinant is zero. There are many occasions when the determinant of a matrix must be evaluated. (See Chapter 7 for numerous examples.)

The determinant of a matrix A of order N is denoted by det (A) and defined by

$$\det (A) = \sum (\pm) a_{i1} a_{j2} a_{k3}, \ldots, a_{rN} \tag{6.8.1}$$

where the summation is extended over all permutations of the first subscripts of a, and ($\pm$) takes plus if the permutation is even and minus if it is odd.*

For a 2×2 matrix, the determinant of A is calculated as

$$\det(A) = \det \begin{bmatrix} a_{11} & a_{12} \\ a_{21} & a_{22} \end{bmatrix} = a_{11}a_{22} - a_{12}a_{21} \qquad (6.8.2)$$

For a 3×3 matrix, the determinant is

$$\det(A) = \det \begin{bmatrix} a_{11} & a_{12} & a_{13} \\ a_{21} & a_{22} & a_{23} \\ a_{31} & a_{32} & a_{33} \end{bmatrix}$$

$$= a_{11}a_{22}a_{33} + a_{21}a_{32}a_{13} + a_{31}a_{12}a_{23}$$

$$- a_{11}a_{32}a_{23} - a_{21}a_{12}a_{33} - a_{31}a_{22}a_{13} \qquad (6.8.3)$$

One may memorize the rule for calculating the determinant of a 3×3 matrix as the *spaghetti rule*. In Figure 6.2, each of three solid lines is connecting three numbers. The products along the solid lines have positive signs in Eq. (6.8.3). The products of three numbers along dotted lines have negative signs in Eq. (6.8.3). This rule cannot be extended to a matrix of 4×4 or greater.

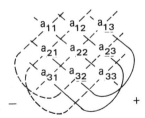

$-$ $+$ **Figure 6.2** Spaghetti rule to calculate determinant of a 3×3 matrix

As the order of a matrix exceeds 3, direct calculation of the determinant by Eq. (6.8.1) becomes impractical because the amount of computations increases very rapidly. Indeed, a matrix of order N has N-factorial permutations, so the determinant of a 5×5 matrix, for example, has 120 terms each of which needs four multiplications. The determinant of 10×10 matrix has more than 2×10^8 terms each requiring nine multiplications.

* The sequence of the first subscript is $(i, j, k, \ldots, r)$ and is called *permutation*. A permutation is odd or even if $(i, j, k, \ldots, r)$ is obtained by changing the order of any two consecutive numbers in $(1, 2, 3, \ldots, N)$ an odd or even number of times, respectively. For example, $(3, 2, 1, 4, \ldots, N)$ is obtained through exchanges of the first three numbers as $123 \rightarrow 213 \rightarrow 231 \rightarrow 321$ (i.e., three times). So, the permutation of $(3, 2, 1, 4, \ldots, N)$ is odd. It turns out, however, that the exchanges of two numbers do not have to be between two consecutive numbers but can be between any pair of numbers. In the present example, $(3, 2, 1, 4, \ldots, N)$ is obtained by exchanging 1 and 3 in $(1, 2, 3, \ldots, N)$. The number of exchange is one so the permutation $(3, 2, 1, 4, \ldots, N)$ is odd.

A practical way of calculating the determinant is to use the forward elimination process of Gauss elimination or, alternatively, the LU decomposition described in Section 6.7. We first look at two important rules of determinants:

Rule 1: $\det (BC) = \det (B) \det (C)$

which means that the determinant of a product of matrices is the product of the determinants of all the matrices.

Rule 2: $\det (M) =$ the product of all diagonal elements of M if M is an upper
or lower triangular matrix.

For example, if all the diagonal elements of a triangular matrix are unity, the determinant is unity.

If no pivoting is used, calculation of the determinant using the LU decomposition is straightforward. According to Rule 1, the determinant can be written as

$$\det (A) = \det (LU) = \det (L) \det (U) = \det (U) \tag{6.8.4}$$

where $\det (L) = 1$ because L is a lower triangular matrix and all its diagonal elements are unity. The $\det (U)$ is the product of all the diagonal elements of U, which equals the $\det (A)$:

$$\det (A) = \prod_{i=1}^{N} u_{ii} \tag{6.8.5}$$

When pivoting is used in the LU decomposition, its effect should be taken into consideration. First, we recognize that the LU decomposition with pivoting is equivalent to performing two separate processes: (1) transform A to A' by performing all shuffling of rows, and (2) then decompose A' to LU with no pivoting. The former step can be expressed by

$$A' = PA, \text{ or equivalently } A = P^{-1}A' \tag{6.8.6}$$

where P is called a *permutation matrix* and represents the pivoting operation. The second process is

$$A' = LU \tag{6.8.7}$$

Therefore, L and U are related to A by

$$A = P^{-1}LU \tag{6.8.8}$$

The determinant of A may now be written as

$$\det (A) = \det (P^{-1}) \det (L) \det (U)$$

or equivalently

$$\det (A) = \gamma \det (U) \tag{6.8.9}$$

where $\det (L) = 1$ is used and $\gamma = \det (P^{-1})$ equals -1 or $+1$ depending on whether the number of pivotings is odd or even, respectively.

The present algorithm of computing determinant is incorporated in PROGRAM 6–3.

Example 6.10

Compute the determinant of the matrix in Example 6.9.

⟨**Solution**⟩

Using the result of the *LU* decomposition in Example 6.9,

$$\det (L) = 1 \quad \text{and} \quad \det (U) = (2)(3.5)(1.5714) = 11$$

Since no pivoting is used, $\gamma = 1$. Thus, the determinant is

$$\det (A) = \gamma \det (U) = 11$$

The determinant of a matrix may also be calculated during the process of Gauss elimination. This is because when the forward elimination is completed, the original matrix has been transformed to the U matrix of the LU decomposition. Therefore, the determinant can be calculated by taking the product of all the terms along the diagonal line and then multiplying by 1 or -1 depending on whether the number of pivoting operations performed is even or odd respectively. This is the algorithm implemented in PROGRAM 6–1 in computing the determinant.

Example 6.11

(a) Calculate the determinant of the coefficient matrix in Eq. (6.3.1).
(b) Calculate the determinant of the matrix in Example 6.7.

⟨**Solution**⟩

(a) The coefficient matrix for Eq. (6.3.1) is

$$A = \begin{bmatrix} 0 & 10 & 1 \\ 1 & 3 & -1 \\ 2 & 4 & 1 \end{bmatrix}$$

Referring to Eq. (6.3.5), the matrix after completion of forward elimination is

$$\begin{bmatrix} 2 & 4 & 1 \\ 0 & 10 & 1 \\ 0 & 0 & -16/10 \end{bmatrix}$$

where two times of pivoting are applied.

Therefore,

$$\det (A) = (-1)^2(2)(10)(-16/10) = -32$$

(b) The matrix [see Eq. (A) in Example 6.7] is defined by

$$A = \begin{bmatrix} 2 & 1 & -3 \\ -1 & 3 & 2 \\ 3 & 1 & -3 \end{bmatrix}$$

After the forward elimination, the upper triangular matrix [see Eq. (B) in Example 6.7] is

$$\begin{bmatrix} 2 & 1 & -3 \\ 0 & 3.5 & 0.5 \\ 0 & 0 & 1.5714 \end{bmatrix}$$

No pivoting is used. Therefore, we get

$$\det (A) = (2)(3.5)(1.5714) = 11$$

SUMMARY OF THIS SECTION

(a) The determinant may easily be calculated for 2×2 and 3×3 matrices by hand calculations.

(b) For larger matrices, Gauss elimination or LU decomposition is used.

(c) The two rules discussed in this section are often very useful in evaluating the determinant.

6.9 ILL-CONDITIONED PROBLEMS

Unsolvable problems are relatively easy to identify. And if such a problem is run, the computer program will stop anyway.

However, a number of problems are solvable, yet their solutions become very inaccurate because of severe round-off errors. Problems of this type are named *ill-conditioned problems.*

Small round-off errors or changes in coefficients can cause significant errors in the solution for an ill-conditioned problem. Although the effect of round-off errors increases as the size of the equations becomes larger, it can still be illustrated with two equations for simplicity of explanations. Consider a set of equations, for illustration,

$$0.12065x + 0.98775y = 2.01045 \quad \text{(line A)}$$
$$0.12032x + 0.98755y = 2.00555 \quad \text{(line B)} \tag{6.9.1}$$

where the two equations are very close to each other. The solution will be denoted by (x_1, y_1) and is given by

$$x_1 = 14.7403$$
$$y_1 = 0.23942$$

To simulate the effect of an error in the coefficients, we artificially increase the inhomogeneous term of the first equation (line A) by 0.001, so Eq. (6.9.1) is now altered to

$$0.12065x + 0.98775y = 2.01145 \quad \text{(line A')}$$
$$0.12032x + 0.98755y = 2.00555 \quad \text{(line B)} \tag{6.9.2}$$

The solution of Eq. (6.9.2) denoted by (x_2, y_2) becomes

$$x_2 = 17.9756$$
$$y_2 = -0.15928$$

which is utterly different from (x_1, y_1).

The amounts of differences between (x_1, y_1) and (x_2, y_2) are significant and astonishing particularly when compared to the amount of the change made on the inhomogeneous term of first equation. Small changes in other coefficients can cause similar effects. Errors in the coefficients can occur by round-off in the process of solving the equations.

A graphical analysis may help understand why such a large change in the solution results. In Figure 6.3, the two equations in Eq. (6.9.1) and the first equation in Eq. (6.9.2) are shown as A, B, and A', respectively. Line A intersects with B with a very acute angle (a characteristic symptom of the ill-conditioned problem). Line A' is parallel to A but slightly higher than A. Since gradients of Line A and B are very close, any small change in the gradient or height of one of the lines causes a serious shift in the location of the intersection of the two lines.

Although a set of only two equations is used in the preceding illustration, similar effects occur with a larger set of ill-conditioned equations. The reader might graphically explain how the solution of an ill-conditioned set of three linear equations can be affected by errors in the coefficients. The effect of round-off errors becomes more pronounced as the number of equations increases.

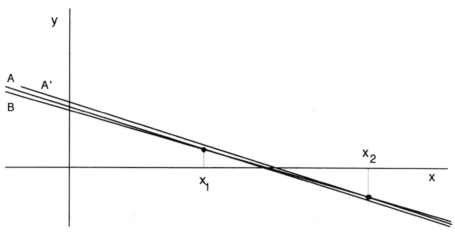

Figure 6.3 Qualitative plotting of Lines A, A' and B

The matrix A of an ill-conditioned problem has the following symptoms:

(a) A small change of coefficients (or matrix elements) causes significant changes in the solution.

(b) Diagonal elements of the coefficient matrix tend to be smaller than off-diagonal elements.

(c) Computed $\det (A) \det (A^{-1})$ significantly deviates from 1.

(d) Computed $(A^{-1})^{-1}$ becomes significantly different from A.

(e) Computed AA^{-1} deviates significantly from the identity matrix.

(f) Computed $A^{-1}(A^{-1})^{-1}$ deviates from the identity matrix more significantly than does AA^{-1}.

Pivoting discussed earlier enhances accuracy of the solution if the problem is mildly ill-conditioned, but for severely ill-conditioned problems, pivoting alone cannot save the accuracy. The best remedy is to increase precision of computation (see Example 6.4 and Chapter 1, where precision is discussed).

Example 6.12

A Hilbert matrix [Morris] is defined by

$$A = [a_{i,j}]$$

where

$$a_{i,j} = \frac{1}{i + j - 1}$$

which is known to be ill-conditioned even for a small order. Compute (a) $A^{-1}(A^{-1})^{-1}$, and (b) det (A) det (A^{-1}) for the 4×4 Hilbert matrix. Use single precision.

⟨Solution⟩

The 4×4 Hilbert matrix is

$$
\begin{bmatrix}
1 & 1/2 & 1/3 & 1/4 \\
1/2 & 1/3 & 1/4 & 1/5 \\
1/3 & 1/4 & 1/5 & 1/6 \\
1/4 & 1/5 & 1/6 & 1/7
\end{bmatrix}
$$

The following results are obtained by single precision on the VAX 8550:

(a) $A^{-1}(A^{-1})^{-1} =$

$$
\begin{bmatrix}
1.0001183 & -0.0014343 & 0.0032959 & -0.0021362 \\
-0.0000019 & 1.0000000 & -0.0001221 & 0.0000610 \\
0.0000000 & 0.0000000 & 0.9999390 & 0.0000305 \\
0.0000000 & -0.0000305 & 0.0000610 & 0.9999390
\end{bmatrix}
$$

(b) det (A) det $(A^{-1}) = (1.6534499E - 07)(6047924.) = 0.99999393$

The product of the determinants deviates from unity as the order of the matrix increases. However, deviation of $A^{-1}(A^{-1})^{-1}$ from the identity matrix detects ill-conditioned matrices more clearly than does the product of the determinants.

SUMMARY OF THIS SECTION

(a) Whether the coefficient matrix of a set of linear equations is ill-conditioned or not cannot be easily found by examining the solution of linear equations.

(b) Methods for examining ill-conditioned matrices include computing $A^{-1}(A^{-1})^{-1}$ and det (A) det (A^{-1}).

6.10 SOLUTION OF M EQUATIONS WITH N UNKNOWNS

In the previous sections we studied how to compute the unique solution of $Ax = y$ by using Gauss elimination or other methods. The necessary condition for existence of a unique solution is that A is a square matrix and det $(A) \neq 0$. If det $(A) = 0$, we called the matrix singular and gave up on finding the solution. However, this is not because there is no solution but because there is no unique solution [Strang]. If det $(A) = 0$, at least one equation is linearly dependent and can be eliminated. After the elimination, the number of equations becomes less than the number of unknowns.

In general, the number of equations, m, can be less than the number of unknowns, n. The equation for such a problem can be written in the form of $Ax = y$ where the matrix A is not square but rectangular. For $m < n$ the number of solutions is infinite, but numerical values of a solution cannot be determined uniquely. In this section we pursue nonunique solutions of m equations with n unknowns, where $m < n$.

As an example of linear equations of $m < n$, we consider

$$x + y = 1 \tag{6.10.1}$$

where $m = 1$ and $n = 2$. The solution may be written as

$$x = 1 - y$$

or

$$y = 1 - x$$

In the first equation, y on the right side is a *free variable*, and x on the left side is a *basic variable*. In the second equation, x is a free variable and y is a basic variable. Whichever is the form of the solution chosen, the solution for the basic variable is given in terms of the free variable. In case the number of equations is insufficient, (a) the solution is given in an equation form, rather than numerically, and (b) the number of solutions is infinite because the free parameters can take any value.

If we have m linearly independent equations for n unknowns and $m < n$, we can find m basic variables and $n - m$ free variables. By placing the basic variables on the left side of equations and all the free variables on the right side, the set of m equations may be solved for the basic variables in terms of free variables.

The only requirement in choosing basic variables is that the m equations for the basic variables are not singular. As a result, the selection of basic variables is not always unique. Which variables can become basic variables is not apparent at a glance. Nevertheless, they can be found systematically by using Gauss (or Gauss-Jordan) elimination devised for $m \times n$ matrices as explained next.

Although we assume in the preceding paragraph that m given equations are linearly independent, we waive this restriction now. This is because, through Gauss elimination, nonindependent equations will be automatically eliminated from the set so that the remaining equations become linearly independent.

Let us consider

$$
\begin{aligned}
-1u + 2v + 2w + x - 2y &= 2 \\
3u - 6v - w + 5x - 4y &= 1 \\
2u - 4v - 1.5w + 2x - y &= -0.5
\end{aligned}
\tag{6.10.2}
$$

where $m = 3$ and $n = 5$. For simplicity, the equations above are rewritten in an augmented array form,

u	v	w	x	y	RHS
−1	2	2	1	−2	2
3	−6	−1	5	−4	1
2	−4	−1.5	2	−1	−0.5

The first task on this set is pivoting because the coefficient of u in the second row is greater than that in the first row:

u	v	w	x	y	RHS
3	−6	−1	5	−4	1
−1	2	2	1	−2	2
2	−4	−1.5	2	−1	−0.5

Now, the coefficient of u in the second and third rows is eliminated by subtracting $-1/3$ and $2/3$ times the first row, respectively:

u	v	w	x	y	RHS
3	−6	−1	5	−4	1
0	0	1.666667	2.666667	−3.333333	2.333333
0	0	−0.833333	−1.333333	1.666667	−1.166667

In the foregoing array, the coefficients of v in the second and third rows vanished automatically. If we are working for Gauss elimination, as described in Section 6.2 or 6.3, we would give up here. However, we proceed to the third row.

Considering the coefficient of w in the second row as a pivot, we eliminate the coefficient of w in the third row by subtracting $-0.833333/1.666667$ times the second row:

u	v	w	x	y	RHS
3	−6	−1	5	−4	1
0	0	1.666667	2.666667	−3.333333	2.333333
0	0	0	0	0	0

In this array, the last row has entirely vanished, which indicates that the third equation was not independent. Now, we consider that the number of equations is $m = 2$ and $n = 5$.

The equations represented by the foregoing array are explicitly written as

$$3u - 6v - 1w + 5x - 4y = 1$$
$$1.666667w + 2.666667x - 3.333333y = 2.333333 \tag{6.10.3}$$

The coefficients of the leading terms after the forward elimination, such as 3 in the first row and 1.666667 in the second row, are pivots. The corresponding variables, u and w, are called *basic variables*. Other variables are *free variables*. By moving all free variables to the right side, we write

$$
\begin{aligned}
3u - \quad w &= \quad 1 + 6v - \quad 5x + \quad 4y \\
1.666667w &= 2.333333 \quad - 2.666667x + 3.333333y
\end{aligned}
\tag{6.10.4}
$$

Obviously the coefficient matrix

$$
\begin{bmatrix} 3 & -1 \\ 0 & 1.666667 \end{bmatrix}
$$

is nonsingular, so the solution for the basic variables may be obtained in terms of free variables. By applying the backward substitution, the final solution is

$$
\begin{aligned}
u &= 0.8 + 2v - 2.2x + 2y \\
w &= 1.4 \quad\quad - 1.6x + 2y
\end{aligned}
\tag{6.10.5}
$$

In developing a program to solve $m \times n$ equations, Gauss-Jordan elimination tends to be more advantageous than Gauss elimination. To illustrate the application of the Gauss-Jordan elimination, we consider another problem:

u	v	w	x	y	RHS
2	3	1	4	1	6
2	3	1	1	−1	1
4	6	−1	1	2	5

After pivoting, the array becomes

$$
\begin{array}{cccccc}
4 & 6 & -1 & 1 & 2 & 5 \\
2 & 3 & 1 & 1 & -1 & 1 \\
2 & 3 & 1 & 4 & 1 & 6
\end{array}
$$

The first row is normalized by dividing by 4, and then the first coefficients in other rows are eliminated:

$$
\begin{array}{cccccc}
1 & 1.5 & -0.25 & 0.25 & 0.5 & 1.25 \\
0 & 0 & 1.5 & 0.5 & -2 & -1.5 \\
0 & 0 & 1.5 & 3.5 & 0 & 3.5
\end{array}
$$

where the second coefficients in the second and third rows have been automatically eliminated. The second row is normalized by dividing by 1.5. Then the third coefficients in the first and third rows are eliminated by subtracting a multiple of the second row from the first and third rows, respectively:

$$
\begin{array}{cccccc}
1 & 1.5 & 0 & 0.333333 & 0.166667 & 1 \\
0 & 0 & 1 & 0.333333 & -1.333333 & -1 \\
0 & 0 & 0 & 3 & 2 & 5
\end{array}
$$

The third row is now normalized by dividing by 3. Then the fourth coefficients of the first and second rows are eliminated by subtracting a multiple of the third row from the first and second rows, respectively:

$$
\begin{array}{cccccc}
1 & 1.5 & 0 & 0 & -0.055556 & 0.444444 \\
0 & 0 & 1 & 0 & -1.555556 & -1.555556 \\
0 & 0 & 0 & 1 & 0.666667 & 1.666667
\end{array}
$$

In the foregoing array, the coefficient of each basic variable is unity, and it is the only nonzero coefficient in each column. The equations obtained are written explicitly as

$$
\begin{aligned}
1u + 1.5v + 0w + 0x - 0.055556y &= 0.444444 \\
1w + 0x - 1.555556y &= -1.555556 \\
1x + 0.666667y &= 1.666667
\end{aligned}
\tag{6.10.6}
$$

or, after moving the free variables to the right side,

$$
\begin{aligned}
u &= 0.444444 - 1.5v + 0.055556y \\
w &= -1.555556 + 1.555556y \\
x &= 1.666667 - 0.666667y
\end{aligned}
\tag{6.10.7}
$$

In a more general way, we consider n equations with m unknowns:

$$
\begin{aligned}
a_{1,1}x_1 + a_{1,2}x_2 + \cdots a_{1,n}x_n &= y_1 \\
a_{2,1}x_1 + a_{2,2}x_2 + \cdots a_{2,n}x_n &= y_2 \\
&\ \ \vdots \\
a_{m,1}x_1 + a_{m,2}x_2 + \cdots a_{m,n}x_n &= y_m
\end{aligned}
\tag{6.10.8}
$$

where we assume $m \leqslant n$ including the case of $m = n$. The equations are expressed in the augumented array form:

$$
\begin{array}{cccc}
a_{1,1} & a_{1,2} & \cdots & a_{1,n} \vdots y_1 \\
a_{2,1} & a_{2,2} & \cdots & a_{2,n} \vdots y_2 \\
& & \cdots & \\
a_{m,1} & a_{m,2} & \cdots & a_{m,n} \vdots y_m
\end{array}
\tag{6.10.9}
$$

Application of Gauss-Jordan elimination to the above array will lead to a form such as, for example,

$$
\begin{array}{cccccccc}
1 & x & 0 & 0 & x & x & : & x' \\
0 & 0 & 1 & 0 & x & x & : & x' \\
0 & 0 & 0 & 1 & x & x & : & x' \\
0 & 0 & 0 & 0 & 0 & 0 & : & 0 \\
0 & 0 & 0 & 0 & 0 & 0 & : & 0
\end{array}
\qquad (6.10.10)
$$

where x denotes nonzero values and the pivots are all unity and the "x" symbols correspond to the coefficients of the free variables. In the process of Gauss-Jordan elimination, rows are exchanged (pivoting) whenever necessary. The null rows represent linearly dependent equations that have been eliminated. Rewriting this array to the equation form is easy as illustrated previously. The solution scheme is implemented in PROGRAM 6–4.

The solution algorithm just explained is a universal one because, in addition to computing the solution of an $m \times n$ equation:

(a) It can be applied to solve the unique solution of an $n \times n$ equation.

(b) It finds the solution even if the coefficient matrix is singular.

(c) It can be used to compute the inverse of a square matrix. To find the inverse of an $n \times n$ matrix, consider an augmented array of $n \times (2n + 1)$, in which the first n columns are the square matrix, the next n columns are an identity matrix, and the last column corresponding to the inhomogeneous terms are set to zero.

SUMMARY OF THIS SECTION

(a) When the rank of linear equations is less than the number of unknowns, the solutions for basic variables are given in terms of free variables.

(b) The solution algorithms for $n \times m$ equations are universal and applicable to solving equations as well as finding the inverse of a matrix.

PROGRAMS

PROGRAM 6–1 Gauss Elimination

(A) Explanations

A set of linear equations may be written as

$$ Ax = y \qquad (A) $$

where A is a square matrix, x is a vector representing unknowns, and y is a vector representing inhomogeneous terms. PROGRAM 6–1 solves Eq. (A) by Gauss elimination with pivoting (see Section 6.3). Double precision is used throughout the program. Gauss elimination is performed in gauss(). Subroutine gauss() is used also in PROGRAMS 5–1 and 8–1. In the arguments, n is an integer while a is a pointer (an array variable written in the argument is a pointer).

The function gauss() also calculates the determinant of the coefficient matrix (see Example 6.11 for the method of calculating the determinant). The machine epsilon (see Section 1.3.4), calculated in gauss(), is used to clean up very small numbers that result from subtractions. This operation becomes important if the coefficient matrix is singular. For a singular problem, the determinant of a singular matrix should become zero, but in the real computations the determinant of a singular matrix may not become zero because of round-off errors. The machine epsilon is used to minimize this possibility.

The user should define the number of equations, n, the coefficients, and the inhomogeneous terms in an augmented form in the declaration statements early in main(). No input is given interactively.

(B) List

```
/* CSL/c6-1.c    Gauss Elimination */
#include <stdio.h>
#include <stdlib.h>
#include <math.h>
#define TRUE 1
/*              a[i][j] : matrix element, a(i,j)
                n : order of matrix
                eps : machine epsilon
                det : determinant                                    */
main()
{
int i, j, _i, _r;
static n = 3;
static float a_init[10][11] = {{ 0,-1,  2,   0},
                               {-2,  2,-1,   0},
                               {-2,  4,  3,   1}};
double a[10][11];
void gauss();
static int _aini = 1;
   printf( "\nCSL/C6-1    Gauss Elimination \n\n" );
   printf( "Augmented matrix\n" );
   for( i = 1; i <= n; i++ ){
      for( j = 1; j <= n+1; j++ ) {
         a[i][j]=a_init[i-1][j-1];  printf(  "  %12.5e", a[i][j] );
      }
      printf(  "\n" );
   }
   gauss( n, a );
   printf( " Solution\n" );
   printf( "-------------------------------------------\n" );
   printf( "        i              x(i)\n" );
   printf( "-------------------------------------------\n" );
   for( i = 1; i <= n; i++ ) printf( "  %5d  %16.6e\n", i, a[i][n+1] );
   printf( "-------------------------------------------\n\n" );
   exit(0);
}
```

```
void gauss(n, a)
int n;   double a[][11];
{
int i, j, jc, jr, k, kc, nv, pv;
double det, eps, ep1, eps2, r, temp, tm, va;
    eps = 1.0; ep1 = 1.0 ;              /* eps = Machine epsilon */
    while( ep1  > 0 ){
        eps = eps/2.0; ep1 = eps*0.98 + 1; ep1 = ep1 - 1;
    }
    eps = eps*2;        eps2 = eps*2;
    printf( "                  Machine epsilon=%g \n", eps );
    det = 1;                           /* Initialization of determinant */
    for( i = 1; i <= (n - 1); i++ ){
        pv = i;
        for( j = i + 1; j <= n; j++ ){
            if( fabs( a[pv][i] ) < fabs( a[j][i] ) )    pv = j;
        }
        if( pv != i ){
            for( jc = 1; jc <= (n + 1); jc++ ){
                tm = a[i][jc];  a[i][jc] = a[pv][jc];     a[pv][jc] = tm;
            }
            det = -det;
        }
        if( a[i][i] == 0 ){         /* Singular matrix */
            printf( "Matrix is singular.\n" );    exit(0);
        }
        for(jr = i + 1; jr <= n; jr++){ /* Elimination of below-diagonal.*/
            if( a[jr][i] != 0 ){
                r = a[jr][i]/a[i][i];
                for( kc = i + 1; kc <= (n + 1); kc++ ){
                    temp = a[jr][kc];
                    a[jr][kc] = a[jr][kc] - r*a[i][kc];
                    if( fabs( a[jr][kc] ) < eps2*temp ) a[jr][kc] = 0.0;
/*                              If the result of subtraction is smaller than
 *                              2 times machine epsilon times the original
 *                              value, it is set to zero. */

                }
            }
        }
    }
    for( i = 1; i <= n; i++ ) {
        det = det*a[i][i];              /*  Determinant is calculated. */
    }
    if( det == 0 ){
        printf( "Matrix is singular.\n" );    exit(0);
    }
    else{                              /* Backward substitution starts. */
        a[n][n+1] = a[n][n+1]/a[n][n];
        for( nv = n - 1; nv >= 1; nv-- ){
            va = a[nv][n+1];
            for( k = nv + 1; k <= n; k++ ) {va = va - a[nv][k]*a[k][n+1];}
            a[nv][n+1] = va/a[nv][nv];
        }
        printf( "                  Determinant = %g \n", det );
        return;
    }
}
```

(C) Sample Output

```
CSL/C6-1     Gauss Elimination

Augmented matrix
  0.00000e+00   -1.00000e+00    2.00000e+00    0.00000e+00
 -2.00000e+00    2.00000e+00   -1.00000e+00    0.00000e+00
 -2.00000e+00    4.00000e+00    3.00000e+00    1.00000e+00
                 Machine epsilon=2.775558e-17
                 Determinant = -16
  Solution
------------------------------------------
      i            x(i)
------------------------------------------
      1          2.187500e+00
      2          1.750000e+00
      3          1.250000e-01
------------------------------------------
```

(D) Discussions

The solution for the equation

$$\begin{pmatrix} 0 & -1 & 2 \\ -2 & 2 & -1 \\ -2 & 4 & 3 \end{pmatrix}\begin{pmatrix} x_1 \\ x_2 \\ x_3 \end{pmatrix} = \begin{pmatrix} 0 \\ 0 \\ 1 \end{pmatrix}$$

is found in the foregoing output. The value of the determinant indicates that the equation is well behaved. (If the determinant is extremely small, it indicates that the matrix is near singular, so the solution may not be reliable.) The machine epsilon is 2.7e-17 because the program uses double precision and is executed on VAX (see Section 1.3.4).

PROGRAM 6–2 Matrix Inversion

(A) Explanations

This program finds the inverse of a matrix A^{-1} for a nonsingular square matrix A. The inverse of a matrix can be obtained as follows: Write A and I (identity matrix) in an augmented form as $[A, I]$. Then the inverse is found where originally the identity matrix was written. Pivoting does not affect the inverse matrix computed.

     In the present program, Gauss elimination is used rather than the Gauss-Jordan elimination. With Gauss elimination, each column of the identity matrix originally in the augmented matrix is viewed as a set of inhomogeneous terms. Gauss elimination for each column is done not separately but simultaneously. When the Gauss elimination is completed, the columns for the original identity matrix are filled with the solution of the Gauss elimination, which exactly comprises the inverse matrix.

Gauss elimination in this program is performed in double precision by gauss() that is essentially same as gauss() in PROGRAM 6–1.

The matrix for inversion and an identity matrix in an augmented form should be written in the declaration statement for a_init[ ][ ], which are later copied to a [ ][ ].

(B) List

```
/* CSL/c6-2.c      Matrix Inversion   */
#include <stdio.h>
#include <stdlib.h>
#include <math.h>
/*              a[i][j] : matrix element
                n : order of matrix
                eps : machine epsilon                        */
main()
{
int i, j, _i, _r;
static n = 3;
static float a_init[11][21]= {{ 2,  1,-3,   1,  0,  0},
                              {-1,  3, 2,   0,  1,  0},
                              { 3,  1,-3,   0,  0,  1}};

double a[11][21];
void gauss();
static int _aini = 1;
        {     /* Initialization of matrix elements*/
              for( j = 1; j <= 2*n; j++ )
              {    for ( i = 1 ; i <= n; i++ )   a[i][j] = a_init[i-1][j-1];
              }

        }
    printf( "\nCSL/C6-2      Matrix Inversion \n\n" );
    printf( "Original Matrix\n" );
    for( i = 1; i <= n; i++ )
    {    for( j = 1; j <= 3; j++ )   printf(  " %12.5e ", a[i][j] );
         printf( "\n" );
    }
    gauss( n, a );
    printf( "Inverse Matrix\n" );
    for( i = 1; i <= n; i++ )
    {    printf( " " );
         for( j = n + 1; j <= (n*2); j++ )   printf( " %12.5e ", a[i][j] );
         printf( "\n" );
    }
    printf( "\n\n" );   exit(0);
}

void gauss(n, a)
int n;   double a[][21];
{
int i, j, jc, jr, k, kc, m, nv, pv;
double det, eps, eps1, eps2, r, temp, tm, va;
    eps = 1.0; eps1 = 1.0;
    while( eps1 > 0 ) {
        eps = eps/2.0; eps1 = eps*0.98 + 1.0; eps1=eps1 - 1;
    }
```

```
eps = eps*2;
printf( "            Machine epsilon = %11.5e \n", eps );
eps2 = eps*2;
det = 1.0;              /*  Initialization of determinant */
for( i = 1; i <= (n - 1); i++ ){
    pv = i;
    for( j = i + 1; j <= n; j++ ){
        if( fabs( a[pv][i] ) < fabs( a[j][i] ) )  pv = j;
    }
    if( pv != i ){
        for( jc = 1; jc <= (n*2); jc++ ){
            tm = a[i][jc]; a[i][jc] = a[pv][jc]; a[pv][jc] = tm;
        }
        det = -det;
    }
    if( det == 0 ){
        printf( "Matrix is singular.\n" ); exit(0);
    }
    for( jr = i + 1; jr <= n; jr++ ){
        if( a[jr][i] != 0 ){
            r = a[jr][i]/a[i][i];
            for( kc = i + 1; kc <= (n*2); kc++ ){
                temp = a[jr][kc];
                a[jr][kc] = a[jr][kc] - r*a[i][kc];
                if( fabs( a[jr][kc] ) < eps2*temp )  a[jr][kc] = 0.0;
            }
        }
    }
}
for( i = 1; i <= n; i++ )   det = det*a[i][i];
printf( "         Determinant=%11.5e \n", det );
if( a[n][n] != 0 ){
    for( m = n + 1; m <= (n*2); m++ ){
        a[n][m] = a[n][m]/a[n][n];
        for( nv = n - 1; nv >= 1; nv-- ){
            va = a[nv][m];
            for(k=nv+1; k<=n; k++) va=va - a[nv][k]*a[k][m];
            a[nv][m] = va/a[nv][nv];
        }
    }
}   return;
}
}
```

(C) Sample Output

```
CSL/C6-2        Matrix Inversion

Original Matrix
  2.00000e+00    1.00000e+00   -3.00000e+00
 -1.00000e+00    3.00000e+00    2.00000e+00
  3.00000e+00    1.00000e+00   -3.00000e+00
            Machine epsilon = 2.77556e-17
            Determinant=1.10000e+01
Inverse Matrix
 -1.00000e+00    0.00000e+00    1.00000e+00
  2.72727e-01    2.72727e-01   -9.09091e-02
 -9.09091e-01    9.09091e-02    6.36364e-01
```

(D) Discussions

Both the original matrix and its inverse are printed out. The students who are learning inversion of a matrix are encouraged to compute AA^{-1} using the computed results to make sure the results are correct and accurate. This check can be done easily by adding a routine to PROGRAM 6–2 to compute the product of A and A^{-1}.

Quiz: The inverse matrix A^{-1} is supposed to satisfy $AA^{-1} = A^{-1}A = I$ by definition. In reality, however, a devastating situation can occur; that is, $AA^{-1} = A^{-1}A = I$ is not always satisfied if the computed inverse is used. Why does this happen and what should you do if it does?

PROGRAM 6–3 *LU* Decomposition

(A) Explanations

This program decomposes a matrix A into the LU form with pivoting. Denoting the pivoting matrix by P, the LU decomposition can be written as

$$PA = LU$$

where L is a lower triangular matrix, U is an upper triangular matrix, and P is a matrix that represents the pivoting operations performed during the decomposition. The 3×3 matrix A, for example, is written as

$$A = \begin{pmatrix} a_{1,1} & a_{1,2} & a_{1,3} \\ a_{2,1} & a_{2,2} & a_{2,3} \\ a_{3,1} & a_{3,2} & a_{3,3} \end{pmatrix}$$

Then, L and U will be in the form

$$L = \begin{pmatrix} 1 & 0 & 0 \\ L_{2,1} & 1 & 0 \\ L_{3,1} & L_{3,2} & 1 \end{pmatrix}, \quad U = \begin{pmatrix} U_{1,1} & U_{1,2} & U_{1,3} \\ 0 & U_{2,2} & U_{2,3} \\ 0 & 0 & U_{3,3} \end{pmatrix}$$

The matrices L and U are printed out in the combined form; that is

$$\begin{matrix} U_{1,1} & U_{1,2} & U_{1,3} \\ L_{2,1} & U_{2,2} & U_{2,3} \\ L_{3,1} & L_{3,2} & U_{3,3} \end{matrix}$$

The compact form is saved in the array el[ ][ ]. This program also calculates the determinant of the original matrix.

To reduce the memory space requirement, one of the array variables, a or el, may be easily eliminated from the program. For example, one can delete the declaration statement for el[][], and replace all el[][] by a[][] throughout the program.

(B) List

```
/* CSL/c6-3.c    LU Decomposition with Pivoting */
#include <stdio.h>
#include <stdlib.h>
#include <math.h>
/*              a[i][j]: matrix element a(i,j)
                el[i][j]: triangular matrices, L and U, in compact form
                ip[i] : i-th element of permutation
                ipc : counting of the pivoting operations
                de : determinant                                          */
main()
{
int i, ip[20], ipc, j, k, m, np, _i, _r;
float de, el[20][20], l, s;
static float a[20][20];
void pivot();
static n = 3;
static float _itmp0[20][20] = {{ 2,  1,-3},
                               {-1,  3,  2},
                               { 3,  1,-3}};
    for(i = 1; i<=3; i++){
        for( j = 1, _r = 0; j <= 3; j++ ) a[i][j] = _itmp0[i-1][j-1];
    }
    printf( "\nCSL/C6-3    LU Decomposition\n" );
    printf( "Original matrix\n" );
    for( i = 1; i <= n; i++ ){
        for( j = 1; j <= n; j++ ) printf(" %12.5e  ",a[i][j]);
        printf( "\n" );
    }
    printf( "If pivoting is desired, type 1, or 0 otherwise. \n" );
    scanf( "%d", &np );
    ipc = 1;      /*--Initialization of pivoting and matrix el */
    for( i = 1; i <= n; i++ ){
        ip[i] = i;
        for( j = 1; j <= n; j++ ) el[i][j] = 0;
    }
    j = 1;
    if( np == 1 )  pivot( n, a, el, j, ip, &ipc );
    for( j = 1; j <= n; j++ ) {el[1][j] = a[1][j];}          /*First row */
    for( i = 2; i<=n; i++) {el[i][1] = a[i][1]/el[1][1];} /*First column */
    for( m = 2; m <= n; m++ ){
        if( np == 1 )  pivot( n, a, el, m, ip, &ipc );
        for( j = m; j <= n; j++ ){    /*         M-th row */
            s = 0;
            for( k = 1; k <= (m - 1); k++ ) s = s + el[m][k]*el[k][j];
            el[m][j] = a[m][j] - s;
        }
        for( i = m + 1; i <= n; i++ ){    /*  M-th column */
            s = 0;
            for( k = 1; k <= m - 1; k++) s = s + el[i][k]*el[k][m];
            el[i][m] = (a[i][m] - s)/el[m][m];
        }
```

```
      }
      printf( "Permutation\n" );
      for( i = 1; i <= n; i++ )    printf( "%d", ip[i] );
      printf( "\n" );
      printf( "LU Matrices in compact form\n" );
      for( i = 1; i <= n; i++ ){
          printf( " " );
          for( m = 1; m <= n; m++ ) printf( "%12.5e", el[i][m] );
          printf( "\n" );
      }
      de = 1;              /* Initializes determinant */
      for( i = 1; i <= n; i++ )   de = de*el[i][i];
      if ( ipc == (long)( ipc/2 )*2 )    de = -de;
      printf( "Determinant =   %12.5e  \n\n\n", de );  exit(0);
}

void pivot(n, a, el, j, ip, ipc)
int n;   float a[][20], el[][20];
int j, ip[], *ipc;
{
int it, jj, k, m;
float t;
    t = 0;
    for( k = j; k <= n; k++ ){
        if( fabs( a[k][j] ) > t ) { jj = k; t = fabs( a[k][j] );}
    }
    if( jj != j ){
        *ipc = *ipc + 1;
        for( m = 1; m <= n; m++ ){
            t = a[j][m];    a[j][m] = a[jj][m];    a[jj][m] = t;
            t = el[j][m];   el[j][m] = el[jj][m];  el[jj][m] = t;
        }
        it = ip[j]; ip[j] = ip[jj];  ip[jj] = it;
        printf( "Number of pivoting =%d\n", *ipc );
    }
    return;
}
```

(C) Sample Output

```
CSL/C6-3     LU Decomposition

Original matrix
   2.00000e+00     1.00000e+00     -3.00000e+00
  -1.00000e+00  .  3.00000e+00      2.00000e+00
   3.00000e+00     1.00000e+00     -3.00000e+00
If pivoting is desired, type 1, or 0 otherwise.
1
Number of pivoting =2
Permutation
321
LU Matrices in compact form
   3.00000e+00 1.00000e+00-3.00000e+00
  -3.33333e-01 3.33333e+00 1.00000e+00
   6.66667e-01 1.00000e-01-1.10000e+00
Determinant =    1.10000e+01
```

(D) Discussions

The original matrix is

$$A = \begin{pmatrix} 2 & 1 & -3 \\ -1 & 3 & 2 \\ 3 & 1 & -3 \end{pmatrix}$$

The LU matrix in the compact form should be interpreted as

$$L = \begin{pmatrix} 1 & 0 & 0 \\ -0.33333 & 1 & 0 \\ 0.66667 & 0.1 & 1 \end{pmatrix}, \quad U = \begin{pmatrix} 3 & 1 & -3 \\ 0 & 3.3333 & 1 \\ 0 & 0 & -1.1 \end{pmatrix}$$

Now, if LU is computed, it becomes

$$LU = \begin{pmatrix} 3 & 1 & -3 \\ -1 & 3 & 2 \\ 2 & 1 & -3 \end{pmatrix}$$

which equals PA, where P is

$$P = \begin{pmatrix} 0 & 0 & 1 \\ 0 & 1 & 0 \\ 1 & 0 & 0 \end{pmatrix}$$

and can be written easily by using the permutation printed out.

PROGRAM 6-4 *M* **Equations with** *N* **Unknowns**

(A) Explanations

A set of m equations with n unknowns ($m < n$, or $m = n$) may be written as

$$Ax = y \tag{A}$$

where A is a rectangular ($m < n$) or a square ($m = n$) matrix, x is a unknown vector and y is a known vector. If $m = 3$ and $n = 4$, for example,

$$A = \begin{pmatrix} a_{1,1} & a_{1,2} & a_{1,3} & a_{1,4} \\ a_{2,1} & a_{2,2} & a_{2,3} & a_{2,4} \\ a_{3,1} & a_{3,2} & a_{3,3} & a_{3,4} \end{pmatrix},$$

$$x = \begin{pmatrix} x_1 \\ x_2 \\ x_3 \\ x_4 \end{pmatrix}, \quad y = \begin{pmatrix} y_1 \\ y_2 \\ y_3 \end{pmatrix}$$

In the program the equation is represented by an augmented array,

$$
\begin{array}{ccccc}
a_{1,1} & a_{1,2} & a_{1,3} & a_{1,4} & y_1 \\
a_{2,1} & a_{2,2} & a_{2,3} & a_{2,4} & y_2 \\
a_{3,1} & a_{3,2} & a_{3,3} & a_{3,4} & y_3
\end{array}
$$

This program reduces the augmented array to the echelon form.

For the reduction of the equation, Gauss-Jordan elimination is used. If the system is inconsistent, the program stops with a message indicating the inconsistency. If $m = n$ and there is a unique solution, the coefficient array becomes an identity matrix. If $m < n$, the coefficients of the basic variables selected by the program become unity, so the solution may be easily written in terms of free variables from the reduced array.

For the sake of simplicity and compactness of the program, elimination of coefficients is done without separating forward and backward eliminations. This means that once a nonzero pivot coefficient is found, all the coefficients along the same column, above and below the pivot, are eliminated before going to the next pivot. Pivoting is done before elimination.

The user defines the matrix and the inhomogeneous terms in the declaration statement for a_ini[][], which is copied to a[][]. No interactive input is necessary.

(B) List

```
/* CSL/c6-4.c      M Equations with N Unknowns    */
#include <stdio.h>
#include <stdlib.h>
#include <math.h>
#define TRUE 1
struct t_bc { double a[10][20]; int m, n;} bc;
/*           a[i][j] : matrix element a(i,j)
             eps : machine epsilon
             rank : rank of matrix
             bc.m: number of equations
             bc.n: number of unknowns                       */
main()
{
int i, j, jc, jr, k, kc, konsis, l, pv, rank, _i, _r, flag=0;
double det, eps, eps1, eps2, r, tm, z;
void list();
static a_init[10][20] = {{2,  3,  1,  4,  1,    6},
                         {2,  3,  1,  1,-1,    1},
                         {4,  6,-1,  1,  2,    5}};
   bc.m = 3;
   bc.n = 5;
   for( j = 1, _r = 0; j <= bc.n + 1; j++ ){
      for ( i = 1; i<= bc.m; i++) bc.a[i][j] = a_init[i-1][j-1];
   }
   printf( "\nCSL/C6-4    M Equations with N Unknowns\n" );
   eps = 1.0; eps1=2.0;     /*  Machine epsilon  */
```

```
for( l = 1; l <= 100; l++ ){
   if( eps1  <= 0 )  break;
     eps = eps/2; eps1= eps*0.98 + 1; eps1 = eps1 -1;
}
eps = eps*2;
printf( " Machine epsilon= %12.5e  \n", eps );
eps2 = eps*2;
printf( " Number of equations =%d\n", bc.m );
printf( " Number of unknowns  =%d\n", bc.n );
printf( " Augmented matrix (last column is inhomogeneous term)\n" );
list();
det = 1.0;
i = 0;
for( k = 1; k <= bc.m; k++ ){    /* Gauss Jordan scheme starts*/
   pv = k;
   while( TRUE ){
      i = i + 1;
      if( i > bc.n )  goto L_600;
      for( j = k + 1; j <= bc.m; j++ ){
         if( fabs( bc.a[pv][i] ) < fabs( bc.a[j][i] ) ) pv = j;
      }
      if( !(pv == k && bc.a[k][i] == 0) )  break;
   }
   if( pv != k ){
      for( jc = i; jc <= (bc.n + 1); jc++ ){    /* Pivoting */
         tm = bc.a[k][jc];
         bc.a[k][jc] = bc.a[pv][jc];  bc.a[pv][jc] = tm;
      }
      det = -det;
   }
   z = bc.a[k][i]; /* Elimination starts */
   det = det*z;
   rank = k;
   for( j = i; j <= (bc.n + 1); j++ ){
      bc.a[k][j] = bc.a[k][j]/z;
   }
   for( jr = 1; jr <= bc.m; jr++ ){
      if( jr != k ){
         if( bc.a[jr][i] != 0 ){
            r = bc.a[jr][i];
            for( kc = i; kc <= (bc.n + 1); kc++ ){
               bc.a[jr][kc] = bc.a[jr][kc] - r*bc.a[k][kc];
               if( fabs( bc.a[jr][kc]/bc.a[k][kc] ) < eps2 )
                       bc.a[jr][kc] = 0;
            }
         }
      }
   }
}
flag=0;
for( j = i; j <= bc.n; j++ ){
   if( bc.a[bc.m][j] != 0 ){
      for( jr = j; jr <= (bc.n + 1); jr++ ){
         bc.a[bc.m][jr] = bc.a[bc.m][jr]/bc.a[bc.m][j];
      }
      bc.a[bc.m][j] = 1.0;    break;
   }
}
```

```
L_600:
    if( rank != k ){    /* Examining consistency  */
        konsis = 0;
        for( j = rank + 1; j <= bc.m; j++ ){
            if( bc.a[j][bc.n+1] != 0 )  konsis = konsis + 1;
        }
    }
    printf( " Rank of matrix =%d\n", rank );
    printf( " Determinant  = %12.5e  \n", det );
    printf( " Reduced matrix:\n" );
    list();
    if(konsis!=0) printf("WARNING: The matrix is not consistent.\n" );
    printf( "\n\n"); exit(0);
}

void list()
{
int i, j;
    for( i = 1; i <= bc.m; i++ ){
        for(j=1;j<= (bc.n+1);j++) printf(" %12.5e",bc.a[i][j] );
        printf(  "\n" );
    }
    printf ("\n");
    return;
}
```

(C) Sample Output

```
CSL/C6-4     M Equations with N Unknowns

Machine epsilon=  2.77556e-17
Number of equations =3
Number of unknowns  =5
Augmented matrix (last column is inhomogeneous term)
2.00000e+00   3.00000e+00   1.00000e+00   4.00000e+00   1.00000e+00   6.00000e+00
2.00000e+00   3.00000e+00   1.00000e+00   1.00000e+00  -1.00000e+00   1.00000e+00
4.00000e+00   6.00000e+00  -1.00000e+00   1.00000e+00   2.00000e+00   5.00000e+00

Rank of matrix =3
Determinant  = -1.80000e+01
Reduced matrix:
1.00000e+00   1.50000e+00   0.00000e+00   0.00000e+00  -5.55556e-02   4.44444e-01
0.00000e+00   0.00000e+00   1.00000e+00   0.00000e+00  -1.55556e+00  -1.55556e+00
0.00000e+00   0.00000e+00   0.00000e+00   1.00000e+00   6.66667e-01   1.66667e+00
```

(D) Discussions

The computed results for the reduced array are interpreted as the transformation of the original equation to

$$\begin{pmatrix} 1 & 1.5 & 0 & 0 & -0.055556 \\ 0 & 0 & 1 & 0 & -1.5556 \\ 0 & 0 & 0 & 1 & 0.66667 \end{pmatrix} \begin{pmatrix} u \\ v \\ w \\ x \\ y \end{pmatrix} = \begin{pmatrix} 0.44444 \\ -1.5556 \\ 1.6667 \end{pmatrix}$$

In the preceding equation, u and w and x are found to be basic variables, while v and y are free variables. The coefficient matrix is in the echelon form. By moving the free variables to the right, the equations may be rewritten as

$$\begin{pmatrix} 1 & 0 & 0 & u \\ 0 & 1 & 0 & w \\ 0 & 0 & 1 & x \end{pmatrix} = \begin{pmatrix} 0.44444 - 1.5v + 0.055556y \\ -1.5556 + 1.5556y \\ 1.6667 - 0.66667y \end{pmatrix}$$

or equivalently

$$u = 0.44444 - 1.5v + 0.055556y$$

$$w = -1.5556 + 1.5556y$$

$$x = 1.6667 - 0.66667y$$

PROGRAM 6–4 can of course solve any usual problem of m equations with m unknowns, even when it is singular. So if the reader understands how to read the output, this program is a universal program to solve linear equations, including singular problems.

PROBLEMS

(6.1) Solve the following set of equations by Gauss elimination in the array form using a hand calculator (without using pivoting).

(a) $2x_1 + x_2 - 3x_3 = -1$
$-x_1 + 3x_2 + 2x_3 = 12$
$3x_1 + x_2 - 3x_3 = 0$

(b) $0.1x_1 - 0.6x_2 + x_3 = 0$
$-2x_1 + 8x_2 + 0.3x_3 = 1$
$x + 6x_2 + 4x_3 = 2$

(6.2) Solve the following sets of equations by Gauss-Jordan elimination:

(a) $4x + y - z = 9$
$3x + 2y - 6z = -2$
$x - 5y + 3z = 1$

(b) $x - y = 0$
$-x + 2y - z = 1$
$- y + 1.1z = 0$

(6.3) Repeat (6.1) with pivoting in the array form using a hand calculator.

(6.4) Solve the following set of equations without pivoting, and then with pivoting:

$$6.122x + 1500.5y = 1506.622$$
$$2000x + 3y = 2003$$

Round off numbers after the sixth significant figure.

(6.5) Solve the following equations by Gauss elimination without pivoting, and then with pivoting. To simulate the effect of round-off, cut off each number after the fourth significant figure.

$$1.001x + 1.5y = 0$$
$$2x + 3y = 1$$

(6.6) The following sets of linear equations have common coefficients but different right-side terms:

(a)
$$x + y + z = 1$$
$$2x - y + 3z = 4$$
$$3x + 2y - 2z = -2$$

(b)
$$x + y + z = -2$$
$$2x - y + 3z = 5$$
$$3x + 2y - 2z = 1$$

(c)
$$x + y + z = 2$$
$$2x - y + 3z = -1$$
$$3x + 2y - 2z = 4$$

The coefficients and the three sets of the right-side terms may be combined into an array

$$
\begin{matrix}
1 & 1 & 1 & 1 & -2 & 2 \\
2 & -1 & 3 & 4 & 5 & -1 \\
3 & 2 & -2 & -2 & 1 & 4
\end{matrix}
$$

If we apply Gauss-Jordan scheme to this array and reduce the first three columns to the unit matrix form, the solutions for the three problems are automatically obtained in the fourth, fifth, and the sixth columns when the elimination is completed. Calculate the solution in this way.

(6.7) Calculate $C \equiv A + B$, $D \equiv A - B$, $E \equiv AB$, where

$$A = \begin{bmatrix} 1 & 2 & 3 \\ 0 & 1 & 4 \\ 3 & 0 & 2 \end{bmatrix}$$

$$B = \begin{bmatrix} 4 & 1 & 2 \\ 3 & 2 & 1 \\ 0 & 1 & 2 \end{bmatrix}$$

(6.8) Calculate $B^t A^t$ and $(AB)^t$ using the definitions in the previous problem, and show that the results are identical.

(6.9) Calculate $E = AB$, where

$$A = \begin{bmatrix} 1 & 2 & 3 \\ 0 & 1 & 4 \\ 3 & 0 & 2 \end{bmatrix}$$

$$B = \begin{bmatrix} 3 \\ 5 \\ 1 \end{bmatrix}$$

(6.10) Calculate $D = A + A'$, $E = A - A'$, $F = AB$, $G = BA$, and $H = BC$ where

$$A = \begin{bmatrix} 1 & 2 & 3 & 1 \\ 0 & 1 & 4 & 2 \\ 3 & 0 & 2 & 3 \end{bmatrix} \quad A' = \begin{bmatrix} 2 & 3 & 0 & 1 \\ 0 & 1 & 0 & 1 \\ 2 & 1 & 5 & 0 \end{bmatrix}$$

$$B = \begin{bmatrix} 4 & 1 & 2 \\ 3 & 2 & 1 \\ 0 & 1 & 2 \\ 3 & 1 & 0 \end{bmatrix}$$

$$C = \begin{bmatrix} 7 \\ 1 \\ 4 \end{bmatrix}$$

(6.11) Compute $E = B + CD$, where

$$B = \begin{bmatrix} 3 & 2 & 1 \\ 0 & 4 & 3 \\ 0 & 0 & 6 \end{bmatrix} \quad C = \begin{bmatrix} 1 & 0 & 2 \\ -1 & 1 & 0 \\ 0 & 3 & 2 \end{bmatrix} \quad D = \begin{bmatrix} 1 & 0 & 0 \\ -2 & 1 & 0 \\ 5 & 2 & 7 \end{bmatrix}$$

(6.12) Calculate the inverse of

$$A = \begin{bmatrix} 7 & 1 \\ 4 & 5 \end{bmatrix}$$

(6.13) Calculate the inverse of

$$A = \begin{bmatrix} 1 & -1 & 0 & 0 \\ -1 & 2 & -1 & 0 \\ 0 & -1 & 2 & -1 \\ 0 & 0 & -1 & 2 \end{bmatrix}$$

$$B = \begin{bmatrix} 1 & 4 & 5 \\ 2 & 1 & 2 \\ 8 & 1 & 1 \end{bmatrix}$$

(6.14) Find the inverse of the following matrix:

$$\begin{bmatrix} 3 & 1 & 0 \\ 1 & 2 & 1 \\ 0 & 1 & 1 \end{bmatrix}$$

(6.15) Find the inverse of the following matrix with pivoting:

$$\begin{bmatrix} 0 & 5 & 1 \\ -1 & 6 & 3 \\ 3 & -9 & 5 \end{bmatrix}$$

(6.16) Decompose the following matrices into L and U matrices by using a calculator, and then verify the decomposition by calculating the product LU.

(a) $\begin{bmatrix} 2 & -1 & 0 \\ -1 & 2 & -1 \\ 0 & -1 & 2 \end{bmatrix}$

(b) $\begin{bmatrix} 2 & -1 & 0 \\ -3 & 4 & -1 \\ 0 & -1 & 2 \end{bmatrix}$

(6.17) Solve the following equations by using the LU decomposition:

(a) $\begin{bmatrix} 2 & -1 & 0 \\ -1 & 2 & -1 \\ 0 & -1 & 2 \end{bmatrix} \begin{bmatrix} x_1 \\ x_2 \\ x_3 \end{bmatrix} = \begin{bmatrix} 1 \\ 2 \\ 3 \end{bmatrix}$

(b) $\begin{bmatrix} 2 & -1 & 1 \\ -3 & 4 & -1 \\ 1 & -1 & 1 \end{bmatrix} \begin{bmatrix} x_1 \\ x_2 \\ x_3 \end{bmatrix} = \begin{bmatrix} 4 \\ 5 \\ 6 \end{bmatrix}$

(6.18) Find the determinant of the following matrices:

$$A = \begin{bmatrix} 1 & 4 \\ 3 & 2 \end{bmatrix}$$

$$B = \begin{bmatrix} 3 & 2 \\ 1 & 3 \end{bmatrix}$$

$$C = \begin{bmatrix} 4 & -1 & 2 \\ 1 & 2 & -3 \\ 0 & 3 & 1 \end{bmatrix}$$

$$D = \begin{bmatrix} -1 & 1 & 2 & -3 \\ 2 & -1 & 3 & 2 \\ 0 & 2 & 4 & 1 \\ 5 & 1 & 1 & -1 \end{bmatrix}$$

(6.19) Calculate the determinant of

$$A = \begin{bmatrix} 8 & 1 & 3 & 2 \\ 2 & 9 & -1 & -2 \\ 1 & 3 & 2 & -1 \\ 1 & 0 & 6 & 4 \end{bmatrix}$$

which may be decomposed to the product of

$$L = \begin{bmatrix} 1 & 0 & 0 & 0 \\ 0.25 & 1 & 0 & 0 \\ 0.125 & 0.328 & 1 & 0 \\ 0.125 & -0.0143 & 2.545 & 1 \end{bmatrix}$$

$$U = \begin{bmatrix} 8 & 1 & 3 & 2 \\ 0 & 8.75 & -1.75 & -2.5 \\ 0 & 0 & 2.2 & -0.4285 \\ 0 & 0 & 0 & 4.8052 \end{bmatrix}$$

(6.20) Evaluate the determinant of A^{-1} where

$$A = BCD$$

and

$$B = \begin{bmatrix} 3 & 2 & 1 \\ 0 & 4 & 3 \\ 0 & 0 & 6 \end{bmatrix} \quad C = \begin{bmatrix} 1 & 0 & 2 \\ -1 & 1 & 0 \\ 0 & 3 & 2 \end{bmatrix} \quad D = \begin{bmatrix} 1 & 0 & 0 \\ -2 & 1 & 0 \\ 5 & 2 & 7 \end{bmatrix}$$

(6.21) Evaluate the determinant of the transpose of the matrices of the previous problem, and show that the determinant of A equals the determinant of A^t.

(6.22) Matrix A is the 5×5 Hilbert matrix given by

$$A = [a_{i, j}] \quad \text{where} \quad a_{i, j} = \frac{1}{i + j - 1}$$

Compute (a) A^{-1}, (b) $A^{-1}A$, (c) $(A^{-1})^{-1}A^{-1}$

(6.23) Expand the determinant of the following matrix into a polynomial form:

$$A = \begin{bmatrix} 2 - s & 4 & 6 \\ 1 & -1 - 9 & 5 \\ 2 & 0 & 1 - s \end{bmatrix}$$

(6.24) Find the general solution of

$$4x + y - z = 9$$
$$3x + 2y - 6z = -2$$

(6.25) Find basic variables and free variables for the following equations. Then, find the general solution.

$$\begin{bmatrix} 1 & 3 & 3 & 2 \\ 2 & 6 & 9 & 5 \\ -1 & 3 & 3 & 0 \end{bmatrix} \begin{bmatrix} u \\ v \\ w \\ y \end{bmatrix} = \begin{bmatrix} 1 \\ 5 \\ 5 \end{bmatrix}$$

REFERENCES

Dongarra, J. J., J. R. Bunch, C. B. Moler, and G. W. Stewart, *LINPACK User's Guide*, SIAM, 1979.

Forsythe, G. E., and C. B. Moler, *Computer Solution of Linear Algebra System*, Prentice-Hall, 1967.

Forsythe, G. E., M. A. Malcolm, and C. B. Moler, *Computer Methods for Mathematical Computations*, Prentice-Hall, 1977.

Jennings, A., *Matrix Computations for Engineers and Scientists*, Wiley, 1977.

Lang, S., *Linear Algebra*, Springer-Verlag, 1987.

Morris, J. L., *Computational Methods in Elementary Numerical Analysis*, Wiley, 1983.

Strang, G., *Linear Algebra and Its Applications*, 2nd ed., Academic Press, 1980.

7
Computations of
Matrix Eigenvalues

7.1 INTRODUCTION

Homogeneous linear equations are often associated with systems that sustain harmonic oscillations without external forces. The vibration of string, a membrane, or some other structural system belongs to this category. When the dynamic nature or stability of such systems is studied, the solution of homogeneous equations (particularly, finding characteristic values of homogeneous equations) becomes necessary.

Homogeneous linear equations are also important in various mathematical analyses. For example, in solving a set of ordinary differential equations, it is necessary to solve a homogeneous linear equation set. Another example is that the stability of numerical solution methods for partial differential equations is related to characteristic values of homogeneous linear equations.

In the beginning of Section 6.2, we pointed out that a set of linear equations with at least one nonzero value on the right side is an inhomogeneous set. All the linear equations discussed in Chapter 6 are inhomogeneous equations. On the other hand, when the right-hand side of each equation is zero, the set is called a homogeneous set. For example,

$$3x - 2y + z = 0$$
$$x + y + 2z = 0 \qquad (7.1.1)$$
$$4x - y + 3z = 0$$

The solution of a homogeneous set is quite different from that for the inhomogeneous linear equations. To explain why, suppose that a solution of Eq. (7.1.1) exists and is written as $x = a$, $y = b$, and $z = c$. Then, the $x = ka$, $y = kb$, and $z = kc$ also satisfy Eq. (7.1.1) where k is an arbitrary constant. This means that we can fix one of the unknowns to an arbitrary value, say $x = \beta$, and then solve the system for the remainder of unknowns.

By fixing x to an arbitrary value β, however, Eq. (7.1.1) becomes

$$
\begin{aligned}
-2y + z &= -3\beta \\
+y + 2z &= -\beta \\
-y + 3z &= -4\beta
\end{aligned}
\tag{7.1.2}
$$

Here we have three equations for two unknowns. If a different combination of two equations yields a different solution, no solution exists to the whole set. The set of equations has a solution only if one of the three equations is identical to another or is a linear combination of others (namely, one equation can be eliminated by adding or subtracting multiples of other equations).

When at least one of the equations in Eq. (7.1.2) is linearly dependent, the determinant of the coefficient matrix of Eq. (7.1.1) becomes zero. Hence, the necessary condition for the solution of a homogeneous set of linear equations to exist is that its determinant be zero. (The situation is contrary to the inhomogeneous set of equations, because an inhomogeneous set of linear equations has a unique solution only if the determinant is nonzero.)

In the case of Eq. (7.1.1), its determinant happens to be zero:

$$
\det(A) = \det \begin{bmatrix} 3 & -2 & 1 \\ 1 & 1 & 2 \\ 4 & -1 & 3 \end{bmatrix} = 0
\tag{7.1.3}
$$

and the solution can be written as

$$
\begin{aligned}
x &= \beta \\
y &= \beta \\
z &= -\beta
\end{aligned}
$$

or equivalently

$$
\begin{bmatrix} x \\ y \\ z \end{bmatrix} = \beta \begin{bmatrix} 1 \\ 1 \\ -1 \end{bmatrix}
\tag{7.1.4}
$$

where β is an arbitrary constant.

One standard form of a homogeneous equation set (with three unknowns, for example) is as follows:

$$(a_{11} - \lambda)x_1 + a_{12}x_2 + a_{13}x_3 = 0$$
$$a_{21}x_1 + (a_{22} - \lambda)x_2 + a_{23}x_3 = 0 \qquad (7.1.5a)$$
$$a_{31}x_1 + a_{32}x_2 + (a_{33} - \lambda)x_3 = 0$$

where λ is called a *characteristic value*, or *eigenvalue*. Equation (7.1.5a) can be equivalently expressed as

$$Ax = \lambda x \qquad (7.1.5b)$$

where

$$A = \begin{bmatrix} a_{11} & a_{12} & a_{13} \\ a_{21} & a_{22} & a_{23} \\ a_{31} & a_{32} & a_{33} \end{bmatrix}, \quad x = \begin{bmatrix} x_1 \\ x_2 \\ x_3 \end{bmatrix}$$

Equation (7.1.5a) or Eq. (7.1.5b) has a nontrivial solution only when the characteristic value satisfies

$$\det \begin{bmatrix} a_{11} - \lambda & a_{12} & a_{13} \\ a_{21} & a_{22} - \lambda & a_{23} \\ a_{31} & a_{32} & a_{33} - \lambda \end{bmatrix} = 0 \qquad (7.1.6a)$$

or more compactly

$$f(\lambda) \equiv \det(A - \lambda I) = 0 \qquad (7.1.6b)$$

The function $f(\lambda)$ is a characteristic function and is a polynomial of λ. The order of the polynomial is equal to the order of the matrix. Eigenvalues are roots of the characteristic equation. Equation (7.1.5b) is called a *matrix eigenvalue problem*.

Once solutions of the characteristic equation are found, the solution of the homogeneous equation can be found for each eigenvalue. Each solution is then called an *eigenvector*.

Another form of eigenvalue problems is given by

$$Ax = \lambda Bx \qquad (7.1.7)$$

where A and B are matrices. The characteristic equation for Eq. (7.1.7) is written as

$$f(\lambda) \equiv \det(A - \lambda B) = 0$$

Example 7.1

Consider a vertical system consisting of masses and springs. The notations in the figure are:

$$k_{0,1},\ k_{1,2},\ k_{2,3},\ \text{and}\ k_{3,4}: \quad \text{spring constants}$$

$$m_i,\ i = 1,\ 2\ \text{and}\ 3: \quad \text{masses}$$

$$y_i: \quad \text{displacement of mass } i \text{ from the static position}$$

Assuming there are no frictions, the differential equations for displacements of the masses are given by

$$m_1 \frac{d^2}{dt^2} y_1(t) = -(k_{01} + k_{12})y_1 + k_{12}y_2$$

$$m_2 \frac{d^2}{dt^2} y_2(t) = k_{12}y_1 - (k_{12} + k_{23})y_2 + k_{23}y_3 \qquad \text{(A)}$$

$$m_3 \frac{d^2}{dt^2} y_3(t) = k_{23}y_2 - (k_{23} + k_{34})y_3$$

Derive the eigenvalue problem associated with a harmonic oscillation (see Figure E7.1). Assume that all the masses are identical, and that $m_1 = m_2 = m_3 = m$.

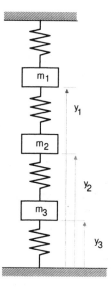

Figure E7.1 A vertical system of masses and springs

⟨**Solution**⟩

For a harmonic oscillation, the solution may be written as

$$y_i = \exp(j\omega t)f_i, \quad i = 1,\ 2,\ 3 \qquad \text{(B)}$$

where ω is an undetermined angular velocity, f_i are unknowns, and $j = \sqrt{-1}$.
Introducing Eq. (B) into Eq. (A) yields

$$-\omega^2 f_1 = -(1/m)(k_{01} + k_{12})f_1 + (1/m)k_{12}f_2$$
$$-\omega^2 f_2 = (1/m)k_{12}f_1 - (1/m)(k_{12} + k_{23})f_2 + (1/m)k_{23}f_3 \qquad \text{(C)}$$
$$-\omega^2 f_3 = (1/m)k_{23}f_2 - (1/m)(k_{23} + k_{34})f_3$$

where equations are divided through by m. In matrix notations, Eq. (C) is written
as

$$Af - \lambda f = 0 \qquad \text{(D)}$$

where

$$\lambda = \omega^2 \qquad \text{(E)}$$

$$A = \begin{bmatrix} (k_{01} + k_{12})/m, & -k_{12}/m, & 0 \\ -k_{12}/m, & (k_{12} + k_{23})/m, & -k_{23}/m \\ 0, & -k_{23}/m, & (k_{23} + k_{34})/m \end{bmatrix} \qquad \text{(F)}$$

and

$$f = \begin{bmatrix} f_1 \\ f_2 \\ f_3 \end{bmatrix} \qquad \text{(G)}$$

In computing eigenvalues of a matrix we should note the following facts:

(a) Eigenvalues of a symmetric matrix are all real.
    (If all the eigenvalues of a symmetric matrix are positive, the matrix is called a
    positive definite matrix.)

Table 7.1 Numerical methods of eigenvalue calculations

Method	Result	Real/Complex	Form[a]	Comments
Method of interpolation	Polynomial	R, C	A, B	Newton forward interpolation (for small matrices only).
Power method/ inverse power method/shifted inverse power method	Eigenvalues	R	A	One eigenvalue is computed at a time.
Householder/ tridiagonal matrix	Eigenvalues	R	A	*Symmetric matrix only.*
Householder/ QR iteration	Eigenvalues	R, C	A	*Nonsymmetric matrix.*

[a] A: $\det(A - \lambda I)$
    B: $\det(A - \lambda B)$

(b) A nonsymmetric matrix consisting of real elements can have complex eigenvalues, all of which appear in complex conjugate pairs.

In the remainder of this chapter, we will concentrate our attention on fundamental numerical methods of calculating eigenvalues. The solution methods discussed in this chapter are summarized in Table 7.1. General references for eigenvalue computations are listed at the end of this chapter.

7.2 METHOD OF INTERPOLATION

We start with the method of interpolation [Faddeeva], which is a primitive algorithm but easy to understand. In this approach the characteristic function is reduced to a power series in λ. Then, the roots of the power series are found by Bairstow's method (described in Chapter 3).

The reduction procedure here consists of the two steps:

(a) To transform the characteristic function to a Newton forward polynomial.
(b) To convert the Newton forward polynomial to a power series.

For a matrix of order N, the characteristic function is a polynomial of order N. As such, if a function table of $f(\lambda)$ for $N + 1$ equispaced values of λ is developed, then $f(\lambda)$ can be expressed in the Newton forward interpolation polynomial of order N (see Chapter 2):

$$f(\lambda) = g(s) = \sum_{n=0}^{N} \binom{s}{n} \Delta^n f_0 \qquad (7.2.1)$$

with

$$f_i = f(\lambda_i), \quad i = 0, 1, 2, \ldots, N$$
$$s = (\lambda - \lambda_0)/\Delta\lambda$$

where λ_i are equispaced values of λ, $\lambda_i = \lambda_{i-1} + \Delta\lambda$. The values of $f_i \equiv f(\lambda_i)$, $i = 0$, $1, 2, \ldots, N$ are evaluated by direct calculation of the determinant of $(A - \lambda_i I)$ (see Section 6.8). Although the increment $\Delta\lambda$ is arbitrary, a too-small value and a too-large value both cause round-off errors in developing the difference table. Because λ_0 is also arbitrary, we set it to $\lambda_0 = 0$, so that s becomes

$$s = \lambda/\Delta\lambda$$

The binomial coefficient $\binom{s}{n}$ may be expressed by

$$\binom{s}{n} = \frac{s(s-1)(s-2)\cdots(s-n)}{n!}$$

$$= \sum_{i=1}^{n} c_{n,i} s^i, \quad n \geqslant 1 \tag{7.2.2}$$

where $c_{n,i}$ are called *Markov coefficients*. Some of their numerical values are listed in Table 7.2. Introducing Eq. (7.2.2) into Eq. (7.2.1) and rearranging terms yield

$$g(\lambda) = f_0 + \sum_{n=1}^{N} \sum_{i=1}^{n} c_{n,i} s^i \Delta^n f_0$$

$$= f_0 + \sum_{i=1}^{N} \left(\sum_{n=i}^{N} c_{n,i} \Delta^n f_0 \right) s^i$$

$$= f_0 + \sum_{i=1}^{N} b_i s^i \tag{7.2.3}$$

where

$$b_i = \sum_{n=i}^{N} c_{n,i} \Delta^n f_0 \tag{7.2.4}$$

Thus, by using $s = \lambda/\Delta\lambda$, Eq. (7.2.3) may be rewritten in terms of λ:

$$g(\lambda) = f_0 + \sum_{i=1}^{N} b_i \left(\frac{\lambda}{\Delta\lambda} \right)^i \tag{7.2.5}$$

This is the desired form of the characteristic equation in a power series. PROGRAM 7–1 transforms the characteristic equation to a power series.

Table 7.2 Markov coefficients, $c_{n,i}$

	$i = 1$	$i = 2$	$i = 3$	$i = 4$	$i = 5$	$i = 6$
$n = 1$	1					
$n = 2$	-0.5	0.5				
$n = 3$	0.33333	-0.5	0.16666			
$n = 4$	-0.25	0.45833	-0.25	0.04167		
$n = 5$	0.2	-0.41667	0.29167	-0.08333	0.00833	
$n = 6$	-0.166666	0.38056	-0.31250	0.11806	-0.02083	0.00139

The present method is applicable to the characteristic equation in the form, det $(A - \lambda B) = 0$, also.

Example 7.2

Find the power series form of the following characteristic equation using a Newton interpolation polynomial:

$$f(\lambda) = \det \begin{bmatrix} 3 - \lambda & 4 & -2 \\ 3 & -1 - \lambda & 1 \\ 2 & 0 & 5 - \lambda \end{bmatrix}$$

Then, calculate the characteristic values by Bairstow's method.

⟨**Solution**⟩

The values of $f(\lambda)$ for four different values of λ, namely, 0, 0.5, 1.0, 1.5, are calculated by direct evaluation of the determinant:

$$\lambda = 0 \ : \text{determinant} = -71$$
$$\lambda = 0.5: \text{determinant} = -68.875$$
$$\lambda = 1.0: \text{determinant} = -64$$
$$\lambda = 1.5: \text{determinant} = -57.125$$

The forward difference table then becomes

i	λ_i	f_i	Δf_i	$\Delta^2 f_i$	$\Delta^3 f_i$
0	0	−71	2.125	2.75	−0.75
1	0.5	−68.875	4.875	2.0	
2	1.0	−64	6.875		
3	1.5	−57.125			

Using the differences along the first line, the Newton forward interpolation formula is written as

$$g(\lambda) = -71 + 2.125s + \frac{2.75}{2} s(s - 1) - \frac{0.75}{6} s(s - 1)(s - 2)$$

where $s = \lambda/0.5$. Using the Markov coefficients, this equation is converted to

$$g(\lambda) = -71 + [(1)(2.125) + (-0.5)(2.75) + (0.333333)(-0.75)]s$$
$$+ [(0.5)(2.75) + (-0.5)(-0.75)]s^2$$
$$+ [(0.166666)(-0.75)]s^3$$

Introducing $s = \lambda/0.5$ and reorganizing the terms yield

$$g(\lambda) = -71 + \lambda + 7\lambda^2 - \lambda^3$$

The roots of the foregoing equation found by PROGRAM 3–7 are

$$4.875 \pm 1.431i, \quad -2.750$$

SUMMARY OF THIS SECTION

(a) Characteristic function is transformed to a Newton interpolation formula, which is then rewritten to a power series form by using Markov coefficients.

(b) The roots of the power series are computed by the Bairstow method.

7.3 HOUSEHOLDER METHOD FOR A SYMMETRIC MATRIX

A symmetric matrix can be transformed to a tridiagonal matrix by the Householder method, which consists of a series of similarity transformations. The eigenvalues of a tridiagonal matrix are calculated by the bisection method.

In the remainder of this section, we study the Householder/bisection method in two steps—the Householder transformation and the bisection method—to calculate eigenvalues of a symmetric tridiagonal matrix.

7.3.1 Transformation of a Symmetric Matrix to a Tridiagonal Matrix

The original matrix A given is now denoted by $A^{(1)}$:

$$A^{(1)} = A = \begin{bmatrix} x & x & x & \cdot & x \\ x & x & x & \cdot & \cdot \\ x & x & x & \cdot & \cdot \\ x & x & x & \cdot & \cdot \\ \cdot & \cdot & \cdot & \cdot & \cdot \\ x & x & x & \cdot & x \end{bmatrix} \tag{7.3.1}$$

The first step in reducing $A^{(1)}$ to a tridiagonal matrix is to transform it to the following form denoted by $A^{(2)}$:

$$A^{(2)} = \begin{bmatrix} x & x & 0 & \cdot & 0 \\ x & x & x & \cdot & x \\ 0 & x & x & \cdot & x \\ 0 & x & x & \cdot & x \\ \cdot & \cdot & \cdot & \cdot & \cdot \\ 0 & x & x & \cdot & x \end{bmatrix} \tag{7.3.2}$$

This transformation is performed by both premultiplying and postmultiplying A by a transformation matrix P as

$$A^{(2)} = P A^{(1)} P \tag{7.3.3}$$

In the foregoing equation, P is defined by

$$P = I - \frac{uu^T}{h} \tag{7.3.4}$$

with

$$u = \text{col} (0, a_{2,1} + G, a_{3,1}, a_{4,1}, \ldots, a_{N,1}) \tag{7.3.5}$$

where $a_{i,j}$ are matrix elements of $A^{(1)}$, and

$$G = \left[\sum_{i=2}^{N} (a_{i,1})^2 \right]^{1/2} \text{sign} (a_{2,1}) \tag{7.3.6}$$

$$h = G^2 + Ga_{2,1} \tag{7.3.7}$$

Here, sign $(a) = +1$ if $a \geqslant 0$, or sign $(a) = -1$ if $a < 0$. The transformation matrix P has the following properties:

$$P = P^{-1}$$
$$P^T = P \tag{7.3.8}$$
$$PP = I$$

Matrix A of order N is reduced to a tridiagonal matrix by repeating transformations of the same kind $N - 2$ times. The matrix $A^{(m)}$ has a principal submatrix of order m in the tridiagonal form at the top left corner. In general, the transformation from $A^{(m)}$ to $A^{(m+1)}$ is written as

$$A^{(m+1)} = PA^{(m)}P \tag{7.3.9}$$

where

$$P = I - \frac{uu^T}{h} \tag{7.3.10}$$

$$u = \text{col} (0, 0, 0, \ldots, 0, a_{m+1,m} + G, a_{m+2,m}, \ldots, a_{N,m}) \tag{7.3.11}$$

where $a_{i,j}$ are elements of $A^{(m)}$ and

$$G = \left[\sum_{i=m+1}^{N} (a_{i,m})^2 \right]^{1/2} \text{sign} (a_{m+1,m}) \tag{7.3.12}$$

$$h = G^2 + Ga_{m+1,m} \tag{7.3.13}$$

The transformation matrix P in each stage satisfies Eq. (7.3.8).

7.3.2 Eigenvalues of a Tridiagonal Matrix

The tridiagonal matrix denoted here by M is symmetric and written as

$$M = \begin{bmatrix} a_1, & b_1 & \cdot & & \cdot & \\ b_1, & a_2, & b_2 & & \cdot & \\ \cdot & b_2, & a_3, & b_3 & & \cdot \\ \cdot & & \cdot & \cdot & & \cdot \\ \cdot & & & \cdot & b_{N-1}, & a_N \end{bmatrix} \tag{7.3.14}$$

We define a series of polynomials by

$$p_0(\lambda) = 1$$
$$p_1(\lambda) = a_1 - \lambda$$
$$p_2(\lambda) = [a_2 - \lambda]p_1(\lambda) - [b_1]^2 \tag{7.3.15}$$
$$\vdots$$
$$p_i(\lambda) = [a_i - \lambda]p_{i-1}(\lambda) - [b_{i-1}]^2 p_{i-2}(\lambda)$$
$$\vdots$$

Notice that p_i in Eq. (7.3.15) is the determinant of the principal submatrix in M including the first diagonal element down to the ith diagonal element. For a tridiagonal matrix of order N, $p_N(\lambda) = 0$ is the characteristic equation to be solved.

The series of polynomials defined in Eq. (7.3.15) has an important property, that is, a root of $p_k(\lambda)$ always separates a pair of two consecutive roots of $p_{k+1}(\lambda)$. In other words, each root of $p_{k+1}(\lambda)$ is between two consecutive roots of $p_k(\lambda)$ except the minimum and maximum roots of the former. The relation among the roots of consecutive polynomials is illustrated in Figure 7.1. Because of this property, the roots of $p_{k+1}(\lambda)$ can be found by the bisection method.

PROGRAM 7–2 transforms a symmetric matrix to a symmetric tridiagonal matrix, and then finds all the eigenvalues by the bisection method.

Figure 7.1 Relation among roots of $p_k(\lambda) = 0$

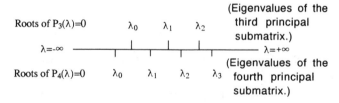

Roots of $P_3(\lambda)=0$ λ_0 λ_1 λ_2 (Eigenvalues of the third principal submatrix.)

$\lambda=-\infty$ ———————— $\lambda=+\infty$

Roots of $P_4(\lambda)=0$ λ_0 λ_1 λ_2 λ_3 (Eigenvalues of the fourth principal submatrix.)

Example 7.3

Calculate eigenvalues of the following matrix by the Householder/bisection method:

$$A^{(1)} = A = \begin{bmatrix} 3 & 1 & 4 \\ 1 & 7 & 2 \\ 4 & 2 & 0 \end{bmatrix}$$

⟨**Solution**⟩

Because $N = 3$, only one transformation is necessary. Elements of u given by Eq. (7.3.11) are calculated as

$$u_1 = 0$$
$$G = \sqrt{1^2 + 4^2} \, \text{sign} \, (1) = 4.1231$$
$$a_{2,1} = 1$$
$$h = (4.1231)^2 + (4.1231)(1) = 21.1231$$
$$u_2 = a_{2,1} + \sqrt{17} = 1 + 4.1231 = 5.1231$$
$$u_3 = a_{3,1} = 4$$

Thus, Eq. (7.3.11) becomes

$$u = \text{col} \, [0, 5.123106, 4]$$

By introducing u and h into Eq. (7.3.14), the transformed matrix becomes

$$A^{(2)} = \left(I - \frac{uu^T}{h} \right) A^{(1)} \left(I - \frac{uu^T}{h} \right)$$
$$= \begin{bmatrix} 3.0000 & -4.1231 & 0.0 \\ -4.1231 & 1.3529 & 3.4118 \\ 0.0 & 3.4118 & 5.6471 \end{bmatrix}$$

The eigenvalue of the first principal submatrix is immediately 3.000. The second principal submatrix is

$$\begin{bmatrix} 3.0000 & -4.1231 \\ -4.1231 & 1.3529 \end{bmatrix}$$

Its eigenvalues found by the bisection method are -2.0280 and 6.3810. (Notice that these two eigenvalues are separated by the eigenvalue of the first principal submatrix, 3.000) Eigenvalues of the whole tridiagonal matrix are -2.8941, 4.3861, and 8.508.

SUMMARY OF THIS SECTION

(a) The Householder method is a series of similarity transformations that transforms a symmetric matrix into a symmetric tridiagonal matrix.

(b) Eigenvalues of the tridiagonal matrix are obtained by the bisection method.

7.4 POWER METHODS

There are two reasons why power methods are important. First, these methods are easy means to compute eigenvalues. Second, they are closely related to the QR iteration discussed in the next section.

The power methods include three versions. First is the *regular* power method based on the power of the matrix, which finds the largest eigenvalue iteratively. Second is the *inverse* power method that is based on the inverse power of the matrix and that finds the smallest eigenvalue. Third is the *shifted* inverse power method. In the remainder of this section, we call the regular power method simply the *power method* and the inverse power method as the *inverse power method*.

POWER METHOD. Consider a $N \times N$ matrix A. The eigenvalues and eigenvectors satisfy

$$Au_i = \lambda_i u_i \tag{7.4.1}$$

where λ_i is the ith eigenvalue and u_i is the ith eigenvector. If A is a symmetric matrix, all the eigenvalues are real. If A is a nonsymmetric matrix, some eigenvalues may become complex values. We assume that the largest eigenvalue is real and separate (it is not a double eigenvalue) and that eigenvalues are numbered in the increasing order,

$$|\lambda_1| \leqslant |\lambda_2| \cdots \leqslant |\lambda_{N-1}| < |\lambda_N| \tag{7.4.2}$$

The power method starts with an initial guess for the eigenvector, $u^{(0)}$, which can be any no-null vector. The first iterative approximation is

$$u^{(1)} = Au^{(0)}$$

and subsequent iteratives are

$$u^{(k+1)} = \frac{1}{\lambda^{(k)}} Au^{(k)} \tag{7.4.3}$$

with

$$\lambda^{(k)} = \frac{(u^{(k)}, u^{(k)})}{\left(u^{(k)}, \frac{1}{\lambda^{(k-1)}} u^{(k-1)}\right)} \tag{7.4.4}$$

where (a, b) denotes a scalar product of two vectors a and b, and k is the iteration number. As iteration proceeds, $\lambda^{(k)}$ converges to the largest eigenvalue and $u^{(k)}$ converges to the corresponding eigenvector.

Example 7.4

Find the largest eigenvalue and the corresponding eigenvector by the power method for

$$A = \begin{bmatrix} 3 & 1 & 4 \\ 1 & 7 & 2 \\ 4 & 2 & 0 \end{bmatrix}$$

⟨**Solution**⟩

We set the initial guess as

$$u^{(0)} = \text{col} \ (1 \ 1 \ 1)$$

The iterative solution after each iteration cycle is shown in Table 7.3.

Table 7.3 Iterative solution

k	$\lambda^{(k)}$	Three elements of $u^{(k)}$		
1	8.333333	0.9600000	1.200000	0.7200000
2	8.453462	0.8233314	1.277583	0.7381592
3	8.492605	0.7889469	1.323826	0.6886570
4	8.503810	0.7579303	1.344461	0.6824518
5	8.506876	0.7462269	1.355852	0.6724730
10	8.507998	0.7319254	1.366952	0.6655880
15	8.508000	0.7314215	1.367357	0.6653084
18	8.508000	0.7314049	1.367370	0.6652992
19	8.508000	0.7314036	1.367371	0.6652986
20	8.508000	0.7314029	1.367372	0.6652982

The eigenvalue converges with approximately 15 iteration steps. The convergence of the eigenvector is slower than that of the eigenvalue.

The reason the power method converges is now discussed. The initial vector may be expanded in terms of eigenvectors of A,

$$u^{(0)} = \sum_{i=1}^{N} a_i u_i \tag{7.4.5}$$

where a_i are expansion coefficients and u_i is the ith eigenvector of A. Introducing Eq. (7.4.5) into Eq. (7.4.3) yields

$$u^{(k)} = \frac{(\lambda_N)^k}{\lambda^{(1)} \cdots \lambda^{(k-1)}} \left[\left(\frac{\lambda_1}{\lambda_N} \right)^k u_1 + \left(\frac{\lambda_2}{\lambda_N} \right)^k u_2 + \cdots u_N \right] \tag{7.4.6}$$

As k increases, all the terms in the brackets of Eq. (7.4.6) vanish except u_N. As $u^{(k)}$ converges to u_N, it is easy to see that Eq. (7.4.4) converges to λ_N.

Before completing the explanation of the power method, we point out that Eq. (7.4.3) may be written as

$$u^{(k+1)} = \frac{1}{\displaystyle\prod_{l=1}^{k} \lambda^{(l)}} A^{k+1} u^{(0)} \tag{7.4.7}$$

Therefore, the essence of the power method is to multiply the initial guess by a power of A. The factor $1/\lambda^{(l)}$ is to normalize the iterative vectors. If not normalized, the magnitude of the iterative vector may increase or decrease unboundedly and cause overflow or underflow.

INVERSE POWER METHOD. The inverse power method is identical to the power method except that the inverse of A^{-1} is used in place of A. Because eigenvalues of A^{-1} are reciprocals of A, the power method applied to A^{-1} will find the smallest eigenvalue of A. Of course, we must assume that the smallest eigenvalue of A is real and separate:

$$|\lambda_1| < |\lambda_2| \cdots \leqslant |\lambda_{N-1}| \leqslant |\lambda_N| \tag{7.4.8}$$

Otherwise, the method does not work.

The first iteration step is

$$A u^{(1)} = u^{(0)} \tag{7.4.9}$$

or equivalently

$$u^{(1)} = A^{-1} u^{(0)} \tag{7.4.10}$$

where $u^{(0)}$ is a nonnull vector as an initial guess. The subsequent iterative steps are

$$A u^{(k+1)} = \lambda^{(k)} u^{(k)} \tag{7.4.11}$$

with

$$\lambda^{(k)} = \frac{(u^{(k)}, \lambda^{(k-1)} u^{(k-1)})}{(u^{(k)}, u^{(k)})} \tag{7.4.12}$$

Equations (7.4.10) and (7.4.11) may be evaluated directly by using A^{-1} if the matrix is small. However, if A is a large sparse matrix, Gauss elimination or LU decomposition is used at every iteration cycle rather than saving A^{-1}.

It is easy to explain why the inverse power method converges. Because the initial vector can be expanded as Eq. (7.4.5), introducing it into Eq. (7.4.11) yields

$$u^{(k)} = \frac{\lambda^{(1)} \cdots \lambda^{(k-1)}}{(\lambda_1)^k} \left[u_1 + \left(\frac{\lambda_1}{\lambda_2}\right)^k u_2 + \cdots + \left(\frac{\lambda_1}{\lambda_N}\right)^k u_N \right]$$

Because λ_1 is the smallest eigenvalue, all the terms in the brackets in the foregoing equation except for u_1 vanishes as k increases.

SHIFTED INVERSE POWER METHOD. The shifted inverse power method is known also as the Wielandt method [Wachspress]. This method can find any eigenvalue and eigenvector provided it is real and separate. The essence of the method is to compute the eigenvalue of a shifted matrix given by

$$A' = A - \alpha I$$

The eigenvalue of A' is shifted by α from those of A, that is

$$\lambda_i' = \lambda_i - \alpha$$

Therefore, if the inverse power method is applied to A', the iterative vector converges to the eigenvalue λ_i' that is closest to zero. Thus, a desired eigenvalue of A is computed by setting α to an estimate of that eigenvalue. This method is applied in Sections 10.7 and 10.8.

7.5 QR ITERATION

The QR iteration is an iterative sequence of similarity transformations. Each iteration step consists of decomposition of the matrix into the QR form and the similarity transformation. Denoting the initial matrix by $A_0 = A$ where A is the original matrix of which eigenvalues are to be found. Matrix A_0 is decomposed by

$$A_0 = Q_0 R_0 \tag{7.5.1}$$

where Q_0 is a orthonormal matrix and R_0 is a upper triangular matrix. The similarity transformation is now written as

$$A_1 = Q_0^{-1} A_0 Q_0$$

or equivalently

$$A_1 = R_0 Q_0 \tag{7.5.2}$$

Subsequent iteration steps are essentially identical and written as

$$A_k = Q_k R_k \tag{7.5.3}$$

$$A_{k+1} = R_k Q_k$$

To improve computational efficiency, two modifications are made to the equations of Eq. (7.5.3). First, the iterative procedure is changed to

$$A_k - \alpha_k I = Q_k R_k \tag{7.5.4}$$

$$A_{k+1} = R_k Q_k + \alpha_k I$$

This change is called *shift* because subtracting $\alpha_k I$ from A_k shifts the eigenvalues of the right side by α_k as well as the eigenvalues of $R_k Q_k$. Adding $\alpha_k I$ in the second equation shifts the eigenvalues of A_{k+1} back to the original values. However, the shifts accelerate convergence of the eigenvalues close to α_k. Second, instead of applying the decomposition to the original matrix, the original matrix is first transformed to the Hessenberg form. When A_0 is in the Hessenberg form, all the subsequent A_k are also in the same form. This decomposition may be done by Gram-Schmit algorithm, but a more efficient process is used. See Morris for explanations for the convergence of QR iteration. For programming, Martin/Peters/Wilkinson and Shoup are also recommended.

If the Householder method is applied to a nonsymmetric matrix, the matrix is reduced to the form

$$\begin{bmatrix}
x & x & x & x & \cdots & x \\
x & x & x & x & & x \\
0 & x & x & x & & x \\
0 & 0 & x & x & & x \\
0 & 0 & 0 & x & & x \\
\cdot & \cdot & \cdot & \cdot & \cdots & \cdot \\
0 & 0 & 0 & 0 & x & x
\end{bmatrix} \tag{7.5.5}$$

which is called an upper Hessenberg form. The algorithm of the Householder transformation explained in the previous section works without any change.

The QR iteration transforms any real matrix in the Hessenberg form to a upper-block-triangular matrix such as

$$\begin{bmatrix}
D & x & x & x & x & x & x \\
0 & D & x & x & x & x & x \\
0 & 0 & D & D & x & x & x \\
0 & 0 & D & D & x & x & x \\
0 & 0 & 0 & 0 & D & D & x \\
0 & 0 & 0 & 0 & D & D & x \\
0 & 0 & 0 & 0 & 0 & 0 & D
\end{bmatrix} \tag{7.5.6}$$

where D and x both represent nonzero elements: D may be considered to belong to 1×1 or 2×2 diagonal blocks (square submatrices). The entire matrix has a form of upper-block-triangular matrix.

Eigenvalues of an upper-block-triangular matrix are equal to those of the block-diagonal matrix that is obtained by setting all the elements above the diagonal blocks to zero:

$$
\begin{bmatrix}
D & 0 & 0 & 0 & 0 & 0 & 0 \\
0 & D & 0 & 0 & 0 & 0 & 0 \\
0 & 0 & D & D & 0 & 0 & 0 \\
0 & 0 & D & D & 0 & 0 & 0 \\
0 & 0 & 0 & 0 & D & D & 0 \\
0 & 0 & 0 & 0 & D & D & 0 \\
0 & 0 & 0 & 0 & 0 & 0 & D
\end{bmatrix}
\tag{7.5.7}
$$

Eigenvalues of a block-diagonal matrix are equal to eigenvalues of the diagonal blocks. Single diagonal elements (such as indicated by D at the first, second, and last positions) are eigenvalues by themselves. Each 2×2 diagonal submatrix has a pair of real or complex conjugate eigenvalues. Eigenvalues of 2×2 blocks are calculated as roots of a quadratic polynomial.

PROGRAM 7–3 contains the Householder transformation and QR iteration, so it finds all eigenvalues of a nonsymmetric matrix at once.

Example 7.5

Find eigenvalues of the following matrix by Householder/QR iteration:

$$
\begin{bmatrix}
5.3 & 2.3 & 4.6 & 2.7 & 1.6 & 2.2 \\
2.4 & 7.8 & 5.7 & 8.4 & 3.4 & 4.2 \\
3.4 & 5.6 & 2.4 & 1.7 & 7.4 & 3.9 \\
8.3 & 7.5 & 9.2 & 6.1 & 5.2 & 7.9 \\
4.3 & 5.9 & 7.2 & 2.6 & 4.9 & 0.8 \\
0.9 & 2.7 & 4.9 & 4.8 & 6.7 & 4.8
\end{bmatrix}
$$

⟨**Solution**⟩

The Householder transformation reduces the foregoing matrix to

$$
\begin{bmatrix}
5.3000E+00 & -5.1043E+00 & 2.4620E+00 & 2.5047E+00 & -1.4576E+00 & -7.9177E-01 \\
-1.0272E+01 & 1.9500E+01 & -1.2706E+01 & -3.5993E+00 & 3.9338E+00 & 4.9378E+00 \\
0.0 & -1.1336E-01 & 4.3307E+01 & -4.1250E+02 & -1.1056E+00 & -8.6185E-01 \\
0.0 & 0.0 & 1.9893E+00 & -3.8121E+00 & -5.4740E+01 & 3.2432E+01 \\
0.0 & 0.0 & 0.0 & 2.4664E+00 & 3.0625E+00 & 3.9123E+00 \\
0.0 & 0.0 & 0.0 & 0.0 & -3.0033E+00 & 2.9188E+00
\end{bmatrix}
$$

The QR iteration transforms the matrix further to

$$
\begin{bmatrix}
28.3953 & -1.0980 & -4.5150 & 4.5051 & -4.6998 & 2.4398 \\
0.0 & -2.7794 & -0.4306 & 1.9669 & 2.4996 & 0.0536 \\
0.0 & 2.1144 & -3.3709 & -4.5749 & -0.3379 & 0.9949 \\
0.0 & 0.0 & 0.0 & 3.1160 & 0.7233 & 0.1572 \\
0.0 & 0.0 & 0.0 & 0.0 & 3.3789 & -3.7079 \\
0.0 & 0.0 & 0.0 & 0.0 & 3.3854 & 2.5600
\end{bmatrix}
$$

Two 2 × 2 diagonal blocks are found in the foregoing matrix:

$$
\begin{bmatrix} -2.7794 & -0.4306 \\ 2.1144 & -3.3709 \end{bmatrix} \text{ and } \begin{bmatrix} 3.3789 & -3.7079 \\ 3.3854 & 2.56 \end{bmatrix}
$$

Eigenvalues of these submatrices are respectively $-3.0751 \pm 0.9142j$ and $2.9694 \pm 3.5193j$ where $j = \sqrt{-1}$. Other diagonal elements of the matrix are eigenvalues. Therefore, eigenvalues of the given matrix are

$$
28.3953, -3.0751 \pm 0.9142j, 3.1160, 2.9694 \pm 3.5193j
$$

One important application of the QR iteration is to find roots of a polynomial in the power series form.

The following form of a matrix is called a *Frobenius matrix*:

$$
P = \begin{bmatrix}
-p_{N-1} & -p_{N-2} & -p_{N-3} & \cdots & -p_0 \\
1 & 0 & 0 & & \\
0 & 1 & 0 & & \\
\cdot & \cdot & \cdot & \cdots & \cdot \\
0 & & & 1 & 0
\end{bmatrix}
\tag{7.5.8}
$$

Its characteristic function is written as

$$
f(\lambda) = \det \begin{bmatrix}
-p_{N-1} - \lambda & -p_{N-2} & -p_{N-3} & \cdots & -p_0 \\
1 & -\lambda & 0 & & \\
0 & 1 & -\lambda & & \\
\cdot & \cdot & \cdot & \cdots & \\
0 & & & 1 & -\lambda
\end{bmatrix}
\tag{7.5.9}
$$

Expansion of Eq. (7.5.9) to a power series yields

$$
\begin{aligned}
f(\lambda) &= |P - \lambda I| \\
&= (-1)^N [\lambda^N + p_{N-1}\lambda^{N-1} + \cdots + p_0]
\end{aligned}
\tag{7.5.10}
$$

On the other hand, any polynomial equation,

$$
a_N x^N + \cdots + a_2 x^2 + a_1 x + a_0 = 0
\tag{7.5.11}
$$

can be expressed in the form of Eq. (7.5.8) with the definitions

$$p_0 = a_0/a_N$$

$$p_1 = a_1/a_N$$

$$p_2 = a_2/a_N \qquad\qquad (7.5.12)$$

$$\vdots$$

$$p_{N-1} = a_{N-1}/a_N$$

The Frobenius matrix is already in the Hessenberg form. Therefore, its eigenvalues may be calculated by the QR iteration. The Frobenius matrix may be used directly as input to a program such as PROGRAM 7–3.

Example 7.6

Calculate roots of the following polynomial equation by the QR iteration:

$$f(x) = x^5 - 0.2x^4 + 7x^3 + x^2 - 3.5x + 2 = 0$$

⟨**Solution**⟩

We regard $f(x)$ as the characteristic function for a Frobenius matrix:

$$M = \begin{bmatrix} 0.2 & -7 & -1 & 3.5 & -2 \\ 1 & 0 & 0 & 0 & 0 \\ 0 & 1 & 0 & 0 & 0 \\ 0 & 0 & 1 & 0 & 0 \\ 0 & 0 & 0 & 1 & 0 \end{bmatrix}$$

PROGRAM 7–3 is used to find eigenvalues. The results are

$$-0.9085, \quad 0.1403 \pm 2.7314i, \quad 0.4139 \pm 0.3506i$$

You can verify the results with Bairstow method, PROGRAM 3–7.

SUMMARY OF THIS SECTION

(a) The QR iteration transforms a matrix in the Hessenberg form to an upper-block-triangular form consisting of at most 2×2 blocks.

(b) A pair of eigenvalues that are complex conjugate or real are found from a 2×2 block in the diagonal position. Diagonal elements that appear without making 2×2 a block are real eigenvalues.

(c) One application of QR iteration is in finding roots of a polynomial. This is possible because a polynomial may be transformed to a Frobenius matrix, which is already in the Hessenberg form, so that QR iteration is directly applicable. The eigenvalues of the Frobenius matrix are the roots of the polynomial.

PROGRAMS

PROGRAM 7-1 Method of Interpolation

(A) Explanations

This program transforms the characteristic function of a matrix to a power series by the method of interpolation described in Section 7.2. The characteristic function is defined by

$$f(\lambda) = \det (A - \lambda I)$$

where A is a square matrix, I is an identity matrix and "det" means the determinant. The user defines the matrix elements in the declaration statement for a_ini[ ][ ]. The results of the computation are the coefficients of the power series representing $f(\lambda)$.

In main(), $f(\lambda) = \det (A - \lambda I)$ is evaluated for $n + 1$ values of λ. Actual calculation of the determinant is performed in determinant() by the forward elimination process of Gauss elimination.

The difference table for the Newton forward interpolation is developed in main(). The Markov coefficients as well as the power coefficients are then computed. No interactive input is necessary.

(B) List

```
/* CSL/c7-1.c      Method of Interpolation */
#include <stdio.h>
#include <stdlib.h>
#include <math.h>
#define TRUE 1
/*                 a[i][j]: matrix elements
                   n: order of matrix
                   de: increment of Lambda
                   ra[i]: i-th values of Lambda
                   df[i][j]: difference table
                   cc[i]: power coefficient
                   mv[l]: Markov coefficient
main()
{
int i, j, jj, k, l, m, _i, _r;
float a[20][20], cc[21], de, df[21][21], ff[21], mv[21], ra[21], s;
static float b[20][20];
void determinant();
static int n = 4;                           /* Defines the order of
matrix. */
static float a_ini[20][20]= {{4,  3,  2,  1},
                             {3,  3,  2,  1},
                             {2,  2,  2,  1},
                             {1,  1,  1,  1}};
    printf( "\n\nCSL/C7-1      Method of Interpolation \n\n" );
    printf( "Order of matrix =%d\n", n );
    printf( "Matrix:\n" );
```

```
for( i = 1; i <= n; i++ ){
   for( j = 1; j <= n; j++ ) {
      b[i][j] = a_ini[i-1][j-1];   printf(   "%12.5e ", b[i][j] );
   }
   printf(   "\n" );
}
while( TRUE ){   /* Increment of lambda for difference table. */
   printf( "\n Delta lambda ?   " );   scanf( "%f", &de );
   for( jj = 0; jj <= n; jj++ ){
      ra[jj] = jj*de;
      printf( "Lambda =   %g \n", ra[jj] );
      for( j = 1; j <= n; j++ ){
         for( i = 1; i <= n; i++ ){
            a[i][j] = b[i][j];
         }
         a[j][j] = a[j][j] - ra[jj];
      }
      determinant( n, a, &s );
      ff[jj] = s;
   }
   /*--Next part calculates forward difference table. */
   for( i = 0; i <= n; i++ ){
      df[i][0] = ff[i]; /* Initialization of difference table */
   }
   m = n;
   for( j = 1; j <= n; j++ ){   /* Difference table */
      m = m - 1;
      for( i = 0; i <= m; i++ )
             df[i][j] = df[i + 1][j - 1] - df[i][j - 1];
   }
   printf( "\nDifference table for determinants \n" );
   for( i = 0; i <= n; i++ ){
      printf( "%8.4f", ra[i] );
      for( j=0; j<=(n-i);j++ ) printf("%11.3e",df[i][j]);
      printf("\n");
   }
   /*Next is to compute power coefficients using Markov coefficients*/
   printf( "\nMarkov coefficients\n" );
   for( i = 0; i <= n; i++ )
   {
      cc[i] = 0; mv[i] = 0;
   }
   mv[1] = 1; /* Markov coefficients initialized */
   cc[0] = df[0][0];             cc[1] = df[0][1];
   for( k = 2; k <= n; k++ ){
      for( l = k; l >= 1; l-- ){
         mv[l] = (mv[l - 1] - (k - 1)*mv[l])/k;
         cc[l] = cc[l] + mv[l]*df[0][k]; /* Markov coefficients */
      }
      for( l = 1;l<= k;l++ ) printf( "%10.5f ",mv[l]);
   printf("\n" );
   }
   printf( "\n----Final result----\n" );
   printf( "\nPower       Coefficients\n" );
   for( i = 0; i <= n; i++ ){
      cc[i] = cc[i]/pow(de, i);
      printf( " %3d      %10.4f\n", i, cc[i] );
   }
```

```
      printf( "\nType 1 to continue, or 0 to stop. \n");scanf("%d", &k);
      if( k != 1 )  exit(0);
   }
}

void determinant(n, a, det)
int n;  float a[][20], *det;
{
int i, j, jc, jr, kc, pc, pv;
float r, tm;
   /* Determinant calculations */
   pc = 0; /* Pivoting counter initialization */
   for( i = 1; i <= (n - 1); i++ ){/* Forward elimination begins. */
      pv = i;
      for( j = i + 1; j <= n; j++ ){
         if( fabs( a[pv][i] ) < fabs( a[j][i] ) ) pv = j;
      }
      if( pv != i ){
         for( jc = 1; jc <= n; jc++ ){
            tm = a[i][jc];
            a[i][jc] = a[pv][jc];
            a[pv][jc] = tm;
         }
         pc = pc + 1;
      }
      for( jr = i + 1; jr <= n; jr++ ){
         if( a[jr][i] != 0 )
         {
            r = a[jr][i]/a[i][i];
            for( kc = i + 1; kc <= n; kc++ ){
               a[jr][kc] = a[jr][kc] - r*a[i][kc];
            }           /* End of forward elimination */
         }
      }
   }
   if( a[n][n] == 0 )  printf( "Matrix is singular.\n" );
   else {
      *det = 1; /* Determinat initialization */
      for( i = 1; i <= n; i++ ) *det = *det*a[i][i];
      if( pc != ( pc/2 )*2 )  *det = -*det;
      printf( " Determinant =    %g \n", *det );
      printf( " No. of pivoting = %d\n", pc );
   }
   return;
}
```

(C) Sample Output

```
CSL/C7-1     Method of Interpolation

Order of matrix =4
Matrix:
   4.00000e+00  3.00000e+00  2.00000e+00  1.00000e+00
   3.00000e+00  3.00000e+00  2.00000e+00  1.00000e+00
   2.00000e+00  2.00000e+00  2.00000e+00  1.00000e+00
   1.00000e+00  1.00000e+00  1.00000e+00  1.00000e+00
```

```
Delta lambda ?  0.5
Lambda =  0
 Determinant =     1
 No. of pivoting = 0
Lambda =  0.5
 Determinant =       6.250000e-02
 No. of pivoting = 1
Lambda =  1
 Determinant =       -8.940698e-08
 No. of pivoting = 0
Lambda =  1.5
 Determinant =      -4.4375
 No. of pivoting = 1
Lambda =  2
 Determinant =      -17
 No. of pivoting = 1


Difference table for determinants
 0.0000   1.000e+00 -9.375e-01  8.750e-01 -5.250e+00  1.500e+00
 0.5000   6.250e-02 -6.250e-02 -4.375e+00 -3.750e+00
 1.0000  -8.941e-08 -4.438e+00 -8.125e+00
 1.5000  -4.438e+00 -1.256e+01
 2.0000  -1.700e+01


Markov coefficients
 -0.50000    0.50000
  0.33333   -0.50000    0.16667
 -0.25000    0.45833   -0.25000    0.04167


----Final result----

Power           Coefficients
    0               1.0000
    1              -7.0000
    2              15.0000
    3             -10.0000
    4               1.0000
```

(D) Discussions

In the output the notation "Lambda" is used to denote the Greek letter λ. With a 4 × 4 sample matrix, this program expands the characteristic function defined by

$$f(\lambda) = \det\left[\begin{pmatrix} 4 & 3 & 2 & 1 \\ 3 & 2 & 2 & 1 \\ 2 & 2 & 2 & 1 \\ 1 & 1 & 1 & 1 \end{pmatrix} - \lambda I\right]$$

in the power series form using the method of interpolation. Because $f(\lambda)$ is a polynomial of the fourth order in this case, the interpolation polynomial of the fourth order fitted to the five values of $f(\lambda)$ will become $f(\lambda)$ exactly. The interpolation polynomial is first generated in the form of Newton's forward interpolation formula, which is then converted to a power series using the Markov coefficients. Looking at the output, one can easily write

$$f(\lambda) = 1 - 7\lambda + 15\lambda^2 - 10\lambda^3 + \lambda^4$$

The roots of the $f(\lambda)$, namely the eigenvalues of the matrix, can be found by using Bairstow's method or other methods to compute the zeros of a nonlinear function. This method is effective only for small matrices.

PROGRAM 7-2 Householder-Bisection Method

(A) Explanations

This program finds eigenvalues of a symmetric real matrix A by transforming the matrix to a symmetric tridiagonal matrix and then computing the eigenvalues of the tridiagonal matrix by the bisection method. Eigenvalues of the matrix A satisfy

$$\det (A - \lambda I) = 0$$

where λ is an eigenvalue of A.

The order and elements of the matrix are specified in the variable declaration statements. The reduction of the given matrix to a tridiagonal form is performed in main(). The bisection scheme is coded in bisec(). The determinant of a tridiagonal matrix is calculated in determ().

(B) List

```
/* CSL/c7-2.c   Householder/Bisection */
#include <stdio.h>
#include <stdlib.h>
#include <math.h>
#define TRUE 1
/*              n: order of matrix
                a[i][j]: matrix elements
                s: s
                ssr: square root of s
                uau: (u-transpose)A(u)
                t[i][j]: temporary memory for matrix
                ei[l][j]: j-th eigenvalue of l-th principal
                          tridiagonal matrix
                xl: lower bound of eigenvlaue estimate
                xh: upper bound of eigenvalue estimate        */
struct t_bc { int l; } bc;
```

```
main()
{
int i, ir, j, k, km, n, _i, _r;
double b23, dumm, ei[11][11], g[11], h, s, ssr,
    t[11][11], u[11], uau, w, xh, xl, xm;
static double a[11][11];
void bisec();
static n_order=4;                              /* order of matrix */
static double a_ini[11][11]= {{4, 3, 2, 1},
                              {3, 3, 2, 1},
                              {2, 2, 2, 1},
                              {1, 1, 1, 1}};
    printf( "\nCSL/C7-2      Householder/Tridiagonal\n" );
    printf( "              (Symmetric matrix only)\n" );
    /*---This program calculates eigenvalues of a symmetric matrix in two
     *    steps: Step 1:
     *           Reduction of a matrix to tridiagonal form by Householder
     *           scheme.
     *           Step 2:
     *           Calculation of eigenvalue of the tridiagonal matrix
     *           by bisection method combined with linear interpolation.*/
    n = 4; /* Defines order of matrix. */
    printf( "         Reduction of a matrix to Hessenberg or \n" );
    printf( "         tridiagonal form by householder scheme.\n" );
    printf( "Original matrix (This matrix must be symmetric.)\n" );
    for( i = 1; i <= n; i++ ){
      for( j = 1; j <= n; j++ ){
        a[i][j] = a_ini[i-1][j-1];   printf( "%g ", a[i][j] );
      }   printf( "\n" );
    }
    for( ir = 1; ir <= (n - 2); ir++ ){ /* Householder scheme starts. */
      s = 0;
      for( i = 1; i <= n; i++ ){
        u[i] = 0;
        if( i > ir + 1 )  u[i] = a[i][ir];
        if( i > ir )  s = s + a[i][ir]*a[i][ir];
      }
      w = 1;
      if( a[ir + 1][ir] < 0 )  w = -1;
      ssr = sqrt( s );
      h = s + fabs( a[ir + 1][ir] )*ssr;
      u[ir + 1] = a[ir + 1][ir] + ssr*w;
      uau = 0;
      for( i = 1; i <= n; i++ ){
        for( j = 1; j <= n; j++ ){
          uau = uau + u[i]*a[i][j]*u[j];
          if( (i <= ir) && (j <= ir) )  t[i][j] = a[i][j];
          else if( (j == ir) && (i >= ir + 2) )  t[i][j] = 0;
          else{
            b23 = 0;
            for( k = 1; k <= n; k++ )
                b23 = b23 - (u[i]*a[k][j] + a[i][k]*u[j])*u[k];
            t[i][j] = a[i][j] + b23/h;
          }
        }
      }
      uau = uau/h/h;
      for( i = 1; i <= n; i++ ){
```

```
         for( j = 1; j <= n; j++ ){
            a[i][j] = t[i][j] + uau*u[i]*u[j];
            if( fabs( a[i][j] ) < .000001 )  a[i][j] = 0;
         }
      }
   }
   printf( "Hessenberg or tridiagonal matrix\n" );
   for( i = 1; i <= n; i++ ){
      printf( " " );
      for( j = 1; j <= 4; j++ )   printf( "%12.5e", a[i][j] );
      printf( "\n" );
   }
   printf( "To continue, type 1 and hit RETURN key.\n" );
   scanf( "%f", &dumm );
   km = n;
   for( bc.l = 1; bc.l <= km; bc.l++ ){
      if( bc.l == 1 )  ei[1][1] = a[1][1];
      else{
          for( j = 1; j <= bc.l; j++ ){
             xl = ei[bc.l - 1][j - 1];  xh = ei[bc.l - 1][j];
             bisec( g, a, &xl, &xh, &xm );  ei[bc.l][j] = xm;
          }
      }
      ei[bc.l][0] = -99;  ei[bc.l][bc.l + 1] = 99;
      if( bc.l == n ){
          printf( "Final result (eigenvalues of the whole matrix)\n" );
          printf("-------------------------------------------------\n");
      }
      else  printf(  "%d%ld \n", bc.l, bc.l );
      printf( " " );
      for( i = 1; i <= bc.l; i++ )  printf( "%12.5e", ei[bc.l][i] );
      printf( "\n" );
   }
   printf( "-------------------------------------------------\n" );
   exit(0);
}

void bisec(g, a, xl, xh, xm)   /* Bisection scheme*/
double g[], a[][11], *xl, *xh, *xm;
{
int ka;
double dx, xb, yh, yl, ym;
void determ();
   ka = 0;          /* Finds roots of a characteristic equation */
   determ( g, a, xl, &yl );
   determ( g, a, xh, &yh );
   while( TRUE ){
      ka = ka + 1;
      if( ka > 99 )  return;
      dx = *xh - *xl;
      if( dx < .0000001 )  return;
      if( dx > 1 ){
         *xm = (*xl + *xh)/2; /* Bisection scheme */
         determ( g, a, xm, &ym );
      }
      else{
         xb = *xm;
         *xm = (*xl*yh - *xh*yl)/(yh - yl); /* Linear interpolation */
```

```
{
char g[8][11];
int i, ir, it, j, k, l, m, na, nn, _i, _r;
double b23, dummy, f[11], h, hh, im[11], ma, p, q,
    r, rl[11], M_eps,e, e_, s, ssr, t[11][11], rd, rd_cnv,
    u[11], uau, w, x, y, z;
static double a[11][11];
static n = 6;          /* order of matrix */
static double a_ini[6][6] ={{5.3,  2.3,  4.6,  2.7,  1.6,  2.2},
                            {2.4,  7.8,  5.7,  8.4,  3.4,  4.2},
                            {3.4,  5.6,  2.4,  1.7,  7.4,  3.9},
                            {8.3,  7.5,  9.2,  6.1,  5.2,  7.9},
                            {4.3,  5.9,  7.2,  2.6,  4.9,  0.8},
                            {0.9,  2.7,  4.9,  4.8,  6.7,  4.8}};
    printf( "\n\nCSL/C7-3     Householder / QR iteration \n\n") ;
    printf( "Step 1-- Reduction of a matrix to Hessenberg or \n" );
    printf( "           tridiagonal form by Householder scheme.\n" );
    printf( "\nOriginal matrix: \n" );
    r = 1;
    for( i = 1; i <= n; i++ ){
       for( j = 1;j<=n;j++ ) {
          a[i][j] = a_ini[i-1][j-1];
          printf( "%12.5e ",a[i][j]);
       }
       printf("\n" );
    }
    for( ir = 1; ir <= (n - 2); ir++ ){
       s = 0;
       for( i = 1; i <= n; i++ ){
          u[i] = 0;
          if( i > ir + 1 )  u[i] = a[i][ir];
          if( i > ir )        s = s + a[i][ir]*a[i][ir];
       }
       w = 1;
       if( a[ir + 1][ir] < 0 ) w = -1;
       ssr = sqrt ( s );
       h = s + fabs( a[ir + 1][ir] )*ssr;
       u[ir + 1] = a[ir + 1][ir] + ssr*w;
       uau = 0;
       for( i = 1; i <= n; i++ ){
          for( j = 1; j <= n; j++ ){
             uau = uau + u[i]*a[i][j]*u[j];
             if( i <= ir && j <= ir )          t[i][j] = a[i][j];
             else if( j == ir && i >= ir + 2 ) t[i][j] = 0;
             else{
                b23 = 0;
                for( k = 1; k <= n; k++ )
                b23 = b23 - (u[i]*a[k][j] + a[i][k]*u[j])* u[k];
                t[i][j] = a[i][j] + b23/h;
             }
          }
       }
       uau = uau/h/h;
       for( i = 1; i <= n; i++ ){
          for( j = 1; j <= n; j++ ){
             a[i][j] = t[i][j] + uau*u[i]*u[j];
             if( fabs( a[i][j] ) < .000001 ) a[i][j] = 0;
          }
       }
    }
```

```
printf( "\nHessenberg or tridiagonal matrix: \n" );
for( i = 1; i <= n; i++ ){
   for( j = 1;j<= 6;j++ ) printf("%12.5e ",a[i][j]); printf(   "\n" );
}
printf( "\nTo continue, type any number and hit RETURN key.\n" );
scanf( "%f", &dummy );
printf( "\nStep 2-- QR iteration to find eigenvlaues\n" );
ma = 1.0; /* Machine epsilon is calculated. */
e = 1.0;
while( e>0 ){
   ma = ma/2;   e=ma*0.98+1; e=e-1; if (e>0) M_eps=e ;
}
ma = M_eps*M_eps  ; /* square of the machine epsilon. */
printf( " Convergence criterion =%g \n\n", ma );
nn = n; rd_cnv = 0.99;
while( nn != 0 ){
   na = nn - 1;
   for (it=1; it<=90; it++) {  /* iteration loop*/
      l=1;
      for( m = nn; m >= 2; m-- ){
         rd =   fabs(a[m][m-1])/(fabs(a[m-1][m-1]) + fabs(a[m][m]));
         if(rd<ma) {
            l=m;
            rd_cnv = rd;
         }
      }
      printf(  "Iter. no.=%3d    nn =%3d   l=%3d  rd_cnv =%12.3e\n",
                           it, nn, l, rd_cnv  );
      x = a[nn][nn];
      if( l == nn ){       /*---------Single root is found.   */
         rl[nn] = x;    im[nn] = 0;
         nn = nn - 1;
         goto L_1202;
      }
      y = a[na][na];
      r = a[nn][na]*a[na][nn];
      if( l == nn-1 ){       /*---------A pair of roots is found.*/
         p = (y - x)/2;  q = p*p + r;  y = sqrt( fabs( q ) );
         if( q < 0 ){   /*-----------complex pair */
            rl[nn - 1] = x + p;           rl[nn] = x + p;
            im[nn - 1] = y;   im[nn] = -y;
            nn = nn - 2;
         }
         else{                  /*-----------They are real roots. */
            if( p < 0 )   y = -y;
            y = p + y;
            rl[nn - 1] = x + y;  rl[nn] = x - r/y;
            im[nn - 1] = 0;      im[nn] = 0;
            nn = nn - 2;
         }
         goto L_1202;
      }
      if( it == 90 ){
            printf( "Iteration limit exceeded.\n" );
            goto L_765;
      }
      if( it == 10 || it == 20 ){
         y = fabs( a[nn][nn-1] ) + fabs( a[nn-1][nn - 2] );
         s = 1.5*y;
         y = y*y;
```

```
        }
        else{
           s = x + y;
           y = x*y - r;
        }
        for( m = nn - 2; m >= 1; m-- ){
           x = a[m][m];
           r = a[m + 1][m];
           z = a[m + 1][m + 1];
           p = x*(x - s) + y + r*a[m][m + 1];
           q = r*(x + z - s);
           r = r*a[m + 2][m + 1];
           w = fabs( p ) + fabs( q ) + fabs( r );
           p = p/w; q = q/w; r = r/w;
           if( m == 1 )  break;
           hh = fabs( a[m][m - 1] )*(fabs( q ) + fabs( r ));
              if( hh < ma*fabs( p )*(fabs( a[m - 1][m - 1] ) +
                              fabs( x ) + fabs( z )) )  break;
        }
        for( i = m + 2; i <= nn; i++ )      a[i][i - 2] = 0;
        for( i = m + 3; i <= nn; i++ )      a[i][i - 3] = 0;
        for( k = m; k <= nn-1; k++ ){
           if( k != m ){
              p = a[k][k-1]; q = a[k+1][k-1]; r = a[k+2][k-1];
              if( nn-1 == k )  r = 0;
                 x = fabs( p ) + fabs( q ) + fabs( r );
                 if( x == 0 )  break;
                 p = p/x; q = q/x; r = r/x;
           }
           s = sqrt( p*p + q*q + r*r );
           if( p < 0 )    s = -s;
           if( k != m )    a[k][k - 1] = -s*x;
           if( l != m )    a[k][k - 1] = -a[k][k - 1];
           p = p + s; x = p/s; y = q/s;
           z = r/s;    q = q/p; r = r/p;
           for( j = k; j <= nn; j++ ) {
              p = a[k][j] + q*a[k + 1][j];
              if( nn-1 != k ){
                 p = p + r*a[k + 2][j];
                 a[k + 2][j] = a[k + 2][j] - p*z;
              }
                 a[k + 1][j] = a[k + 1][j] - p*y;
                 a[k][j] = a[k][j] - p*x;
           }
           j = nn;    if( k + 3 < nn )    j = k + 3;
           for( i = 1; i <= j; i++ ){
              p = x*a[i][k] + y*a[i][k + 1];
              if( nn-1 != k ){
                 p = p + z*a[i][k + 2];
                 a[i][k + 2] = a[i][k + 2] - p*r;
              }
              a[i][k + 1] = a[i][k + 1] - p*q;
              a[i][k] = a[i][k] - p;
           }
        }
     }
  }
L_1202:;
  }
```

```
L_765:    printf( "\nEigenvalues: \n" );
    printf( "-----------------------------------------------\n" );
    printf( "No.      Real part      Imaginary part \n" );
    printf( "-----------------------------------------------\n" );
    for( i = 1; i <= n; i++ )
        printf( "%3d      %12.5e    %12.5e \n", i, rl[i], im[i] );
    printf( "-----------------------------------------------\n" );
    printf( "\nReduced matrix:\n" );
    for( j = 1; j <= n; j++ ) {
        for( i = j+1; i <= n; i++ ) {
            if ( fabs(a[i][j]) < fabs(a[j][j])*1.0e-10 ) a[i][j] = 0.0;
        }
    }
    for( i = 1; i <= n; i++ ){
        for( j = 1; j <= n; j++ )  printf( "%12.5e ", a[i][j] );
        printf( "\n" );
    }
    printf("\n\n\n");
    exit(0);
}
```

(C) Sample Output

```
CSL/C7-3     Householder / QR iteration

Step 1-- Reduction of a matrix to Hessenberg or
         tridiagonal form by Householder scheme.

Original matrix:
  5.30000e+00  2.30000e+00  4.60000e+00  2.70000e+00  1.60000e+00  2.20000e+00
  2.40000e+00  7.80000e+00  5.70000e+00  8.40000e+00  3.40000e+00  4.20000e+00
  3.40000e+00  5.60000e+00  2.40000e+00  1.70000e+00  7.40000e+00  3.90000e+00
  8.30000e+00  7.50000e+00  9.20000e+00  6.10000e+00  5.20000e+00  7.90000e+00
  4.30000e+00  5.90000e+00  7.20000e+00  2.60000e+00  4.90000e+00  8.00000e-01
  9.00000e-01  2.70000e+00  4.90000e+00  4.80000e+00  6.70000e+00  4.80000e+00


Hessenberg or tridiagonal matrix:
  5.30000e+00 -5.10426e+00  2.46195e+00  2.50473e+00 -1.45763e+00 -7.91762e-01
 -1.02718e+01  1.95001e+01 -1.27062e+01 -3.59933e+00  3.93381e+00  4.93779e+00
  0.00000e+00 -1.13363e+01  4.33072e+00 -4.12490e-02 -1.10558e+00 -8.61841e-01
  0.00000e+00  0.00000e+00  1.98931e+00 -3.81215e+00 -5.47416e-01  3.24329e-01
  0.00000e+00  0.00000e+00  0.00000e+00  2.46643e+00  3.06251e+00  3.91227e+00
  0.00000e+00  0.00000e+00  0.00000e+00  0.00000e+00 -3.00331e+00  2.91877e+00

Step 2-- QR iteration to find eigenvlaues
    Convergence criteria =7.703720e-34
  Iteration no.=  1    nn=   6
  Iteration no.=  2    nn=   6
  Iteration no.=  3    nn=   6
  Iteration no.=  4    nn=   6
  Iteration no.=  5    nn=   6
  Iteration no.=  1    nn=   4
  Iteration no.=  2    nn=   4
  Iteration no.=  3    nn=   4
  Iteration no.=  4    nn=   4
  Iteration no.=  1    nn=   3
  Iteration no.=  2    nn=   3
```

```
Eigenvalues:
-------------------------------------------------
No.       Real part        Imaginary part
-------------------------------------------------
 1       2.83953e+01        0.00000e+00
 2      -3.07518e+00        9.14233e-01
 3      -3.07518e+00       -9.14233e-01
 4       3.11612e+00        0.00000e+00
 5       2.96947e+00        3.51931e+00
 6       2.96947e+00       -3.51931e+00
-------------------------------------------------
```

```
Reduced matrix:
 2.83953e+01   4.01039e+00  -2.34910e+00  -4.50410e+00  -2.86934e+00  -4.45069e+00
 0.00000e+00  -3.53696e+00  -2.06736e+00  -1.22093e+00   1.80744e-01   2.49369e+00
 0.00000e+00   5.07438e-01  -2.61340e+00  -4.82801e+00  -1.02226e+00  -2.43248e-01
 0.00000e+00   0.00000e+00   0.00000e+00   3.11612e+00  -8.87623e-02   7.34865e-01
 0.00000e+00   0.00000e+00   0.00000e+00   0.00000e+00   2.59727e+00  -3.31195e+00
 0.00000e+00   0.00000e+00   0.00000e+00   0.00000e+00   3.78147e+00   3.34167e+00
```

(D) Discussions

Eigenvalues of the 6×6 matrix defined within the program are computed. Among the six eigenvalues found, two are real and four are complex.

PROBLEMS

(7.1) Transform the following characteristic function to a Newton forward polynomial by evaluating $f(\lambda)$ for $\lambda = -1, 0$, and 1:

$$f(\lambda) = \det \begin{bmatrix} 2 - \lambda & -2 \\ 1 & 3 - \lambda \end{bmatrix}$$

Then reduce the Newton interpolation polynomial to the power series by using the Markov coefficients. Compare your result with the power series obtained by direct expansion of the determinant.

(7.2) Transform the following characteristic equation to a power series by using the Newton interpolation polynomial:

$$f(\lambda) = \det (A - \lambda I)$$

where

$$A = \begin{bmatrix} 3 & 2 & 1 \\ 2 & 1 & 1 \\ 1 & 1 & 1 \end{bmatrix}$$

(*Hint:* Evaluate $f(\lambda)$ for $\lambda = 0, 1, 2$, and 3.)

(7.3) Repeat Problem (7.2) for the following matrices:

(a) $\begin{bmatrix} 2 & -1 & 0 \\ -1 & 2 & -1 \\ 0 & -1 & 2 \end{bmatrix}$

(b) $\begin{bmatrix} 2 & -1 & 0 \\ -1.5 & 2 & -0.5 \\ 0 & -1 & 1 \end{bmatrix}$

(7.4) Find eigenvalues of the following matrices by the method of interpolation:

(a) $\begin{bmatrix} 5 & 2 & 7 \\ 3 & 1 & 5 \\ 2 & 6 & 2 \end{bmatrix}$

(b) $\begin{bmatrix} 1 & 1 & 1 & 1 \\ 1 & 2 & 2 & 2 \\ 1 & 2 & 3 & 3 \\ 1 & 2 & 3 & 4 \end{bmatrix}$

(7.5) Transform the characteristic equation for the matrix given by

$$\begin{bmatrix} 5.3 & 2.3 & 4.6 & 2.7 \\ 2.4 & 7.8 & 5.7 & 8.4 \\ 3.4 & 5.6 & 2.4 & 1.7 \\ 8.3 & 7.5 & 9.2 & 6.1 \end{bmatrix}$$

to a power series by the method of interpolation, and then find the eigenvalues by using the Bairstow scheme:

(7.6) Find eigenvalues of the following matrices by using the Householder/tridiagonal scheme:

$$\begin{bmatrix} 3 & 2 & 1 \\ 2 & 2 & 1 \\ 1 & 1 & 1 \end{bmatrix}$$

(7.7) Find eigenvalues of the following symmetric matrix by using the Householder/tridiagonal scheme:

$$M = \begin{bmatrix} 4 & 3 & 2 & 1 \\ 3 & 3 & 2 & 1 \\ 2 & 2 & 2 & 1 \\ 1 & 1 & 1 & 1 \end{bmatrix}$$

(7.8) Transform the following matrices to the Hessenberg form by using a similarity transformation by hand calculations.

(a) $\begin{bmatrix} 3 & 1 & 2 \\ 1 & 2 & 1 \\ 2 & 1 & 4 \end{bmatrix}$

(b) $\begin{bmatrix} 4 & 1 & 2 \\ 3 & 2 & 1 \\ 0 & 1 & 2 \end{bmatrix}$

(7.9) Find eigenvalues of the following matrices by using the Householder/tridiagonal scheme:

$$\begin{bmatrix} 5.3 & 2.3 & 4.6 & 2.7 \\ 2.3 & 7.8 & 5.7 & 8.4 \\ 4.6 & 5.7 & 2.4 & 1.7 \\ 2.7 & 8.4 & 1.7 & 6.1 \end{bmatrix}$$

(7.10) Find eigenvalues of the following matrix by using the Householder/QR scheme:

$$\begin{bmatrix} 5.3 & 2.3 & 4.6 & 2.7 \\ 2.4 & 7.8 & 5.7 & 8.4 \\ 3.4 & 5.6 & 2.4 & 1.7 \\ 8.3 & 7.5 & 9.2 & 6.1 \end{bmatrix}$$

(7.11) Find eigenvalues of the matrix given in Problem (7.9) by using the Householder/QR scheme.

(7.12) Calculate the roots of the following polynomials by using the QR scheme:

(a) $y = 1.1 + 4x + 3x^2 + x^3$

(b) $y = -2.202 - 5.301x - 3x^2 + 1.1x^3 + x^4$

(*Hint:* Transform the polynomial to an equivalent matrix eigenvalue problem by using the Frobenius matrix, and then apply the QR scheme.)

(7.13) As multiplicity of an eigenvalue increases, accuracy of the QR scheme becomes poorer. This fact must be kept in mind also in applying the scheme to find roots of a polynomial. (The situation is very similar to the Bairstow scheme; see Problem 3.30). The following polynomial has sextuple roots of $x = 1$:

$$x^6 - 6x^5 + 15x^4 - 20x^3 + 15x^2 - 6x + 1 = 0$$

Try to find all the roots by using the QR or Householder/QR scheme.

REFERENCES

Cowell, W. R., ed., *Sources and Development of Mathematical Software*, Prentice-Hall, 1984.

Faddeeva, V. N., *Computational Methods of Linear Algebra*, Dover, 1959.

Garbow, B. S., J. M. Boyle, J. J. Dongarra, and C. B. Moler, Matrix Eigensystem Routines— EISPACK Guide Extension, *Lecture Notes in Computer Science*, Vol. 51, Springer-Verlag, 1977.

IMSL Library Reference Manual, IMSL, Inc., 7500 Bellaire Boulvard, Houston, Texas 77036.

Jennings, A., *Matrix Computations for Engineers and Scientists*, Wiley, 1977.

Lawson, C. R. H., D. Kincaid, and F. Krogh, *Matrix Eigensystem Routines–EISPACK Guide Extension*, *Lecture Notes in Computer Science*, Vol. 51, Springer-Verlag, 1977.

Martin, R. S., P. Peters, and J. H. Wilkinson, "The QR Algorithm for Real Hessenberg Matrices," *Numerische Math.*, Vol. 14, 1970.

Morris, J. L., *Computational Methods in Elementary Numerical Analysis*, Wiley, 1983.

Shoup, E., *Applied Numerical Methods for the Micro-Computer*, Prentice-Hall, 1984.

Smith, B. T., J. M. Boyle, J. J. Dongarra, B. S. Garbow, Y. Ikebe, V. C. Klema, and C. B. Moler, Matrix Eigensystem Routines—EISPACK Guide, *Lecture Notes in Computer Science*, Vol. 6, Springer-Verlag, 1976.

Stewart, G. W., *Introduction to Matrix Computations*, Academic Press, 1974.

Wachspress, E. L., *Iterative Solution of Elliptic Systems*, Prentice-Hall, 1966.

8

Curve Fitting to Measured Data

8.1 INTRODUCTION

Measured data fluctuate. Fluctuations may be a result of random error of the measurement system or the intrinsic stochastic behavior of the system measured. Whatever the reason, the need to fit a function to measured data often occurs. For example, a researcher may want to develop an empirical formula for the system measured, or an economist may want to fit a curve to a current economic trend to forecast the future.

If the number of data points equals the order of the polynomial plus one, we can exactly fit a polynomial to the data points (this is the polynomial interpolation described in Chapter 2). In fitting a function to measured data, however, a much larger number of measured data than the order of a polynomial is used. Indeed, as the number of data points increases, accuracy of the fitted curve increases.

How, then, can we fit a function to data points? The best we can do is to consider a function with a small number of free parameters and determine the parameters so that deviation of the function from the data points is minimal. The minimization of deviation is achieved by means of the least square method.

8.2 LINE REGRESSION

Suppose we desire to find a linear function that fits the data in Table 8.1 (see also Figure 8.1) with minimal deviations. The linear function determined in this manner

Table 8.1 A set of
measured data

i	x	y
1	0.1	0.61
2	0.4	0.92
3	0.5	0.99
4	0.7	1.52
5	0.7	1.47
6	0.9	2.03

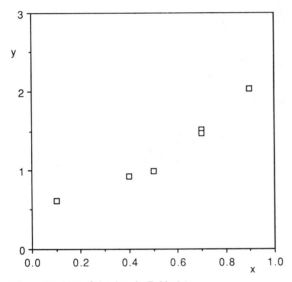

Figure 8.1 Plot of the data in Table 8.1

is called a regression line. The linear function is now expressed by

$$g(x) = a + bx \tag{8.2.1}$$

where a and b are undetermined constants. Deviation of the line from each data point
is defined by

$$r_i = y_i - g(x_i) = y_i - (a + bx_i), \quad i = 1, 2, \ldots, L \tag{8.2.2}$$

where L is the total number of data points, which is 6 in the present example, and a
and b are the constants to be determined.

The square total of the deviations is given by

$$R = \sum_{i=1}^{L} (r_i)^2 = \sum_{i=1}^{L} (y_i - a - bx_i)^2 \tag{8.2.3}$$

Because a and b are arbitrary parameters, they are determined so as to minimize R. The minimum of R occurs if partial derivatives of R with respect to a and b become zero:

$$\frac{\partial R}{\partial a} = -2 \sum_{i=1}^{L} (y_i - a - bx_i) = 0$$

$$\frac{\partial R}{\partial b} = -2 \sum_{i=1}^{L} x_i(y_i - a - bx_i) = 0 \tag{8.2.4}$$

which may be rewritten after dividing by -2 as

$$\begin{bmatrix} A_{1,1} & A_{1,2} \\ A_{2,1} & A_{2,2} \end{bmatrix} \begin{bmatrix} a \\ b \end{bmatrix} = \begin{bmatrix} Z_1 \\ Z_2 \end{bmatrix} \tag{8.2.5}$$

where

$A_{1,1} = L$
$A_{1,2} = \sum x_i$
$Z_1 = \sum y_i$
$A_{2,1} = \sum x_i$
$A_{2,2} = \sum (x_i)^2$
$Z_2 = \sum x_i y_i$

In the foregoing equations the summation is over i from 1 to L. Notice that $A_{2,1}$ equals $A_{1,2}$. The solution of Eq. (8.2.5) is written as

$$a = \frac{A_{2,2}Z_1 - A_{1,2}Z_2}{d}$$

$$b = \frac{A_{1,1}Z_2 - A_{2,1}Z_1}{d} \tag{8.2.6}$$

where

$$d = A_{1,1}A_{2,2} - A_{1,2}A_{2,1}$$

PROGRAM 8–1 included in this chapter performs the line regression.

Example 8.1

Determine the regression line for the data in Table 8.1.

⟨**Solution**⟩

We calculate the coefficients of Eq. (8.2.5) as in Table 8.2.

Table 8.2

Purpose	A12, A21	Z1	A22	Z2
i	x_i	y_i	x_i^2	$x_i y_i$
1	.1	.61	.01	.061
2	.4	.92	.16	.368
3	.5	.99	.25	.495
4	.7	1.52	.49	1.064
5	.7	1.47	.49	1.029
6	.9	2.03	.81	1.827
Total	3.3	7.54	2.21	4.844

From the results in Table 8.2 we get

$$A11 = L = 6, \quad A12 = 3.3, \quad Z1 = 7.54$$
$$A21 = 3.3, \quad A22 = 2.21, \quad Z2 = 4.844$$

Thus, Eq. (8.2.5) becomes

$$\begin{bmatrix} 6 & 3.3 \\ 3.3 & 2.21 \end{bmatrix} \begin{bmatrix} a \\ b \end{bmatrix} = \begin{bmatrix} 7.54 \\ 4.844 \end{bmatrix}$$

The solution is

$$a = 0.2862, \quad b = 1.7645$$

The regression line therefore becomes

$$g(x) = 0.2862 + 1.7645x$$

which is plotted in Figure E8.1 with the given data points.
We now evaluate deviation of the fitted line as in Table 8.3.

Table 8.3

i	$x(i)$	$y(i)$	$g = a + bx$	Deviation
1	0.1	0.61	0.46261	0.14738
2	0.4	0.92	0.99198	−0.07198
3	0.5	0.99	1.16843	−0.17844
4	0.7	1.52	1.52135	−0.00135
5	0.7	1.47	1.52135	−0.05135
6	0.9	2.03	1.87426	0.15574

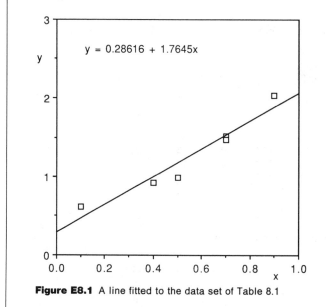

Figure E8.1 A line fitted to the data set of Table 8.1

The line regression method may be applied even to fitting nonlinear functions to a data set in some cases. We consider here fitting a power form

$$g(x) = Cx^b \tag{8.2.7}$$

to a data set (x_i, y_i), where C and b are constants to be determined.

By taking logarithm of the preceding formula we get

$$\log(g) = \log(C) + b\log(x) \tag{8.2.8}$$

With the definitions

$$Y = \log(g), \quad X = \log(x), \quad a = \log(C) \tag{8.2.9}$$

Eq. (8.2.8) is reduced to

$$Y = a + bX \qquad (8.2.10)$$

which is a linear function. The data set may be transformed to $(\log(x_i), \log(y_i))$. Then determining a and b is a line regression problem.

Example 8.2

The following data set is plotted in Figure E8.2A. Fit the function, $g(x) = Cx^b$, to the data set.

i	$x(i)$	$y(i)$
1	0.1500	4.4964
2	0.4000	5.1284
3	0.6000	5.6931
4	1.0100	6.2884
5	1.5000	7.0989
6	2.2000	7.5507
7	2.4000	7.5106
8	2.7000	8.0756
9	2.9000	7.8708
10	3.5000	8.2403
11	3.8000	8.5303
12	4.4000	8.7394
13	4.6000	8.9981
14	5.1000	9.1450
15	6.6000	9.5070
16	7.6000	9.9115

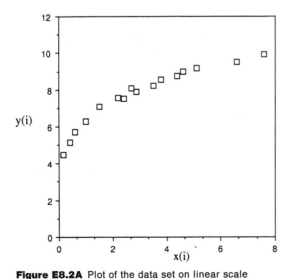

Figure E8.2A Plot of the data set on linear scale

⟨**Solution**⟩

We first convert the data to log (x_i) versus log (y_i) as follows (natural logarithm is used):

i	log (x_i)	log (y_i)
1	−1.8971	1.5033
2	−0.9163	1.6348
3	−0.5108	1.7393
4	0.0100	1.8387
5	0.4055	1.9599
6	0.7885	2.0216
7	0.8755	2.0163
8	0.9933	2.0889
9	1.0647	2.0632
10	1.2528	2.1090
11	1.3350	2.1436
12	1.4816	2.1678
13	1.5261	2.1970
14	1.6292	2.2132
15	1.8871	2.2520
16	2.0281	2.2937

By the line regression using PROGRAM 8–8, we find that the line fitted to the log-log data (see Figure E8.2B) is

$$G(X) = 1.8588 + 0.2093X$$

or

$$\log (C) = 1.8588, \quad b = 0.2093$$

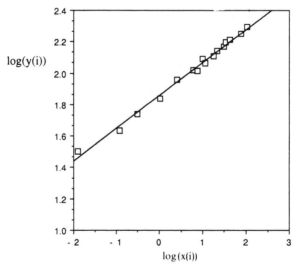

Figure E8.2B Plot of the data and fitted line on log scale

Therefore, the fitted curve on the x-y coordinates becomes

$$g(x) = (6.4160)x^{0.2093}$$

where $C = e^{1.8588} = 6.4160$ is used.

SUMMARY OF THIS SECTION

(a) The purpose of line regression is to fit a linear function to data points by the least square method.

(b) Linear regression is also applicable to fitting a power law form.

(c) PROGRAM 8–1 may be used for a line regression.

8.3 CURVE FITTING WITH A HIGHER-ORDER POLYNOMIAL

Line regression as explained in the previous section works well if the measured data are intrinsically linear or if the range of abscissa is small. For other cases, however, better results may be obtained by fitting a higher-order polynomial to the data set.

The principle of least square can be extended to fitting a polynomial of any order to measured data. An Nth order polynomial is first written as

$$g(x) = a_0 + a_1 x + a_2 x^2 + \cdots a_N x^N \tag{8.3.1}$$

Deviation of the curve from each data point is

$$r_i = y_i - g(x_i), \quad i = 1, 2, \ldots, L \tag{8.3.2}$$

where L is the number of data points. The square total of deviations is as follows:

$$R = \sum_{i=1}^{L} (r_i)^2 \tag{8.3.3}$$

We set the partial derivatives of R with respect to the coefficients of the polynomial to zero to minimize R:

$$\frac{\partial R}{\partial a_k} = 0, \quad k = 0, 1, 2, \ldots, N \tag{8.3.4}$$

or equivalently

$$\sum_{n=0}^{N} \left[\sum_{i=1}^{L} x_i^{n+k} \right] a_n = \sum_{i=1}^{L} x_i^k y_i \quad \text{for } k = 0, 1, 2, \ldots, N \tag{8.3.5}$$

which can be written more explicitly as

$$
\begin{bmatrix}
L & \sum x_i & \sum x_i^2 & \cdots & \sum x_i^N \\
\sum x_i & \sum x_i^2 & \sum x_i^3 & \cdots & \sum x_i^{N+1} \\
& & \cdots & & \\
\sum x_i^N & \sum x_i^{N+1} & \sum x_i^{N+2} & \cdots & \sum x_i^{2N}
\end{bmatrix}
\begin{bmatrix}
a_0 \\
a_1 \\
\\
a_N
\end{bmatrix}
=
\begin{bmatrix}
\sum y_i \\
\sum x_i y_i \\
\\
\sum x_i^N y_i
\end{bmatrix}
$$

The coefficients, a_n, $n = 0, 1, 2, \ldots, N$, are determined by solving Eq. (8.3.5) simultaneously (using Gauss elimination).

Example 8.3

Fit a quadratic polynomial to the data in Table 8.1.

⟨**Solution**⟩

The equation for the coefficients in matrix notations becomes

$$
\begin{bmatrix}
6.0000 & 3.3000 & 2.2100 \\
3.3000 & 2.2100 & 1.6050 \\
2.2100 & 1.6050 & 1.2245
\end{bmatrix}
\begin{bmatrix}
a_0 \\
a_1 \\
a_2
\end{bmatrix}
=
\begin{bmatrix}
7.5400 \\
4.8440 \\
3.5102
\end{bmatrix}
$$

Power coefficients determined are:

Power n	Coefficient a_n
0	0.587114
1	0.059102
2	1.729537

So, the quadratic polynomial determined is

$$y = 0.587114 + 0.059102x + 1.729537x^2$$

Error evaluation is shown in Table 8.4. The data and the fitted polynomials are depicted in Figure E8.3.

Table 8.4

i	$x(i)$	$y(i)$	Polynomial	Deviation
1	0.1	0.61	0.6103198	−0.00032
2	0.4	0.92	0.8874811	0.03252
3	0.5	0.99	1.049050	−0.05905
4	0.7	1.52	1.475959	0.04404
5	0.7	1.47	1.475959	−0.00596
6	0.9	2.03	2.041231	−0.01123

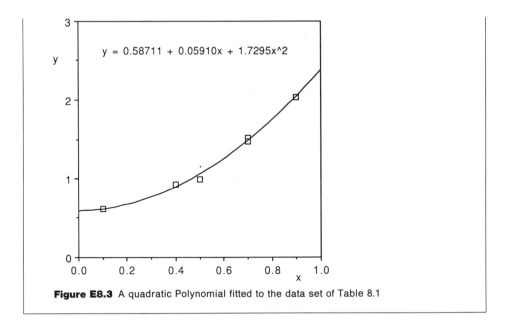

Figure E8.3 A quadratic Polynomial fitted to the data set of Table 8.1

Linear equations that arise in curve fitting are often ill-conditioned when the coefficients of the linear equations become a mix of very small and very large numbers. The spread becomes intensified as the range of x values of data points and the order of the polynomial both increase. Therefore, it is desirable to use double precision in solving the linear equations (see PROGRAM 8–1).

SUMMARY OF THIS SECTION

(a) Polynomial fitting is an extension of line fitting, and is based on least square method.

(b) The number of points fitted is generally much greater than the order of the polynomial.

(c) The linear equation associated with polynomial fitting is often highly vulnerable to round-off errors. Using double precision is recommended.

8.4 CURVE FITTING BY A LINEAR COMBINATION OF KNOWN FUNCTIONS

In fitting a function to data points, a linear combination of any known functions may be used instead of polynomials.

The curve fitting to data points is written here as

$$g(x) = a_1 f_1(x) + a_2 f_2(x) + a_3 f_3(x) + \cdots + a_N f_N(x)$$

$$= \sum_{n=1}^{N} a_n f_n(x) \tag{8.4.1}$$

where $f_1, f_2, \ldots$ are prescribed functions, $a_1, a_2, \ldots$ are undetermined coefficients, and N is the total number of prescribed functions.

The deviation of the curve from each data point is defined by

$$r_i = y_i - \sum_{n=1}^{N} a_n f_n(x_i), \quad i = 1, 2, \ldots, L \tag{8.4.2}$$

where L is the total number of data points. The square total of the deviations is

$$R = \sum_{i=1}^{L} [r_i]^2$$

$$= \sum_{i=1}^{L} \left[y_i - \sum_{n=1}^{N} a_n f_n(x_i) \right]^2 \tag{8.4.3}$$

By setting the partial derivatives of R with respect to undetermined coeficients to zero, we have

$$\frac{\partial R}{\partial a_k} = 0, \quad k = 1, 2, \ldots, N \tag{8.4.4}$$

or equivalently

$$\sum_{m=1}^{N} \left[\sum_{i=1}^{L} f_m(x_i) f_k(x_i) \right] a_m = \sum_{i=1}^{L} y_i f_k(x_i) \quad \text{for } k = 1 \text{ to } N \tag{8.4.5}$$

where the equation has been divided by 2. Equation (8.4.5) has N equations with N unknowns. Thus, the equations can be solved by Gauss elimination.

Example 8.4

Determine the coefficients of the function

$$g(x) = a_1 + a_2 x + a_3 \sin(x) + a_4 \exp(x)$$

fitted to the data in the following table:

x	y
0.1	0.61
0.4	0.92
0.5	0.99
0.7	1.52
0.7	1.47
0.9	2.03

⟨**Solution**⟩

Written in a matrix form, Eq. (8.4.5) becomes

$$
\begin{bmatrix}
6.0000E+00 & 3.3000E+00 & 3.0404E+00 & 1.0733E+01 \\
3.3000E+00 & 2.2100E+00 & 2.0124E+00 & 6.5645E+00 \\
3.0404E+00 & 2.0124E+00 & 1.8351E+00 & 6.0030E+00 \\
1.0733E+01 & 6.5645E+00 & 6.0030E+00 & 2.0325E+01
\end{bmatrix}
\begin{bmatrix}
a_1 \\ a_2 \\ a_3 \\ a_4
\end{bmatrix}
=
\begin{bmatrix}
7.5400E+00 \\
4.8440E+00 \\
4.4102E+00 \\
1.4693E+01
\end{bmatrix}
$$

Coefficients determined are as follows:

Function n	Coefficient a_n
1	−5.265785
2	−19.706939
3	13.496765
4	5.882096

Error evaluation is given in Table 8.5. The data and the fitted functions are both depicted in Figure E8.4.

Table 8.5

i	$x(i)$	$y(i)$	Curve fitted	Deviation
1	0.1	0.61	0.6117	−0.0017
2	0.4	0.92	0.8824	0.0376
3	0.5	0.99	1.0494	−0.0594
4	0.7	1.52	1.4793	0.0407
5	0.7	1.47	1.4793	−0.0093
6	0.9	2.03	2.0380	−0.0080

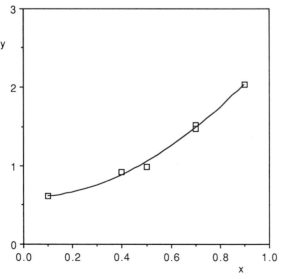

Figure E8.4 Linear combination of functions fitted to the data of Table 8.1

SUMMARY OF THIS SECTION

(a) Curve fitting including line and polynomial fitting is based on the least square method.

(b) Functions in a linear combination can be selected from experience as well as theoretical reasons.

PROGRAM

PROGRAM 8–1 Curve Fitting

(A) Explanations

This program includes the curve fitting schemes of line regression, polynomial fitting, and linear combination of functions in one program.

The program analyzes deviations of the fitted curve. The polynomial can be any order less than 11. If the order is set to 1 (or $M = 2$ equivalently), the program performs the line regression analysis.

When executed, the computer asks which of two curve fitting approaches is desired: either polynomial fitting or a linear combination of functions. Input 0 selects the former, and 1 selects the latter. However, if linear combination is selected, the number of functions to be combined as well as the functions to be used must have been defined in fun().

To increase the accuracy of the computations, double precision is used throughout the program. The linear equations are solved by gauss().

(B) List

```
/* CSL/c8-1.c   Curve Fitting with Polynomial Fitting or
 *                   Linear Combination by Least Square */
#include <stdio.h>
#include <stdlib.h>
#include <math.h>
#define TRUE 1
/*          x[j], y[j] : data table
            a[i][j] : matrix
            a[i][m+1] : coefficients of the curve
            no_of_data : number of data points              */
main()
{
int i, j, jex, jj, k, k_, lp, m, n, nord, _i, _r;
double fk, gg[100], yy, fljj;
double x[100], y[100];
double a[10][10], power, fun();
void gauss();
double pwr();
static no_of_data = 6;
static int _aini = 1;
```

```
static double x_ini[] = {0.1,  0.4,  0.5,  0.7,  0.7,  0.9};
static double y_ini[] = {0.61,  0.92,  0.99,  1.52,  1.47,  2.03};
L_10:
   for( j = 1; j <= no_of_data; j++ ){
      x[j] = x_ini[j-1];
      y[j] = y_ini[j-1];
   }
   printf( "\nCSL/C8-1       Curve Fitting by Least Square \n" );
   printf( "Type 0 for polynomial fitting\n" );
   printf( "     1 for linear combination\n" );
   scanf( "%d", &lp );
   if( lp == 0 ){   /* Polynomial fitting */
      printf( " Type order of polynomial\n" );     scanf( "%d", &nord );
      m = nord + 1;
      for (k=1; k<=m; k++) {              /* initialization of matrix*/
         for( j = 1; j <= m + 1; j++ )   a[k][j] = 0.0;
      }
      for( k = 1; k <= m; k++ ){
         /* Developing matrix for polynomial fitting*/
         for( i = 1; i <= no_of_data; i++ ){
                              /*no_of_data:number of data points.*/
            for( j = 1; j <= m; j++ ){
               jj = k - 1 + j - 1;
               a[k][j] = a[k][j] + pwr(x[i], jj);
            }
            yy = pwr(x[i], k-1);
            a[k][m + 1] = a[k][m + 1] + y[i]*yy;
         }
      }
   }
   else{   /* Linear combination */
      printf( " Type the number of functions in linear combination\n" );
      printf( "     (Currently only four functions are available)\n" );
                  /* m is the number of functions linearly combined.*/
      scanf( "%d", &m );
      for (k=1; k<=m; k++) {              /* initialization of matrix*/
         for( j = 1; j <= m + 1; j++ )   a[k][j] = 0.0;
      }
      for( k = 1; k <= m; k++ ){
         for( i = 1; i <= no_of_data; i++ ){
            fk = fun( k, x[i] );
            for( j=1; j<= m; j++ ) {
               a[k][j]=a[k][j]+fk*fun(j,x[i]);
            }
            a[k][m + 1] = a[k][m + 1] + y[i]*fk;
         }
      }
   }
   for( i = 1; i <= m; i++ ){     /* printing matrix */
      for( j = 1; j <= (m + 1); j++ ){
         printf( " %11.4e ", a[i][j] );
      }
      printf( "\n" );
   }
   n = m;
   gauss( m, a );
   printf( "Power coefficients determined \n" );
   printf( "----------------------------------------\n" );
```

```
   if( lp == 1 ){
      printf( "   Function        Coefficient        \n" );
   }
   else{
      printf( "   Power          Coefficient\n" );
   }
   printf( "-----------------------------------------\n" );
   for( i = 1; i <= m; i++ ){
      if( lp == 1 ){
         printf( "   %3d              %12.5e \n", i, a[i][m + 1] );
      }
      else{
         printf( "   %3d              %12.5e \n", i - 1, a[i][m + 1] );
      }
   }
   printf( "-----------------------------------------\n" );
   for( i = 1; i <= no_of_data; i++ ){
      gg[i] = 0.0;
      for (k=1; k<=m; k++) {
         if( lp == 1 ){
            gg[i] = gg[i] + a[k][m + 1]*fun( k, x[i] );
         }
         else {
            gg[i] = gg[i] + a[k][m + 1]*pwr(x[i], k-1);
         }
      }
   }
   printf( "Error evaluation\n" );
   printf( "-----------------------------------------------------\n" );
   printf( "   i        x(i)        y(i)        Curve fitted   Deviation\n" );
   printf( "-----------------------------------------------------\n" );
   for( i = 1; i <= no_of_data; i++ ){
      printf( "%4d   %11.4e %11.4e %11.4e %11.4e \n",
               i, x[i], y[i], gg[i], y[i] - gg[i] );
   }
   printf( "-----------------------------------------------------\n" );
   printf("\n Type 0 to stop, 1 to continue.\n");  scanf("%d", &k);
   if (k != 0 ) goto L_10;
   exit(0);
}

double pwr(x,n)    /* Calculates of n-th power of x */
double x; int n;
{
double f; int i;
   f=1.0;
   if (n > 0){
      for (i=1; i<=n; i++) f=f*x;
   }
   return f;
}

void  gauss(n, a)  /* Solves linear equations by Gauss elimination */
int n;  double a[][10];
{
int i, ipv, j, jc, jr, k, kc, nv;
double r, tm, va;
```

```
for( i = 1; i <= (n - 1); i++ ){
    ipv = i;
    for( j = i + 1; j <= n; j++ )
        if( fabs( a[ipv][i] ) < fabs( a[j][i] ) )   ipv = j;
    if( ipv != i ){
        for( jc = 1; jc <= (n + 1); jc++ ){
            tm = a[i][jc];   a[i][jc] = a[ipv][jc];   a[ipv][jc] = tm;
        }
    }
    for( jr = i + 1; jr <= n; jr++ ){
        if( a[jr][i] != 0 ){
            if( a[i][i] == 0.0 ){
                printf( "Matrix is singular!\n" );     exit(0);
            }
            r = a[jr][i]/a[i][i];
            for( kc = i + 1; kc <= (n + 1); kc++ )
                        a[jr][kc] = a[jr][kc] - r*a[i][kc];
        }
    }
}
if( a[n][n] == 0.0 ){
    printf( "Matrix is singular!\n" );     exit(0);
}
a[n][n + 1] = a[n][n + 1]/a[n][n];
for( nv = n - 1; nv >= 1; nv-- ){
va = a[nv][n + 1];
for( k = nv + 1; k <= n; k++ )   va = va - a[nv][k]*a[k][n + 1];
a[nv][n + 1] = va/a[nv][nv];
}
return;
}

double fun(k, x)
int k;   double x;
{
double f;
    if( k == 1 )   f = 1; /*  First function */
    if( k == 2 )   f = x; /*  Second function */
    if( k == 3 )   f = sin( x ); /*  ect. */
    if( k == 4 )   f = exp( x );
    return( f );
}
```

(C) Sample Output

```
(Case 1)
CSL/C8-1          Curve Fitting by Least Square

Type 0 for polynomial fitting
     1 for linear combination
0
 Type order of polynomial
2
    6.0000e+00    3.3000e+00    2.2100e+00    7.5400e+00
    3.3000e+00    2.2100e+00    1.6050e+00    4.8440e+00
    2.2100e+00    1.6050e+00    1.2245e+00    3.5102e+00
```

```
Power coefficients determined
-----------------------------------------
   Power            Coefficient
-----------------------------------------
     0               5.87116e-01
     1               5.90918e-02
     2               1.72955e+00
-----------------------------------------
Error evaluation
-----------------------------------------------------------
  i      x(i)       y(i)      Curve fitted    Deviation
-----------------------------------------------------------
  1   1.0000e-01  6.1000e-01  6.1032e-01   -3.2085e-04
  2   4.0000e-01  9.2000e-01  8.8748e-01    3.2519e-02
  3   5.0000e-01  9.9000e-01  1.0490e+00   -5.9049e-02
  4   7.0000e-01  1.5200e+00  1.4760e+00    4.4041e-02
  5   7.0000e-01  1.4700e+00  1.4760e+00   -5.9587e-03
  6   9.0000e-01  2.0300e+00  2.0412e+00   -1.1232e-02
-----------------------------------------------------------


(Case 2)
CSL/C8-1       Curve Fitting by Least Square
Type 0 for polynomial fitting
     1 for linear combination
1
 Type the number of functions in linear combination
   (Currently only four functions are available)
3
  6.0000e+00    3.3000e+00    3.0404e+00    7.5400e+00
  3.3000e+00    2.2100e+00    2.0124e+00    4.8440e+00
  3.0404e+00    2.0124e+00    1.8351e+00    4.4102e+00
Power coefficients determined
-----------------------------------------
   Function          Coefficient
-----------------------------------------
     1               5.22529e-01
     2               7.90342e+00
     3              -7.12938e+00
-----------------------------------------
Error evaluation
-----------------------------------------------------------
  i      x(i)       y(i)      Curve fitted    Deviation
-----------------------------------------------------------
  1   1.0000e-01  6.1000e-01  6.0112e-01    8.8794e-03
  2   4.0000e-01  9.2000e-01  9.0758e-01    1.2416e-02
  3   5.0000e-01  9.9000e-01  1.0562e+00   -6.6230e-02
  4   7.0000e-01  1.5200e+00  1.4620e+00    5.7952e-02
  5   7.0000e-01  1.4700e+00  1.4620e+00    7.9521e-03
  6   9.0000e-01  2.0300e+00  2.0510e+00   -2.0969e-02
-----------------------------------------------------------
```

(D) Discussions

Two cases of runs are shown in the sample output. In Case 1, a second-order polynomial is fitted to the data set specified within the code. In Case 2, a linear combination of three functions defined in fun() are fitted to the same data. The values of the fitted functions are compared to the data and deviations of the curve from the data are also calculated.

PROBLEMS

(8.1) Determine a linear function fitted to the following data points by the least square method. (Work with a hand calculator first, and then verify the answer with PROGRAM 8–1.)

i	x_i	y_i
1	1.0	2.0
2	1.5	3.2
3	2.0	4.1
4	2.5	4.9
5	3.0	5.9

(8.2) Determine a linear function fitted to the following data points by the least square method. (Answer the question by using PROGRAM 8–1.)

i	x_i	y_i
1	0.1	9.9
2	0.2	9.2
3	0.3	8.4
4	0.4	6.6
5	0.5	5.9
6	0.6	5.0
7	0.7	4.1
8	0.8	3.1
9	0.9	1.9
10	1.0	1.1

(8.3) Fit a quadratic polynomial to the following data set.

x	y
0	1
1	0
2	0
3	2

(8.4) (a) Fit a quadratic polynomial to the following data set:

i	x_i	y_i
1	0	0
2	1	2.3
3	2	4.2
4	3	5.7
5	4	6.5
6	5	6.9
7	6	6.8

(b) Evaluate the deviations of the polynomial from the data.

(8.5) Repeat the previous problem with first-order and third-order polynomials.

(8.6) Fit first-, second-, and third-order polynomials to the following data and compare deviations of the three polynomials:

$x(i)$	$y(i)$
0	0
0.002	0.618
0.004	1.1756
0.006	1.618
0.008	1.9021

(8.7) Fit a quadratic function to the following data and plot the fitted curve with the data points:

i	x_i	y_i
1	0	0
2	0.2	7.78
3	0.4	10.68
4	0.6	8.37
5	0.8	3.97
6	1	0

(8.8) Fit a cubic polynomial to the data in the previous problem.

(8.9) Fit

$$g(x) = a_0 + a_1 x + a_2 \sin(\pi x) + a_3 \sin(2\pi x)$$

to the following table:

i	$x(i)$	$y(i)$
1	0.1	0
2	0.2	2.1220
3	0.3	3.0244
4	0.4	3.2568
5	0.5	3.1399
6	0.6	2.8579
7	0.7	2.5140
8	0.8	2.1639
9	0.9	1.8358

REFERENCES

Daniel, A., and F. S. Wood, *Fitting Equations to Data*, Wiley-Interscience, 1971.

Dongarra, J. J., J. R. Bunch, C. B. Moler, and G. W. Stewart, *LINPACK User's Guide*, SIAM, 1979.

Forsythe, G. E., M. A. Malcolm, and C. B. Moler, *Computer Methods for Mathematical Computations*, Prentice-Hall, 1977.

Jennings, A., *Matrix Computations for Engineers and Scientists*, Wiley, 1977.

Robinson, E. A., *Least Squares Regression Analysis in Terms of Linear Algebra*, Goose Pond Press, 1981.

9

Initial Value Problems of Ordinary Differential Equations

9.1 INTRODUCTION

Problems of solving ordinary differential equations (ODE) are classified into initial value problems and boundary value problems. Many initial value problems are time-dependent, in which all the conditions for the solution are specified at the initial time. The numerical methods for initial value problems are significantly different from those for boundary value problems. Therefore, the present chapter discusses the numerical solution methods for the former type only, and Chapter 10 describes the numerical methods for the latter.

The initial value problem of a first-order ODE may be written in the form

$$y'(t) = f(y, t), \quad y(0) = y_0 \qquad (9.1.1)$$

where $f(y, t)$ is a function of y and t, and the second equation is an initial condition. In Eq. (9.1.1), the first derivative of y is given by a known function of y and t, and we desire to compute the unknown function y by numerically integrating $f(y, t)$. If f were independent of y, the computation would be one of the straightforward integrations discussed in Chapter 4. However, the fact that f is a function of the unknown function y makes the integration different.

The initial condition is always a part of the problem definition because the solution of an initial value problem can be uniquely determined only if an initial condition is given.

More examples for initial value problems of first-order ordinary differential equations follow:

(a) $y'(t) = 3y + 5$, $y(0) = 1$

(b) $y'(t) = ty + 1$, $y(0) = 0$

(c) $y'(t) = -\dfrac{1}{1 + y^2}$, $y(0) = 1$

(e) $y' = z$, $z' = -y$, $y(0) = 1$, $z(0) = 0$

Numerical methods for ordinary differential equations calculate the solution on the points, $t_n = t_{n-1} + h$, where h is the step size (or time interval).

Three types of numerical integration methods for initial value problems are described in this chapter: Euler methods, Runge-Kutta methods, and predictor-corrector methods. Major aspects of the methods are summarized in Table 9.1.

Table 9.1 Summary of the methods for initial value problems of ODEs

Name of methods	Relevant formula	Error Local	Error Global	Other features[a]
Nonstiff Equation:				
Euler methods				
Forward	Forward difference	$O(h^2)$	$O(h)$	SS, EC
Modified	Trapezoidal rule	$O(h^3)$	$O(h^2)$	SS, EC, NL
Backward	Backward difference	$O(h^2)$	$O(h)$	SS, EC, NL
Runge-Kutta				
Second-order	Trapezoidal rule	$O(h^3)$	$O(h^2)$	SS, EC
Third-order	Simpson's 1/3 rule	$O(h^4)$	$O(h^3)$	SS, EC
Fourth-order	Simpson's 1/3 or 3/8	$O(h^5)$	$O(h^4)$	SS, EC
Predictor-Corrector				
Second-order	(identical with second-order Runge-Kutta)			SS, EC
Third-order	Newton backward	$O(h^4)$	$O(h^3)$	NS, DC
Fourth-order	Newton backward	$O(h^5)$	$O(h^4)$	NS, DC
Stiff Equations:				
Implicit Methods	Backward difference: Gear method			SS/NS
Exponential Transformation	Exponential transformation			SS

[a] SS: self-starting capability.

NS: no self-starting capability.

EC: step size can be changed easily in the middle of solution.

DC: difficult to change the step size.

NL: solution of nonlinear equations may be necessary in each step.

Once we learn the numerical methods to solve first-order differential equations, they can be easily applied to higher-order ODEs because a higher-order ODE can be decomposed to a set of first-order differential equations. For example, consider

$$y'''' + ay''' + by'' + cy' + ey = g \tag{9.1.2}$$

where a, b, c, e, and g are constants or known functions of t. The initial conditions are given as

$$y(0) = y_0, \quad y'(0) = y'_0,$$
$$y''(0) = y''_0, \quad y'''(0) = y'''_0$$

and where y_0, y'_0, y''_0, and y'''_0 are prescribed values. By defining u, v, and w as

$$u = y', \quad v = y'', \quad w = y'''$$

Eq. (9.1.2) can be written as

$$w' + aw + bv + cu + ey = g \tag{9.1.3}$$

So, Eq. (9.1.2) is equivalent to the set of four first-order ordinary differential equations:

$$y' = u, \qquad\qquad\qquad y(0) = y_0$$
$$u' = v, \qquad\qquad\qquad u(0) = y'_0$$
$$v' = w, \qquad\qquad\qquad v(0) = y''_0$$
$$w' = g - aw - bv - cu - ey, \quad w(0) = y'''_0$$

The numerical methods for first-order ordinary differential equations are then applicable to the foregoing set.

The numerical methods may be applied to integro-differential equations, too. For example, consider the equation given by

$$y'' + ay + \int_0^t y(s)\,ds = g, \quad y(0) = y_0, \quad y'(0) = y'_0 \tag{9.1.4}$$

By defining u and v as

$$u = y', \quad v = \int_0^t y(s)\,ds$$

Eq. (9.1.4) becomes

$$u' = -ay - v + g, \quad u(0) = y'_0$$
$$v' = y, \qquad\qquad\quad v(0) = 0 \tag{9.1.5}$$
$$y' = u, \qquad\qquad\quad y(0) = y_0$$

The foregoing set of first-order ordinary differential equations can be solved by a numerical method.

9.2 EULER METHODS

We start our study with the Euler methods, which are suitable for a quick programming because of their great simplicity. It should be pointed out that, as the system of equations becomes more complicated, the Euler methods are more often used. Indeed, a large fraction of numerical methods for parabolic and hyperbolic partial differential equations, which are far more complicated than ordinary differential equations, are based on Euler methods rather than the Runge-Kutta or predictor-corrector methods.

Euler methods consist of three versions: (a) forward Euler, (b) modified Euler, and (c) backward Euler methods.

9.2.1 Forward Euler Method

The forward Euler method for $y' = f(y, t)$ is obtained by rewriting the forward difference approximation,

$$\frac{y_{n+1} - y_n}{h} \simeq y'_n \tag{9.2.2}$$

to

$$y_{n+1} = y_n + hf(y_n, t_n) \tag{9.2.3}$$

where $y'_n = f(y_n, t_n)$ is used. Using Eq. (9.2.3), y_n is recursively calculated as

$$y_1 = y_0 + hy'_0 = y_0 + hf(y_0, t_0)$$
$$y_2 = y_1 + hf(y_1, t_1)$$
$$y_3 = y_2 + hf(y_2, t_2)$$
$$\vdots$$
$$y_n = y_{n-1} + hf(y_{n-1}, t_{n-1})$$

Example 9.1

(a) Solve $y' = -20y + 7 \exp(-0.5t)$, $y(0) = 5$, using the forward Euler method with $h = 0.01$ for $0 < t \leqslant 0.02$. Do this part by hand calculation.

(b) Repeat the same for $h = 0.01$, 0.001, and 0.0001 on a computer for $0 \leqslant t \leqslant 0.09$. Evaluate errors of the three calculations by comparison to the analytical solution given by

$$y = 5e^{-20t} + (7/19.5)(e^{-0.5t} - e^{-20t})$$

⟨**Solution**⟩

(a) The first few steps of calculations with $h = 0.1$ are shown next:

$t_0 = 0$, $\quad y_0 = y(0) = 5$

$t_1 = 0.01$, $\quad y_1 = y_0 + hy_0' = 5 + (0.01)(-20(5) + 7 \exp(0))$

$\qquad\qquad = 4.07$

$t_2 = 0.02$, $\quad y_2 = y_1 + hy_1' = 4.07 + (0.01)(-20(4.07) + 7 \exp(-0.005))$

$\qquad\qquad = 3.32565$

$$\vdots$$

$t_n = nh$, $\quad y_n = y_{n-1} + hy_{n-1}'$

The computational results for selected values of t with three values of time interval (grid spacing) are shown in Table 9.2.

Table 9.2 Forward Euler method

t	$h = 0.01$	$h = 0.001$	$h = 0.0001$
0.01	4.07000 (8.693)[a]	4.14924 (0.769)[a]	4.15617 (0.076)[a]
0.02	3.32565 (14.072)	3.45379 (1.259)	3.46513 (0.124)
0.03	2.72982 (17.085)	2.88524 (1.544)	2.89915 (0.153)
0.04	2.25282 (18.440)	2.42037 (1.684)	2.43554 (0.167)
0.05	1.87087 (18.658)	2.04023 (1.722)	2.05574 (0.171)
0.06	1.56497 (18.125)	1.72932 (1.690)	1.74454 (0.168)
0.07	1.31990 (17.119)	1.47496 (1.613)	1.48949 (0.169)
0.08	1.12352 (15.839)	1.26683 (1.507)	1.28041 (0.150)
0.09	0.96607 (14.427)	1.09646 (1.387)	1.10895 (0.138)
0.10	0.83977 (12.979)	0.95696 (1.261)	0.96831 (0.126)

[a](error) × 100

Comments: Accuracy of the forward Euler method increases with a decrease in time interval h. In effect, magnitudes of errors are approximately proportional to h. However, further reduction of h without using double precision is not advantageous because it increases numerical error caused by round-off (see Chapter 1).

Although the forward Euler method is simple, it has to be used carefully for two kinds of errors. The first is the truncation errors that we have already seen in Example 9.1. The second is a potential of instability, which occurs when the time constant of the equation is negative unless time interval h is sufficiently small. A typical equation for a diminishing solution is $y' = -\alpha y$, with $y(0) = y_0 > 0$, where $\alpha > 0$. The exact solution is $y = y_0 \exp(-\alpha t)$ that approaches zero as t increases. The forward Euler method for this problem becomes

$$y_{n+1} = (1 - \alpha h)y_n$$

If $\alpha h < 1$, the numerical solution is diminishing and positive, but if $\alpha h > 1$, the sign of the solution alternates. Furthermore, if $\alpha h > 2$, the magnitude of the solution increases after each step, and the solution oscillates. This is the *instability*.

The forward Euler method is applicable to a set of first-order ODEs. Consider a set of first-order ODEs given by

$$y' = f(y, z, t), \quad y(0) = y_0$$
$$z' = g(y, z, t), \quad z(0) = z_0$$
(9.2.4)

The Euler method for Equation (9.2.4) is written as

$$y_{n+1} = y_n + hy'_n = y_n + hf(y_n, z_n, t_n)$$
$$z_{n+1} = z_n + hz'_n = z_n + hg(y_n, z_n, t_n)$$
(9.2.5)

A higher-order ordinary differential equation may be broken into a set of coupled first-order differential equations as mentioned earlier.

Example 9.2

Using the forward Euler method with $h = 0.5$, find the values of $y(1)$ and $y'(1)$ for

$$y''(t) - 0.05y'(t) + 0.15y(t) = 0, \quad y'(0) = 0, \quad y(0) = 1$$

⟨Solution⟩

Let $y' = z$, then the second-order ODE becomes

$$y' = z, \qquad\qquad y(0) = 1$$
$$z' = 0.05z - 0.15y, \quad z(0) = 0$$

We will denote $y_n = y(nh)$ and $z_n = z(nh)$. The initial conditions are expressed as $y_0 = y(0) = 1$ and $z_0 = y'(0) = 0$. Using the forward Euler method, y and z at $n = 1$ and $n = 2$ become:

$t = 0.5$:

$$y'_0 = z_0 = 0$$
$$z'_0 = 0.05z_0 - 0.15y_0 = -0.15$$
$$y_1 = y_0 + hy'_0 = 1 + (0.5)(0) = 1$$
$$z_1 = z_0 + hz'_0 = 0 + (0.5)(-0.15) = -0.075$$

$t = 1$:

$$y'_1 = z_1 = -0.075$$
$$z'_1 = 0.05z_1 - 0.15y_1 = (0.05)(-0.075) - (0.15)(1) = -0.15375$$
$$y_2 = y_1 + hy'_1 = 1 + (0.5)(-0.075) = 0.96250$$
$$z_2 = z_1 + hz'_1 = -0.075 + (0.5)(-0.15375) = -0.15187$$

Therefore

$$y(1) = y_2 = 0.96250$$
$$y'(1) = z(1) = z_2 = -0.15187$$

Example 9.3

Solve the following set of first-order ODEs by the forward Euler method with $h = 0.005\pi$ and $h = 0.0005\pi$:

$$y' = z, \qquad y(0) = 1$$
$$z' = -y, \quad z(0) = 0 \tag{A}$$

⟨**Solution**⟩

The calculations for the first few steps with $h = 0.0005\pi$ are shown next.

$t_0 = 0$: $y_0 = 1$

                $z_0 = 0$

$t_1 = 0.0005\pi$: $y_1 = y_0 + hz_0 = 1 + (0.0005\pi)(0) = 1.0$

                $z_1 = z_0 - hy_0 = 0 - (0.0005\pi)(1) = -0.00157$

$t_2 = 0.001\pi$: $y_2 = y_1 + hz_1 = 1 + (0.0005\pi)(-0.00157) = 0.99999$

                $z_2 = z_1 - hy_1 = -0.00157 - (0.0005\pi)(1) = -0.00314$

In Table 9.3 the results of the present calculations for selected values of t are compared to the exact solution, $y = \cos(t)$ and $z = -\sin(t)$.

Table 9.3

	Exact		$h = 0.005\pi$		$h = 0.0005\pi$	
t	$y = \cos(t)$	$z = -\sin(t)$	y	z	y	z
0.5π	0	-1	$1.32E - 4$	-1.01241	$2.62E - 6$	-1.00123
π	-1	0	-1.02497	$-2.67E - 4$	-1.00247	$-5.25E - 6$
1.5π	0	1	$-4.01E - 4$	1.03770	$-7.88E - 6$	1.00371
2π	1	0	1.05058	$5.48E - 4$	1.00495	$1.05E - 5$
3π	-1	0	-1.07682	$-8.43E - 4$	-1.00743	$-1.58E - 5$
6π	1	0	1.15954	$1.82E - 3$	1.01491	$3.19E - 5$
8π	1	0	1.21819	$2.54E - 3$	1.01994	$4.27E - 5$

It is observed in the results shown in Table 9.3 that the error of y increases with increase in t, and in proportion to h. (See the y values for $t = \pi$, 2π, 3π, 6π and 8π: z values for these t values do not follow the same trend because, when z is close to zero, the errors of z are significantly affected by phase shift.)

9.2.3 Modified Euler Method

The motivation for the modified Euler method is twofold. First, the modified Euler method is more accurate than the forward Euler method. Second, it is more stable than the forward Euler method.

The modified Euler method is derived by applying the trapezoidal rule to integrating $y' = f(y, t)$:

$$y_{n+1} = y_n + \frac{h}{2}\left[f(y_{n+1}, t_{n+1}) + f(y_n, t_n)\right] \tag{9.2.6}$$

If f is linear in y, both Eqs.(9.2.6) may be easily solved for y_{n+1}. For example, if the ODE is given by

$$y' = ay + \cos(t)$$

Eq. (9.2.6) becomes

$$y_{n+1} = y_n + \frac{h}{2}\left[ay_{n+1} + \cos(t_{n+1}) + ay_n + \cos(t_n)\right]$$

Therefore, solving for y_{n+1} yields

$$y_{n+1} = \frac{1 + ah/2}{1 - ah/2}y_n + \frac{h/2}{1 - ah/2}\left[\cos(t_{n+1}) + \cos(t_n)\right] \tag{9.2.7}$$

If f is a nonlinear function of y, Eq. (9.2.6) becomes a nonlinear function of y_{n+1}, so an algorithm to solve nonlinear equations is necessary. A widely used method for solving nonlinear equations is the successive substitution method (Section 3.6):

$$y_{n+1}^{(k)} - y_n = \frac{h}{2}\left[f(y_{n+1}^{(k-1)}, t_{n+1}) + f(y_n, t_n)\right] \tag{9.2.8}$$

where $y_{n+1}^{(k)}$ is the kth iterative approximation for y_{n+1}, and $y_{n+1}^{(0)}$ is an initial guess for y_{n+1}. The above iteration is terminated when $\left|y_{n+1}^{(k)} - y_{n+1}^{(k-1)}\right|$ becomes less than a prescribed tolerance. The initial guess is set to y_n. Then, the first iteration step becomes identical with the forward Euler method. If only one more iteration step is used, the scheme becomes the second-order Runge-Kutta method or, equivalently, the Euler predictor-corrector method. But, in the modified Euler method, iteration is continued until the tolerance of convergence is satisfied.

Example 9.4 shows an application of the modified Euler method to a nonlinear first-order ODE.

Example 9.4

By the modified Euler method with $h = 0.1$, solve

$$y' = -y^{1.5} + 1, \quad y(0) = 10$$

for $0 \leqslant t \leqslant 1$. Print the results up to $t = 1$.

⟨Solution⟩

The modified Euler method becomes

$$y_{n+1} = y_n + \frac{h}{2}\left[-(y_{n+1})^{1.5} - (y_n)^{1.5} + 2\right] \tag{A}$$

For $n = 0$:

$$y_1 = y_0 + \frac{h}{2}\left[-(y_1)^{1.5} - (y_0)^{1.5} + 2\right]$$

The best estimate for y_1 on the right side is y_0. By introducing $y_1 \simeq y_0$ to the right side, the equation becomes

$$y_1 \simeq 10 + \frac{0.1}{2}[-(10)^{1.5} - (10)^{1.5} + 2] = 6.93772$$

Introducing the above result to y_1 on the right side of Eq. (A) again yields

$$y_1 \simeq 10 + \frac{0.1}{2}[-(6.93772)^{1.5} - (10)^{1.5} + 2] = 7.60517$$

Repeating the substitution a few more times, we get

$$y_1 \simeq 10 + \frac{0.1}{2}[-(7.60517)^{1.5} - (10)^{1.5} + 2] = 7.47020$$

$$y_1 \simeq 10 + \frac{0.1}{2}[-(7.47020)^{1.5} - (10)^{1.5} + 2] = 7.49799$$

$$y_1 \simeq 10 + \frac{0.1}{2}[-(7.49799)^{1.5} - (10)^{1.5} + 2] = 7.49229$$

$$\vdots$$

$$y_1 = 10 + \frac{0.1}{2}[-(7.49326)^{1.5} - (10)^{1.5} + 2] = 7.49326$$

The computed results for ten time steps follow:

t	y
0.0	10.0
0.1	7.4932
0.2	5.8586
0.3	4.7345
0.4	3.9298
0.5	3.3357
0.6	2.8859
0.7	2.5386
0.8	2.2658
0.9	2.0487
1.0	1.8738

Why is the accuracy of the modified Euler method higher than that of the forward Euler method? To explain the reason analytically, let us consider a test equation, $y' = \alpha y$. Equation (9.2.6) for this problem may then be written as

$$y_{n+1} = y_n + \frac{\alpha h}{2}(y_{n+1} + y_n) \tag{9.2.9}$$

or equivalently, solving for y_{n+1},

$$y_{n+1} = \left(1 + \frac{\alpha h}{2}\right)\left(1 - \frac{\alpha h}{2}\right)^{-1} y_n$$

Expanding the coefficient of the foregoing equation yields

$$y_{n+1} = \left(1 + \alpha h + \frac{1}{2}(\alpha h)^2 + \frac{1}{4}(\alpha h)^3 + \cdots\right) y_n$$

Comparing this expansion to the Taylor expansion of the exact solution $y(t_{n+1}) = \exp(\alpha h)y_n$, it is found that Eq. (9.2.9) is accurate to the second-order term. Thus, the modified Euler method is a second-order (accurate) method. On the other hand, a similar analysis for the forward Euler method indicates that the forward Euler method is first-order accurate.

The local error (error generated in each step) of the forward Euler method is proportional to h^2 and its global error is proportional to h, whereas the local error of the modified Euler method is proportional to h^3 and its global error is proportional to h^2. The order of errors of the backward Euler method is the same as in the forward Euler method.

If the modified Euler method is applied to a set of ODEs, the whole equations must be solved simultaneously or "implicitly." However, the advantage of the implicit solution is that the method is more stable than the forward Euler method and thus allows a larger time step.

9.2.4 Backward Euler Method

The backward Euler method is based on the backward difference approximation and is written as

$$y_{n+1} = y_n + hf(y_{n+1}, t_{n+1}) \tag{9.2.10}$$

The accuracy of this method is the same as that of the forward Euler method. Besides, if f is a nonlinear function of y, an iterative scheme has to be used in each step just as in the modified Euler method. However, the advantages are (a) the method is stable for stiff problems, and (b) positivity of solution is guaranteed when the exact solution is positive. See applications of the backward Euler method in Section 9.5 and Chapter 12.

SUMMARY OF THIS SECTION

(a) The forward Euler method is based on the forward difference approximation. Its error in one interval is proportional to h^2 and its global error to h. The forward Euler method may become unstable if the ODE has a negative time constant unless a small h is used.

(b) The modified Euler method is based on the trapezoidal rule. If the ODE is not linear, an iterative method is necessary for each interval. Its error in one interval is proportional to h^3 and its global error to h^2.

(c) The backward Euler method is based on the backward difference approximation. Its errors are comparable to those of the forward Euler method. The method is stable so it is used to solve stiff problems that are difficult to solve by other methods.

9.3 RUNGE-KUTTA METHODS

A major drawback of the Euler methods is that the orders of accuracy are low. This disadvantage is twofold. To maintain a high accuracy requires very small h, which increases computational time and causes round-off errors.

In Runge-Kutta methods, the order of accuracy is increased by using intermediate points in each step interval. A higher accuracy also implies that errors decrease more quickly than in lower-order accuracy methods when h is reduced.

Consider an ordinary differential equation

$$y' = f(y, t), \quad y(0) = y_0 \tag{9.3.1}$$

To calculate y_{n+1} at $t_{n+1} = t_n + h$ with a known value of y_n, we integrate Eq. (9.3.1) over the interval $[t_n, t_{n+1}]$ as

$$y_{n+1} = y_n + \int_{t_n}^{t_{n+1}} f(y, t)\, dt \tag{9.3.2}$$

Runge-Kutta methods are derived by applying a numerical integration method to the integral on the right side of Eq. (9.3.2) [Fox/Mayers]. In the remainder of this section, the second-, third-, and fourth-order Runge-Kutta methods are explained.

9.3.1 Second-Order Runge-Kutta Method

Here we examine an application of the trapezoidal rule to the right side of Eq. (9.3.2):

$$\int_{t_n}^{t_{n+1}} f(y, t)\, dt \simeq \frac{1}{2} h [f(y_n, t_n) + f(y_{n+1}, t_{n+1})] \tag{9.3.3}$$

In Eq. (9.3.3) y_{n+1} is not known, so the second term is approximated by $f(\bar{y}_{n+1}, t_{n+1})$, where $\bar{y}_{n+1}$ is the first estimate for y_{n+1} calculated by the forward Euler method. The scheme derived here is called the second-order Runge-Kutta method and summarized as

$$\bar{y}_{n+1} = y_n + hf(y_n, t_n)$$

$$y_{n+1} = y_n + \frac{h}{2} [f(y_n, t_n) + f(\bar{y}_{n+1}, t_{n+1})]$$

or in a more standard form as

$$k_1 = hf(y_n, t_n)$$

$$k_2 = hf(y_n + k_1, t_{n+1})$$

$$y_{n+1} = y_n + \frac{1}{2}[k_1 + k_2]$$

(9.3.4)

The second-order Runge-Kutta method is identical to the Euler predictor-corrector method, which is the simplest predictor-corrector method (See Section 9.4). It is also equivalent to the modified Euler method applied with only one iteration step.

Example 9.5

The circuit shown in Figure E9.5 has a self-inductance of $L = 50H$, a resistance of $R = 20$ ohm, and a voltage source of $V = 10$ volt. If the switch is closed at $t = 0$, the current $I(t)$ satisfies

$$L\frac{d}{dt}I(t) + RI(t) = E, \quad I(0) = 0 \tag{A}$$

Find the electric current for $0 < t \leqslant 10$ sec by using the second-order Runge-Kutta method with $h = 0.1$.

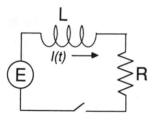

Figure E9.5 An electric circuit

⟨Solution⟩

We first rewrite Eq. (A) as

$$\frac{d}{dt}I = -\frac{R}{L}I + \frac{E}{L} \equiv f(I, t)$$

Then, the second-order Runge-Kutta method becomes

$$k_1 = h\left[-\frac{R}{L}I_n + \frac{E}{L}\right]$$

$$k_2 = h\left[-\frac{R}{L}(I_n + k_1) + \frac{E}{L}\right]$$

$$I_{n+1} = I_n + \frac{1}{2}(k_1 + k_2)$$

The calculations for the first two steps are shown next:

$n = 0$ $(t = 0.1)$: $k_1 = 0.1[(-0.4)(0) + 0.2] = 0.02$

$k_2 = 0.1[(-0.4)(0 + 0.02) + 0.2] = 0.0192$

$$I_1 = I_0 + \frac{1}{2}(k_1 + k_2) = 0 + \frac{1}{2}(0.02 + 0.0192) = 0.0196$$

$n = 1$ $(t = 0.2)$: $k_1 = 0.1[(-0.4)(0.0196) + 0.2] = 0.019216$

$k_2 = 0.1[(-0.4)(0.0196 + 0.019216) + 0.2] = 0.018447$

$$I_2 = I_1 + \frac{1}{2}(k_1 + k_2)$$

$$= 0.0196 + \frac{1}{2}(0.019216 + 0.018447) = 0.038431$$

The final result of computation (at multiples of 10 steps) is as follows:

t (sec)	I (amp)
0	0
1	0.1648
2	0.2752
3	0.3493
4	0.3990
5	0.4332
6	0.4546
7	0.4695
8	0.4796
9	0.4863
10	0.4908
(∞)	(0.5000)

The accuracy of the second-order Runge-Kutta method may be analyzed by using the test equation $y' = \alpha y$ as described toward the end of Section 9.2. However, to show a more formal approach, we consider here a generic form, $y' = f(y, t)$. We first expand the exact value of y_{n+1} in the Taylor series:

$$y_{n+1} = y_n + hf + \frac{h^2}{2}[f_t + f_y f]$$

$$+ \frac{h^3}{6}[f_{tt} + 2f_{ty}f + f_{yy}f^2 + f_t f_y + f_y^2 f] + 0(h^4) \qquad (9.3.5)$$

where all the derivatives of y are expressed in terms of f and the partial derivatives of f at t_n.

Next, we expand the third equation in Eq. (9.3.4) in a Taylor series:

$$y_{n+1} = y_n + hf + \frac{h^2}{2}[f_t + f_y f] + \frac{h^3}{4}[f_{tt} + 2f_{ty}f + f_{yy}f^2] + 0(h^4) \qquad (9.3.6)$$

By comparing Eq. (9.3.6) to Eq. (9.3.5), Eq. (9.3.4) is found to be accurate to the order of h^2 and the discrepancy (error generated in one step) is proportional to h^3. Notice that the second-order Runge-Kutta method is identical to the modified Euler method given by Eq. (9.2.8) with two iteration steps. However, the order of accuracy of the former is identical to that of the latter, which requires iterative convergence. This indicates that strict convergence of the iteration in the modified Euler method is meaningless. (Indeed, using the second-order Runge-Kutta method with a smaller h is far more effective in improving accuracy than using the modified Euler method with strict iterative convergence.) The foregoing analysis may be done more easily by considering the test equation $y' = \alpha y$, but this approach is left as a student's exercise.

Application of the second-order Runge-Kutta method to a higher-order ordinary differential equation is easy. For illustration, we consider the second-order differential equation:

$$y''(t) + ay'(t) + by(t) = q(t), \quad y(0) = 1, \, y'(0) = 0 \tag{9.3.7}$$

where a and b are coefficients and $q(t)$ is a known function, and two initial conditions are given. By defining

$$z(t) = y'(t) \tag{9.3.8}$$

Eq. (9.3.7) can be reduced to coupled first-order differential equations:

$$\begin{aligned} y' &= f(y, z, t) \equiv z, & y(0) &= 1 \\ z' &= g(y, z, t) \equiv -az - by + q, & z(0) &= 0 \end{aligned} \tag{9.3.9}$$

The second-order Runge-Kutta method for the foregoing equations is written as

$$\begin{aligned} k_1 &= hf(y_n, z_n, t_n) = hz_n \\ l_1 &= hg(y_n, z_n, t_n) = h(-az_n - by_n + q_n) \\ k_2 &= hf(y_n + k_1, z_n + l_1, t_{n+1}) = h(z_n + l_1) \\ l_2 &= hg(y_n + k_1, z_n + l_1, t_{n+1}) = h(-a(z_n + l_1) - b(y_n + k_1) + q_{n+1}) \end{aligned} \tag{9.3.10}$$

$$y_{n+1} = y_n + \frac{1}{2}(k_1 + k_2)$$

$$z_{n+1} = z_n + \frac{1}{2}(l_1 + l_2)$$

Example 9.6

A cubic material of mass $M = 0.5$ kg is fixed to the lower end of a massless spring. The upper end of the spring is fixed to a structure at rest. The cube receives resistance $R = -B\,dy/dt$ from the air, where B is a damping constant.

(See Figure E9.6.) The equation of motion is

$$M \frac{d^2}{dt^2} y + B \frac{d}{dt} y + ky = 0, \quad y(0) = 1, \ y'(0) = 0 \tag{A}$$

where y is the displacement from the static position, k is the spring constant equal to 100 kg/s², and $B = 10$ kg/s.

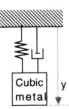

Figure E9.6 A spring-mass system

  (a) Calculate $y(t)$ for $0 < t < 0.05$ using the second-order Runge-Kutta method with $h = 0.025$ by hand calculations.
  (b) Calculate $y(t)$ for $0 < t < 10$ sec using the second-order Runge-Kutta method with $h = 0.001$.
  (c) Repeat (b) for $B = 0$.

⟨**Solution**⟩

  Equation (A) may be written as

$$y' = z \equiv f(y, z, t), \qquad\qquad y(0) = 1$$
$$z' = -\frac{B}{M} z - \frac{k}{M} y \equiv g(y, z, t), \quad z(0) = 0 \tag{B}$$

By setting $a = B/M = 20$ and $b = k/M = 200$ and $g = 0$, the second-order Runge-Kutta method for Eq. (A) becomes the form of Eq. (9.3.9).
  (a) For $n = 1$: $t = 0.025$

$$k_1 = hf(y_0, z_0, t_0) = hz_0 = 0.025(0) = 0$$
$$l_1 = hg(y_0, z_0, t_0) = h(-20z_0 - 200y_0) = 0.025(-20(0) - 200(1)) = -5$$
$$k_2 = hf(y_0 + k_1, z_0 + l_1, t_0) = h(z_0 + l_1) = 0.025(0 - 5) = -0.125$$
$$l_2 = hg(y_0 + k_1, z_0 + l_1, t_1) = h[-20(z_0 + l_1) - 200(y_0 + k_1)]$$
$$= 0.025[-20(0 - 5) - 200(1 + 0)] = -2.5$$

$$y_1 = y_0 + \frac{1}{2}(0 - 0.125) = 0.9375$$

$$z_1 = z_0 + \frac{1}{2}(-5 - 2.5) = -3.75$$

For $n = 2$: $t = 0.05$

$$k_1 = hf(y_1, z_1, t_1) = hz_1 = 0.025(-3.75) = -0.09375$$
$$l_1 = hg(y_1, z_1, t_1) = h(-20z_1 - 200y_1)$$
$$= 0.025[-20(-3.75) - 200(0.9375)] = -2.8125$$

$$k_2 = hf(y_1 + k_1, z_1 + l_1, t_1) = h(z_1 + l_1)$$

$$= 0.025(-3.75 - 2.8125) = 0.1640625$$

$$l_2 = hg(y_1 + k_1, z_1 + l_1, t_1) = h[-20(z_1 + l_1) - 200(y_1 + k_1)]$$

$$= 0.025[-20(-3.75 - 2.8125) - 200(0.9375 - 0.093750)]$$

$$= -0.9375$$

$$y_2 = y_1 + \frac{1}{2}(-0.09375 - 0.1640625) = 0.80859$$

$$z_2 = z_1 + \frac{1}{2}(-2.8125 - 0.9375) = -5.625$$

(b) and (c) This part of the computations was performed by using PROGRAM 9–1. The computational results after every 50 steps up to 0.75 sec are shown below:

	(b)	(c)
t (sec)	y (meter)	y (meter)
	(B = 10)	(B = 0)
0	1.000	1.000
0.05	0.823	0.760
0.1	0.508	0.155
0.15	0.238	−0.523
0.2	0.066	−0.951
0.25	−0.016	−0.923
0.3	−0.042	−0.45
0.35	−0.038	0.235
0.4	−0.025	0.810
0.45	−0.013	0.996
0.5	−0.004	0.705
0.55	0.000	0.075
0.6	0.001	−0.590
0.65	0.001	−0.973
0.7	0.001	−0.889
0.75	0.000	−0.378

9.3.2 Third-Order Runge-Kutta Method

A Runge-Kutta method that is more accurate than the second-order Runge-Kutta method may be derived by using a higher-order numerical integration scheme for the second term of Eq. (9.3.2). Using the Simpson's 1/3 rule, Eq. (9.3.2) is approximately

$$y_{n+1} = y_n + \frac{h}{6}\left[f(y_n, t_n) + 4f(\bar{y}_{n+\frac{1}{2}}, t_{n+\frac{1}{2}}) + f(\bar{y}_{n+1}, t_{n+1})\right] \qquad (9.3.11)$$

where $\bar{y}_{n+1}$ and $\bar{y}_{n+\frac{1}{2}}$ are estimates because $y_{n+\frac{1}{2}}$ and y_{n+1} are not known.

The estimate $\bar{y}_{n+\frac{1}{2}}$ is obtained by the forward Euler method as

$$\bar{y}_{n+\frac{1}{2}} = y_n + \frac{h}{2}f(y_n, t_n) \qquad (9.3.12)$$

The estimate $\bar{y}_{n+1}$ may be obtained by

$$\bar{y}_{n+1} = y_n + hf(y_n, t_n)$$

or

$$\bar{y}_{n+1} = y_n + hf(\bar{y}_{n+\frac{1}{2}}, t_{n+\frac{1}{2}})$$

or a linear combination of both

$$\bar{y}_{n+1} = y_n + h[\theta f(y_n, t_n) + (1 - \theta)f(\bar{y}_{n+\frac{1}{2}}, t_{n+\frac{1}{2}})] \qquad (9.3.13)$$

Here θ is an undetermined parameter, which will be determined to maximize accuracy of the numerical method. With Eq. (9.3.13), the whole scheme is written in the following form:

$$k_1 = hf(y_n, t_n)$$

$$k_2 = hf\left(y_n + \frac{1}{2}k_1, t_n + \frac{h}{2}\right)$$

$$k_3 = hf(y_n + \theta k_1 + (1 - \theta)k_2, t_n + h) \qquad (9.3.14)$$

$$y_{n+1} = y_n + \frac{1}{6}(k_1 + 4k_2 + k_3)$$

To optimize θ, we expand k_1, k_2, and k_3 in Taylor series as

$$k_1 = hf \qquad (9.3.15a)$$

$$k_2 = hf + \frac{1}{2}h^2(f_t + f_y f) + \frac{1}{8}h^3(f_{tt} + 2f_{ty}f + f_{yy}f^2) \qquad (9.3.15b)$$

$$k_3 = hf + h^2(f_t + f_y f) + \frac{1}{2}h^3[f_{tt} + 2f_{ty}f$$

$$+ f_{yy}f^2 + (1 - \theta)(f_t + f_y f)f_y] \qquad (9.3.15c)$$

where f and its derivatives are evaluated at t_n. By introducing Eq. (9.3.15) into Eq. (9.3.14) and comparing it to Eq. (9.3.5), we find that $\theta = -1$ is the optimum because Eq. (9.3.14) then agrees with Eq. (9.3.5) to the third-order term.

The foregoing derivation may be more easily understood if it is applied to the test equation $y' = \alpha y$.

In summary, the third-order-accurate Runge-Kutta method is written as

$$k_1 = hf(y_n, t_n)$$

$$k_2 = hf\left(y_n + \frac{1}{2}k_1, t_n + \frac{h}{2}\right)$$

$$k_3 = hf(y_n - k_1 + 2k_2, t_n + h) \qquad (9.3.16)$$

$$y_{n+1} = y_n + \frac{1}{6}(k_1 + 4k_2 + k_3)$$

9.3.3 Fourth-Order Runge-Kutta Method

Derivation of the fourth-order Runge-Kutta method is similar to that of the third-order method except one more intermediate step of evaluating the derivative is used. There are several alternative choices for the numerical integration scheme to be used in Eq. (9.3.2). The fourth-order Runge-Kutta method is accurate to the fourth-order term of the Taylor expansion, so the local error is proportional to h^5.

The following two versions of the fourth-order Runge-Kutta method are most popularly used. The first version is based on the Simpson's 1/3 rule and is written as

$$k_1 = hf(y_n, t_n)$$

$$k_2 = hf\left(y_n + \frac{k_1}{2}, t_n + \frac{h}{2}\right)$$

$$k_3 = hf\left(y_n + \frac{k_2}{2}, t_n + \frac{h}{2}\right) \qquad (9.3.17)$$

$$k_4 = hf(y_n + k_3, t_n + h)$$

$$y_{n+1} = y_n + \frac{1}{6}[k_1 + 2k_2 + 2k_3 + k_4]$$

The second version is based on the Simpson's 3/8 rule and is written as

$$k_1 = hf(y_n, t_n)$$

$$k_2 = hf\left(y_n + \frac{k_1}{3}, t_n + \frac{h}{3}\right)$$

$$k_3 = hf\left(y_n + \frac{k_1}{3} + \frac{k_2}{3}, t_n + \frac{2h}{3}\right) \qquad (9.3.18)$$

$$k_4 = hf(y_n + k_1 - k_2 + k_3, t_n + h)$$

$$y_{n+1} = y_n + \frac{1}{8}[k_1 + 3k_2 + 3k_3 + k_4]$$

Example 9.7

Calculate $y(1)$ by solving

$$y' = -1/(1 + y^2), \quad y(0) = 1$$

using the fourth-order Runge-Kutta method with $h = 1$.

⟨**Solution**⟩

We set

$$f(y, t) = -\frac{1}{1 + y^2}$$

and $y_0 = 1$ and $t_0 = 0$. Because we have only one interval, the whole calculations are:

$$k_1 = hf(y_0, t_0) = -\frac{1}{(1 + 1)} = -\frac{1}{2}$$

$$k_2 = hf\left(y_0 + \frac{k_1}{2}, t_0 + \frac{h}{2}\right) = -\frac{1}{(1 + (0.75)^2)} = -0.64$$

$$k_3 = hf\left(y_0 + \frac{k_2}{2}, t_0 + \frac{h}{2}\right) = -\frac{1}{(1 + (0.68)^2)} = -0.6838$$

$$k_4 = hf(y_0 + k_3, t_0 + h) = -\frac{1}{(1 + (0.3161)^2)} = -0.9091$$

$$y_1 = y_0 + \frac{1}{6}[k_1 + 2k_2 + 2k_3 + k_4]$$

$$= 1 + \frac{1}{6}[-0.5 - 2(0.64) - 2(0.6838) - 0.9091] = 0.3238$$

Example 9.8

Solve

$$y' = ty + 1, \quad y(0) = 0$$

using the fourth-order Runge-Kutta method, Eq. (9.3.17), with $h = 0.2, 0.1$, and 0.05, respectively, and evaluate the error for each h at $t = 1, 2, 3, 4$, and 5.

⟨**Solution**⟩

Computations for this example were performed by using PROGRAM 9–2. The results follow:

	$h = 0.2$		$h = 0.1$		$h = 0.05$	
t	y	p.e.	y	p.e.	y	p.e.[a]
1	1.41067	(0.00)	1.41069	(0.00)	1.41068	(0.00)
2	8.83839	(0.01)	8.83937	(0.00)	8.83943	(0.00)
3	112.394	(0.11)	112.506	(0.01)	112.514	(0.00)
4	3716.42	(0.52)	3734.23	(0.04)	3735.72	(0.00)
5	330549.	(1.71)	335798.	(0.15)	336273.	(0.01)

[a] p.e.: percentage error

A comparison of these results to those of Euler methods reveals that the error of the fourth-order Runge-Kutta method with $h = 0.1$ is comparable to that of the modified Euler method with $h = 0.01$. Also, the fourth-order Runge-Kutta method with $h = 0.2$ is comparable to the forward Euler method with $h = 0.001$.

Application of the fourth-order Runge-Kutta method to a set of ordinary differential equations is very similar to that of the second-order Runge-Kutta method. For simplicity of explanation, we consider a set of two equations:

$$y' = f(y, z, t)$$
$$z' = g(y, z, t)$$

(9.3.19)

The fourth-order Runge-Kutta method for the set of two equations becomes

$$k_1 = hf(y_n, z_n, t_n)$$
$$l_1 = hg(y_n, z_n, t_n)$$
$$k_2 = hf\left(y_n + \frac{k_1}{2}, z_n + \frac{l_1}{2}, t_n + \frac{h}{2}\right)$$
$$l_2 = hg\left(y_n + \frac{k_1}{2}, z_n + \frac{l_1}{2}, t_n + \frac{h}{2}\right)$$
$$k_3 = hf\left(y_n + \frac{k_2}{2}, z_n + \frac{l_2}{2}, t_n + \frac{h}{2}\right)$$
$$l_3 = hg\left(y_n + \frac{k_2}{2}, z_n + \frac{l_2}{2}, t_n + \frac{h}{2}\right)$$
$$k_4 = hf(y_n + k_3, z_n + l_3, t_n + h)$$
$$l_4 = hg(y_n + k_3, z_n + l_3, t_n + h)$$

(9.3.20)

$$y_{n+1} = y_n + \frac{1}{6}[k_1 + 2k_2 + 2k_3 + k_4]$$

(9.3.21)

$$z_{n+1} = z_n + \frac{1}{6}[l_1 + 2l_2 + 2l_3 + l_4]$$

(9.3.22)

Even when the number of equations in a set is greater than two, the derivation of the fourth-order Runge-Kutta method is essentially the same. A program to solve a set of equations using the fourth-order Runge-Kutta method is given as PROGRAM 9–3.

Example 9.9

Repeat the problem in Example 9.3 by using the fourth-order Runge-Kutta method with $h = 0.2\pi$ and $h = 0.05\pi$.

⟨Solution⟩

PROGRAM 9–3 is used to produce the following results:

	Exact		$h = 0.2\pi$		$h = 0.05\pi$	
t	$y = \cos(t)$	$z = -\sin(t)$	y	z	y	z
0.5π	0	-1	$1.23E-4$	-0.99997	$1.32E-6$	-0.99999
π	-1	0	-0.99993	$-2.48E-4$	-0.99999	$-2.65E-6$
1.5π	0	1	$-3.72E-4$	0.99990	$-3.96E-6$	0.99999
2π	1	0	0.99987	$4.95E-4$	0.99999	$5.29E-6$
3π	-1	0	-0.99989	$-7.43E-4$	-0.99999	$-7.94E-6$
6π	1	0	0.99960	$1.49E-3$	0.99999	$1.57E-5$
8π	1	0	0.99947	$1.98E-3$	0.99999	$2.11E-5$

Comparing these values to the results of the forward Euler solution in Example 9.3, the accuracy of the fourth-order Runge-Kutta method even with $h = 0.2\pi$ is significantly better than the forward Euler method with $h = 0.01\pi$.

9.3.5 Error, Stability, and Grid Interval Optimization

The Runge-Kutta methods are subject to two kinds of errors—truncation error and instability. As discussed earlier, the truncation error is due to the discrepancy between the Taylor expansion of the numerical method and the Taylor expansion of the exact solution. The amount of error decreases as the order of the method becomes higher. On the other hand, instability is an accumulated effect of the local error such that the error of the solution grows unboundedly as the time steps are advanced.

To analyze the instability of a Runge-Kutta method, let us consider the test equation

$$y' = \alpha y \tag{9.3.23}$$

where $\alpha < 0$. For a given value of y_n, the exact value for y_{n+1} is analytically given as

$$y_{n+1} = \exp(\alpha h)y_n \tag{9.3.24}$$

Notice that because $\alpha < 0$, $|y_{n+1}|$ decreases as n (or time) increases.

The numerical solution of Eq. (9.3.23) by the fourth-order Runge-Kutta method becomes

$$k_1 = \alpha h y_n$$

$$k_2 = \alpha h\left(y_n + \frac{k_1}{2}\right) = \alpha h\left(1 + \frac{1}{2}\alpha h\right)y_n$$

$$k_3 = \alpha h\left(y_n + \frac{k_2}{2}\right) = \alpha h\left(1 + \frac{1}{2}\alpha h\left(1 + \frac{1}{2}\alpha h\right)\right)y_n \tag{9.3.25}$$

$$k_4 = \alpha h(y_n + k_3) = \alpha h\left(1 + \alpha h\left(1 + \frac{1}{2}\alpha h\left(1 + \frac{1}{2}\alpha h\right)\right)\right)y_n$$

$$y_{n+1} = \left[1 + \alpha h + \frac{1}{2}(\alpha h)^2 + \frac{1}{6}(\alpha h)^3 + \frac{1}{24}(\alpha h)^4\right]y_n \tag{9.3.26}$$

Equation (9.3.26) equals the first five terms of the Taylor expansion for the right side of Eq. (9.3.24) about t_n. The factor

$$\gamma = 1 + \alpha h + \frac{1}{2}(\alpha h)^2 + \frac{1}{6}(\alpha h)^3 + \frac{1}{24}(\alpha h)^4 \tag{9.3.27}$$

in Eq. (9.3.26) is approximating $\exp(\alpha h)$ of Eq. (9.3.24), so the truncation error and instability of Eq. (9.3.26) both originate in this approximation.

Equation (9.3.27) and $\exp(\alpha h)$ are plotted together in Figure 9.1 for comparison. The figure indicates that if $\alpha < 0$ and the modulus (absolute value) of αh increases, the deviation of γ from $\exp(\alpha h)$ increases, so that the error of the Runge-Kutta method increases. Particularly, if $\alpha h \leqslant -2.785$, the method becomes unstable because the modulus of the numerical solution grows in each step whereas the modulus of the true solution decreases by a factor, $\exp(\alpha h)$, in each step.

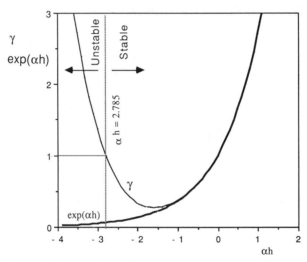

Figure 9.1 Domain of stability for the fourth-order Runge-Kutta method

In practical applications of the Runge-Kutta method, the size of an optimal grid interval can be determined in the following way. For illustration purposes, suppose we desire to keep the local error of the third-order Runge-Kutta method less than ξ. The local error of the third-order Runge-Kutta method for a test interval h is proportional to h^4, so we express the local error in the form

$$E_h = Bh^4 \tag{9.3.28}$$

where B is a constant that depends on the given problem. If we apply the same Runge-Kutta method in two steps with $h/2$ as the time interval, the error becomes proportional to $(h/2)^4$ times 2, where the factor 2 is due to the accumulation of error in two steps. Thus, we have another relation,

$$2E_{h/2} = 2B\left(\frac{h}{2}\right)^4 = \frac{1}{8}Bh^4 \tag{9.3.29}$$

By subtracting Eq. (9.3.29) from Eq. (9.3.28), we get

$$E_h - 2E_{h/2} = Bh^4 - \frac{1}{8}Bh^4 = \frac{7}{8}Bh^4 \qquad (9.3.30)$$

The left side of the foregoing equation may be evaluated by a numerical experiment—that is, by running the scheme twice starting from the same initial value. In the first run, only one time step is advanced using a trial value for h as the time interval. We denote the result of this calculation as $[y_1]_h$. In the second run, $[y_2]_{h/2}$ is calculated in two time steps using $h/2$ as the time interval. Using the results of those two calculations, the left side of Eq. (9.3.30) is evaluated as

$$E_h - 2E_{h/2} = [y_1]_h - [y_2]_{h/2} \qquad (9.3.31)$$

Introducing Eq. (9.3.31) into Eq. (9.3.30) and solving for B yields

$$B = \frac{8}{7}([y_1]_h - [y_2]_{h/2})/h^4 \qquad (9.3.32)$$

Once B is determined, the maximum (or optimum) h that satisfies the criterion $E_h \leqslant \xi$ may be found by introducing $E_h = \xi$ into Eq. (9.3.28) and solving for h, as follows:

$$h = \left(\frac{\xi}{B}\right)^{0.25} \qquad (9.3.33)$$

The theory we have just described is reminiscent of the Romberg integration explained in Section 3.2.

Example 9.10

Assume that a fourth-order Runge-Kutta method is applied to

$$y' = -\frac{y}{1 + t^2}, \quad y(0) = 1$$

find an optimal step interval satisfying $E_h \leqslant 0.00001$.

⟨Solution⟩

For the fourth-order Runge-Kutta method, the local error is expressed by

$$E_h = Bh^5 \qquad (A)$$

The approach is very similar to Eqs. (9.3.28) through (9.3.33) except that the order of error is five. The error accumulated in two steps using $h/2$ is $2E_{h/2} = 2B(h/2)^5$. The difference between the errors of one-step and two-step calculations, namely $E_h - 2E_{h/2}$, is numerically evaluated by

$$2E_h - 2E_{h/2} = [y_1]_h - [y_2]_{h/2}. \qquad (B)$$

In Eq. (B), $[y_1]_h$ is the result of the fourth-order Runge-Kutta method for only one step with h, and $[y_2]_{h/2}$ is the result of the same for two steps with $h/2$. Introducing Eq. (A) into Eq. (B) and solving for B, we have

$$B = \frac{16}{15}([y_1]_h - [y_1]_{h/2})/h^5 \qquad (C)$$

Now we actually run the fourth-order Runge-Kutta method for only one step with $h = 1$ starting with the given initial condition. Then we run it for two steps with $h/2 = 1/2$. The results are

$$[y_1]_1 = 0.4566667 \quad \text{(one interval only)}$$
$$[y_2]_{1/2} = 0.4559973 \quad \text{(two intervals)}$$

From Eq. (C), we obtain B as

$$B = \frac{16}{15}(0.4566667 - 0.4559973)/(1)^5 = 6.3 \times 10^{-4} \qquad (D)$$

By introducing this into Eq. (A), the local error for any h is expressed by

$$E_h = 6.3 \times 10^{-4}h^5$$

The maximum h that satisfies the given criterion, $E_h < 0.00001$, is

$$h = (0.00001/6.3 \times 10^{-4})^{1/5} = 0.44 \qquad (E)$$

SUMMARY OF THIS SECTION

(a) The Runge-Kutta methods are derived by integrating the first order ODE with numerical integration methods. The second-order Runge-Kutta method is identical to the modified Euler method with two iteration cycles as well as to the second-order predictor-corrector method.

(b) A higher-order ODE can be solved by a Runge-Kutta method after it is transformed to a set of first-order ODEs.

(c) Each Runge-Kutta method becomes unstable if α is negative and $|\alpha h|$ exceeds a certain criterion.

(d) The local error of a Runge-Kutta method can be found by running the same method twice: the first time for one interval with a value of h, and the second time for two intervals with $h/2$.

9.4 PREDICTOR-CORRECTOR METHODS

9.4.1 Third-Order Adams Predictor-Corrector Method

A predictor-corrector method consists of a predictor step and a corrector step in each interval. The predictor estimates the solution for the new point, and then the corrector improves its accuracy. Predictor-corrector methods use the solutions for previous points instead of using intermediate points in each interval.

y_{n-3}	y_{n-2}	y_{n-1}	y_n	y_{n+1}

t_{n-3}	t_{n-2}	t_{n-1}	t_n	t_{n+1}
$z = -3h$	$z = -2h$	$z = -h$	$z = 0$	$z = h$

Figure 9.2 Grid points used in predictor-corrector methods

To explain the methods, let us consider an equispaced time interval and assume that the solution has been calculated up to time point n so that the values of y and y' on the previous time points may be used for the calculation of y_{n+1}.

Both predictor and corrector formulas are derived by introducing an appropriate polynomial approximation for $y'(t)$ into Eq. (9.3.2). The most primitive member of the predictor-corrector methods is the second-order predictor-corrector method, which is identical to the second-order Runge-Kutta method.

Let us derive a third-order predictor by approximating $y' = f(y, t)$ with a quadratic interpolation polynomial fitted to f'_n, y'_{n-1} and y'_{n-2}:

$$y'(z) = \frac{1}{2h^2} \left[(z + h)(z + 2h)y'_n - 2z(z + 2h)y'_{n-1} + z(z + h)y'_{n-2} \right] + E(z) \quad (9.4.1)$$

where z is a local coordinate defined by

$$z = t - t_n$$

and $E(z)$ is the error (see Section 2.3). Equation (9.4.1) is the Lagrange interpolation fitted to the values y'_n, y'_{n-1} and y'_{n-2}. The error of the polynomial is

$$E(z) = \frac{1}{3!} z(z + h)(z + 2h)y^{(iv)}(\xi), \quad t_{n-2} < \xi < t_{n+1} \quad (9.4.2)$$

Here, the derivative in the error term is of the fourth order because a quadratic polynomial is fitted to y'.

Equation (9.3.2) can be rewritten in terms of the local coordinate $z = t - t_n$ as

$$y_{n+1} = y_n + \int_0^h y'(z)\,dz \quad (9.4.3)$$

Introducing Eq. (9.4.1) into Eq. (9.4.3) yields

$$y_{n+1} = y_n + \frac{h}{12}(23y'_n - 16y'_{n-1} + 5y'_{n-2}) + 0(h^4) \quad (9.4.4)$$

Equation (9.4.4) is called the third-order Adams-Bashforth predictor formula. The error of Eq. (9.4.4) is attributable to Eq. (9.4.2) and is evaluated by integrating Eq.

(9.4.2) in $[0, h]$, as follows:

$$0(h^4) = \frac{3}{8}h^4 y^{(iv)}(\xi), \quad t_{n-2} < \xi < t_{n+1}$$

In deriving Eq. (9.4.4), notice that Eq. (9.4.1) has been used as an extrapolation. As pointed out in Section 2.9, extrapolation is less accurate than interpolation (See Section 2.9 and Appendix A). Therefore, Eq. (9.4.4) is used only as a predictor and is written as

$$\bar{y}_{n+1} = y_n + \frac{h}{12}(23y'_n - 16y'_{n-1} + 5y'_{n-2}) + 0(h^4) \tag{9.4.5}$$

where the overbar indicates a predictor.

To derive a corrector formula, a predicted value of y'_{n+1} denoted by $\bar{y}'_{n+1}$ is necessary, which is calculated by introducing $\bar{y}_{n+1}$ into $y'(t) = f(y, t)$ as

$$\bar{y}'_{n+1} = f(\bar{y}_{n+1}, t_{n+1})$$

The quadratic polynomial fitted to $\bar{y}'_{n+1}$, y'_n, and y'_{n-1} is written as

$$y'(z) = \frac{1}{2h^2}\left[z(z + h)\bar{y}'_{n+1} - 2(z - h)(z + h)y'_n + z(z - h)y'_{n-1}\right] + E(z) \tag{9.4.6}$$

where z is the local coordinate defined after Eq. (9.4.1). The error of this equation is

$$E(z) = \frac{1}{3!}(z - h)z(z + h)y^{(iv)}(\xi), \quad t_{n-1} < \xi < t_{n+1}$$

Introducing Eq. (9.4.6) into Eq. (9.4.3) yields a corrector formula as

$$y_{n+1} = y_n + \frac{h}{12}(5\bar{y}'_{n+1} + 8y'_n - y'_{n-1}) + 0(h^4) \tag{9.4.7}$$

The error is

$$0(h^4) = -\frac{1}{24}h^4 y^{(iv)}(\xi), \quad t_{n-1} < \xi < t_{n+1}$$

Equation (9.4.7) is named the *Adams-Moulton corrector formula of order 3*. The set of Eqs. (9.4.5) and (9.4.7) is called the *third-order Adams predictor-corrector method*.

As seen from the preceding derivation, numerous formulas can be derived by changing the choice of the extrapolating and interpolating polynomials.

In discussing the predictor-corrector methods, we have assumed that the solutions for previous points are available. The third-order predictor-corrector method needs three previous values of y as explained earlier. Therefore, to start up the method, the solutions for $n = 0$, $n = 1$, and $n = 2$ are necessary, the first of which is given by an initial condition, but the second and third should be provided by some other means than the predictor-corrector method, such as a Runge-Kutta method.

Example 9.11

Repeat the problem of Example 9.8 using the third-order Adams predictor-corrector method with $h = 0.1$ and 0.01.

⟨**Solution**⟩

Since the predictor-corrector methods cannot be self-started, we use the fourth-order Runge-Kutta method to provide the solution for the first two intervals. For the present computation, PROGRAM 9–3 is modified by incorporating the predictor-corrector method, so y_1 and y_2 are calculated by the Runge-Kutta method, and then the remainder is calculated by the predictor-corrector method. The program used is given as PROGRAM 9–4. The computational results are as follows:

	$h = 0.1$		$h = 0.01$	
t	y	(p.e.)^a	y	(p.e.)^a
1	1.41091	(−0.01)	1.41069	(0.0000)
2	8.84404	(−0.05)	8.83943	(−0.0001)
3	112.644	(−0.12)	112.514	(−0.0004)
4	3740.07	(−0.11)	3736.00	(0.0004)
5	335593	(0.22)	336344.	(0.0009)

^{a} p.e.: percentage error

9.4.2 Fourth-Order Adams Predictor-Corrector Method

The interpolation polynomial fitted to y' at points, $n, n - 1, n - 2, \ldots, n - m$ may be written in the Newton backward formula [see Eq. (2.4.14)] as

$$g_m(t) = \sum_{k=0}^{m} (-1)^k \binom{s + k - 1}{k} \Delta^k y'_{n-k} \qquad (9.4.8)$$

where

$$s = \frac{t - t_n}{h}$$

By introducing Eq. (9.4.8) into Eq. (9.3.2) as an approximation for $f(y, t)$, we obtain the Adams-Bashforth predictor formula of order $m + 1$:

$$\bar{y}_{n+1} = y_n + h[b_0 y'_n + b_1 \Delta y'_{n-1} + \cdots + b_m \Delta^m y'_{n-m}] \qquad (9.4.9)$$

where

$$b_k = \int_0^1 \binom{s + k - 1}{k} ds \qquad (9.4.10)$$

The first few values of b_k follow:

$$b_0 = 1$$

$$b_1 = \frac{1}{2}$$

$$b_2 = \frac{5}{12}$$

$$b_3 = \frac{3}{8}$$

$$b_4 = \frac{251}{720}$$

If we set $m = 2$ in Eq. (9.4.9) for example, we obtain the third-order predictor given by Eq. (9.4.4). By following the same procedure for $m = 3$, the fourth-order predictor formula is derived as

$$\bar{y}_{n+1} = y_n + \frac{9h}{24}(55y'_n - 59y'_{n-1} + 37y'_{n-2} - 9y'_{n-3}) + 0(h^5) \qquad (9.4.11)$$

where

$$0(h^5) = \frac{251}{720} h^5 y^{(v)}(\xi), \quad t_{n-3} < \xi < t_{n+1}$$

The corrector formulas may be derived by using the polynomial fitted to y' at grid points, $n + 1, n, n - 1, \ldots, n - m + 1$. The Newton backward interpolation formula fitted to y' at these points (see Section 2.5) is

$$g_m(t) = \sum_{k=0}^m \binom{s + k - 2}{k} \Delta^k y'_{n+1-k} \qquad (9.4.12)$$

Introducing this equation into Eq. (9.3.2) yields the Adams-Moulton corrector formula:

$$y_{n+1} = y_n + h[c_0 y'_{n+1} + c_1 \Delta y'_n + \cdots + c_m \Delta^m y'_{n-m}] \qquad (9.4.13)$$

where

$$c_k = \int_0^1 \binom{s + k - 2}{k} \, ds$$

The first few values of c_k follow:

$$c_0 = 1$$

$$c_1 = -\frac{1}{2}$$

$$c_2 = -\frac{1}{12}$$

$$c_3 = -\frac{1}{24}$$

$$c_4 = -\frac{19}{720}$$

By setting $m = 3$ in Eq. (9.4.13), we obtain the fourth-order Adams-Moulton corrector:

$$y_{n+1} = y_n + \frac{h}{24}(9\bar{y}'_{n+1} + 19y'_n - 5y'_{n-1} + y'_{n-2}) + 0(h^5) \qquad (9.4.14)$$

where $y'_n = f(y_n, t_n)$ and

$$0(h^5) = -\frac{19}{720}h^5 y^{(v)}(\xi), \quad t_{n-2} < \xi < t_{n+1}$$

The set of Eqs. (9.4.11) and (9.4.14) is called the *fourth-order Adams predictor-corrector method*.

9.4.3 Advantages and Disadvantages of Predictor-Corrector Methods

An advantage of predictor-corrector methods is their computational efficiency: they use information from previous steps. Indeed, the function $f(y, t)$ is evaluated only twice in each step regardless of the order of the predictor-corrector method, whereas the fourth-order Runge-Kutta method evaluates $f(y, t)$ four times in each interval. Another advantage is that the local error can be detected at each step with a small additional computing effort. The technique of detecting local error is discussed in Subsection 9.4.4. On the other hand, there are some disadvantages as follows:

(a) The method cannot be started by itself because of the use of previous points. Until the solutions for enough points are determined, another method such as a Runge-Kutta method must be used.

(b) Because previous points are used, changing the interval size in the middle of solution is not easy. Although the predictor-corrector formulas may be derived on nonuniformly spaced points, the coefficients of the formulas change for each interval, so programming becomes very cumbersome.

(c) The predictor-corrector method cannot be used if y' becomes discontinuous. This can happen when one of the coefficients of the differential equation changes discontinuously in the middle of the domain.

However, the last two difficulties can be overcome as follows: Because the predictor-corrector program must contain a self-starting method such as a Runge-Kutta method anyway, the computation can be restarted whenever the step interval has to be changed or when y' becomes discontinuous.

9.4.4 Analysis of Local Error and Instability of Predictor-Corrector Methods

One advantage of the predictor-corrector methods is that local error may be evaluated easily by observing the difference between the predictor and the corrector in each step. For illustration of analysis, we consider the third-order Adams predictor-corrector method. Equations (9.4.5) and (9.4.7) indicate that, assuming that $y_n, y_{n-1}, y_{n-2}, \ldots$ are given, the predictor and corrector values become

$$\bar{y}_{n+1} = y_{n+1,\,\text{exact}} - \frac{3}{8}h^4 y^{(iv)}(\xi) \qquad (9.4.15)$$

$$y_{n+1} = y_{n+1,\,\text{exact}} + \frac{1}{24}h^4 y^{(iv)}(\xi) \qquad (9.4.16)$$

If we assume further that the values of the fourth derivative in Eqs. (9.4.15) and (9.4.16) take the same value, then subtracting Eq. (9.4.16) from Eq. (9.4.15) yields

$$\bar{y}_{n+1} - y_{n+1} = -\frac{10}{24}h^4 y^{(iv)}(\xi) \qquad (9.4.17)$$

Backsubstituting Eq. (9.4.17) into Eq. (9.4.16) yields

$$y_{n+1,\,\text{exact}} - y_{n+1} = \frac{1}{10}(\bar{y}_{n+1} - y_{n+1}) \qquad (9.4.18)$$

The right side of Eq. (9.4.18) is the local error of the corrector. Because it is expressed in terms of the difference between the predictor and the corrector, the calculation is simple. By using this algorithm at every step interval, the local error of the method can be automatically monitored in a program.

Now we study the stability of a predictor-corrector method by considering again the third-order Adams predictor-corrector method given by Eq. (9.4.5) and Eq. (9.4.7). Suppose we apply the method to the test equation given by

$$y' = \alpha y \tag{9.4.19}$$

Introducing Eq. (9.4.19) into Eqs. (9.4.5) and (9.4.7) yields

$$\bar{y}_{n+1} = y_n + \frac{\alpha h}{12}(23y_n - 16y_{n-1} + 5y_{n-2})$$

$$y_{n+1} = y_n + \frac{\alpha h}{12}(5\bar{y}_{n+1} + 8y_n - y_{n-1})$$

Eliminating $\bar{y}_{n+1}$ in the foregoing two equations and reorganizing the terms lead to

$$y_{n+1} = -a_2 y_n - a_1 y_{n-1} - a_0 y_{n-2} \tag{9.4.20}$$

where

$$a_2 = -(1 + 13b + 115b^2)$$
$$a_1 = b + 80b^2$$
$$a_0 = -25b^2$$
$$b = \frac{\alpha h}{12}$$

Equation (9.4.20) may be considered as the initial value problem of a difference equation, for which the analytical solution may be obtained in a similar way as for a third-order linear ordinary differential equation. Indeed, the analytical solution of Eq. (9.4.20) may be found in the form,

$$y_n = c\gamma^n \tag{9.4.21}$$

where γ is a characteristic value and c is a constant. By introducing Eq. (9.4.21) into Eq. (9.4.20), we get the characteristic equation:

$$\gamma^3 + a_2\gamma^2 + a_1\gamma + a_0 = 0 \tag{9.4.22}$$

Equation (9.4.22) is a third-order polynomial equation, so it has three roots although two of them can be complex values. We denote the three roots by

$$\gamma_1, \quad \gamma_2, \quad \text{and} \quad \gamma_3$$

Because each of γ_1, γ_2, and γ_3 satisfies Eq. (9.4.20), a linear combination of all the solutions is also a solution of Eq. (9.4.20). The general solution of Eq. (9.4.20) may now be written as

$$y_n = c_1(\gamma_1)^n + c_2(\gamma_2)^n + c_3(\gamma_3)^n \tag{9.4.23}$$

where c_1, c_2, and c_3 are determined when the initial values of y_0, y_1, and y_2 are given (remember that the third-order predictor-corrector method needs three starting values).

The exact solution to the original problem, Eq. (9.4.19), is given by

$$y_n = y(0) \exp(\alpha n h) \tag{9.4.24}$$

where $y(0)$ is the initial value of $y(t)$. The question is how each of the three terms in Eq. (9.4.23) is related to Eq. (9.4.24). The answer is that one term in Eq. (9.4.23) is an approximation for Eq. (9.4.24), but the other two are irrelevant to the true solution and constitute a part of the error of the scheme. We assume that the first term is the approximation and that the second and the third are the errors. Instability of the method is then related to the second and third terms. If these error terms vanish as n increases, there is no instability. If the magnitude of these error terms becomes greater than unity, erratic behavior of the numerical solution occurs. This is the instability, and it happens if

$$|\gamma_2| > 1 \quad \text{or} \quad |\gamma_3| > 1 \quad \text{or both}$$

When the predictor-corrector method is applied to Eq. (9.4.19), both α and h affect the instability. However, because α and h always appear as a product [see Eq. (9.4.20)], we can consider αh as one parameter. The roots of Eq. (9.4.22) for various values of αh are shown in Table 9.2.

Table 9.2 Characteristic values of the third-order predictor-corrector method applied to $y'(t) = \alpha y(t)$

αh	$\exp(\alpha h)$	γ_1	Percentage Error	γ_2, γ_3	$|\gamma_2|, |\gamma_3|$
0.1	1.1051	1.1051	0	$0.006 \pm 0.039j$	0.040
0.2	1.2214	1.2214	0	$0.014 \pm 0.074j$	0.075
0.5	1.6487	1.6477	0.06	$0.047 \pm 0.155j$	0.162
1.0	2.7183	2.6668	1.90	$0.108 \pm 0.231j$	0.255
1.5	4.4816	4.1105	8.3	$0.155 \pm 0.266j$	0.308
2.0	7.3891	5.9811	23.	$0.190 \pm 0.283j$	0.341
2.5	12.1825	8.2705	32.	$0.215 \pm 0.292j$	0.362
−0.1	0.9048	0.9048	0	$-0.003 \pm 0.043j$	0.043
−0.2	0.8187	0.8189	0.02	$-0.002 \pm 0.092j$	0.092
−0.3	0.7408	0.7416	0.1	$-0.003 \pm 0.145j$	0.145
−0.4	0.6703	0.6732	0.43	$-0.011 \pm 0.203j$	0.203
−0.5	0.6065	0.6147	1.35	$0.022 \pm 0.265j$	0.266
−1.0	0.3678	0.4824	31.2	$0.116 \pm 0.588j$	0.600
−1.5	0.2231	0.4944	121.	$0.338 \pm 0.821j$	0.889
−2.0	0.1353	0.5650	419.	$0.731 \pm 0.833j$	1.109

$j = \sqrt{-1}$

In Table 9.2, γ_1 is the root relevant to the true solution; indeed, it is an approximation to exp (αh). The other two γs are irrelevant roots. The last column shows the magnitude of the second and the third roots. It is seen that when $\alpha h > 0$ (that is, $\alpha > 0$), the magnitudes of γ_2 and γ_3 are always smaller than γ_1. Therefore, as n increases, the magnitude of the second and third terms relative to the first term diminishes. Thus, there is no instability if $\alpha > 0$.

In the second half of Table 9.2 where $\alpha < 0$, the relevant root γ_1 is always smaller than 1, and it decreases as αh becomes more negative. When the magnitude of αh is very small, the irrelevant roots are smaller than the relevant root, but the magnitude of the irrelevant roots keeps increasing and exceeds the relevant roots before αh reaches -1. The magnitude of the irrelevant root exceeds unity approximately when $\alpha h = -1.8$. If the magnitude of the irrelevant root exceeds unity, the second and third terms of Eq. (9.4.20) will show an erratic behavior. That is, while the first term approaches 0, the second and the third terms diverge with oscillatory behavior. Therefore, instability of the third-order predictor-corrector occurs if $\alpha h < -1.8$.

Table 9.2 also provides important information on the accuracy of the third-order Adams predictor-corrector method. As described previously, the first root γ_1 in Table 9.2 approximates exp (αh). The discrepancy between the γ_1 and exp (αh) is a direct measure of the local error. The table shows that, for $\alpha > 0$, the percentage of error is small until αh reaches 0.5. For $\alpha < 0$, the percentage of error increases quickly as $|\alpha h|$ increases, and for $\alpha h = -0.5$, which is still far from the instability domain, the percentage of error is significant.

SUMMARY OF THIS SECTION

(a) A predictor-corrector method consists of a predictor and a corrector.

(b) The predictors of the Adams predictor-corrector methods are named Adams-Bashforth predictors. They are derived by integrating a polynomial extrapolation of y' for the previous points.

(c) The correctors of the Adams predictor-corrector methods are named Adams-Moulton predictors, and they are derived by integrating a polynomial interpolation of y' for the previous points plus $\bar{y}'$ (the predicted value for the new point).

(d) The second-order predictor-corrector method is identical with the second-order Runge-Kutta method.

(e) The third- and fourth-order predictor-corrector methods cannot be self-started. However, once started, their computational efficiency is higher than that of the Runge-Kutta method. Error check in each interval is easier than for the Runge-Kutta method.

9.5 MORE APPLICATIONS

In this section, five applications of the numerical methods for initial value problems are shown. Although the fourth-order Runge-Kutta method is used throughout this section, it can be replaced by any other method for ordinary differential equations described in this chapter.

Example 9.12

A metal piece of 0.1 kg mass and 200° C (or 473° K) is suddenly placed in a room of temperature 25° C, where it is subject to both natural convection cooling and radiation heat transfer. Assuming that the temperature distribution in the metal is uniform, the equation for the temperature may be written as

$$\frac{dT}{dt} = \frac{A}{\rho c v}[\varepsilon\sigma(297^4 - T^4) + h_c(297 - T)], \quad T(0) = 473 \tag{A}$$

where T is the temperature in degrees Kelvin, and we assume that the constants are given by

$\rho = 300 \text{ kg/m}^3$	(density of the metal)
$v = 0.001 \text{ m}^3$	(volume of the metal)
$A = 0.25 \text{ m}^2$	(surface area of the metal)
$c = 900 \text{ J/kgK}$	(specific heat of the metal)
$h_c = 30 \text{ J/m}^2\text{K}$	(heat transfer coefficient)
$\varepsilon = 0.8$	(emissivity of the metal)
$\sigma = 5.67 \times 10^{-8} \text{ w/m}^2\text{K}^4$	(Stefan-Boltzmann constant)

⟨**Solution**⟩

This problem can be solved by modifying PROGRAM 9–2, which uses the fourth-order Runge-Kutta method. Temperatures calculated by the fourth-order Runge-Kutta method with $h = 1$ sec follow for selected values of t:

t (sec)	T (° K)
0	473
10	418.0
20	381.7
30	356.9
60	318.8
120	300.0
180	297.4

Example 9.13

The electric current of the circuit shown in Figure E9.13a satisfies the integro-differential equation

$$L\frac{di}{dt} + Ri + \frac{1}{C}\int_0^t i(t')\,dt' + \frac{1}{C}q(0) = E(t), \quad t > 0 \tag{A}$$

where the switch is closed at $t = 0$; $i = i(t)$ is the current (amp); R is a resistance (ohm); L, C, and E are given by

$$L = 200 \text{ henry}$$
$$C = 0.001 \text{ farad}$$
$$E(t) = 1 \text{ volt for } t > 0$$

Initial conditions are $q(0) = 0$ (capacitor's initial charge) and $i(0) = 0$. Calculate the current for $0 \leqslant t \leqslant 5$ sec after closing the switch ($t = 0$) for the following four values of R:

(a) $R = 0$ ohm
(b) $R = 50$ ohm
(c) $R = 100$ ohm
(d) $R = 300$ ohm

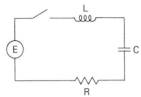

R   **Figure E9.13a** Electric circuit

⟨**Solution**⟩

We first define

$$q(t) = \int_0^t i(t')\, dt' \tag{B}$$

Differentiating Eq. (B) yields

$$\frac{d}{dt} q(t) = i(t), \quad q(0) = 0 \tag{C}$$

Introducing Eq. (B) into Eq. (A) and rewriting give

$$\frac{d}{dt} i(t) = -\frac{R}{L} i(t) - \frac{1}{LC} q(t) + \frac{1}{LC} q(0) + \frac{E(t)}{L}, \quad i(0) = 0 \tag{D}$$

Thus, Eq. (A) is transformed to a set of two first-order ODEs, Eqs. (C) and (D). PROGRAM 9–4 was modified for the present problem in two respects (see note below). The result of the computation is shown in a graphic form in Figure E9.13b.

Note: (a) To perform the calculations for all four cases in one run, four pairs of coupled first-order ODEs are incorporated, the first pair corresponding to the first case, the second pair to the second case, and so on. This is possible because not all equations in the program have to be mathematically coupled. (b) A graphic plotting routine is added, so all four cases are plotted on one graphic output.

temperature is 273° K. Using the following constants, determine the temperature distribution in the axial direction:

$k = 60$ W/mK   (thermal conductivity)

$Q = 50$ W/m   (heat generation rate per unit length of the bar)

$\sigma = 5.67 \times 10^{-8}$ W/m²K⁴   (Stefan-Boltzmann constant)

$A = 0.0001$ m²   (the cross sectional area)

$P = 0.01$ m   (perimeter of the rod)

⟨**Solution**⟩

The heat conduction equation in the axial direction x is written as

$$-Ak \frac{d^2}{dx^2} T + P\sigma(T^4 - 273^4) = Q \quad 0 < x < 1.0 \tag{A}$$

with the boundary conditions

$$T(0) = T(1.0) = 273 \text{ K}$$

where T is temperature in degrees Kelvin.

The present problem is a boundary value problem (boundary conditions are specified at $x = 0$ and $x = 1$), but it can be solved as an initial value problem on the trial-and-error basis. By defining y_1 and y_2 as

$$y_1(x) = T(x)$$
$$y_2(x) = T'(x)$$

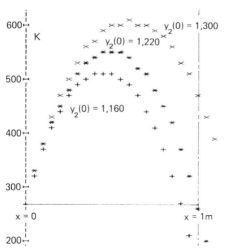

Figure E9.15 Computed results of the shooting method

Eq. (A) may be rewritten as a set of two first-order ODEs as

$$y'_1 = y_2 \qquad\qquad\qquad\qquad\qquad\text{(B)}$$

$$y'_2 = \frac{P}{Ak}\sigma(y^4 - 273^4) - \frac{Q}{kA}$$

Only one initial condition, $y_1(0) = 273$, is known from the boundary conditions (but $y_2(0)$ is not known). So we solve Eq. (A) with trial values for $y_2(0)$ until the boundary condition for the right end, namely $y_1(1) = 273$, is satisfied. This approach is called the *shooting method* [Rieder/Busby].

For the present example, PROGRAM 9–3 is used with some modifications. The results are directly plotted on a printer and shown in Figure E9.15. It is seen that $y_2(0) = 1160$ is too small as an initial guess, whereas $y_2(0) = 1300$ is too large. Some $y_2(0)$ in between these values should give the best result. After a few more trials, $y_2(0) = 1220$ is found to satisfy almost exactly the right boundary condition.

Example 9.16

The temperature of a perfectly insulated iron bar 55 cm long is initially at 200° C. The temperature of the left edge is suddenly reduced and fixed to 0° C at $t = 0$ sec. Calculate the temperature distribution at every 100 sec until 1000 sec is reached. The property constants are

$$k = 80.2 \text{ W/mK} \qquad \text{(thermal conductivity)}$$
$$\rho = 7870 \text{ kg/m}^3 \qquad \text{(density)}$$
$$c = 0.447 \text{ kJ/kg}^\circ\text{K} \qquad \text{(specific heat unit)}$$

⟨Solution⟩

We first divide the rod into eleven control volumes as shown in Figure E9.16a. Denoting the average temperature of control volume i by $T_i(t)$, the heat balance equation for control volume i is written as

$$\rho c \Delta x A (dT_i/dt) = (q_{i-1} - q_i)A \qquad\qquad\qquad\text{(A)}$$

In Eq. (A), q_i is the heat flux (rate of conduction of heat transfer per unit cross-sectional area) at the boundary of the control volumes i and $i + 1$, and written

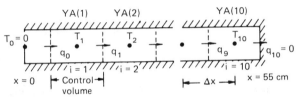

Figure E9.16a An insulated bar

```
INPUT PRINTING OR PLOTTING INTERVAL(P.I.)
 100
INPUT NUMBER OF STEPS IN ONE P.I. OF X
 5
INPUT MAXIMUM X TO STOP CALCULATION
 1005
```

H= 20	T_0	T_1	T_2	....						T_{10}
t= 100 (sec)	0	109	170	192	198	200	200	200	200	200	200
200	0	81	141	175	191	197	199	200	200	200	200
300	0	67	122	160	182	193	197	199	200	200	200
400	0	58	108	146	172	186	194	198	199	200	200
500	0	52	99	136	163	180	190	195	198	199	200
600	0	48	91	127	155	173	186	193	196	198	199
700	0	44	85	120	147	167	181	190	195	197	198
800	0	41	80	114	141	162	176	186	192	196	197
900	0	39	76	108	135	156	172	183	190	194	196
1000	0	37	72	104	130	152	168	179	187	192	194

Figure E9.16b Result of computations

by

$$q_i = -\frac{k}{\Delta x}(T_{i+1} - T_i) \quad \text{for } i = 0, 1, 2, \ldots, 9 \tag{B}$$

and

$$q_{10} = 0 \tag{C}$$

Introducing Eq. (B) into Eq. (A) and rearranging yield

$$\frac{dT_i}{dt} = \frac{k}{\rho c \Delta x^2}(T_{i-1} - 2T_i + T_{i+1}) \tag{D-1}$$

for $i = 1, 2, 3, \ldots, 9$, and

$$\frac{dT_{10}}{dt} = \frac{k}{\rho c \Delta x^2}(T_9 - T_{10}) \tag{D-2}$$

Equation (D) may be considered as a set of first-order ODEs and solved by using one of the Runge-Kutta methods. The set of equations is solved by PROGRAM 9–3 with some modifications. The computed results are shown in Figure E9.16b.

Notes:

(a) Equation (D) may be viewed as a semidifference approximation for the heat conduction equation (parabolic partial differential equation)

$$\frac{\partial}{\partial x} k \frac{\partial}{\partial x} T(x, t) = \rho c \frac{\partial T(x, t)}{\partial t}$$

with the initial condition, $T(x, 0) = 200°\,$C, and the boundary conditions, $T(0, t) = T'(55, t) = 0$.

(b) The present solution technique for the partial differential equation using a numerical method for ODEs is called the *method of lines*.

(c) Space-dependent and time-dependent thermal conductivity can be implemented with a minor change: that is, to recalculate k for each boundary of the control volumes in each time step.

(d) The author's study indicates that the computations using $h = 50$ sec agree well with that of $h = 1$ sec, but the solution scheme becomes unstable with $h = 100$ sec.

9.6 STIFF ODEs

9.6.1 Why Stiff Equations Are Difficult

Stiffness refers to a very short time constant of an ODE. Consider, for example,

$$y' = -\alpha y + s(t), \quad y(0) = y_0 \tag{9.6.1}$$

where $\alpha > 0$. The solution of this equation is, if $s = 0$,

$$y(t) = y_0 e^{-\alpha t} \tag{9.6.2a}$$

and if $s(t) \neq 0$, then

$$y(t) = y_0 e^{-\alpha t} + e^{-\alpha t} \int_0^t s(\xi) e^{\alpha \xi} \, d\xi \tag{9.6.2b}$$

The response of the system to the initial condition as well as to the changes of $s(t)$ is characterized by $1/|\alpha|$ that is called the *time constant*.

Solution of a stiff problem with a standard Runge-Kutta or predictor-corrector method is difficult or, sometimes, impossible. For example, if the fourth-order Runge-Kutta method is used for Eq. (9.6.2a and b), the computation becomes unstable unless $h < 2.785/|\alpha|$ (see Subsection 9.3.5). As the time constant becomes shorter, one has to use a progressively smaller time step. For $\alpha = -100000$ sec^{-1} as an example, h must be smaller than $2.785/100000 = 0.000002785$ sec just to maintain stability. The predictor-corrector methods discussed earlier in this chapter are subject to similar constraints.

When very fast transients of a system are computed, the necessity of small time steps is understandable. On the other hand, when $s(t)$ is a slowly varying function or a constant, the solution changes vary slowly, so naturally we desire to use larger time steps. Nonetheless, the same small time are necessary to assure stability of the numerical solution, no matter how slow the actual change of the solution is.

Stiffness is particularly serious for a set of ODEs [Gear (1971); Shampine/Gear (1979); Hall/Watt; Fertziger; Kuo]. If the set of equations contains only one stiff equation, the stability of a numerical method is governed by the short time constant of the stiffest equation.* For example, in the case of two equations,

$$y' = -y + z + 3$$
$$z = -10^7 z + y$$

(9.6.3)

the second equation has a significantly shorter time constant than the first.

A number of numerical methods that allow a large time step have been proposed including the implicit Runge-Kutta method and the rational Runge-Kutta method. Two such methods are introduced in the remainder of this section.

9.6.2 Implicit Methods

For simplicity, let us consider a set of two ODEs:

$$\frac{d}{dt} y = f(y, z, t)$$
$$\frac{d}{dt} z = g(y, z, t)$$

(9.6.4)

Using the backward difference approximation to the left side, we can write an implicit scheme as

$$y_{n+1} - y_n = hf(y_{n+1}, z_{n+1}, t_{n+1}) \equiv hf_{n+1}$$
$$z_{n+1} - z_n = hg(y_{n+1}, z_{n+1}, t_{n+1}) \equiv hg_{n+1}$$

(9.6.5)

where the f and g terms on the right side have unknowns, y_{n+1} and z_{n+1}.

If f and g are nonlinear functions, Eq. (9.6.5) cannot be solved in a closed form. However, the iterative solution explained in Subsection 9.2.3 can be applied very easily [Hall/Watt]. Indeed, the reader is encouraged to try it. Unfortunately, it is not computationally efficient for a large system of ODEs. A more efficient approach is to linearize the equations by Taylor expansions [Kubicek; Constantinites]. The Taylor expansion of $f_{k, n+1}$ about t_n becomes

$$f_{n+1} = f_n + f_y \Delta y + f_z \Delta z + f_t h$$
$$g_{n+1} = g_n + g_y \Delta y + g_z \Delta z + g_t h$$

(9.6.6)

---

* More strictly speaking, the time constant is an eigenvalue of the system and therefore not a value associated with any single equation in the set. Nonetheless, if one of the equations is much stiffer than the others, that equation determines the smallest time constant and there is very little influence from other equations.

where

$$\Delta y = y_{n+1} - y_n, \quad \Delta z = z_{n+1} - z_n \tag{9.6.7}$$

Introducing Eq. (9.6.6) into Eq. (9.6.5) and using Eq. (9.6.7) yields

$$\begin{bmatrix} 1 - hf_y & -hf_z \\ -hg_y & 1 - hg_z \end{bmatrix} \begin{bmatrix} \Delta y \\ \Delta z \end{bmatrix} = \begin{bmatrix} hf_n + h^2 f_t \\ hg_n + h^2 g_t \end{bmatrix} \tag{9.6.8a}$$

or more compactly

$$(I - hJ)\Delta \bar{y} = RHS \tag{9.6.8b}$$

where

RHS = vector on the right side of Eq. (9.6.8a)
  J = the Jacobian matrix defined by

$$J = \begin{bmatrix} f_y & f_z \\ g_y & g_z \end{bmatrix}$$

  I = identity matrix
$\Delta \bar{y}$ = col $(\Delta y, \Delta z)$

Equation (9.6.8) is solved by Gauss elimination. The implicit method is unconditionally stable unless nonlinear effects cause instability.

The method has been extended to a larger set of coupled ODEs. The Gear methods [Gear, 1971], which are available in NAG library [NAG], use higher-order backward difference approximations with variable time steps.

9.6.3 Exponential Method

Exponential transformation and exponential fitting have been proposed and used by various researchers to solve stiff ODEs. This subsection gives only a brief introduction of the basic ideas in exponential methods.

To explain the principle, consider a single first-order ODE:

$$y' = f(y, t) \tag{9.6.9}$$

where, for simplicity of the discussions, we assume f does not include t explicitly.

Adding cy to both terms of Eq. (9.6.9) yields

$$y' + cy = f(y, t) + cy \tag{9.6.10}$$

where c is a constant. Using e^{-ct} as an integrating factor, Eq. (9.6.10) is integrated in the interval $[t_n, t_{n+1}]$ as

$$y(t_{n+1}) = y_n e^{-ch} + \int_0^h [f(y(t_n + \xi), t_n + \xi) + cy(t_n + \xi)]e^{c(\xi - h)} d\xi \tag{9.6.11}$$

where $t_{n+1} = t_n + h$. Equation (9.6.11) is exact regardless of the choice of c.

Several different numerical schemes may be derived by introducing an approximation for $f + cy$ in the integrand. However, the accuracy of an approximate integration is then affected by the value of c. To find an appropriate value of c, we write y as

$$y(t) = y_n + \delta y(t) \tag{9.6.12}$$

Introducing Eq. (9.6.12) into Eq. (9.6.9) yields

$$\delta y' = f(y_n + \delta y)$$
$$= f_n + (f_y)_n \delta y + 0(\delta y^2) \tag{9.6.13}$$

By ignoring the second-order error term, Eq. (9.6.13) can be equivalently written as

$$y' - (f_y)_n y = f_n - (f_y)_n y_n \tag{9.6.14}$$

which is a linearized approximation for Eq. (9.6.9) about $t = t_n$. If c in Eq. (9.6.10) is set to

$$c = -(f_y)_n \tag{9.6.15}$$

then, Eq. (9.6.10) becomes identical with Eq. (9.6.14). By using

$$y'' = f' = f_y y' = f_y f \tag{9.6.16}$$

Eq. (9.6.15) may also be expressed by

$$c = -(f'/f)_n \tag{9.6.17}$$

An explicit numerical scheme is obtained by setting the terms in the brackets of Eq. (9.6.11) by

$$[f(y, t_n + \xi) + cy(t_n + \xi)] \simeq f_n + cy_n \tag{9.6.18}$$

Because the right side of Eq. (9.6.18) is constant, Eq. (9.6.11) reduces to

$$y_{n+1} = y_n e^{ch} + (1/c)(1 - e^{-ch})[f_n + cy_n]$$

$$= y_n + hf_n \left[\frac{1 - e^{-ch}}{ch} \right] \qquad (9.6.19)$$

which is known as the *exponentially fitted method* [Bui; Oran; Hetric; Fergason/ Hansen]. Not only this method is unconditionally stable but also positivity of the solution is guaranteed whenever the exact solution is expected to be positive.

The errors of Eq. (9.6.19) come from the approximation of Eq. (9.6.18). A more accurate method using an iterative procedure is developed in the remainder of this subsection. Based on Eq. (9.6.19), a predictor for $y(t)$ for $t_n < t < t_{n+1}$ can be set

$$\bar{y}(t) = y_n + \left[\frac{1 - e^{-c\xi}}{c} \right] f_n, \quad \xi = t - t_n \qquad (9.6.20)$$

and for t_{n+1},

$$\bar{y}_{n+1} = y_n + \left[\frac{1 - e^{-ch}}{c} \right] f_n$$

By introducing Eq. (9.6.20) into Eq. (9.6.11) we obtain

$$y_{n+1} = \bar{y}_n + \int_{\xi=0}^{h} [f(\bar{y}(t_n + \xi), t_n + \xi) - f_n + cy(t_n + \xi) - cy_n] e^{c(\xi - h)} d\xi \qquad (9.6.21)$$

The second term of Eq. (9.6.21) is a correction of Eq. (9.6.20), and can be evaluated by any one of the following:

(a) Analytical integration if it is possible.

(b) Approximating the terms in the brackets by a linear interpolation.

(c) Integrating by the trapezoidal rule.

Approach (a) is not easy unless f is a simple function, so we do not consider it any further. To pursue (b), the linear interpolation of the bracketed part is written as

$$[f(\bar{y}(t_n + \xi)) - f_n + cy(t_n + \xi) - cy_n] \cong B\xi \qquad (9.6.22)$$

where

$$B = \frac{f_{n+1} - f_n + c(y_{n+1} - y_n)}{h}$$

Introducing Eq. (9.6.22) into Eq. (9.6.21), the corrector becomes

$$y_{n+1} = \bar{y}_{n+1} + \frac{Bh^2}{ch}\left(\frac{1 - e^{-ch}}{ch - 1}\right) \tag{9.6.23}$$

If the trapezoidal rule is used, the corrector becomes

$$y_{n+1} = \bar{y}_{n+1} + \frac{Bh^2}{2} \tag{9.6.24}$$

which agrees with Eq. (9.6.23) in the limit of $ch \rightarrow 0$.

The second term of Eq. (9.6.23) or Eq. (9.6.24) is a correction to Eq. (9.6.19). To compute the second term, Eq. (9.6.19) is first evaluated, and then the second term is computed.

An extension of the exponential method to a set of nonlinear equations is straightforward, and the procedure is essentially the same as for a single equation. That is, Eq. (9.6.19) is independently evaluated for all the equations. Once the predictors for all the variables are obtained, then the second term of Eq. (9.6.23) or Eq. (9.6.24) is evaluated.

SUMMARY OF THIS SECTION

(a) An ODE becomes stiff if its time constant is short and $f''/f < 0$ (if there is no homogeneous term, the solution approaches zero). If a standard numerical method such as one of the Runge-Kutta or predictor-corrector methods is used, a very small time step is required even when the solution is slowly changing.

(b) To alleviate the difficulty of the stiff ODEs, two methods, including an implicit method and exponential method, are introduced.

PROGRAMS

PROGRAM 9–1 Second-order Runge-Kutta Method

(A) Explanations

This program solves a second-order ordinary differential equation in Example 9.6,

$$M\frac{d^2}{dt^2}y(t) + B\frac{d}{dt}y(t) + ky = 0, \quad y(0) = 1, \quad y'(0) = 0 \tag{A}$$

which represents the motion of a mass attached to one end of a spring-damper system that is hung from the ceiling as shown in Fig. E9.6. In the foregoing equation, M,

B, and *k* may be interpreted as mass, damping constant, and spring constant, respectively. To apply the second-order Runge-Kutta method to the second-order ODE, Eq. (A) is first split into two first-order equations as

$$\frac{d}{dt}\,y(t) = f(y, z, t), \quad y(0) = 1$$

$$\frac{d}{dt}\,z(t) = g(y, z, t), \quad z(0) = 0 \tag{B}$$

where

$$f(y, z, t) = z(t)$$

$$g(y, z, t) = -(B/M)z(t) - (k/M)y(t)$$

The constants and initial conditions are specified within the code. No input is necessary.

(B) List

```
/* CSL/c9-1.c        Second Order Runge-Kutta Scheme
 *                   (Solving the problem II of Example 9.6) */
#include <stdio.h>
#include <stdlib.h>
#include <math.h>
/*        time : t
          y,z: y,y'
          kount: number of steps between two lines of printing
          k, m, b: k, M(mass), B(damping coefficient) in Example 9.6 */
main()
{
int kount, n, kstep=0;
float bm, k1, k2, km, l1, l2;
static float time,   k = 100.0, m = 0.5, b = 10.0, z = 0.0;
static float y = 1.0,   h = 0.001;
    printf( "CSL/C9-1    Second Order Runge-Kutta Scheme \n" );
    printf( "       t              y              z\n" );
    printf( "    %12.6f    %12.5e   %12.5e \n", time, y, z );
    km = k/m;
    bm = b/m;
    for( n = 1; n <= 20; n++ ){
       for( kount = 1; kount <= 50; kount++ ){
          kstep=kstep+1; time = h*kstep ;
          k1 = h*z;                  l1 = -h*(bm*z + km*y);
          k2 = h*(z + l1);           l2 = -h*(bm*(z + l1) + km*(y + k1));
          y = y + (k1 + k2)/2;       z = z + (l1 + l2)/2;
       }
       printf( "    %12.6f    %12.5e   %12.5e \n", time, y, z );
    }
    exit(0);
}
```

(C) Sample Output

```
CSL/C9-1    Second Order Runge-Kutta Scheme
       t              y                z
    0.000000       1.00000e+00      0.00000e+00
    0.050000       8.23049e-01     -5.81545e+00
    0.100000       5.08312e-01     -6.19085e+00
    0.150000       2.38353e-01     -4.45118e+00
    0.200000       6.67480e-02     -2.46111e+00
    0.250000      -1.66253e-02     -9.82537e-01
    0.300000      -4.22529e-02     -1.40603e-01
    0.350000      -3.88646e-02      2.11764e-01
    0.400000      -2.58300e-02      2.77157e-01
    0.450000      -1.32004e-02      2.17147e-01
    0.500000      -4.55050e-03      1.29208e-01
    0.550000       1.17297e-05      5.76674e-02
    0.600000       1.68646e-03      1.38587e-02
```

(D) Discussions

The physical meaning of y is displacement of the mass and z is the velocity. The initial conditions $y(0) = 1$ and $z(0) = 0$ mean that initially the mass is held at $y = 1$ with zero velocity, and released at $t = 0$. The computed results show that the mass starts moving to the negative direction with increasing negative velocity, but the direction of motion is changed between $t = 0.3$ and 0.35. The motion of the mass is oscillatory but dies away quickly because of a strong damping effect.

If B is set to zero, the damping effect is removed and the motion theoretically becomes harmonic. If the program is run with $B = 0$, the computed results will show a more sustained oscillation. However, numerical error tends to damp the oscillation. The reader is encouraged to run the program with varying time steps. The accuracy of calculating a harmonic oscillation is significantly improved by a higher-order numerical method such as the fourth-order Runge-Kutta method.

PROGRAM 9-2 Fourth-order Runge-Kutta Scheme

(A) Explanations

This program integrates a first-order ordinary differential equation,

$$y' = f(y, t), \quad y(0) = y_0$$

When executed, the user is asked interrogatively for the number of steps, nstep_pr, in each printing time interval denoted by t_pr. The step interval h for the Runge-Kutta scheme is then set to $h = $ t_pr/nsteps_pr. The user is also asked for the maximum t to which the solution must proceed. The initial condition for y is set in the program.

(B) List

```
/* CSL/c9-2.c       Fourth-Order Runge-Kutta Scheme
 *                       (See Example 9.8) */
#include <stdio.h>
#include <stdlib.h>
#include <math.h>
#define TRUE 1
/*              time : t
                h : time step interval
                k1, k2, k3, and k4 : k1, k2, k3 and k4
                t_new and t_old : t for the new point and old point, resp.
                y : solution
                nstep_pr : number of steps in one printing t_pr
                t_pr : time interval of printing output              */
main()
{
int nstep_pr, j, j_, k;
float _f0, h, hh, k1, k2, k3, k4, t_old, t_limit,
                                t_mid, t_new, t_pr, y, ya, yn;
double fun();
    printf( "\n CSL/C9-2  Fourth-Order Runge-Kutta Scheme \n" );
    while(TRUE){
        printf( "Interval of t for printing ?\n" );     scanf( "%f", &t_pr );
        printf( "Number of steps in one printing interval?\n" );
        scanf( "%d", &nstep_pr );
        printf( "Maximum t?\n" );              scanf( "%f", &t_limit );
        y = 0;      /* Setting the initial value of the solution  */
        h = t_pr/nstep_pr;
        printf( "h=%g \n", h );
        t_new = 0;                            /* Time is initialized. */
        hh = h/2;
        printf( "------------------------------------------\n" );
        printf( "       t                    y\n" );
        printf( "------------------------------------------\n" );
        printf( " %12.5f      %15.6e \n", t_new, y );
        do{
            for( j = 1; j <= nstep_pr; j++ ){
                t_old = t_new;
                t_new = t_new + h;
                yn = y;
                t_mid = t_old + hh;
                yn = y;            k1 = h*fun( yn, t_old );
                ya = yn + k1/2; k2 = h*fun( ya, t_mid );
                ya = yn + k2/2; k3 = h*fun( ya, t_mid );
                ya = yn + k3  ; k4 = h*fun( ya, t_new );
                y = yn + (k1 + k2*2 + k3*2 + k4)/6;
            }
            printf( " %12.5f      %15.6e \n", t_new, y );
        } while( t_new <= t_limit );
        printf( "------------------------------------------\n" );
        printf( " Maximum t limit exceeded \n" );
        printf( "Type 1 to continue, or 0 to stop.\n" );
        scanf( "%d", &k );
        if( k != 1 )  exit(0);

    }
}
```

```
double fun(y, t)
float y, t;
{
float fun_v;
    fun_v = t*y +1;        /* Definition of f(y,t)    */
    return( fun_v );
}
```

(C) Sample Output

```
CSL/C9-2   Fourth-Order Runge-Kutta Scheme

Interval of t for printing ?
1
Number of steps in one printing interval?
10
Maximum t?
5
h=0.1
-------------------------------------
        t                    y
-------------------------------------
     0.00000             0.000000e+00
     1.00000             1.410686e+00
     2.00000             8.839369e+00
     3.00000             1.125059e+02
     4.00000             3.734231e+03
     5.00000             3.357972e+05
     6.00000             8.194355e+07
-------------------------------------

  Maximum t limit exceeded
```

(D) Discussions

The equation solved in the foregoing output is

$$y'(t) = -\frac{y(t)}{t^2 + y^2(t)}$$

with the initial condition $y(0) = 1$. The interactive input specifies that the printing interval is 1, which is divided into ten time steps for numerical integration.

PROGRAM 9–3 Fourth-order Runge-Kutta Method
for a Set of ODEs

(A) Explanations

This program is designed to solve by the fourth-order Runge-Kutta method a set of an arbitrary number of first-order ordinary differential equations written as

$$\frac{dy_i(t)}{dt} = f_i(y_j, j = 1, \ldots, I; t), \quad i = 1, 2, \ldots, I$$

where I is the number of equations coupled, and the initial conditions are

$$y_i(0) = y_{i,0}, \quad i = 1, \ldots, I$$

The functions on the right side of the differential equations are hard-coded in func(). Initial conditions are also specified within the program.

The printing time interval for the solution, the number of steps for integration per one printing time interval, and the maximum time for integration are given interactively by input.

(B) List

```
/* CSL/c9-3.c         Fourth-Order Runge-Kutta Scheme
 *                    for a Set of Equations
 *                    (See Example 9.9) */
#include <stdio.h>
#include <stdlib.h>
#include <math.h>
#define TRUE 1
/*              k1[], k2[], k3[], k4[] : k1, k2, k3 and k4 for j-th eq.
                t_new : t for the new point
                t_old : t for the previous point
                t_mid : t for the midpoint
                y[j] : solution for the j-th equation              */
main()
{
int i, No_of_eqs, j, k, n, ns;
float k1[11], k2[11], k3[11], k4[11],
    h, hh, pi, ta, t_old, t_limit, t_mid, t_new,
    y[11], ya[11], yn[11];
void f();
    printf( "\n CSL/C9-3      Fourth-Order Runge-Kutta Scheme \n" );
    printf( "              for a Set of Equations \n" );
    while (TRUE){
        No_of_eqs = 2;          /* Number of equations */
        y[1] = 1;          /* Initial condition for y1 at t=0. */
        y[2] = 0;          /* Initial condition for y2 at y=0. */
        printf( "Interval of t for printing ?  " );    scanf( "%f", &pi );
        printf( "Number of steps in one print interval ? " );
        scanf( "%d", &ns );
        printf("Maximum t to stop calculations ?");scanf("%f", &t_limit);
        h = pi/ns;
        printf( " h= %g \n", h );
        t_new = 0;
        hh = h/2;
        printf( "     t          y(1),              y(2), .....\n" );
        printf( "  %10.4f ", t_new );
        for( i = 1; i <= No_of_eqs; i++ ) printf(  "%12.5e ", y[i] );
        printf(  "\n" );
        do{
            for( n = 1; n <= ns; n++ ){
                t_old = t_new; /* Old time */
                t_new = t_new + h; /* New time */
                t_mid = t_old + hh; /* Midpoint time */
                for( i = 1; i <= No_of_eqs; i++ )    ya[i] = y[i];
                  f( k1,  ya, &t_old, &h );
                for( i = 1; i <= No_of_eqs; i++ )    ya[i] = y[i] + k1[i]/2;
                  f( k2, ya, &t_mid, &h );
```

```
        for( i = 1; i <= No_of_eqs; i++ )    ya[i] = y[i] + k2[i]/2;
          f( k3, ya, &t_mid, &h );
        for( i = 1; i <= No_of_eqs; i++ )    ya[i] = y[i] + k3[i];
          f( k4, ya, &t_new, &h );
        for( i = 1; i <= No_of_eqs; i++ )
          y[i] = y[i] + (k1[i] + k2[i]*2 + k3[i]*2 + k4[i])/6;
      }
      printf(  "   %10.4f ", t_new );
      for( i = 1; i <= No_of_eqs; i++ )  printf(  "%12.5e ", y[i] );
      printf(  "\n" );
    }while ( t_new < t_limit );
    printf( "Type 1 to continue, or 0 to stop. \n" );
    scanf( "%d", &k );   if( k != 1 )  exit(0);
  }
}

void f(k, y, t, h)
float k[], y[], *t, *h;
{
  k[1] =  y[2]**h;
  k[2] = -y[1]**h;
  /* More equations come here if the number of equations are greater.*/
  return;
}
```

(C) Sample Output

```
CSL/C9-3       Fourth-Order Runge-Kutta Scheme
                  for a Set of Equations
Interval of t for printing ?  0.5
Number of steps in one print interval ?  2
Maximum t to stop calculations ?  5.1
  h= 0.25
        t             y(1),            y(2), .....
       0.0000   1.00000e+00   0.00000e+00
       0.5000   8.77587e-01  -4.79410e-01
       1.0000   5.40326e-01  -8.41448e-01
       1.5000   7.07842e-02  -9.97482e-01
       2.0000  -4.16083e-01  -9.09312e-01
       2.5000  -8.01083e-01  -5.98526e-01
       3.0000  -9.89959e-01  -1.41212e-01
       3.5000  -9.36474e-01   3.50671e-01
       4.0000  -6.53723e-01   7.56699e-01
       4.5000  -2.10930e-01   9.77470e-01
       5.0000   2.83500e-01   9.58937e-01
```

(D) Discussions

A set of two ODEs given by

$$\frac{d}{dt} y_1(t) = y_2(t), \quad y_1(0) = 1$$

$$\frac{d}{dt} y_2(t) = -y_1(t), \quad y_2(0) = 0$$

is solved by the fourth-order Runge-Kutta method.

Results are printed out for every increment of $\Delta t = 0.5$ as specified by input. The time step interval for the Runge-Kutta method is half of the printing time interval, namely, $h = 0.25$. The computation is stopped when t exceeds 5. If a smaller time step for integration is desired, a larger number can be given for "Number of steps in one print interval."

The problem solved here may be interpreted as a harmonic oscillation problem because, if y_2 is eliminated, the equation for y_1 becomes

$$\frac{d^2 y_1}{dt^2} + y_1 = 0$$

The reader is encouraged to run this program with a finer time interval and plot the solution for a longer period, so a sustaining harmonic oscillation is seen. (See also Example 9.9.)

PROGRAM 9–4 Third-order Predictor-Corrector Method

(A) Explanations

This program solves a first-order ordinary differential equation,

$$\frac{dy}{dt} = f(y, t), \quad y(0) = y_0$$

using the third-order predictor-corrector method.

Since the predictor-corrector method cannot be self-started, the fourth-order Runge-Kutta method is used to start up the predictor-corrector method.

For the first two time steps, the fourth-order Runge-Kutta scheme is used. After then, the program ignores the Runge-Kutta method and starts using the predictor-corrector scheme. This program can be upgraded very easily to the fourth-order predictor-corrector method.

The user defines $f(y, t)$ and the initial condition in the program in a similar manner as PROGRAMS 9–2 and 9–3. Some input data are given interactively.

(B) List

```
/*CSL/c9-4.c   Third-Order Predictor-Corrector Method */
#include <stdio.h>
#include <stdlib.h>
#include <math.h>
#define TRUE 1
/*
                h : step interval
                h_half : 1/2 of h
                f_old : f at the previous step
                f_2step_old : f of 2 steps ago
```

```
                    f_3step-old : f of 3 steps ago
                    time_pr : time interval for printing output
                    nstep_pr : number of steps in one printing interval
                    y : solution
                    yp : predictor                                    */
main()
{
int nstep_pr, j, n;
float   f0, f, fa, f_old, f_2step_old, f_3step_old, g, h, h_half,
       k1, k2, k3, k4, t_old, time_pr, t_mid,
       tmax, t_new, ta, y, ya, yb, yp;
void func();
   printf( "\n\nCSL/C9-4  Third-Order Predictor-Corrector Method \n" );
   printf( "Time interval for printing solution ?\n" );
   scanf( "%f", &time_pr );
   printf( "Number of steps in one print interval ?\n" );
   scanf( "%d", &nstep_pr );
   printf( "Limit of t ?\n" );
   scanf( "%f", &tmax );
   h = time_pr/nstep_pr;                            /*  Time interval */
   h_half = h/2;
   printf( "Step size=%g \n", h );
   y = 0;                  /*  Initial condition for the solution */
   yb = y;
   ta =0;
   t_new = 0;
   f_2step_old = 0;
   f_old = 0;
   g = h/12;
   printf( " Solution:\n" );
   printf( "        t                y\n" );
   func( &fa, ta, y );
   n = 0;              /* Initialize time step counter */
   while(TRUE){
       for( j = 1; j <= nstep_pr; j++ ){
           /* nstep_pr time steps are advanced in this printing cycle */
           n = n + 1;  /* Counting time steps */
           f_3step_old = f_2step_old;  f_2step_old = f_old;   f_old = fa;
           t_old = t_new;  t_new = t_old + h;  t_mid = t_old + h_half;
           if( n <= 2 ){               /*  4-th order Runge-Kutta */
               func( &f, t_old, y );
               k1 = h*f; ya=y+k1/2;     func( &f, t_mid, ya );
               k2 = h*f; ya=y+k2/2;     func( &f, t_mid, ya );
               k3 = h*f; ya=y+k3;       func( &f, t_new, ya );
               k4 = h*f;
               y = y + (k1 + k2*2. + k3*2. + k4)/6.;
           }
           else{
               yp = y + g*(23*f_old - 16*f_2step_old + 5*f_3step_old);
               func( &fa, t_new, yp );
               y = y + g*(5*fa + 8*f_old - f_2step_old);
           }
           func( &fa, t_new, y );
       }
       printf("%10.4f        %12.5e \n", t_new, y );
       if( t_new > tmax )   exit(0);
   }
}
```

```
void  func(f, t, y)
float *f, t, y;
{
    *f = t*y + 1;       /* Defines the differential equation */
    return;
}
```

(C) Sample Output

```
CSL/C9-4  Third-Order Predictor-Corrector Method

Time interval for printing solution ?
1
Number of steps in one print interval ?
10
Limit of t ?
5.0
Step size=0.1
 Solution:
       t                    y
    1.0000              1.41091e+00
    2.0000              8.84414e+00
    3.0000              1.12644e+02
    4.0000              3.74007e+03
    5.0000              3.35593e+05
    6.0000              8.11660e+07
```

(D) Discussions

The problem solved is

$$\frac{dy}{dt} = ty + 1, \quad y(0) = 1$$

The solution approaches infinity because the right side is positive and increases with t.

PROBLEMS

(9.1) Solve the following problems in $0 \leqslant t \leqslant 5$ using the forward Euler method with $h = 0.5$ by hand calculation. Repeat the same with $h = 0.01$ by using a computer (write a short program yourself). Evaluate the errors by comparing to the exact solutions that follow:

(a) $y' + ty = 1,$ $y(0) = 1$

(b) $y' + 3y = e^{-t},$ $y(0) = 1$

(c) $y' = (t^2 - y),$ $y(0) = 0.5$

(d) $y' + y|y| = 0,$ $y(0) = 1$

(e) $y' + |y|^{1/2} = \sin(t),$ $y(0) = 1$

Exact Solution

t	Case a y	b^a y	c y	d y	e y
0	1.0000	1.0000	0.5000	1.0000	1.0000
1	1.3313	0.2088	0.4482	0.5000	0.6147
2	0.7753	0.06890	1.7969	0.3333	0.7458
3	0.4043	2.4955E–2	4.9253	0.2500	0.4993
4	0.2707	9.1610E–3	9.9725	0.2000	−0.2714
5	0.2092	3.3692E–3	16.980	0.1666	−2.2495

^{a}*Hint:* Solution of (b) may oscillate with $h = 0.5$, but you are encouraged to try anyway.

(9.2) Solve

$$y''(t) - 0.05y'(t) + 0.15y(t) = 0, \quad y'(0) = 0, \quad y(0) = 1$$

and find the values of $y(1)$ and $y(2)$ using the forward Euler method with $h = 0.5$.

(9.3) Solve the following problems in $0 \leqslant t \leqslant 5$ using the forward Euler method with $h = 0.1$ and $h = 0.01$ (write your own program).

(a) $y'' + 8y = 0,\ y(0) = 1,\ y'(0) = 0$

(b) $y'' - 0.01(y')^2 + 2y = \sin(t),\ y(0) = 0,\ y'(0) = 1$

(c) $y'' + 2ty' + ty = 0,\ y(0) = 1,\ y'(0) = 0$

(d) $(e^t + y)y'' = t,\ y(0) = 1,\ y'(0) = 0$

Evaluate the errors using the following exact solutions:

Exact Solution

t	Case a y	b y	c y	d y
0	1.0	0.0000	1.0000	1.0000
1	−0.9514	0.8450	0.8773	1.0629
2	0.8102	0.9135	0.5372	1.3653
3	−0.5902	0.1412	0.3042	1.8926
4	0.3128	−0.7540	0.1763	2.5589
5	−0.0050	−0.9589	0.1035	3.2978

(9.4) Solve the following equations for $0 < t < 5$ by the modified Euler method:

$$4y' = -3y + 7z + 2t, \quad y(0) = 1$$
$$7z' = -2y + 8z, \qquad\quad z(0) = 0$$

Use both $h = 0.01$ and 0.001.

(9.5) A conical tank contains water up to 0.5 m high from the bottom. The tank has a hole of 0.02 m radius at the bottom. The radius of the tank at y is given by $r = 0.25y$, where r is the radius and y is the height measured from the bottom. The velocity of the water that drains through the hole is given by $v^2 = 2gy$ where $g = 9.8$ m/sec^2. Using the forward Euler method (use $h = 0.001$ sec), find out how many minutes it will take until the tank becomes empty.

(9.6) A circuit shown in Figure P9.6 has self-inductance of $L = 100$ henry, a resistance of $R = 2$ ohm and a DC voltage source of 10 volt. If the switch is closed at $t = 0$, the current, $I(t)$, changes in accordance with

$$L \frac{d}{dt} I(t) + RI(t) = E, \quad I(0) = 0$$

(a) Find the current I at $t = 1, 2, 3, 4$, and 5 sec by using the forward Euler method with $h = 0.01$.

(b) Evaluate the error by comparing the numerical solution to the analytical solution given by $I(t) = (E/R)(1 - \exp^{-Rt/L})$.

(c) Investigate the effect of h by repeating the foregoing calculations with $h = 0.1$.

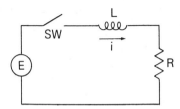

Figure P9.6 Electric circuit

(9.7) A U-tube of 0.05 m radius is initially filled with water but separated by a partition so that the water level of the left vertical part is 0.2 m higher than the water level of the right vertical part. At $t = 0$ the partition is suddenly removed. The water level of left vertical portion, y_A, measured from the midplane between two surfaces, satisfies

$$Ly_A'' = -2gy_A$$

where L is the total length of water in the U-tube, which is assumed to be 1 m, and $g = 9.8$ m/sec^2. Ignoring the friction in the tube, calculate the water level by the forward Euler method for $0 < t < 10$ sec, and find when y_A reaches minimums and maximums. Use $h = 0.001$.

(9.8) Repeat the previous problem assuming that there is friction in the pipe so that the equation of motion is given by

$$Ly_A'' = -2gy_A - \beta y_A'$$

where $\beta = 0.8$ m/sec. Use $h = 0.001$.

(9.9) The number density (number of atoms per cm^3) of iodine-135 (radioisotope) satisfies

$$\frac{d}{dt} N_i(t) = -\lambda_i N_i(t)$$

where $N(t)$ is the number density of iodine-135 and λ_i is its decay constant equal to 0.1044 hr^{-1}. If $N_i(0) = 10^5$ atoms/cm^3 at $t = 0$, compute $N_i(t)$ at $t = 1$ hr by the modified Euler method. Set h to 0.05 hr.

(9.10) The decay product of iodine-135 (considered in the previous problem) is xenon-135; it is also radioactive. Its decay constant is $\lambda_x = 0.0753$ hr^{-1}. The number density of xenon satisfies

$$\frac{d}{dt} N_x(t) = -\lambda_x N_x(t) + \lambda_i N_i(t)$$

where N_x is the number density of xenon and N_i is the number density of iodine defined in the previous problem. Assuming that $N_x(0) = 0$, develop a program to compute N_i and N_x together based on the modified Euler method. (Because the differential equations are linear, use closed form solutions for each time step.) Print out the solution for every 5 hours until 50 hrs is reached. Use $h = 0.1$ hr.

(9.11) Calculate $y(1)$ by solving the following equation by the second-order Runge-Kutta method with $h = 0.5$:

$$y' = -\frac{y}{t + y^2}, \quad y(0) = 1$$

(9.12) Calculate $y(2)$ for the following equation using the second-order Runge-Kutta method with $h = 1$:

$$y'' + 0.2y' + 0.003y \sin(t) = 0, \quad y(0) = 0, \, y'(0) = 1$$

(9.13) Find the value of $y(1)$ by solving

$$y'' - 0.05y' + 0.15y = 0, \quad y(0) = 1, \quad y'(0) = 0$$

Use the second-order Runge-Kutta method with $h = 0.5$.

(9.14) Solve the following differential equation:

$$2y'' + (y')^2 + y = 0, \quad y(0) = 0, \, y'(0) = 1$$

by the second-order Runge-Kutta method with $h = 0.5$ and evaluate $y(1)$ and $y'(1)$.

(9.15) An initial value problem of an ordinary differential equation is given by

$$y''' = -y, \quad y(0) = 1$$
$$y'(0) = y''(0) = 0$$

Using the second-order Runge-Kutta method with $h = 0.2$, calculate $y(0.4)$ and $y(1)$.

(9.16) (a) A 50-gal tank of water contains salt at a concentration of 10 oz/gal. To dilute the salt content, fresh water is supplied at the rate of 2 gal/min. If the tank is well mixed, and the same amount of water leaves the tank every minute, the salt content satisfies

$$y_1'(t) = -\frac{2}{50}y_1, \quad y_1(0) = 10$$

where $y_1(t)$ is the salt concentration in oz/gal, and t is time in minutes. By the second-order Runge-Kutta method with $h = 1$ min, find out how long it takes until the salt concentration reaches $1/10$ of its initial value.

(b) The water that leaves the tank enters another tank of 20 gal, into which fresh water is also poured at the rate of 3 gal/min and well mixed. The salt concentration in this tank satisfies

$$y_2'(t) = -\frac{3}{20}y_2(t) + \frac{2}{20}y_1(t), \quad y_2(0) = 0$$

where $y_1(t)$ is the salt concentration of the 50 gal tank of the previous problem. By using the second-order Runge-Kutta method, find when the salt concentration of the 20 gal tank reaches its maximum. Assume that the second tank has fresh water at $t = 0$.

(9.17) Repeat Problem 9.12 by the third-order Runge-Kutta method.

(9.18) A bullet is shot into the air at a $45°$ angle from the ground at $u = v = 150$ m/sec, where u and v are horizontal and vertical velocities, respectively. The equations of motion are given by

$$\begin{aligned} u' &= -cVu, & u(0) &= 150 \text{ m/sec} \\ v' &= -g - cVv, & v(0) &= 150 \text{ m/sec} \end{aligned} \quad \text{(A)}$$

where u and v are functions of time, $u = u(t)$, and $v = v(t)$, and

$$V = \sqrt{u^2 + v^2}$$
$$c = 0.005 \quad \text{(coefficient of drag)}$$
$$g = 9.8 \text{ m/sec}^2 \quad \text{(gravity)}$$

The equations of motion may be solved by one of the Runge-Kutta methods. The trajectory of the bullet may be calculated by integrating,

$$x' = u \quad \text{and} \quad y' = v$$

or

$$\begin{aligned} x &= \int_0^t u(t')\,dt' \\ y &= \int_0^t v(t')\,dt' \end{aligned} \quad \text{(B)}$$

A program based on the second-order Runge-Kutta method to solve Eq. (A) and evaluate Eq. (B) follows:

```
/* CSL/bullet2.c    Bullet shot to the sky (2nd order R-K) */
#include <stdio.h>
#include <stdlib.h>
#include <math.h>
main()
{
int n;
float K1, K2, L1, L2, u, v, VEL1, VEL2, VEL3;
static float ub=150.0, vb=150.0, h=0.1, c=0.005, t=0.0, x=0.0, y=0.0;
    printf( "\n\n CSL/BLLET2    Bullet shot to the sky\n " );
    printf( " t            u              v              x              y");
```

```
printf(  "\n%12.5e  %12.5e  %12.5e  %12.5e  %12.5e", t, ub, vb, x, y );
for( n = 1; n <= 200; n++ ){
   t = t + h;
   VEL1 = sqrt( pow(ub, 2) + pow(vb, 2) );
   K1 = -c*VEL1*ub*h;
   L1 = (-9.8 - c*VEL1*vb)*h;
   VEL2 = sqrt( pow(ub + K1/2, 2) + pow(vb + L1/2, 2) );
   K2 = -c*VEL2*(ub + K1)*h;
   L2 = (-9.8 - c*VEL2*(vb + L1))*h;
   u = ub + (K1 + K2)/2;
   v = vb + (L1 + L2)/2;
   x = x + 0.5*(u + ub)*h;
   y = y + 0.5*(v + vb)*h;
   ub = u;
   vb = v;
   printf("\n%12.5e  %12.5e  %12.5e  %12.5e  %12.5e",t,ub,vb,x,y);
   if( y < 0 ) break ;
}
printf(  "\n\n" );
}
```

(a) Run the program and plot the trajectory of the bullet.

(b) Rewrite the program using the third-order Runge-Kutta method.

(9.19) Calculate $y(1)$ by solving the following equation using the fourth-order Runge-Kutta method with $h = 1$:

$$y' = -\frac{y}{t + y^2}, \quad y(0) = 1 \text{ for } t = 0$$

(9.20) The solution of $y' = -1(1 + y^2)$ by the second-order Runge-Kutta method is shown for two different h values.

t	h = 0.1 y	h = 0.2 y
0.0	1.0000000	1.0000000
0.1	0.9487188	
0.2	0.894672	0.8947514
0.3	0.8375606	
0.4	0.7770516	0.7772616
0.5	0.7127807	
0.6	0.6443626	0.6447898
0.7	0.5714135	
0.8	0.4935937	0.4943817
0.9	0.4106803	
1.0	0.3226759	0.3240404

(a) Estimate the local error with $h = 0.1$.

(b) Estimate a more accurate value of $y(1)$.

(9.21) By hand calculations, find the solution of

$$y'(t) = -\frac{1}{1 + y^2}, \quad y(0) = 1$$

for $t = 1$ and $t = 2$ using the fourth-order Runge-Kutta method with $h = 0.5$ and $h = 1$.

(9.22) Repeat Problem 9.1 with the fourth-order Runge-Kutta method with $h = 0.1$.

(9.23) For the equation given by

$$y' = 3y + \exp(1 - t), \quad y(0) = 1$$

find an optimal time step for the second-order Runge-Kutta method to satisfy the condition for the local error, $E(h) < 0.0001$. (Run the second-order Runge-Kutta method for one interval with a value of h, and rerun it for two intervals with $h/2$.)

(9.24) Repeat Problem (9.23) for the fourth-order Runge-Kutta method.

(9.25) By repeating the analysis of Eqs. (9.3.23) through (9.3.27), derive the equation corresponding to Eq. (9.3.27) for the third-order Runge-Kutta method.

(9.26) If the third-order Runge-Kutta method is applied to $y' = -\alpha y$, find in what range of h the method is unstable.

(9.27) The initial temperature of the metal piece described in Example 9.12 is now assumed to be 25° C. The metal piece is internally heated electrically at the rate of $q = 3000$ W. The equation for the temperature is written as

$$\frac{dT}{dt} = \frac{1}{\rho c v}\left[q - \epsilon \sigma A(T^4 - 298^4) - h_c A(T - 298)\right], \quad T(0) = 298$$

Calculate the temperature until $t = 10$ min, and print it out for every 0.5 min by the fourth-order Runge-Kutta method with $h = 0.1$ min. (Use the constants given in Example 9.12.)

(9.28) The motion of the mass system shown in Figure P9.28 is given by

$$y'' + 2\zeta\omega y' + \omega^2 y = F(t)/M$$

where

$\omega = (k/M)^{1/2}$ (undamped natural frequency, s^{-1})
$\zeta = c/(2M\omega) = 0.5$ (damping factor)
$k = 3.2$ (spring constant, kg/s^2)
$M = 5$ (mass, kg)
$F(t) = 1$ (kg force) for $0 < t < 1$, or 0 for $t > 1$ (1 kg force = 9.8 Newton)

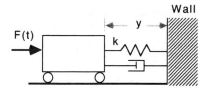

Figure P9.28 Spring-mass system

If $F(t)$ is a step function of magnitude $F_0 = 1$ kg force and time duration 1 sec, determine the motion of the mass for $0 < t < 10$ sec using the fourth-order Runge-Kutta method.

(9.29) Determine the response and dynamic load of the damping mass system of the previous problem subject to a triangular force pulse

$$F(t) = 2F_0 t, \qquad 0 \leqslant t \leqslant 1 \text{ sec}$$
$$= 2F_0(1 - t), \quad 1 \leqslant t \leqslant 2 \text{ sec}$$
$$= 0, \qquad\qquad t > 2 \text{ sec}$$

where $F_0 = 1$ kg (force). Use the fourth-order Runge-Kutta method.

(9.30) The differential equations for the circuit shown in Figure P9.30 is

$$L_1 \frac{d}{dt} i_1 + R_A(i_1 - i_2) + \frac{1}{C} \int_0^t (i_1(t') - i_2(t')) \, dt' = e(t)$$

$$-\frac{1}{C} \int_0^t (i_1(t') - i_2(t')) \, dt' - R_A(i_1 - i_2) + R_B i_2 + L_2 \frac{d}{dt} i_2 = 0$$

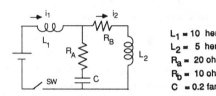

$L_1 = 10$ henry
$L_2 = 5$ henry
$R_a = 20$ ohm
$R_b = 10$ ohm
$C = 0.2$ farad

Figure P9.30 Electric circuit

The initial conditions are

$$i_1(0) = i_2(0) = 0,$$

and $e(t) = 1$. Using the fourth-order Runge-Kutta method with $h = 0.1$ sec, determine i_1 and i_2 for $0 < t < 10$ sec.

(9.31) The problem in Example 9.15 has a geometrical symmetry about $x = 0.5$. Therefore, the problem may be equivalently restated as

$$Ak \frac{d^2}{dx^2} T + P\sigma(T^4 - 273^4) = Q, \quad 0.5 < x < 1$$

$$T'(0.5) = 0$$
$$T(1) = 0$$

Using the fourth-order Runge-Kutta method, solve the foregoing equations by the shooting method. (*Hint:* change $T(0.5)$ by trial and error until $T(1) = 0$ is satisfied.)

(9.32) Repeat (a), (b), and (c) of Problem 9.1 with the Adams third-order predictor-corrector method using $h = 0.5$.

(9.33) Repeat (a), (b), and (c) of Problem 9.1 with the Adams fourth-order predictor-corrector method using $h = 1.0$.

(9.34) Explicitly write down the Adams-Bashforth predictors of orders 2, 3, 4, and 5 respectively.

(9.35) Explicitly write down the Adams-Moulton correctors of orders 2, 3, 4, and 5 respectively.

(9.36) Solve the problem in Example 9.11 by the Adams fourth-order predictor-corrector scheme. (*Hint:* Modify PROGRAM 9–4 so that the first three steps are calculated by the fourth-order Runge-Kutta method and the remainder is then calculated by the fourth-order predictor-corrector method.)

(9.37) The Euler predictor-corrector method is written as

$$\overline{y}_{n+1} = y_n + hy'_n$$

$$y_{n+1} = y_n + \frac{1}{2}h(\overline{y}'_{n+1} + y'_n)$$

If the Euler predictor-corrector method is applied to $y'(t) = \alpha y(t)$ where $\alpha < 0$, instability occurs if $\alpha h < -2$. Prove this.

(9.38) The Milne predictor-corrector method is given by

$$\overline{y}_{n+1} = y_{n-1} + \frac{4}{3}h(2y'_n - y'_{n-1} + 2y'_{n-2}) \quad \text{predictor}$$

$$y_{n+1} = y_{n-1} + \frac{1}{3}h(\overline{y}'_{n+1} + 4y'_n + y'_{n-1}) \quad \text{corrector}$$

Show that the Milne method becomes unstable for $y' = \alpha y$ if $\alpha < 0$ and

$$-\infty < \alpha h < -0.8 \quad \text{or} \quad -0.3 < \alpha h < 0$$

(*Hint:* Calculate the roots of the characteristic equation as illustrated in Section 9.4.4.)

REFERENCES

Bui, T. D., A. K. Oppenheim, and D. T. Pratt, "Recent advances in methods for numerical solution of O.D.E. initial value problems," *J. Comp. Math.*, Vol. 11, p. 283–296, 1984.

Constantinides, A., *Applied Numerical Methods with Personal Computers*, McGraw-Hill, 1987.

Creese, T. M., and R. M. Haralick, *Differential Equations for Engineers*, McGraw-Hill, 1978.

Ferziger, J. H., *Numerical Methods for Engineering Application*, Wiley-Interscience, 1981.

Fox, L., and D. F. Mayers, *Computing Methods for Scientists and Engineers*, Oxford, University Press 1968.

Furgason, D. R., and K. F. Hansen, "Solution of the space dependent reactor kinetics equations in three dimensions," *Nucl. Sci. Eng.*, Vol. 51, p. 189–205, 1973.

Gear, C. W., *Numerical Initial Value Problems in Ordinary Differential Equations*, Prentice-Hall, 1971.

Habib, I. S., *Engineering Analysis Methods*, Lexington Books, 1975.

Hall, G., and J. M. Watt, *Modern Numerical Methods for Ordinary Differential Equations*, Clarendon Press, 1976.

Hetric, D. L., *Dynamics of Nuclear Reactors*, University of Chicago Press, 1971.

Kubicek, M., and V. Hlavacek, *Numerical Solution of Nonlinear Boundary Value Problems with Applications*, Prentice-Hall, 1983.

Kuo, K. K., *Principles of Combustion*, Wiley-Interscience, 1986.

Lapidus, L., and J. H. Seinfeld, *Numerical Solution of Ordinary Differential Equations*, Academic Press, 1971.

NAG Fortran Library (1987), Algorithm Group Inc., 1101 31st, Suite 100, Downers Grove, IL 60515–1263.

Oran, E. S., and J. P. Boris, *Numerical Simulation of Reactive Flow*, Elsevier, 1987.

Rieder, W. G., and H. R. Busby, *Introductory Engineering Modeling*, Wiley, 1986.

Shampine, L. F. and C. W. Gear, "A User's View of Solving Stiff Ordinary Differential Equations," *SIAM Review*, Vol. 21, 1979.

10

Boundary Value Problems of Ordinary Differential Equations

10.1 INTRODUCTION

In a one-dimensional boundary value problem of ordinary differential equations, the solution is required to satisfy boundary conditions at both ends of the one-dimensional domain. Definition of boundary conditions is an important part of a boundary value problem. For example, consider a thin metal rod of length H with each end connected to a different heat source. If heat escapes from the surface of the rod to the air only by convection heat transfer, the equation for the temperature may be written as

$$-A \frac{d}{dx} k(x) \frac{d}{dx} T(x) + h_c P T(x) = h_c P T_\infty + A S(x) \qquad (10.1.1)$$

where $T(x)$ is the temperature at distance x from the left end, A the constant cross sectional area of the rod, k the thermal conductivity, P the perimeter of the rod, h_c the convection heat transfer coefficient, and T_∞ is the bulk temperature of the air, S is the heat source. The boundary conditions are

$$T(0) = T_L$$
$$T(H) = T_R \qquad (10.1.2)$$

where T_L and T_R are the given temperatures of the body at the left and right ends, respectively.

If $\overline{T}$ is defined as

$$\overline{T} = T - T_\infty$$

Eq. (10.1.1) may be written as

$$-\frac{d}{dx} k(x) \frac{d}{dx} \overline{T}(x) + \frac{h_c P}{A} \overline{T}(x) = S(x) \qquad (10.1.3)$$

where the equation has been divided by A. The first term represents the diffusion of heat, the second term is the removal of heat by convection to the air, and the right side is the heat source.

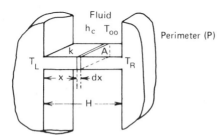

Fig. 10.1 A fin connected to two heat sources

Another example of an ODE in a similar form is the neutron diffusion equation given by

$$-\frac{d}{dx} D(x) \frac{d}{dx} \psi(x) + \Sigma_a \psi(x) = S(x) \qquad (10.1.4)$$

where ψ is neutron flux, D is diffusion coefficient, and S is neutron source. The meaning of the first term is the diffusion of neutrons, the second term is the removal by absorption, and the right side is the neutron source.

In the remainder of this chapter we consider the equation

$$-\frac{d}{dx} p(x) \frac{d}{dx} \phi(x) + q(x)\phi(x) = S(x) \qquad (10.1.5)$$

or similar equations on cylindrical or spherical coordinates. The first term is the diffusion term, the second is the removal term, and the right side is the source term, regardless of the particular physical situation considered.

It should be emphasized that Eq. (10.1.5) is a conservation law of diffusion. Indeed, integrating Eq. (10.1.5) in the interval $[a, b]$ yields

$$Z(b) - Z(a) + \int_a^b q(x)\phi(x)\, dx = \int_a^b S(x)\, dx \qquad (10.1.6)$$

where

$$Z(x) = -p(x) \frac{d}{dx} \phi(x)$$

is heat flux at x if heat conduction is considered, or neutron current if neutron diffusion is considered. In any case, the first and second terms in Eq. (10.1.6) are, respectively, inflow and outflow of the property represented by ϕ, the third term is the total removal in $[a, b]$, and the right side is the total source in $[a, b]$. Thus, Eq. (10.1.6) represents conservation of the property in $[a, b]$.

If Eq. (10.1.1) were an initial value problem, two boundary conditions would be specified at only one boundary, so the numerical solution could proceed from that end to the other, using a numerical method such as the fourth-order Runge-Kutta method. Although the solution methods for initial value problems can be used for boundary value problems as illustrated in Chapter 9, they work only on the trial-and-error basis (known as the shooting method; see Example 9.15). An advantage of the shooting method is that an existing program for initial value problems may be easily used. However, the shooting method often becomes unsuccessful because it may face numerical instability. Furthermore, its application becomes very difficult if the number of end conditions exceeds two [Hall/Watt].

A more general way of solving boundary value problems consists of (a) deriving difference equations and (b) solving all the difference equations simultaneously.

In this chapter, we first study the derivation of difference approximations for boundary value problems and its simultaneous solution. Then we explore the application of the methods to nonlinear boundary value problems as well as to eigenvalue problems. Studying the numerical methods for one-dimensional boundary value problems will help us understand the solution methods for partial differential equations. A brief summary of the methods is given in Table 10.1.

10.2 BOUNDARY VALUE PROBLEMS FOR RODS AND SLABS

In this section, we derive finite difference equations for second-order ordinary differential equations with boundary conditions.

To explain the principle of the method, we consider the equation,

$$-\phi''(x) + q\phi(x) = S(x) \tag{10.2.1}$$

$$0 < x < H$$

with the boundary conditions

$$\phi'(0) = 0 \quad \text{(left B.C.)}$$
$$\phi(H) = \phi_R \quad \text{(right B.C.)} \tag{10.2.2}$$

Table 10.1 Summary of the methods for one-dimensional boundary value problems

Type of Problems and Solution Method	Advantages	Disadvantages
Inhomogeneous Problems		
Shooting method	An existing program for initial value problems may be used.	Trial-and-error basis. Application is limited to a narrow class of problems. Solution may become unstable.
Finite difference method using the tridiagonal solution	No instability problem. No trial and error for linear problems. Applicable to nonlinear problems with iteration.	Program may have to be developed for each particular problem.
Eigenvalue Problems		
Matrix method (see Chapter 7)	All eigenvalues are calculated at once.	Not applicable if the size of the matrix is large.
Inverse power method	Simplicity.	Fundamental eigenvalue only.
Shifted inverse power method	As simple as power method. Any eigenvalue may be calculated.	Trial and error in estimating the eigenvalue; real eigenvalues only.

where q is a constant coefficient. By dividing the domain into N equispaced intervals, we obtain a grid as shown in Figure 10.2, where the grid intervals are $h = H/N$.

Applying the central difference approximation [see (f) of Table 5.2] to the first term of Eq. (10.2.1), the difference equation for grid i is derived as

$$\frac{-\phi_{i-1} + 2\phi_i - \phi_{i+1}}{h^2} + q\phi_i \simeq S_i \qquad (10.2.3)$$

where $\phi_i = \phi(x_i)$ and $S_i = S(x_i)$ and q is assumed to be constant. Multiplying Eq. (10.2.3) by h^2 yields

$$-\phi_{i-1} + (2 + w)\phi_i - \phi_{i+1} = h^2 S_i \qquad (10.2.4)$$

where $w = h^2 q$. Eq. (10.2.4) applies to all the grid points except for $i = 1$ and $i = N + 1$.

Figure 10.2 One-dimensional grid for a slab

The left boundary condition given by Eq. (10.2.2) is equivalent to a symmetry boundary condition called an *adiabatic boundary condition* in heat transfer. If a hypothetical grid point $i = 0$ located at $x = -h$ is considered, Eq. (10.2.4) for $i = 1$ becomes

$$-\phi_0 + (2 + w)\phi_1 - \phi_2 = h^2 S_1$$

In the foregoing equation, ϕ_0 can be set to $\phi_0 = \phi_2$ because of the symmetry. Then, dividing the resulting equation by 2 yields

$$\left(1 + \frac{w}{2}\right)\phi_1 - \phi_2 = \frac{1}{2}h^2 S_1 \tag{10.2.5}$$

Since $\phi_{N+1} = \phi(H) = \phi_R$ at the right boundary, Eq. (10.2.4) for $i = N$ is written as

$$-\phi_{N-1} + (2 + w)\phi_N = h^2 S_N + \phi_R \tag{10.2.6}$$

where all the known terms are brought to the right side.

The set of Eqs. (10.2.4), (10.2.5), and (10.2.6) are written together as

$$
\begin{aligned}
(1 + w/2)\phi_1 - \quad \phi_2 \qquad\qquad\qquad\qquad &= h^2 S_1/2 \\
-\phi_1 + (2 + w)\phi_2 - \quad \phi_3 \qquad\qquad &= h^2 S_2 \\
-\phi_2 + (2 + w)\phi_3 - \phi_4 \qquad &= h^2 S_3 \\
\vdots \qquad\qquad\qquad & \\
-\phi_{N-1} + (2 + w)\phi_N &= h^2 S_N + \phi_R
\end{aligned}
\tag{10.2.7a}
$$

or equivalently in the matrix form as

$$
\begin{bmatrix}
1 + w/2 & -1 & & & \\
-1 & 2 + w & -1 & & \\
& -1 & 2 + w & -1 & \\
& & & \ddots & \\
& & & -1 & 2 + w
\end{bmatrix}
\begin{bmatrix}
\phi_1 \\ \phi_2 \\ \phi_3 \\ \vdots \\ \phi_N
\end{bmatrix}
=
\begin{bmatrix}
h^2 S_1/2 \\ h^2 S_2 \\ h^2 S_3 \\ \vdots \\ h^2 S_N + \phi_R
\end{bmatrix}
\tag{10.2.7b}
$$

The matrix elements of Eq. (10.2.7b) are all zero except along the three diagonal lines. This special form of the matrix is called a *tridiagonal matrix*, and it appears very often in the numerical method for boundary value problems. We call Eq. (10.2.7a) or Eq. (10.2.7b) a *tridiagonal equation*, which is solved by the tridiagonal solution described in the next section.

Table 10.2 Three types of boundary conditions

Type Name	Explanation	Examples
Fixed value boundary condition (Dirichlet type)	Functional value of the solution is given.	$\phi(0) = 0,$ $\phi(0) = 1$
Derivative boundary condition (Neumann type)	Derivative of the solution is given.	$\phi'(0) = 0,$ $\phi'(0) = 1$
Mixed boundary condition (Mixed type)	Functional value is related to the derivative.	$\phi'(0) + \alpha\phi(0) = \beta$

Boundary conditions are classified into the three types as shown in Table 10.2. To discuss implementation of a mixed-type boundary condition, suppose both boundary conditions for Eq. (10.2.1) are given by the mixed type, namely

$$-\phi'(0) + f_L\phi(0) = g_L \tag{10.2.8}$$

$$\phi'(H) + f_R\phi(H) = g_R \tag{10.2.9}$$

where f_L, f_R, g_L, and g_R are constants. We will consider the grid shown in Figure 10.3 (which is the same as Figure 10.2 except that the last grid point is numbered N rather than $N + 1$).

Figure 10.3 One-dimensional grid

The difference equation, Eq. (10.2.4), is unchanged for $i = 2$ through $N - 1$, but the equations for $i = 1$ and N need to be revised because of the new boundary conditions. We consider the left boundary first. Using the forward difference approximation based on an interval of $h/2$ for Eq. (10.2.1) at $x = 0$ yields

$$-\frac{\phi'\left(\dfrac{h}{2}\right) - \phi'(0)}{\dfrac{h}{2}} + q\phi_1 \simeq S_1$$

Here, $\phi'(h/2)$ may be substituted by the central difference approximation,

$$\phi'\left(\frac{h}{2}\right) \simeq \frac{1}{h}(\phi_2 - \phi_1)$$

and $\phi'(0)$ may be eliminated by using Eq. (10.2.8). Thus, we obtain

$$-\frac{\dfrac{1}{h}(\phi_2 - \phi_1) + g_L - f_L\phi_1}{\dfrac{h}{2}} + q\phi_1 = S_1$$

or, equivalently,

$$\left(1 + \frac{w}{2} + hf_L\right)\phi_1 - \phi_2 = \frac{1}{2}h^2 S_1 + hg_L \tag{10.2.10}$$

where $w = qh^2$, and all the known terms are located on the right side.

The difference equation for the right boundary is derived by a similar procedure:

$$-\phi_{N-1} + \left(1 + \frac{w}{2} + hf_R\right)\phi_N = \frac{1}{2}h^2 S_N + hg_R \tag{10.2.11}$$

The set of Eqs. (10.2.10), (10.2.4), and (10.2.11) forms a tridiagonal equation set.

Example 10.1

Derive difference equations for the following boundary value problem:

$$-2y''(x) + y(x) = \exp(-0.2x) \tag{A}$$

with the boundary conditions

$$y(0) = 1$$
$$y'(10) = -y(10)$$

Assume that the grid spacing is unity.

⟨Solution⟩

Consider the grid shown in Figure E10.1. The difference equations for $i = 1$ through 9 are as follows:

$$2(-y_{i-1} + 2y_i - y_{i+1}) + y_i \simeq \exp(-0.2i) \tag{B}$$

where $x_i = i$ is used.

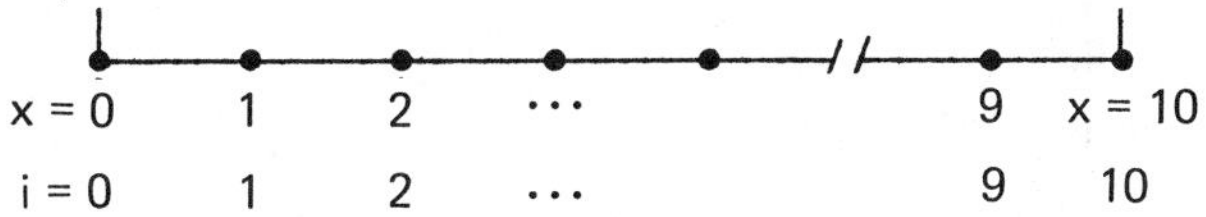

x = 0	1	2	⋯		9	x = 10
i = 0	1	2	⋯		9	10

Figure E10.1

For $i = 1$, the boundary condition $y_0 = y(0) = 1$ is introduced into the foregoing equations to give

$$5y_1 - 2y_2 = \exp(-0.2) + 2 \tag{C}$$

For $i = 10$, we approximate Eq. (A) first by

$$-\frac{2[y'(10) - y'(9.5)]}{\frac{1}{2}} + y(10) \simeq \exp(-2) \qquad \text{(D)}$$

Using the central difference approximation, the term $y'(9.5)$ becomes

$$y'(9.5) \simeq [y(10) - y(9)]/1 \qquad \text{(E)}$$

Introducing Eq. (E) and the right boundary condition $y'(10) = -y(10)$ into Eq. (D) yields

$$-2y_9 + 4.5y_{10} = 0.5 \exp(-2) \qquad \text{(F)}$$

Summarizing the difference equations obtained, we write

$$\begin{aligned}
5y_1 \quad - 2y_2 \quad &= \exp(-0.2) + 2 \\
-2y_{i-1} + 5y_i - 2y_{i+1} &= \exp(-0.2x_i), \quad \text{for } i = 2 \text{ to } 9 \qquad \text{(G)} \\
-2y_9 + 4.5y_{10} &= 0.5 \exp(-2)
\end{aligned}$$

where $x_i = j$ is used, PROGRAM 10-1 solves all the foregoing equations, (G).

SUMMARY OF THIS SECTION

(a) Basic numerical methods for the boundary value problem of the second-order ordinary differential equations are introduced. Two boundary conditions are given, one at the left end and another at the right end of the domain.

(b) The second derivative term is approximated by the central difference approximation.

(c) The set of difference equations for each problem becomes a tridiagonal equation in the matrix form.

10.3 SOLUTION ALGORITHM FOR TRIDIAGONAL EQUATIONS

We write the tridiagonal equation derived in Section 10.2 in the following form:

$$\begin{bmatrix}
B_1 & C_1 & & & & & \\
A_2 & B_2 & C_2 & & & & \\
& A_3 & B_3 & C_3 & & & \\
& & & \ddots & & & \\
& & & A_i & B_i & C_i & \\
& & & & & \ddots & \\
& & & & & A_N & B_N
\end{bmatrix}
\begin{bmatrix}
\phi_1 \\ \phi_2 \\ \phi_3 \\ \vdots \\ \phi_i \\ \vdots \\ \phi_N
\end{bmatrix}
=
\begin{bmatrix}
D_1 \\ D_2 \\ D_3 \\ \vdots \\ D_i \\ \vdots \\ D_N
\end{bmatrix}
\qquad (10.3.1)$$

The solution algorithm for the tridiagonal equation is called the *tridiagonal solution* (a variant of Gauss elimination) and given next:

(a) Initialize the two new variables:

$$B'_1 = B_1 \quad \text{and} \quad D'_1 = D_1$$

(b) Recurrently calculate the following equations in increasing order of i until $i = N$ is reached:

$$R = A_i/B'_{i-1}$$

$$B'_i = B_i - RC_{i-1}$$

$$D'_i = D_i - RD'_{i-1} \quad \text{for } i = 2, 3, \ldots, N.$$

(c) Calculate the solution for the last unknown by

$$\phi_N = D'_N/B'_N$$

(d) Calculate the following equation in decreasing order of i:

$$\phi_i = (D'_i - C_i\phi_{i+1})/B'_i, \quad i = N - 1, N - 2, \ldots, 1$$

In a computer program, the primed variables B'_i and D'_i need not be distinguished from B_i and D_i respectively because B'_i and D'_i are stored in the same memory spaces as for B_i and D_i. Therefore, Step (a) is not necessary in real programming.

The tridiagonal solution is used in PROGRAM 10–1 through PROGRAM 10–4. A subroutine to perform the tridiagonal solution follows:

```
void trdg(a, b, c, d, n)    /* Tridiagonal solution */
float a[], b[], c[], d[];
int n;
{
int i;
float r;
   for( i = 2; i <= n; i++ ){
      r = a[i]/b[i - 1];
      b[i] = b[i] - r*c[i - 1];
      d[i] = d[i] - r*d[i - 1];
   }
   d[n] = d[n]/b[n];
   for( i = n - 1; i >= 1; i-- ){
      d[i] = (d[i] - c[i]*d[i + 1])/b[i];
   }
   return;
}
```

When the computation in the subroutine is completed, the solution is stored in array $d[]$.

SUMMARY OF THIS SECTION. Tridiagonal solution is the most basic numerical method used in solving boundary value problems of ordinary differential equations.

10.4 VARIABLE COEFFICIENTS AND NONUNIFORM GRID SPACING IN THE SLAB GEOMETRY

In many problems, the coefficients of the differential equation are space-dependent and a nonequispaced grid is used. The variable coefficients and nonequispaced grid occur when the geometry consists of different materials, for example.

The second-order ordinary differential equation for the slab geometry with variable coefficients is written here as

$$-(p(x)\phi'(x))' + q(x)\phi(x) = S(x) \tag{10.4.1}$$

with the boundary conditions given by Eqs. (10.2.8) and (10.2.9). The grid spacing between x_i to x_{i+1} will be denoted by h_i. We assume that p, q, and S in each grid interval are constant and denoted by p_i, q_i, and S_i, respectively, as shown in Figure 10.4.

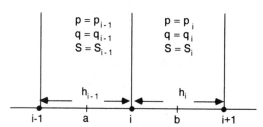

Figure 10.4 Constants in grid intervals

One natural way of deriving difference equations with piecewise-constant coefficients is the method of integration. In this method, Eq. (10.4.1) is integrated from a to b (see Figure 10.4):

$$-\int_a^b (p(x)\phi'(x))' \, dx + \int_a^b q(x)\phi(x) \, dx = \int_a^b S(x) \, dx \tag{10.4.2}$$

where $a = x_i - h_{i-1}/2$, and $b = x_i + h_i/2$ (which are midpoints between $i - 1$ and i, and i and $i + 1$, respectively).

The first term of Eq. (10.4.2) becomes

$$-\int_a^b (p\phi')' \, dx = -(p\phi')_{i+\frac{1}{2}} + (p\phi')_{i-\frac{1}{2}}$$

The derivatives on the right side are approximated by the central difference approximation:

$$(p\phi')_{i-\frac{1}{2}} \simeq p_{i-1}(\phi_i - \phi_{i-1})/h_{i-1}$$
$$(p\phi')_{i+\frac{1}{2}} \simeq p_i(\phi_{i+1} - \phi_i)/h_i$$

where $p(x) = p_i$ for $x_i < x < x_{i+1}$. Thus, the first term of Eq. (10.4.2) becomes

$$-\int_a^b (p\phi')' \, dx \simeq -\frac{p_{i-1}}{h_{i-1}}\phi_{i-1} + \left(\frac{p_{i-1}}{h_{i-1}} + \frac{p_i}{h_i}\right)\phi_i - \frac{p_i}{h_i}\phi_{i+1} \qquad (10.4.3)$$

The second term of Eq. (10.4.2) becomes

$$\int_a^b q(x)\phi(x) \, dx \simeq \frac{1}{2}(q_{i-1}h_{i-1} + q_i h_i)\phi_i \qquad (10.4.4)$$

where $\phi(x)$ in the integrand is approximated by ϕ_i. The right side of Eq. (10.4.1) becomes

$$\int_a^b S(x) \, dx \simeq \frac{1}{2}(S_{i-1}h_{i-1} + S_i h_i) \qquad (10.4.5)$$

Introducing Eqs. (10.4.3), (10.4.4), and (10.4.5) into Eq. (10.4.2) yields

$$-\frac{p_{i-1}}{h_{i-1}}\phi_{i-1} + \left(\frac{p_{i-1}}{h_{i-1}} + \frac{p_i}{h_i}\right)\phi_i - \frac{p_i}{h_i}\phi_{i+1} + \frac{1}{2}(q_{i-1}h_{i-1} + q_i h_i)\phi_i$$

$$= \frac{1}{2}(S_{i-1}h_{i-1} + S_i h_i) \qquad (10.4.6)$$

Equation (10.4.6) can now be written in the form

$$A_i\phi_{i-1} + B_i\phi_i + C_i\phi_{i+1} = D_i \qquad (10.4.7)$$

where

$$A_i = -\frac{p_{i-1}}{h_{i-1}}$$

$$B_i = \frac{p_{i-1}}{h_{i-1}} + \frac{p_i}{h_i} + \frac{1}{2}(q_{i-1}h_{i-1} + q_i h_i)$$

$$C_i = -\frac{p_i}{h_i}$$

$$D_i = \frac{1}{2}(S_{i-1}h_{i-1} + S_i h_i)$$

Assuming that the boundary conditions are given by Eq. (10.2.8) and Eq. (10.2.9), difference equations for the left and right boundary points are also derived by integrating Eq. (10.4.1). Considering the left boundary point, a and b in Eq. (10.4.2) are

$$a = x_1 \qquad \text{(the left boundary point)}$$

$$b = x_1 + h_1/2 \quad \text{(the midpoint between } x_1 \text{ and } x_2)$$

Then, the first term of Eq. (10.4.2) becomes

$$-\int_a^b (p\phi')'\, dx = -(p\phi')_{1+\frac{1}{2}} + (p\phi')_1 \tag{10.4.8}$$

The first term on the right side is approximated by the central difference approximation,

$$-(p\phi')_{1+\frac{1}{2}} \simeq \frac{-p_1(\phi_2 - \phi_1)}{h_1} \tag{10.4.9}$$

The ϕ' in the second term on the right hand side of Eq. (10.4.8) is eliminated by Eq. (10.2.8). Therefore, Eq. (10.4.8) becomes

$$-\int_a^b (p\phi')'\, dx \simeq -\frac{p_1}{h_1}(\phi_2 - \phi_1) + p_1(-g_L + f_L\phi_1) \tag{10.4.10}$$

The second term and right side of Eq. (10.4.2) become, respectively,

$$\int_a^b q(x)\phi(x)\, dx \simeq \frac{1}{2}q_1 h_1 \phi_1 \tag{10.4.11}$$

$$\int_a^b S(x)\, dx \simeq \frac{1}{2}S_1 h_1 \tag{10.4.12}$$

Introducing Eqs. (10.4.10) through (10.4.12) into Eq. (10.4.2) yields

$$\left(\frac{p_1}{h_1} + p_1 f_L + \frac{1}{2}q_1 h_1\right)\phi_1 - \frac{p_1}{h_1}\phi_2 = \frac{1}{2}S_1 h_1 + p_1 g_L \tag{10.4.13}$$

which may be rewritten more compactly as

$$B_1\phi_1 + C_1\phi_2 = D_1$$

The difference equation for the right boundary point can also be derived similarly and written as

$$A_N\phi_{N-1} + B_N\phi_N = D_N \tag{10.4.15}$$

The set of difference equations thus derived, namely Eqs. (10.4.14), (10.4.7), and (10.4.15), becomes exactly the form of Eq. (10.3.1).

The difference equations derived in this section satisfy the conservation law. Before completing this section, conservation property of Eq. (10.4.7) and positivity of the solution are mentioned.

If we define

$$Z_i = -\frac{p_i}{h_i}(\phi_{i+1} - \phi_i) \qquad (10.4.16)$$

$$\overline{B}_i = \frac{1}{2}(q_{i-1}h_{i-1} + q_i h_i)\phi_i \qquad (10.4.17)$$

then Eq. (10.4.7) can be rewritten as

$$Z_i - Z_{i-1} + \overline{B}_i = D_i \qquad (10.4.18)$$

where D_i is defined after Eq. (10.4.7). If we add Eq. (10.4.12) for $i = k, k+1, \ldots, m$, into a single equation, we get

$$Z_m - Z_{k-1} + \sum_{i=k}^{m} \overline{B}_i = \sum_{i=k}^{m} D_i \qquad (10.4.19)$$

Equation (10.4.19) satisfies conservation of the property represented by ϕ in $x_{k-\frac{1}{2}} < x < x_{m+\frac{1}{2}}$ because the first and second terms are inflow and outflow of the property, the third term is the total removal, and the right side is the total source. This statement is true for any arbitrary choice of k and m.

When the set of difference equations satisfies the conservation law, the difference equations are said to be in the *conservation form*. If difference equations are derived in some other way, they may not satisfy the conservation law.

Provided that (a) physically meaningful boundary conditions are imposed and (b) that the coefficient of the removal term is nonnegative, the coefficient matrix of the difference equations in the conservation form satisfies the following conditions when the difference equations are written in the form of Eq. (10.3.1):

(a) The coefficient matrix is symmetric.

(b) The diagonal coefficients are all positive.

(c) The off-diagonal nonzero coefficients are all negative.

(d) The coefficients in each row satisfy

$$B_i \geqslant -A_i - C_i$$

with strict inequality for at least one row.

(e) A_i, B_i, and C_i are all nonzero (A_1 and C_N do not exist).

The inverse of the matrix satisfying all five of the foregoing conditions is shown to be a positive matrix, that is, all the elements of the inverse matrix are positive. This implies that, if $D_i \geqslant 0$ with strict inequality for at least one i, the solution is positive everywhere.

SUMMARY OF THIS SECTION

(a) The coefficients of the ordinary differential equation are considered space-dependent but assumed to be constant in each grid interval.

(b) Difference equations form a tridiagonal equation.

(c) The difference equations derived in this section are in the conservation form.

10.5 BOUNDARY VALUE PROBLEMS FOR CYLINDERS AND SPHERES

Derivation of difference equations for second-order ordinary differential equations for cylindrical and spherical geometries is very similar to that discussed in Section 10.4. The difference equations for these two geometries will have the form of Eq. (10.3.1) again, which is solved by the tridiagonal solution described in Section 10.3. Therefore, we discuss derivation of difference equations.

The second-order ordinary differential equation for cylindrical and spherical geometries may be written in a single expression as

$$-\frac{1}{r^m}\frac{d}{dr}\,p(r)r^m\frac{d}{dr}\,\phi(r) + q(r)\phi(r) = S(r) \tag{10.5.1}$$

where

$$m = 1 \text{ for cylinder}$$

$$m = 2 \text{ for sphere}$$

Notice also that if $m = 0$, Eq. (10.5.1) becomes for slab geometry.

Considering space-dependent coefficients and nonequispaced grid as discussed in the previous section, we derive difference approximations by the method of integration, which is to integrate the equation over a cylindrical volume cell or a spherical cell depending on the geometry. Of course, direct differencing of Eq. (10.5.1) is also possible provided that q and S are averaged appropriately.

Derivation of difference equations is now shown for a cylinder ($m = 1$) using the notations for h, p, q, and S defined in Figure 10.4, where p, q, and S are assumed to be constant between two consecutive points. We multiply Eq. (10.5.1) by r and integrate from $a = r_{i-\frac{1}{2}}$ to $b = r_{i+\frac{1}{2}}$ which are midpoints of $[r_{i-1}, r_i]$ and $[r_i, r_{i+1}]$, respectively:

$$-\int_a^b \left[\frac{d}{dr}\,rp(r)\frac{d}{dr}\,\phi(r)\right]dr + \int_a^b q(r)\phi(r)r\,dr = \int_a^b S(r)r\,dr \tag{10.5.2}$$

Here, $r\,dr$ represents an infinitesimal volume element divided by $2\pi L$ (L is the height of the circular cylinder). The first term of Eq. (10.5.2) becomes

$$p_{i-1}r_{i-\frac{1}{2}}\left[\frac{d}{dr}\,\phi(r)\right]_{i-\frac{1}{2}} - p_i r_{i+\frac{1}{2}}\left[\frac{d}{dr}\,\phi(r)\right]_{i+\frac{1}{2}}$$

and then, using the difference approximation for the derivatives yields

$$p_{i-1}r_{i-\frac{1}{2}}(\phi_i - \phi_{i-1})/h_{i-1} - p_i r_{i+\frac{1}{2}}(\phi_{i+1} - \phi_i)/h_i \qquad (10.5.3)$$

The first term times $2\pi L$ is the total flux of the property through the cylindrical surface at $a = r_{i-\frac{1}{2}}$, and the second is the same for $b = r_{i+\frac{1}{2}}$.

The second term of Eq. (10.5.2) may be approximated by

$$\int_a^b q(r)\phi(r)r\,dr \simeq [v_l q_{i-1} + v_r q_i]\phi_i \qquad (10.5.4)$$

and represents the total removal of the physical property in $[r_{i-\frac{1}{2}}, r_{i+\frac{1}{2}}]$, where

$$v_l = \frac{1}{2}\left[r_i^2 - \left(r_i - \frac{h_{i-1}}{2}\right)^2\right] = \frac{h_{i-1}}{2}\left(r_i - \frac{h_{i-1}}{4}\right) \qquad (10.5.5)$$

$$v_r = \frac{1}{2}\left[\left(r_i + \frac{h_i}{2}\right)^2 - r_i^2\right] = \frac{h_i}{2}\left(r_i + \frac{h_i}{4}\right) \qquad (10.5.6)$$

Notice here that, v_l times $2\pi L$ becomes the volume of a cylindrical cell between $r = r_{i-\frac{1}{2}}$ and $r = r_i$, while v_r becomes the same between r_i and $r_{i+\frac{1}{2}}$. The third term of Eq. (10.5.2) may be approximated similarly as

$$\int_a^b S(r)r\,dr \simeq v_l S_{i-1} + v_r S_i \qquad (10.5.7)$$

Collecting all the terms, the difference approximation for Eq. (10.5.1) becomes the tridiagonal form. Although the cylindrical case has been described in detail, derivation for the spherical geometry is essentially the same.

The difference equations derived in this section are in the conservation form. The coefficient matrix for a cylinder has exactly the same mathematical properties as for the slab geometry (see Section 10.4), so it has a positive inverse matrix.

The conservation property can be lost if the difference equations are derived in a different way. To illustrate this for Eq. (10.5.1), assume, for simplicity, that $m = 1$, and p, q, and s are constant. By rewriting the first term, Eq. (10.5.2) becomes

$$-p\frac{d^2}{dr^2}\phi(r) - \frac{p}{r}\frac{d}{dr}\phi + q\phi(r) = S \qquad (10.5.8)$$

Straightforward differencing of Eq. (10.5.8) on an equispaced grid yields

$$p\frac{-\phi_{i-1} + 2\phi_i - \phi_{i+1}}{h^2} - p\frac{\phi_{i+1} - \phi_{i-1}}{2hr_i} + q\phi_i = S \qquad (10.5.9)$$

The difference equation above violates the conservation law, although it is often used.

In general, the nonconservation form of difference equations should be avoided if possible, because they are less accurate than conservative difference approximations. Indeed, accuracy of the solution for Eq. (10.5.9) becomes increasingly poorer toward the center of the coordinate where r is zero [Smith].

SUMMARY OF THIS SECTION

(a) The difference equations for cylindrical and spherical geometry are derived by integrating the differential equation in space.

(b) The difference equations for cylindrical and spherical geometries are written in the tridiagonal equation form.

10.6 BOUNDARY VALUE PROBLEMS OF NONLINEAR ORDINARY DIFFERENTIAL EQUATIONS

An ordinary differential equation is nonlinear if the unknown appears in a nonlinear form, or if its coefficient(s) depends on the solution. For example, the heat conduction equation for a cooling fin becomes nonlinear if radiation heat transfer from the surface is involved. The diffusion equation for a chemical species is nonlinear if it has a removal term of which the coefficient is dependent on the density of the species. In a nuclear reactor, properties of the materials are significantly affected by the neutron population although indirectly when the power level is high, so the governing equation for the neutron flux becomes nonlinear. See Kubicek/Hlavacek, and also Nishida/Miura/Fujii for numerous examples of nonlinear boundary value problems.

Solution methods for nonlinear boundary value problems require iterative applications of a solution method for linear boundary value problems. Two general methods will be discussed considering a nonlinear diffusion equation given by

$$-\phi'' + 0.01\phi^2 = \exp(-x), \quad 0 < x < H$$
$$\phi(0) = \phi(H) = 0 \tag{10.6.1}$$

Before proceeding to the numerical solution algorithms, however, we note some peculiar aspects of nonlinear boundary value problems. First, unlike a linear boundary value problem, existence of the solution is not guaranteed. Second, a nonlinear boundary value problem can have more than one solution. Indeed, different solutions may be obtained for different initial guesses for an iterative algorithm. Therefore, when a numerical solution is obtained, one must examine it to see if it is physically meaningful.

SUCCESSIVE SUBSTITUTION.   Equation (10.6.1) is now rewritten as

$$-\phi'' + \alpha(x)\phi(x) = \exp(-x) \tag{10.6.2}$$

where

$$\alpha(x) = 0.01\phi(x)$$

The method explained here is an extension of the successive substitution method described in Chapter 3 and proceeds as follows:

(a) Set $\alpha(x)$ to an estimate, for example $\alpha(x) = 0.01$.

(b) Solve Eq. (10.6.2) numerically as a linear boundary value problem (since α is fixed, the equation is linear).

(c) Revise $\alpha(x) = 0.01\phi(x)$ with the updated value of $\phi(x)$ from (b).

(d) Repeat (b) and (c) until $\phi(x)$ in two consecutive solutions agree within a prescribed tolerance.

NEWTON'S METHOD. Suppose an estimate for $\phi(x)$ denoted by $\psi(x)$ is available. The exact solution may then be expressed as

$$\phi(x) = \psi(x) + \delta\psi(x) \tag{10.6.3}$$

where $\delta\psi(x)$ is a correction for the estimate. Introducing Eq. (10.6.3) into Eq. (10.6.1) gives

$$-\delta\psi'' + (0.01)[2\psi\delta\psi + (\delta\psi)^2] = \psi'' - 0.01\psi^2 + \exp(-x) \tag{10.6.4}$$

Ignoring the second-order term $(\delta\psi)^2$ yields

$$-\delta\psi'' + 0.02\psi\delta\psi \simeq \psi'' - 0.01\psi^2 + \exp(-x) \tag{10.6.5}$$

which may be solved as a linear boundary value problem. An approximate solution for Eq. (10.6.1) is then obtained by $\psi(x) + \delta\psi(x)$. The solution may be further improved by repeating the procedure using the most updated result as a new estimate. This procedure is an extension of Newton's method described in Chapter 3.

Example 10.2

Derive linearized difference equations based on the Newton's method for Eq. (10.6.1) in the domain $0 < x < 2$ with the boundary conditions $\phi(0) = \phi(2) = 0$ using 10 grid intervals. Solve the equations.

⟨**Solution**⟩

The linearized form of Eq. (10.6.1) is given by Eq. (10.6.5). With the grid spacing $h = 2/10 = 0.2$, the difference equations for Eq. (10.6.5) are written as

$$-\delta\psi_{i-1} + 2\delta\psi_i - \delta\psi_{i+1} + 0.02h^2\psi_i\delta\psi_i = \psi_{i-1} - 2\psi_i$$

$$+ \psi_{i+1} - 0.01h^2\psi_i^2 + h^2\exp(-ih), \quad i = 1, 2, \ldots, 9$$

where $i = 0$ for $x = 0$, and the equation has been multiplied by h^2. The foregoing equation may be written in the form of Eq. (10.3.1) if we define

$$A_i = -1$$
$$B_i = 2 + 0.02h^2\psi_i$$
$$C_i = -1$$
$$D_i = \psi_{i-1} - 2\psi_i + \psi_{i+1} - 0.01h^2\psi_i^2 + h^2 \exp(-ih)$$

We start the Newton's iteration by setting an estimate as $\psi_i = 0$ for all the grid points. Then, the difference equations for $i = 1, 2, 3, \ldots, 9$ are solved by the tridiagonal solution. The iterative solution for the first five grid points is given in Table 10.3.

Table 10.3

Iteration Number	Grid Points				
	$i = 1$	$i = 2$	$i = 3$	$i = 4$	$i = 5$
1	0.0850	0.1406	0.1720	0.1837	0.1792
2	0.0935	0.1546	0.1891	0.2019	0.1970
3	0.0943	0.1560	0.1908	0.2038	0.1988
4	0.0944	0.1561	0.1910	0.2040	0.1990

Example 10.3

Boundary layer equations [Arpaci/Larsen, Incropera/DeWitt] are usually solved by the shooting method, but it can also be solved as a boundary value problem. As an example, consider

$$f''' + \frac{1}{2}ff'' = 0 \tag{A}$$

with the boundary conditions,

$$f(0) = 0, \quad f'(0) = 0, \quad f'(\infty) = 1$$

Solve Eq. (A) by the difference approximation as a boundary value problem.

⟨Solution⟩

Although Eq. (A) is a third-order ODE, it can be reduced to a second-order ODE as follows. Define

$$g(\eta) = f'(\eta) \tag{B}$$

or equivalently

$$f(\eta) = \int_0^\eta g(t)\, dt \tag{C}$$

which satisfies the boundary condition, $f(0) = 0$. By introducing Eq. (B) into

Eq. (A) yields

$$g'' + \frac{1}{2}fg' = 0 \qquad\qquad (D)$$

Equation (D) is a second-order ODE with the boundary conditions,

$$g(0) = 0, \quad g(\infty) = 1$$

Equation (D) can be solved iteratively as a boundary value problem. The boundary at the infinity is replaced with a finite boundary at $\eta = \eta_{max}$, where η_{max} is a sufficiently large value of η. The finite difference approximation for Eq. (D) is written as

$$\frac{g_{i-1} - 2g_i + g_{i+1}}{(\Delta\eta)^2} + 0.5f_i\,\frac{g_{i+1} - g_{i-1}}{2\Delta\eta} = 0 \qquad\qquad (E)$$

where

$$f_i = \sum_{k=1}^{i} \Delta\eta\,\frac{g_{k-1} + g_k}{2} \qquad\qquad (F)$$

In Eq. (F), the extended trapezoidal rule is applied to Eq. (C) with $g_0 = 0$. Equations (E) is solved iteratively as follows.

(i)   Assume an initial distribution of g_i. Calculate f_i by Eq. (F).

(ii) Solve Eq. (E) by the tridiagonal solution.

(iii) Update f_i by introducing g_i just calculated into (F).

(iv) Repeat (ii) and (iii) until the solution converges.

The result is shown in a graphic form next:

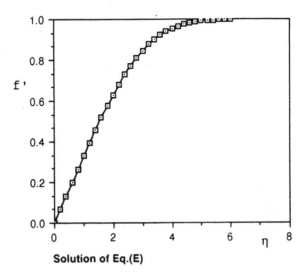

Solution of Eq.(E)

Note: The program used for the present analysis is included in CSL-C.

(a) Nonlinear differential equations are solved iteratively by successive substitution or by Newton's method.

(b) In each iteration step of successive substitution or Newton's method, difference equations are solved by the tridiagonal solution.

10.7 EIGENVALUE PROBLEMS OF ORDINARY DIFFERENTIAL EQUATIONS

A boundary value problem becomes an eigenvalue problem also when (a) the source term (or inhomogeneous term) of the differential equation is zero, and (b) boundary conditions for both ends are in the form of $\phi = 0$ or $\phi' = \gamma\phi$ where γ is a constant, which are called *homogeneous boundary conditions*. The solution of an eigenvalue problem can be normalized by an arbitrary constant. Eigenvalue problems of ordinary differential equations are in close relation with eigenvalue problems of matrices, because a set of finite difference equations for the former becomes the latter.

There are many examples of eigenvalue problems, including the harmonic vibration of rods, strings, and beams; the bending of a beam due to a longitudinal force (see Problem 10.23); and the neutron flux distribution of a nuclear reactor at criticality [Nakamura] (see Problem 10.21 also).

There are two types of solution methods for the difference equations of eigenvalue problems. The first refers to the direct solution methods for matrix eigenvalue problems as described in Chapter 7. The second is the iterative type.

To explain the numerical methods for solving eigenvalue problems, we first consider longitudinal vibration of a rod as an example. The equations for torsional vibration of a rod (or spring) and transverse vibration of a string are very similar to that of the longitudinal vibration of a rod. Therefore, if a computer program for one of them is developed, it can be applied to the other two types by changing the definitions of variables and constants.

For longitudinal vibration of an elastic rod with variable cross-sectional areas, the wave equation is written as

$$\frac{\partial}{\partial x}\left(EA(x)\frac{\partial u}{\partial x}\right) = w(x)\frac{\partial^2 u}{\partial t^2} \tag{10.7.1}$$

where $u = u(x, t)$ is displacement of the rod at x and t, E is the modulus of elasticity, $A(x)$ is the cross sectional area of the rod, $w(x)$ is the mass of the rod per unit length. We will assume that the length of the rod is H and that both ends are fixed.

When the rod is in a sustaining harmonic oscillation, the solution of Eq. (10.7.1) may be written in the form

$$u(x, t) = \sin(2\pi vt + \omega_0)f(x) \tag{10.7.2}$$

where v is the frequency of vibration, $f(x)$ is the spatial mode of oscillation, and ω_0 is the phase. The equation for $f(x)$ is now derived by introducing Eq. (10.7.2) into Eq. (10.7.1) and dividing by $\sin(2\pi vt + \omega_0)$:

$$\frac{d}{dx}\left[EA(x)\frac{d}{dx}f(x)\right] = -(2\pi v)^2 w(x)f(x)$$

where the partial derivatives have been changed to the ordinary derivatives because the equation does not involve t any longer. The foregoing equation may be written more compactly as

$$-[p(x)f'(x)]' = \lambda v(x)f(x) \tag{10.7.3}$$

where $p(x) = EA(x)$, $\lambda = v^2$ and $v(x) = (2\pi)^2 w(x)$. Because both ends are fixed, the boundary conditions are

$$f(0) = f(H) = 0 \tag{10.7.4}$$

We now derive difference equations assuming that the total number of grid points is equal to $N + 2$ including the two end points, and the grid interval is $h = H/(N + 1)$. Considering the three grid points as shown in Figure 10.5, the difference approximation for the left side of Eq. (10.7.3) is

$$
\begin{aligned}
(p(x)f')' &\simeq \frac{p(b)f'(b) - p(a)f'(a)}{h} \\[2mm]
&\simeq \frac{p(b)\dfrac{f_{i+1} - f_i}{h} - p(a)\dfrac{f_i - f_{i-1}}{h}}{h} \\[2mm]
&= \frac{p(a)f_{i-1} - (p(a) + p(b))f_i + p(b)f_{i+1}}{h^2}
\end{aligned}
\tag{10.7.5}
$$

where the right side of the first line is a central difference approximation for the outer derivative of $(p(x)f')'$, the second line is obtained by applying central difference approximations to f', and the third line is obtained by reorganizing the second line.

Figure 10.5 Grid configuration about point i (a and b are midpoints)

The right side of Eq. (10.7.3) may be approximated either by

$$v(x)f(x) \simeq v(x_i)f_i \tag{10.7.6a}$$

or

$$v(x)f(x) \simeq \frac{v(a) + v(b)}{2} f_i \tag{10.7.6b}$$

By introducing Eqs. (10.7.5) and (10.7.6a) into Eq. (10.7.3), the difference equation becomes

$$-p(a)f_{i-1} + [p(a) + p(b)]f_i - p(b)f_{i+1} = \lambda v(x_i)h^2 f_i \qquad (10.7.7)$$

If Eq. (10.7.6b) is used instead of Eq. (10.7.6a), the right side of Eq. (10.7.7) is changed accordingly.

Derivation of difference equations for other one-dimensional coordinates, or other types of physical problems, is essentially the same. In general, eigenvalue problems of second-order ordinary differential equations may be written as

$$-\frac{1}{x^m}(x^m p(x)\phi')' + q(x)\phi(x) = \lambda v(x)\phi(x) \qquad (10.7.8)$$

where $p(x)$, $q(x)$, and $v(x)$ are prescribed coefficients and m is a geometric parameter given as $m = 0$ for slab, $m = 1$ for cylinder, and $m = 2$ for sphere. If $q = 0$ and $m = 0$, Eq. (10.7.8) reduces to the form of Eq. (10.7.3). If $m > 0$, Eq. (10.7.8) should first be multiplied by x^m before difference approximation is derived, as described in Section 10.5.

When the coefficients of Eq. (10.7.8) are real values, all the eigenvalues are real. Particularly if $q \geqslant 0$, all eigenvalues become positive (positive definiteness). The eigenfunction corresponding to the lowest eigenvalue has no zero between the two ends. The second eigenfunction has one zero, the third eigenfunction has two zeroes, and so on as illustrated in Figure 10.6.

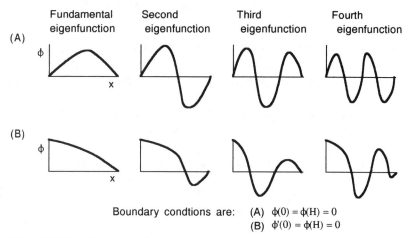

Boundary condtions are: (A) $\phi(0) = \phi(H) = 0$
                          (B) $\phi'(0) = \phi(H) = 0$

Figure 10.6 Distributions of eigenfunctions

The difference equations for Eq. (10.7.8) may be written in the tridiagonal form:

$$B_1 f_1 + C_1 f_2 = \lambda G_1 f_1$$
$$A_2 f_1 + B_2 f_2 + C_2 f_3 = \lambda G_2 f_2$$
$$A_i f_{i-1} + B_i f_i + C_i f_{i+1} = \lambda G_i f_i \tag{10.7.9}$$
$$\vdots$$
$$A_N f_{N-1} + B_N f_N = \lambda G_N f_N$$

Equation (10.7.9) may be solved either directly by the matrix eigenvalue method described in Chapter 7 or iteratively as explained in the remainder of this section. When you choose a solution method, consider the following aspects:

(a) The matrix method can find all the eigenvalues including complex eigenvalues. However, as the number of grid points increases, its computing time rapidly increases.

(b) Accuracy of the difference equation becomes poorer as the number of spatial oscillations increases (or equivalently, the eigenvalue becomes higher). This is because the effect of truncation error of the difference equations increases rapidly as the number of nodes of spatial oscillation becomes greater.

(c) If the coefficients of the tridiagonal equation are symmetric, the eigenvalues are all real.

(d) Generally the lowest and next few eigenvalues and the corresponding eigenfunctions are most frequently asked for in many scientific problems. For those problems, iterative methods are suitable.

INVERSE POWER METHOD. The inverse power method is an iterative method to find the eigenvalue that is closest to zero and the corresponding eigenfunction. When the lowest eigenvalue is real and positive, the fundamental eigenvalue and its eigenfunction are found by the inverse power method.

The inverse power method for Eq. (10.7.9) may be written as

$$A_i f_{i-1}^{(t)} + B_i f_i^{(t)} + C_i f_{i+1}^{(t)} = \lambda^{(t-1)} G_i f^{(t-1)} \tag{10.7.10}$$
$$i = 1, 2, \ldots, N$$

where t is the iteration number and $f_0 = f_{N+1} = 0$. The solution proceeds as follows:

Step 1: $f_i^{(0)}$ for all i are set to an arbitrary initial guess. The initial guess can be zero except for one grid point.

Step 2: $\lambda^{(0)}$ is an initial guess for λ and is set to unity.

Step 3: Eq. (10.7.10) is solved by the tridiagonal solution for $f_i^{(1)}$.

Step 4: The next estimate for λ is calculated by the equation:

$$\lambda^{(1)} = \lambda^{(0)} \frac{\sum_i G_i f_i^{(0)} f_i^{(1)}}{\sum_i G_i [f_i^{(1)}]^2} \tag{10.7.11}$$

Step 5: $\lambda^{(1)}$ and $f_i^{(1)}$ are introduced into the right side of Eq. (10.7.10), which is then solved for $f_i^{(2)}$ by the tridiagonal solution.

Step 6: The operation similar to Step 5 is repeated as the iteration cycle t increases. The calculation of λ after each iteration cycle is

$$\lambda^{(t)} = \lambda^{(t-1)} \frac{\sum_i G_i f_i^{(t-1)} f_i^{(t)}}{\sum_i G_i [f_i^{(t)}]^2} \tag{10.7.12}$$

Step 7: The iteration is stopped when the convergence test

$$\left| \lambda^{(t-1)}/\lambda^{(t)} - 1 \right| < \varepsilon$$

is satisfied, where ε is a prescribed criterion for convergence.

For simplicity, Eq. (10.7.12) may be replaced by

$$\lambda^{(t)} = \lambda^{(t-1)} \frac{\sum_i G_i f_i^{(t-1)}}{\sum_i G_i f_i^{(t)}} \tag{10.7.13}$$

or

$$\lambda^{(t)} = \lambda^{(t-1)} \frac{\sum_i f_i^{(t-1)}}{\sum_i f_i^{(t)}} \tag{10.7.14}$$

However, the convergence rates of the eigenvalue with the two equations above becomes slower than with Eq. (10.7.12).

If $q > 0$, the convergence rate is rapidly slowed down as the values of q increase or as the domain size increases. For one-dimensional eigenvalue problems, the slow iterative convergence can be improved by the shifted inverse power method written later. If $q = 0$ in Eq. (10.7.8) on the other hand, the convergence rate of the inverse power method is not affected by the number of grid points.

SHIFTED INVERSE POWER METHOD. The shifted inverse power method (see also Section 7.3) is derived by a simple modification to the inverse power method, but it significantly accelerates iterative convergence for the fundamental eigenvalue and eigenfunction. It also has the capability of finding higher real eigenvalues and eigenfunctions.

We set the eigenvalue in Eq. (10.7.9) to

$$\lambda = \lambda_e + \delta\lambda \tag{10.7.15}$$

where λ_e is an estimate for the eigenvalue to be found and $\delta\lambda$ is the correction. Then, Eq. (10.7.9) can be written as

$$\begin{aligned}
(2 - \lambda_e)f_1 - f_2 &= \delta\lambda f_1 \\
-f_1 + (2 - \lambda_e)f_2 - f_3 &= \delta\lambda f_2 \\
-f_2 + (2 - \lambda_e)f_3 - f_4 &= \delta\lambda f_3 \\
&\vdots \\
-f_{N-1} + (2 - \lambda_e)f_N \quad\;\; &= \delta\lambda f_N
\end{aligned} \tag{10.7.16}$$

where for simplicity of illustration the coefficients are set to $A = C = -1$, $B = 2$, and $D = 1$. Equation (10.7.16) is solved by the inverse power method considering $\delta\lambda$ as an alternate definition of the eigenvalue. Once $\delta\lambda$ is found, the true eigenvalue is computed by introducing $\delta\lambda$ into Eq. (10.7.15). The convergence rate becomes faster as λ_e becomes closer to the exact eigenvalue. Even with a rough estimate for λ_e, the convergence rate is significantly improved. When no estimate is available for the fundamental eigenvalue, one may obtain it by a few iteration steps with $\lambda_e = 0$.

However, λ_e should not be too close to an exact eigenvalue. If it is, the truncation error in the tridiagonal solution becomes serious, and the numerical solution will be unreliable (the coefficient matrix becomes ill-conditioned). Indeed, if λ_e exactly equals an eigenvalue, the left side of Eq. (10.7.16) becomes singular, and the tridiagonal solution will stop because of an arithmetic error such as overflow or zero division. See PROGRAM 10–4.

Example 10.4

By using PROGRAM 10–4, find the first, second, and third eigenvalues and eigenfunctions.

⟨**Solution**⟩

Let us first find the fundamental eigenvalue and eigenfunction of Eq. (10.7.10) using PROGRAM 10–4. Because there is no prior estimate for the first eigenvalue, we set $\lambda_e = 0$. The successive approximations for the first eigenvalue become:

Iteration Number	
1	0.0819672
2	0.0810257
3	0.0810142
4	0.0810140
Final value of λ is 0.0810140	

The eigenfunction for the first eigenvalue is as follows:

Grid point i	f_i
1	0.28469
2	0.54628
3	0.76359
4	0.91901
5	1.00000
6	1.00000
7	0.91901
8	0.76359
9	0.54628
10	0.28469

Setting $\lambda_e = 0.2$ as the first trial for obtaining the second eigenvalue, the eigenvalue converges to 0.081014, which is still the fundamental eigenvalue. Setting $\lambda_e = 0.5$ as a second trial results in $\lambda = 0.690274$, for which the eigenfunction changes the sign two times. Therefore, $\lambda = 0.690274$ is the third eigenvalue. The second eigenvalue must be between the first and the third eigenvalues. So, we try $\lambda_e = 0.4$ and get $\lambda = 0.317493$, which is the second eigenvalue because its eigenfunction changes sign only once.

The distribution of the second and third eigenfunctions are as follows:

i	Second eigenfunction ($\lambda = 0.317493$) f_i	Third eigenfunction ($\lambda = 0.690274$) f_i
1	−0.54560	0.76386
2	−0.91809	0.99999
3	−0.99923	0.54498
4	−0.76319	−0.28743
5	−0.28475	−0.92291
6	0.28428	−0.92291
7	0.76326	−0.28743
8	0.99999	0.54498
9	0.91918	0.99999
10	0.54638	0.76386

SUMMARY OF THIS SECTIONS

(a) The differential equation and boundary conditions in a eigenvalue problem are both homogeneous.

(b) If the lowest eigenvalue is real and positive, it can be computed using the inverse power method as well as the shifted inverse power method.

10.8 CONVERGENCE ANALYSIS OF THE ITERATIVE METHODS

In this section we explain why the inverse power method and the shifted inverse power method both converge. See Chapter 2 of Nakamura (1986) for iterative convergence of nuclear engineering problems.

Assuming $q(x) \geq 0$ and $v(x) > 0$ in Eq. (10.7.8), the difference equation given by Eq. (10.7.9) has the following properties:

(a) There are N real, positive, and distinctive eigenvalues, where N is the number of unknowns in Eq. (10.7.9).

(b) The fundamental eigenfunction (the eigenfunction associated with the lowest eigenvalue) has no zero except at the two end points.

(c) The second eigenfunction has one zero in the middle of the domain, and the nth eigenfunction has $n - 1$ zeroes in the middle of the domain.

For further discussions, we write Eq. (10.7.9) compactly as

$$Mf = \lambda Gf \tag{10.8.1}$$

where M and G are tridiagonal and diagonal matrices, respectively, and f represents an eigenvector (eigenfunction in a vector form).

We assume that N different eigenvalues of Eq. (10.8.1) are ordered to satisfy

$$0 < \lambda_0 < \lambda_1 < \cdots < \lambda_{N-1}$$

where λ_0 is the fundamental eigenvalue. Denoting the eigenvector corresponding to λ_n by u_n, Eq. (10.8.1) may be written as

$$Mu_n = \lambda_n Gu_n, \quad n = 0, 1, \ldots, N - 1 \tag{10.8.2}$$

When M is a symmetric matrix and all diagonal elements of G are nonzero, the eigenvectors have the following properties.

(a) Different eigenvectors are orthogonal to each other:

$$(u_m)^T Gu_n = 0 \quad \text{if } n \neq m \text{ (orthogonality relation)} \tag{10.8.3}$$

(b) Because all the eigenvectors are independent, any arbitrary vector of order N may be expanded in a linear combination of the eigenvectors as

$$z = \sum_{n=0}^{N-1} a_n u_n \quad \text{(completeness)} \tag{10.8.4}$$

where z is an arbitrary vector of order N, and a_n is a coefficient.

We now write the inverse power method given by Eq. (10.7.10) as

$$Mf^{(t)} = \lambda^{(t-1)} Gf^{(t-1)} \tag{10.8.5}$$

Using the property of Eq. (10.8.4), the initial guess $f^{(0)}$ may be expanded into the eigenvectors as

$$f^{(0)} = \sum_{n=0}^{N-1} c_n u_n \tag{10.8.6}$$

where c_n is a coefficient and can be determined by using the orthogonality relation, Eq. (10.8.3). The initial guess $\lambda^{(0)}$ is set to unity. The solution of Eq. (10.8.5) for $f^{(1)}$ can be written as

$$f^{(1)} = \sum_{n=0}^{N-1} c_n \left(\frac{1}{\lambda_n}\right) u_n \tag{10.8.7}$$

where $\lambda^{(0)} = 1$ has been used in Eq. (10.8.5) for $t = 1$. Equation (10.8.7) may be proven by introducing Eq. (10.8.6) into the right side of Eq. (10.8.5) with $t = 1$ and using Eq. (10.8.2).

For the second iteration cycle, Eq. (10.8.7) is introduced into the right side of Eq. (10.8.5) for $t = 2$ and the equation is solved for $f^{(2)}$. The result may be written as

$$f^{(2)} = \lambda^{(1)} \sum_{n=0}^{N-1} c_n \left(\frac{1}{\lambda_n}\right)^2 u_n \tag{10.8.8}$$

The iterative solution after t iteration cycles becomes

$$f^{(t)} = \lambda^{(1)}\lambda^{(2)} \cdots \lambda^{(t-1)} \sum_{n} c_n \left(\frac{1}{\lambda_n}\right)^t u_n \tag{10.8.9}$$

or equivalently

$$f^{(t)} = \frac{\lambda^{(1)}\lambda^{(2)} \cdots \lambda^{(t-1)}}{(\lambda_0)^t} \left[c_0 u_0 + c_1 \left(\frac{\lambda_0}{\lambda_1}\right)^t u_1 + c_2 \left(\frac{\lambda_0}{\lambda_2}\right)^t u_2 + \cdots \right] \tag{10.8.10}$$

Because λ_0 is the smallest eigenvalue, all the terms with the factor $(\lambda_0/\lambda_n)^t$ will vanish as t increases. Thus, Eq. (10.8.10) will approach

$$f^{(t)} \longrightarrow Cu_0$$

where C is a constant. The rate of convergence is governed by the ratio defined by $\sigma \equiv \lambda_0/\lambda_1$, called the *dominance ratio*, which is the rate that the coefficient of u_1 in Eq. (10.8.10) decreases in one iteration cycle.

The convergence of the eigenvalue is explained next. For simplicity, we first rewrite Eq. (10.7.11) in the matrix and vector form as

$$\lambda^{(t)} = \lambda^{(t-1)} \frac{(f^{(t-1)})^T G f^{(t)}}{(f^{(t)})^T G f^{(t)}} \tag{10.8.11}$$

By using Eq. (10.8.10), the numerator of Eq. (10.8.11) becomes

$$(f^{(t-1)})^T G f^{(t)} = \frac{\left[\lambda^{(1)}\lambda^{(2)}\cdots\lambda^{(t-2)}\right]^2\lambda^{(t-1)}}{(\lambda_0)^{2t-1}}\left[c_0^2 b_0 + c_1^2\left(\frac{\lambda_0}{\lambda_1}\right)^{2t-1} b_1 + \cdots\right] \tag{10.8.12}$$

where

$$b_n = (u_n)^T G u_n$$

and the orthogonality relations given by Eq. (10.8.3) have been used. Similarly, the denominator can be written

$$(f^{(t)})^T G f^{(t)} = \frac{\left[\lambda^{(1)}\lambda^{(2)}\cdots\lambda^{(t-1)}\right]^2}{(\lambda_0)^{2t}}\left[c_0^2 b_0 + c_1^2\left(\frac{\lambda_0}{\lambda_1}\right)^{2t} b_1 + \cdots\right] \tag{10.8.13}$$

Introducing Eqs. (10.8.12) and (10.8.13) into Eq. (10.8.11) and organizing the terms yield

$$\lambda^{(t)} = \lambda_0 \frac{1 + k_1\left(\dfrac{\lambda_0}{\lambda_1}\right)^{2t-1} + k_2\left(\dfrac{\lambda_0}{\lambda_2}\right)^{2t-1} + \cdots}{1 + k_1\left(\dfrac{\lambda_0}{\lambda_1}\right)^{2t} + k_2\left(\dfrac{\lambda_0}{\lambda_2}\right)^{2t} + \cdots} \tag{10.8.14}$$

where

$$k_m = \left(\frac{c_m}{c_0}\right)^2 \frac{b_m}{b_0}$$

Because $\lambda_0/\lambda_1 < 1$, the equation above approaches λ_0 as t increases.

Notice also that the ratio defined by

$$\frac{\lambda^{(t)} - \lambda^{(t-1)}}{\lambda^{(t-1)} - \lambda^{(t-2)}} \tag{10.8.15}$$

approaches the square of the dominance ratio as t increases. The proof is left for the reader's exercise (see Problem 10.25).

The effective dominance ratio for the shifted inverse power method is

$$\frac{\lambda_0 - \lambda_e}{\lambda_1 - \lambda_e} \tag{10.8.16}$$

When $\lambda_e \simeq \lambda_0$, the absolute value of the ratio of Eq. (10.8.16) becomes significantly smaller than λ_0/λ_1. This explains why the shifted inverse power method converges much faster than does the inverse power method. Convergence of the shifted inverse power method to a higher eigenvalue may be explained in a similar manner.

SUMMARY OF THIS SECTION

(a) The reason the inverse power method converges is explained by considering the positive definite problem (all eigenvalues are real and positive).

(b) An initial guess may be expanded into eigenvectors. With the inverse power method, the magnitude of each eigenvector diminishes by inverse power of the corresponding eigenvalue.

(c) The convergence rate of the inverse power method is determined by the dominance ratio.

(d) The shifted inverse power method is a modification of the inverse power method, but its effective dominance ratio is significantly smaller than that of the inverse power method.

10.9 BENDING AND VIBRATION OF A BEAM

Boundary value problems of ordinary differential equations higher than the second order may be solved in a similar manner. For illustration, we show numerical methods for the bending and vibration of a beam as boundary value problems for fourth-order ordinary differential equations. We first consider the bending of a beam under a distributed load $P(x)$, which is an inhomogeneous boundary value problem.

For a distributed load $P(x)$, the equation for the displacement of a beam is written as

$$\frac{d^2}{dx^2}\left(EI\,\frac{d^2y}{dx^2}\right) = P(x) \tag{10.9.1}$$

where E is the modulus of elasticity and I is the moment of inertia of the cross section, y is the displacement and P is the load as illustrated in Figure 10.7. The product EI is called *flexural rigidity*.

We will develop finite difference approximation of Eq. (10.9.1) for a nonuniform beam, which is clamped at the left end but free at the right end, as shown in Figure 10.7.

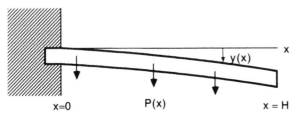

Figure 10.7 A beam clamped at one end

The boundary conditions for the present problem are given by

$$y(0) = 0 \quad \text{and} \quad y'(0) = 0 \text{ for the left boundary}$$

$$M = y''(H) = 0 \quad \text{and} \quad V = y'''(H) = 0 \text{ for the right boundary}$$

where M and V are respectively the bending and shear moments at the right boundary.

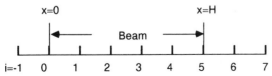

Figure 10.8 A grid for the beam ($i = -1, 6$, and 7 are hypothetical grid points)

To derive difference equations we consider the grid system shown in Figure 10.8. In Figure 10.8, the grid points at $i = -1, 6$, and 7 are hypothetical points. We assume that the grid points are equally spaced. Equation (10.9.1) is now written as

$$EI_i y_i'''' + 2EI_i' y_i''' + EI_i'' y_i'' = P(x) \tag{10.9.2}$$

The derivatives, y'''', y''', and y'', are numerically evaluated by the central difference approximations as

$$y'''' \simeq (y_{i-2} - 4y_{i-1} + 6y_i - 4y_{i+1} + y_{i+2})/h^4$$

$$y''' \simeq (-y_{i-2} + 2y_{i-1} - 2y_{i+1} + y_{i+2})/2h^3 \tag{10.9.3}$$

$$y'' \simeq (y_{i-1} - 2y_i + y_{i+1})/h^2$$

The terms I_i' and I_i'' may be calculated by the finite difference approximations as follows: for $i = 1$ through $N - 1$,

$$I_i' \simeq \frac{I_{i+1} - I_{i-1}}{2h}$$

$$I_i'' \simeq \frac{I_{i+1} - 2I_i + I_{i-1}}{h^2} \qquad (10.9.4)$$

where the central difference approximations are used; for N,

$$I_N' \simeq \frac{3I_N - 4I_{N-1} + I_{N-2}}{(2h)}$$

$$I_N'' \simeq \frac{-2I_N + 5I_{N-1} - 4I_{N-2} + I_{N-3}}{h^2} \qquad (10.9.5)$$

where backward difference approximations are used.

By introducing Eqs. (10.9.3) into Eq. (10.9.2), the difference equations become

$$a_i y_{i-2} + b_i y_{i-1} + c_i y_i + d_i y_{i+1} + e_i y_{i+2} = f_i \quad i = 1, 2, \ldots, N \qquad (10.9.6)$$

where

$a_i = \quad EI_i/h^4 - \quad EI_i'/h^3$

$b_i = -4EI_i/h^4 + 2EI_i'/h^3 + \quad EI_i''/h^2$

$c_i = \quad 6EI_i/h^4 \qquad\qquad + 2EI_i''/h^2$

$d_i = -4EI_i/h^4 - 2EI_i'/h^3 + \quad EI_i''/h^2$

$e_i = \quad EI_i/h^4 + \quad EI_i'/h^3$

$f_i = \quad P(x_i)$

For $i = 1$, Eq. (10.9.6) is written as

$$a_1 y_{-1} + b_1 y_0 + c_1 y_1 + d_1 y_2 + e_1 y_3 = f_1 \qquad (10.9.7)$$

which involves a hypothetical point $i = -1$ outside the domain. In Eq. (10.9.7) we set $y_0 = 0$, since $y(0) = 0$ is given as a boundary condition. The second boundary condition for the left end, $y'(0) = 0$, can be approximated by $(y_1 - y_{-1})/2h = 0$, so we can set $y_{-1} = y_1$. Introducing these relations into Eq. (10.9.7) yields

$$(a_1 + c_1)y_1 + d_1 y_2 + e_1 y_3 = f_1 \qquad (10.9.8)$$

To implement the right boundary conditions, we write the difference approximations of the right boundary conditions:

$$y''' \simeq \frac{-y_{N-2} + 2y_{N-1} - 2y_{N+1} + y_{N+2}}{2h^3} = 0$$

$$y'' \simeq \frac{y_{N-1} - 2y_N + y_{N+1}}{2h^2} = 0 \tag{10.9.9}$$

The equations of Eq. (10.9.9) will be considered as members of simultaneous equations.

Thus, the whole set of equations to be solved is written in a matrix form as

$$
\begin{bmatrix}
a_1 + c_1, & d_1, & e_1 & & & & \\
b_2, & c_2, & d_2, & e_2 & & & \\
a_3, & b_3, & c_3, & d_3, & e_3 & & \\
& a_4, & b_4, & c_4, & d_4, & e_4 & \\
& & a_5, & b_5, & c_5, & d_5, & e_5 \\
& & & -1, & 2, & 0, & -2, & 1 \\
& & & & 1, & -2, & 1, & 0
\end{bmatrix}
\begin{bmatrix}
y_1 \\ y_2 \\ y_3 \\ y_4 \\ y_5 \\ y_6 \\ y_7
\end{bmatrix}
=
\begin{bmatrix}
f_1 \\ f_2 \\ f_3 \\ f_4 \\ f_5 \\ 0 \\ 0
\end{bmatrix}
\tag{10.9.10}
$$

Eq. (10.9.10) is solved by the Gauss elimination or by *LU* decomposition. A Fortran program to compute the solution is available in Nakamura, 1991.

Now we consider harmonic vibration of the same beam [Thomson]. For a harmonic oscillation, Eq. (10.9.1) is replaced by

$$\frac{d^2}{dx^2}\left[EI\frac{d^2}{dx^2}\bar{y}(x, t)\right] = -w(x)\frac{d^2}{dt^2}\bar{y}(x, t) \tag{10.9.11}$$

where $\bar{y}$ is the displacement at x and time t, and $w(x)$ is the mass of the beam per unit length at x. We seek the solution in the form

$$\bar{y}(x, t) = y(x)\exp(j\omega t) \tag{10.9.12}$$

where $y(x)$ is the harmonic mode and ω is the angular velocity of oscillation (2π times the frequency), and $j = \sqrt{-1}$. Introducing Eq. (10.9.12) into Eq. (10.9.11) and dividing the equation through by $\exp(j\omega t)$ yield

$$\frac{d^2}{dx^2}\left[EI\frac{d^2}{dx^2}y(x)\right] = \lambda w(x)y(x) \tag{10.9.13}$$

where $\lambda = \omega^2$ is considered as an eigenvalue.

A Fortran program to compute the solution is available in Nakamura, 1991.

(a) Numerical methods for solving problems of the bending and vibration of a beam are discussed.

(b) The difference equations for the beam bending problem are solved by the *LU* decomposition (since the differential equations are not of second order, difference equations are not tridiagonal any more).

(c) The vibration problem is an eigenvalue problem. Its difference equations are essentially the same as those for the bending problem except that the equations are homogeneous and have eigenvalues. By using the shifted inverse power method, not only the fundamental mode but also higher modes are computed.

PROGRAMS

PROGRAM 10-1 Linear Boundary Value Problem

(A) Explanations

PROGRAM 10-1 solves

$$-2y''(x) + y(x) = e(-0.2x)$$

with the boundary conditions,

$$y(0) = 1, \quad y'(10) = -y(10)$$

The total number of unknowns and the coefficients of difference equations are first computed in the program.

The tridiagonal solution scheme is implemented in the subroutine starting from line 190. When returned from the subroutine, the solution is saved in variable $d[i]$. $d[0] = 1$ is for printing purposes. There is no input to be given interactively.

(B) List

```
/* CSL/c10-1.c    Linear Boundary Value Problem */
#include <stdio.h>
#include <stdlib.h>
#include <math.h>
/*        a[i], b[i], c[i], d[i] : a(i), b(i), c(i), and d(i)
          n: number of grid points                                    */
main()
{
int i, n;
float a[20], b[20], c[20], d[20], x;
```

```
void trdg();
   printf( "\n\nCSL/C10-1    Linear Boundary Value Problem\n" );
   n = 10;                            /* n: Number of grid points */
   for( i = 1; i <= n; i++ ){
      x = i;
      a[i] = -2;                      /* Tridiagonal coefficients */
      b[i] = 5;
      c[i] = -2;
      d[i] = exp( -0.2*x );    /* Source (inhomogeneous) term */
   }
   d[1] = d[1] + 2;                   /* Adjusts with boundary condition */
   d[n] = d[n]*0.5;
   b[n] = 4.5;
   trdg( a, b, c, d, n );
   d[0] = 1;                      /* Setting d[0] for printing purpose */
   printf( "\n Grid point no.        Solution\n" );
   for( i = 0; i <= n; i++ ){
      printf(  "   %3.1d              %12.5e \n", i, d[i] );
   }
   exit(0);
}

void trdg(a, b, c, d, n)     /* Tridiagonal solution */
float a[], b[], c[], d[];
int n;
{
int i;
float r;
   for( i = 2; i <= n; i++ ){
      r = a[i]/b[i - 1];
      b[i] = b[i] - r*c[i - 1];
      d[i] = d[i] - r*d[i - 1];
   }
   d[n] = d[n]/b[n];
   for( i = n - 1; i >= 1; i-- ){
      d[i] = (d[i] - c[i]*d[i + 1])/b[i];
   }
   return;
}
```

(C) Sample Output

```
CSL/C10-1    Linear Boundary Value Problem

Grid point no.         Solution
    0              1.00000e+00
    1              8.46449e-01
    2              7.06756e-01
    3              5.85282e-01
    4              4.82043e-01
    5              3.95162e-01
    6              3.21921e-01
    7              2.59044e-01
    8              2.02390e-01
    9              1.45983e-01
   10              7.99187e-02
```

(D) Discussions

The solution is computed for grid points $i = 1$ through 10. The boundary condition for $i = 0$ is used in the difference equation for $i = 1$.

PROGRAM 10-2 Nonlinear Boundary Value Problem

(A) Explanations

PROGRAM 10–2 illustrates the solution of a boundary value problem of a nonlinear ordinary differential equation given by

$$-\psi'' + 0.01\psi^2 = \exp(-x), \quad 0 < x < H \tag{A}$$

with boundary conditions,

$$\psi(0) = \psi(H) = 0$$

Newton's iteration is used. All parameters for the problem are specified within the program. No interactive input is necessary.

This program illustrates the use of arrayed variables in structure. The algorithm of trdg() that solves the tridiagonal equations is identical to that in PROGRAM 10–1.

(B) List

```
/* CSL/c10-2.c    Nonlinear Boundary Value Problem    */
#include <stdio.h>
#include <stdlib.h>
#include <math.h>
struct tP {
    int n;
    float a[30], b[30], c[30], d[30], ps[20];
       } bc;
/*
                ep : convergence criterion
                w : under relaxation parameter
                h : grid point spacing
                bc.n : number of grid points                        */
main()
{
int i, k, nt;
float ep, h, h2, w;
void trdg();
    printf( "\n CSL/C10-2    Nonlinear Boundary Value Problem \n");
    bc.n = 9;
    h = 0.2;
    h2 = h*h;
    k = 0;
    ep = 0.0001;
    w = 0.9;
    for( i = 0; i <= bc.n+1; i++ )    bc.ps[i] = 0;
    while( 1 ){
```

```
      k = k + 1;
      for( i = 1; i <= bc.n; i++ ){
         bc.a[i] = -1;
         bc.b[i] = 2 + 0.02*h2*bc.ps[i];
         bc.c[i] = -1;
         bc.d[i] = bc.ps[i - 1] - 2*bc.ps[i] + bc.ps[i + 1] -
                   0.01*h2*bc.ps[i]*bc.ps[i] + exp( -0.2*i )*h2;
      }
      trdg();
      nt = 0;
      for( i = 1; i <= bc.n; i++ ){
         if( fabs( bc.d[i] ) > ep )    nt = nt + 1;
      }
      if( nt == 0 )    goto L_270;
      printf( " it. no.=%3d\n", k );
      printf( "   i       psi(i)       Delta_psi(i)\n" );
      for( i = 1; i <= bc.n; i++ ){
         bc.ps[i] = bc.d[i]*w + bc.ps[i];
         printf(  "%3.1d %12.4e  %12.4e \n", i, bc.ps[i], bc.d[i]);
      }
   }
L_270:
   printf( "\n Final Solution:");
   printf( "\n  i       psi(i) \n" );
   bc.ps[0] = 0;
   for( i = 0; i <= bc.n; i++ )
      printf(  "%3.1d   %12.4e \n", i, bc.ps[i] );
   exit(0);
}

void trdg()    /* Tridiagonal solution */
{
int i, N;
float r;
struct {  int n;    float a[30], b[30], c[30], d[30];
      }
   *P = (void*)&bc;
   N=P->n;
   for( i = 2; i <= N; i++ )
   {   r = P->a[i]/P->b[i - 1];
       P->b[i] = P->b[i] - r*P->c[i - 1];
       P->d[i] = P->d[i] - r*P->d[i - 1];
   }
   P->d[N - 1] = P->d[N - 1]/P->b[N - 1];
   for( i = N - 1; i >= 1; i-- )
      P->d[i] = (P->d[i] - P->c[i]*P->d[i+1])/P->b[i];
   return;
}
```

(C) Sample Output

```
CSL/C10-2    Nonlinear Boundary Value Problem
it. no.=  1
 i       psi(i)       Delta_psi(i)
 1    8.4965e-02     9.4405e-02
 2    1.4046e-01     1.5606e-01
 3    1.7181e-01     1.9090e-01
 4    1.8342e-01     2.0380e-01
```

```
     5     1.7884e-01      1.9871e-01
     6     1.6102e-01      1.7892e-01
     7     1.3236e-01      1.4707e-01
     8     9.4826e-02      1.0536e-01
     9     5.6315e-02      6.2572e-02
it. no.=   2
     i        psi(i)       Delta_psi(i)
     1     9.3353e-02      9.3196e-03
     2     1.5429e-01      1.5368e-02
     3     1.8868e-01      1.8744e-02
     4     2.0136e-01      1.9940e-02
     5     1.9626e-01      1.9355e-02
     6     1.7661e-01      1.7313e-02
     7     1.4504e-01      1.4080e-02
     8     1.0371e-01      9.8687e-03
     9     5.6275e-02     -4.3778e-05
it. no.=   3
     i        psi(i)       Delta_psi(i)
     1     9.4303e-02      1.0559e-03
     2     1.5589e-01      1.7848e-03
     3     1.9071e-01      2.2466e-03
     4     2.0360e-01      2.4905e-03
     5     1.9856e-01      2.5566e-03
     6     1.7884e-01      2.4773e-03
     7     1.4709e-01      2.2790e-03
     8     1.0549e-01      1.9832e-03
     9     5.6275e-02     -4.2558e-07
it. no.=   4
     i        psi(i)       Delta_psi(i)
     1     9.4420e-02      1.3042e-04
     2     1.5610e-01      2.2813e-04
     3     1.9098e-01      2.9914e-04
     4     2.0392e-01      3.4838e-04
     5     1.9890e-01      3.7984e-04
     6     1.7919e-01      3.9681e-04
     7     1.4745e-01      4.0188e-04
     8     1.0585e-01      3.9723e-04
     9     5.6275e-02     -7.3823e-09

Final Solution:
     i        psi(i)
     0     0.0000e+00
     1     9.4420e-02
     2     1.5610e-01
     3     1.9098e-01
     4     2.0392e-01
     5     1.9890e-01
     6     1.7919e-01
     7     1.4745e-01
     8     1.0585e-01
     9     5.6275e-02
```

(D) Discussions

In the foregoing output, the solution converges after five iterations. After each iteration, ψ and $\delta\psi$ are printed out.

PROGRAM 10–3 Inverse Power Method

(A) Explanations

An eigenvalue problem of ordinary differential equation is written as

$$-\frac{1}{x^m}\frac{d}{dx}\left(p(x)x^m\frac{d}{dx}f(x)\right) + q(x)f(x) = \lambda v(x)f(x) \quad 0 < x < H \tag{A}$$

where p, q, and v are known coefficients and functions of x, λ is an eigenvalue, and m is a parameter of geometry:

$$m = 0 \quad \text{for slab}$$
$$= 1 \quad \text{for cylinder}$$
$$= 2 \quad \text{for sphere}$$

We assume

$$p(x) = 1, \quad q(x) = 0, \quad v(x) = 1 \quad \text{and} \quad m = 0 \tag{B}$$

The boundary conditions are $f(0) = 0$ for the left boundary, and $f(H) = 0$ for the right boundary. It can be shown that the eigenvalues of the present problem are all real, distinct, and positive.

The inverse power method for difference approximation of Eq. (A) may be expressed by

$$Af^{(t)} = \lambda^{(t-1)}Gf^{(t-1)} \tag{C}$$

where t is iteration times, and $\lambda^{(t)}$ is computed by

$$\lambda^{(t)} = \lambda^{(t-1)}\frac{\langle f^{(t-1)}, Gf^{(t)}\rangle}{\langle f^{(t)}, Gf^{(t)}\rangle} \tag{D}$$

and where $\langle a, b\rangle$ is the inner product of vectors a and b. Equation (C) is solved in each iteration step by the tridiagonal solution scheme.

(B) List

```
/* CSL/c10-3.c    Inverse Power Method */
#include <stdio.h>
#include <stdlib.h>
#include <math.h>
#define TRUE 1
main()
```

```
{
int i, it, k, kstop, n;
float a[20], as[20], b[20], bs[20], c[20], cs[20], d[20], ds[20],
     eb, ei, eigen, ep, f[20], fb[20], s, sb, z;
void trdg();
   printf( "\nCSL/C10-3     Inverse Power Method \n" );
   while( TRUE ){
      k = 0;
      n = 10;        /* Maximum grid number */
      ei = 1;        /* Initialization of eigenvalue */
      s = 1;
      it = 30;       /* Limit of iteration */
      ep = 0.0001; /* Convergence criterion */
      for( i = 1; i <= n; i++ ){
         as[i] = -1.0;
         bs[i] = 2.0;
         cs[i] = -1.0;
         f[i] = 1.0;
         ds[i] = 1.0;
      }
      printf( "\n    It. No.     Eigenvalue\n" );
      while( TRUE ){
         k = k + 1;
         for( i = 1; i <= n; i++ )     fb[i] = f[i]*ei;
         for( i = 1; i <= n; i++ ){
         a[i] = as[i];
            b[i] = bs[i];
            c[i] = cs[i];
            d[i] = ds[i]*fb[i];
         }
         trdg( a, b, c, d, n );
         sb = 0;
         s = 0;
         for( i = 1; i <= n; i++ ){
            f[i] = d[i];
            s = s +    f[i]*f[i];
            sb = sb + f[i]*fb[i];
         }
         eb = ei;
         ei = sb/s;
         printf( "     %3.1d      %12.6e \n", k, ei );
         if( fabs( 1.0 - ei/eb ) <= ep ) break;
         if( k > it ){
            printf( " Iteration limit exceeded.\n" );
            break;
         }
      }
      z = 0;
      for( i = 1; i <= n; i++ ){   /*Normalization of the solution */
         if( fabs( z ) <= fabs( f[i] ) ) z = f[i];
      }
      for( i = 1; i <= n; i++ )    f[i] = f[i]/z;
      eigen = ei;
      printf( "    Eigenvalue = %g \n", eigen );
      printf( "\n  Eigenfunction\n");
      printf( "    i      f(i)\n" );
      for( i = 1; i <= n; i++ )
         printf( "%3.1d     %12.5e \n", i, f[i] );
      printf( " -----------------------------------------\n" );
```

```
        printf( " Type 1 to continue, or 0 to stop. \n" );
        scanf( "%d", &kstop );
        if ( kstop != 1 )    exit(0);
    }
}

void trdg (a, b, c, d, n)     /* Tridiagonal solution */
float a[], b[], c[], d[];
int n;
{
int i;
float r;
    for ( i = 2; i <= n; i++ ) {
        r = a[i]/b[i - 1];
        b[i] = b[i] - r*c[i - 1];
        d[i] = d[i] - r*d[i - 1];
    }
    d[n] = d[n]/b[n];
    for ( i = n - 1; i >= 1; i-- ) {
        d[i] = (d[i] - c[i]*d[i + 1])/b[i];
    }
    return;
}
```

(C) Sample Output

```
CSL/C10-3     Inverse Power Method

    It. No.      Eigenvalue
       1         8.196724e-02
       2         8.102578e-02
       3         8.101425e-02
       4         8.101407e-02
    Eigenvalue = 8.101407e-02

Eigenfunction
 i      f(i)
 1       2.84692e-01
 2       5.46290e-01
 3       7.63595e-01
 4       9.19019e-01
 5       1.00000e+00
 6       1.00000e+00
 7       9.19018e-01
 8       7.63594e-01
 9       5.46290e-01
10       2.84692e-01
```

(D) Discussions

With the conditions given by Eq. (B), eigenvalues of (A) are all positive and distinctive. The inverse power method finds the smallest positive eigenvalue and the corresponding eigenfunction.

The iterative solution converges here in only four iteration steps. This is considered to be very fast. In general, the convergence rate is highest if $b = |a| + |c|$ is satisfied at every point except at the boundaries. The convergence rate is significantly slowed down as b becomes increasingly larger than the sum, $|a| + |c|$. Naturally, the solution for $f[i]$ is geometrically symmetric because the problem is symmetric about the midpoint of the domain.

The eigenfunction computed is obviously the fundamental eigenfunction because there is no change of sign in the solution.

PROGRAM 10–4 Shifted Inverse Power Method

(A) Explanations

PROGRAM 10–4 solves the same equation as PROGRAM 10–3 but uses the shifted inverse power method:

$$(A - \lambda_e I)f^{(t)} = \delta\lambda^{(t-1)}Gf^{(t-1)} \tag{A}$$

where A and G are coefficient matrices, f is the solution vector, t is the iteration count, λ_e is an estimate for the desired eigenvalue, and $\delta\lambda$ is the shifted eigenvalue. Equation (A) is solved for each iteration step by the tridiagonal scheme. This method is also known as the Wielandt method [Nakamura 1986].

There are two major changes from PROGRAM 10–3, however. The first is that the program interactively reads λ_e as input. After the iterative solution is completed for the given λ_e, the program requests for the next λ_e to repeat the solution of the same problem. The second is that $bs[[i]$ is defined in accordance with Eq. (A).

Finding a higher eigenvalue and eigenfunction by using the shifted inverse power method is a trial-and-error procedure, since we generally have no estimates for the higher eigenvalues. The approach is as follows:

(a) Set λ_e to an estimate for the eigenvalue.

(b) Run the shifted inverse power scheme.

(c) Determine which eigenvalue and eigenfunction are obtained by counting the number of zeros in the eigenfunction. If the solution changes the sign J times in the middle of the domain, then the $J + 1$th eigenfunction has been obtained.

(d) If a lower eigenfunction rather than the one obtained in (b) is desired, try a lower estimate for λ_e and repeat (b) and (c). If a higher eigenfunction is desired, try a higher estimate for λ_e and do the same.

(B) List

```
/*  CSL/c10-4.c      Shfited Inverse Power Method   */
#include <stdio.h>
#include <stdlib.h>
#include <math.h>
/*              n : total number of points
                it : iteration limit
                eigen : eigenvalue
```

```
                esti_eign : estimate for eigenvalue
                del_eig : delta-lambda
                eb : previous value of del_eig:x
                ep : convergence criterion       */
main()
{
int i, it, k, kstop, n;
float a[20], as[20], b[20], bs[20], c[20], cs[20], d[20], ds[20],
    eb, del_eig, eigen, ep, esti_eign, f[20], fb[20], s, sb, z;
void trdg();
    printf( "\nCSL/C10-4    Shifted Inverse Power Method \n" );
    while( TRUE ){
    k = 0;  n = 10;  del_eig = 1;  s = 1;  it = 30;  ep = 0.0001;
        printf( "\nType estimate for Lambda ?" );
        scanf( "%f", &esti_eign );
        for( i = 1; i <= n; i++ ){
            as[i] = -1.0;  bs[i] = 2.0 - esti_eign;  cs[i] = -1.0;
            f[i] = 1.0;
            ds[i] = 1.0;
        }
        printf( "\n It. No.    Del-Lamb\n" );
        while( TRUE ){
            k = k + 1;
            for( i = 1; i <= n; i++ )  fb[i] = f[i]*del_eig;
            for( i = 1; i <= n; i++ ){
            a[i] = as[i];             b[i] = bs[i];
                c[i] = cs[i];             d[i] = ds[i]*fb[i];
            }
            trdg( a, b, c, d, n );
            sb = 0;  s = 0;
            for( i = 1; i <= n; i++ ){
            f[i] = d[i];             s = s + f[i]*f[i];
                sb = sb + f[i]*fb[i];
            }
            eb = del_eig;    del_eig = sb/s;
            printf( "%5d    %15.8e \n", k, del_eig );
            if( fabs( 1.0 - del_eig/eb ) <= ep )   break;
            if( k > it ){
                    printf( " Iteration limit exceeded \n" );
                    break;
            }
        }
        z = 0;
        for( i = 1; i <= n; i++ ){
            if( fabs( z ) <= fabs( f[i] ) )    z = f[i];
        }
        for( i = 1; i <= n; i++ )   f[i] = f[i]/z;
        /* Normalization of solution */
        eigen = del_eig + esti_eign;
        printf( "\n           Eigenvalue = %g \n", eigen );
        printf( "\nEigenfunction");
        printf( "\n i          f(I)\n" );
        for( i = 1; i <= n; i++ ) printf("%3d       %12.5e \n", i, f[i] );
        printf( "\n\n Type 1 to continue, or 0 to stop. \n" );
        scanf( "%d", &kstop );
        if( kstop != 1 )   exit(0);
    }
}
```

```
void trdg(a, b, c, d, n)    /* Tridiagonal solution */
float a[], b[], c[], d[];   int n;
{
int i;   float r;
   for( i = 2; i <= n; i++ ){
      r = a[i]/b[i - 1];
       b[i] = b[i] - r*c[i - 1];     d[i] = d[i] - r*d[i - 1];
   }
   d[n] = d[n]/b[n];
   for( i = n - 1; i >= 1; i-- ) d[i] = (d[i] - c[i]*d[i + 1])/b[i];
   return;
}
```

(C) Sample Output

```
CSL/C10-4    Shifted Inverse Power Method

Type estimate for Lambda ? 0.3
 It. No.    Del-Lamb
     1     -1.99321449e-01
     2     -2.13031471e-01
     3     -2.17106611e-01
     4     -2.18393013e-01
     5     -2.18783662e-01
     6     -2.16522843e-01
     7     -7.30994642e-02
     8      1.65595580e-02
     9      1.74870081e-02
    10      1.74929500e-02
    11      1.74929891e-02

        Eigenvalue = 0.317493

Eigenfunction
   i         f(I)
   1       5.46196e-01
   2       9.18981e-01
   3       1.00000e+00
   4       7.63532e-01
   5       2.84655e-01
   6      -2.84591e-01
   7      -7.63473e-01
   8      -9.99951e-01
   9      -9.18946e-01
  10      -5.46178e-01


Type estimate for Lambda ? 0.001
 It. No.    Del-Lamb
     1      8.09463114e-02
     2      8.00252259e-02
     3      8.00141841e-02
     4      8.00140426e-02

        Eigenvalue = 8.101404e-02
```

```
Eigenfunction
     i          f(I)
     1       2.84689e-01
     2       5.46286e-01
     3       7.63591e-01
     4       9.19017e-01
     5       1.00000e+00
     6       1.00000e+00
     7       9.19017e-01
     8       7.63591e-01
     9       5.46286e-01
    10       2.84689e-01

Type estimate for Lambda ? 5.0
 It. No.     Del-Lamb
     1      -4.60952997e+00
     2      -3.74964619e+00
     3      -2.09650660e+00
     4      -1.46248806e+00
     5      -1.35241580e+00
     6      -1.32855117e+00
     7      -1.32138360e+00
     8      -1.31890595e+00
     9      -1.31801128e+00
    10      -1.31768370e+00
    11      -1.31756318e+00

          Eigenvalue = 3.68244

Eigenfunction
     i          f(I)
     1        5.56247e-01
     2       -9.28377e-01
     3        1.00000e+00
     4       -7.56585e-01
     5        2.80560e-01
     6        2.80560e-01
     7       -7.56586e-01
     8        1.00000e+00
     9       -9.28377e-01
    10        5.56247e-01
```

(D) Discussions

Because an estimate of the eigenvalue is given rather randomly, at least in the beginning, which eigenfunction is being solved is not known until the eigenfunction solved is examined. By examination, we find that the computed eigenfunction corresponding to the eigenvalue of 0.31749 is the second eigenfunction because the sign of the solution changes once in the middle of the domain. The next computed solution is the fundamental eigenfunction because it does not change the sign, and it equals the solution that was obtained by PROGRAM 10–3. The last computed solution that has an eigenvalue of 3.6824 must be the next to the highest one because the sign changes for every point except at the center. Symmetry of the solution is slightly off, but this deviation is caused by the round-off error.

QUIZ: The convergence rate of the shifted inverse power method becomes faster as λ_e approaches the true eigenvalue. However, what would happen if λ_e is set very close to or exactly equals one of the eigenvalues? You can answer this theoretically. Try also to run PROGRAM 10–4 with an exact or nearly exact eigenvalue for λ_e.

PROBLEMS

(10.1) Derive difference equations for $i = 1$ and $i = 10$ in Example 10.1 assuming that the boundary conditions are changed to $y'(1) = y(1)$ and $y'(10) = 0$.

(10.2) Derive difference equations for

$$-(p(x)\phi'(x))' + q(x)\phi(x) = S(x), \quad 0 < x < H$$
$$\phi'(0) = \phi(H) = 0$$

The geometry, grid, and constants follow:

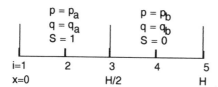

i=1   2   3   4   5
x=0   H/2   H   Figure P10.2

Grid spacings are $h = H/4$ for all the intervals.

(10.3) Repeat the previous problem assuming that the grid spacing for the first two intervals is h_1 and that for the last two intervals is h_2.

(10.4) The differential equation for a flexible cable of 50 m long fixed at both ends is given by

$$y''(x) = w(x)/T \quad y(0) = y(50) = 0$$

where x is in meter, $y(x)$ is displacement of the wire measured from the level of the end points of the wire (positive downward), T is the horizontal component of tension (5000 kg force) and $w(x)$ is the load distribution given by

$$w(x) = 20[1 + \exp(x/25)] \text{ kg weight/m}$$

Determine the shape of the cable. (Use 10 grid intervals.)

(10.5) Consider a cooling fin with a variable cross-sectional area and a variable perimeter. Assuming that the temperature across any cross section perpendicular to the axis is uniform, the temperature in the axial direction is the solution of the equation,

$$-[kA(x)T'(x)]' + P(x)h_c T(x) = P(x)h_c T_\infty$$

where k is the thermal conductivity, $P(x)$ is the perimeter, $A(x)$ is the cross-sectional area, and T_∞ is the temperature of the surrounding environment. The boundary conditions are given by

$$T(0) = 100° \text{ C}$$
$$-kT'(H) = h_c(T(H) - T_\infty)$$

where H is the length of the fin, and h_c is the convection heat transfer coefficient. Solve the foregoing problem assuming the following constants:

$$h_c = 30 \text{ w/m}^2\text{K}$$

$$H = 0.1 \text{ m}, \quad k = 100 \text{ w/mK}, \quad T_\infty = 20° \text{ C}, \quad \text{and}$$

$$A(x) = (0.005)(0.05 - 0.25x) \text{ m}^2, \quad P(x) = A(x)/0.005 + 0.01 \text{ m}$$

(Use 10 grid intervals.)

(10.6) The boundary condition in the form of Eq. (10.2.8) becomes numerically equivalent to $\phi(0) = 0$ if g_L is set to 0 and f_L is set to a very large value such as 10^{10}. What values for g_L and f_L make Eq. (10.2.8) equivalent to $\phi(0) = 2$?

(10.7) Consider a cylindrical unit of fuel cell in a light water nuclear reactor consisting of a fuel pin and moderator. as shown in Figure P10.7. The thermal neutron flux in the cell satisfies the neutron diffusion equation given by

$$-\frac{1}{r}\frac{d}{dr} Dr \frac{d}{dr} \phi(r) + \Sigma_a\phi(r) = S(r)$$

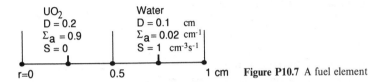

UO$_2$
D = 0.2
$\Sigma_a = 0.9$
S = 0

Water
D = 0.1   cm
$\Sigma_a = 0.02$ cm^{-1}
S = 1 cm^{-3}s^{-1}

r=0   0.5   1 cm   **Figure P10.7** A fuel element

where D is the diffusion coefficient, Σ_a is the absorption cross section, and S is the neutron source. The constants for UO$_2$ and H$_2$O are shown in the figure. The boundary conditions are

$$\phi'(0) = \phi'(1) = 0$$

(a) Using five grid points for the whole domain with a constant interval of 0.25 cm, derive difference equations for each grid point.

(b) Solve the difference equations derived in **(a)** by the tridiagonal solution

(10.8) A beam of 3.5 m long is supported at two points, one at 0.5 m and another at 2.5 m from the left end as shown in Figure P10.8. Assuming that the beam is weightless, the load distribution on the beam is given by

$$W(x) = (x - 0.5)\sqrt{1.5 - x} \quad \text{(N/m)}, \quad \text{for} - 0.5 < x < 1.5$$

$$= 0 \qquad\qquad\qquad\qquad\qquad \text{for } 1.5 < x$$

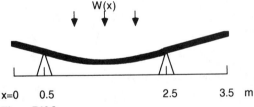

W(x)

x=0   0.5   2.5   3.5 m
Figure P10.8

The reactions of the supports at a and b are given by

$$R_a = \int_{0.5}^{1.5} (x - x_b)w(x)\,dx/(x_a - x_b), \quad (N)$$

$$R_b = \int_{0.5}^{1.5} (x - x_a)w(x)\,dx/(x_b - x_a), \quad (N)$$

(a) The bending moment of the beam satisfies

$$\frac{d}{dx} M(x) = -w(x) + R_a\delta(x - x_a) + R_b\delta(x - x_b)$$

where $\delta(x)$ is the delta function and $M(x)$ is the bending moment distribution. The boundary conditions are $M(0) = M(3.5) = 0$. Find the bending moment distribution by solving the foregoing differential equation by the finite difference method. (Set a grid in such way that grid points are located at the supports. Then, consider the reactions as forces concentrated at the grid points located at the supports. The integrals for the reactions may be evaluated by using the extended trapezoidal rule or the extended Simpson's 1/3 rule.)

(b) The deflection of the beam is related to the moment by

$$EI\,\frac{d}{dx} y(x) = M(x)$$

Determine the deflection between $x = 0.5$ and $x = 2.5$ by solving the equation by the finite difference approximation. Assume $EI = 1000$ Nm^2.

(10.9) Take summation of all the equations in Eq. (10.2.7a), and explain how the result of the summation can be directly derived from Eq. (10.2.1).

(10.10) Take summation of Eq. (10.4.7) for $i = j - 1$, $i = j$, and $i = j + 1$, and cancel as many terms as possible by using the definitions of A_i, B_i, C_i, and D_i. Explain the physical meaning of the summation.

(10.11) For a slab material of thickness of 0.2 cm, the left side is perfectly insulated, but the right surface temperature is fixed to $0°$ C. The slab has a distributed heat source. The temperature equation is given by $-T''(x) = q(x)/k$. Develop a program to compute the temperature distribution using 10 grid intervals. Assuming the thermal conductivity is $k = 30$ w/m^{2}K, run the program for the following two heat source distributions:

(a) $q(x) = 200$ kw/m^3

(b) $q(x) = 100 \exp(-10x)$ kw/m^3

Compare the results with the analytical solutions shown next:

(a) $T(x) = (10/3)(0.04 - x^2)$

(b) $T(x) = 0.033(e^{-2} + 2 - 10x - e^{-10x})$

(10.12) Consider the equation

$$-\phi''(r) - \phi'(r)/r = S(r), \quad a < r < b$$

for a cylindrical coordinate with the boundary conditions

$$\phi'(a) = 0, \quad \phi'(b) = k(\phi_\infty - \phi(b))$$

(a) Using a uniformly spaced grid, derive difference equations by central difference approximation for ϕ'' and ϕ', and show that the set of difference equations do not have the conservation property.

(b) Derive differential equations in the conservation form by rewriting the differential equation.

(10.13) The diffusion equation for a cylindrical geometry is given by

$$-\frac{1}{r}[p(r)r\phi'(r)]' + q(r)\phi(r) = S(r)$$

Considering the three grid points as shown in Figure P10.13, difference equations may be derived by integrating the equation from the midpoint between $i-1$ and i to the midpoint between i and $i+1$. Assuming that the coefficients are constants as illustrated in Figure P10.13 and that the grid spacings are not uniform, derive the difference equations by integrating in the volume between a and b.

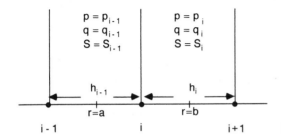

i - 1   i   i+1   **Figure P10.13**

(10.14) The equation for the displacement of a circular membrane loaded with a constant pressure P is given by

$$y''(r) + \frac{1}{r}y'(r) = -P/T, \quad 0.2\ \text{m} \leqslant r \leqslant 0.5\ \text{m}$$

where r is the radial coordinate, y is the displacement of the membrane (positive downward), T is tension (400 kg/m), and the pressure is given as $P = 800$ kg/m^2. The boundary conditions are $y(0.2) = y(0.5) = 0$. Determine the displacement of the membrane, $y(r)$. Use 10 intervals.

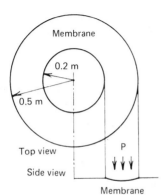

Figure P10.14 A membrane under pressure

(10.15) The surface of a perfectly spherical material of radius 0.05 m is uniformly irradiated by a gamma radiation. The gamma ray penetrates the material and then is absorbed. Thus, we assume that the heat source distribution due to the radiation is expressed by

$$S(r) = 300 \exp\left[20(r - 0.05)\right]$$

where r is radius in meter, and the unit of S is W/m³. The surface of the sphere is exposed to air. Heat escapes to the surrounding air by convection with the heat transfer coefficient, $h_c = 20$ W/m²K. At steady state, the temperature distribution is the solution of the equation,

$$-\frac{1}{r^2}\frac{d}{dr}r^2 k\frac{d}{dr}T(r) = S(r)$$

The boundary conditions are

$$T'(0) = 0$$
$$kT'(R) = h_c(T_\infty - T(R)), \quad T_\infty = 20°\,\text{C}, \qquad k = 1$$

(a) Write the difference equations for the temperature using four equally spaced grid intervals.

(b) Solve the difference equations by using the tridiagonal solution ($T_\infty = 20°\,\text{C}$).

(10.16) One end of a rectangular cooling fin of length $H = 0.1$ m is attached to a heat source of 500° C. The fin transfers heat by both radiation and convection to the environment of 20° C. Assuming that the fin and environment are both black bodies, the temperature of the fin satisfies the nonlinear diffusion equation

$$-AkT''(x) + Ph_c(T(x) - T_\infty) + P\sigma(T^4(x) - T_\infty^4) = 0$$

where

$k = 120$ W/mK (thermal conductivity)
$A = 1.5 \times 10^{-4}$ m² (cross-sectional area of the fin)
$P = 0.106$ m (perimeter of the fin)
$h_c = 100$ W/m²K (convection heat transfer coefficient)
$\sigma = 5.67 \times 10^{-8}$ W/m²K⁴ (Stefan-Boltzmann constant)
$T_\infty = 293$ K (temperature of the environment)

The boundary conditions are given by

$$T(0) = 500 + 273°\,\text{K}$$
$$T'(H) = 0$$

where the right end of the fin is assumed to be perfectly insulated.

(a) Derive the difference equation for the foregoing differential equation using 10 equally spaced grid intervals.

(b) Solve the difference equation by means of the successive substitution.

(c) Repeat (b) by using Newton's method.

(10.17) Solve the following equation by Newton's method:

$$-\phi''(x) + [2 + \sin(\phi(x))]\phi(x) = 2, \quad \phi(0) = \phi(2) = 0$$

Use 20 mesh intervals.

(10.18) In a chemical reactor, the density of a material is governed by

$$-\phi''(x) + 0.1\phi'(x) = \exp(1 + 0.05\phi), \quad 0 < x < 2$$

Boundary conditions are $\phi(0) = 0$ and $\phi'(2) = 0$. With 10 equally spaced grid intervals, solve the equation by (a) successive substitution, and (b) Newton's method.

(10.19) The displacement in the axisymmetric vibration of a circular membrane of radius 0.5 m is the solution of

$$y''(r) + \frac{1}{r} y'(r) = -\lambda y(r) \tag{A}$$

where λ is the eigenvalue and boundary conditions are $y'(0) = 0$ and $y(0.5) = 0$. The eigenvalue in the foregoing equation is related to the frequency by

$$\lambda = \omega^2 \rho T = (2\pi v)^2 \rho T$$

where ω is the angular velocity, v is the frequency, ρ is the mass per unit area of the membrane, and T is the tension.

(a) Find the fundamental eigenvalue of Eq. (A) by the power method using 11 grid points including those for both boundaries.

(b) Repeat (i) for the boundary conditions $y(0) = y(0.5) = 0$.

(10.20) Using the shifted inverse power method, find the first to third eigenvalues of the vibration problems (i) and (ii) in Problem (10.19).

(10.21) Consider a slab nuclear reactor shown in Figure P10.21. According to the mono-energetic neutron model, the neutron flux distribution in a critical slab reactor satisfies the neutron diffusion equation given by

$$-\frac{d}{dx} D \frac{d}{dx} \phi(x) + \Sigma_a \phi(x) = \lambda \Sigma_f \phi(x)$$

where λ is the reciprocal of the effective multiplication factor and an eigenvalue of the equation. Develop a computer program to find the fundamental eigenvalue and eigenfunction of the foregoing equation using the shifted inverse power method. The boundary conditions are $\phi'(0) = 0$ and $\phi(30) = 0$.

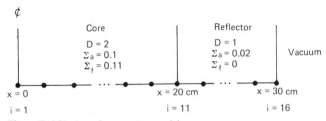

Figure P10.21 A nuclear reactor model

(10.22) The mono-energetic neutron diffusion equation for a cylindrical reactor is given by

$$-\frac{1}{r} \frac{d}{dr} Dr \frac{d}{dr} \phi(r) + \Sigma_a \phi(r) = \lambda \Sigma_f \phi(r)$$

Solve this equation assuming that the constants and dimensions are the same as in Problem 10.21 (x in Figure P10.21 is interpreted as r). Use the same boundary conditions as in Problem 10.21.

(10.23) A beam 1 m long shown in Figure P10.23 is subject to an axial force P. The deflection of the beam is the solution of the equation,

$$EIy''(x) = M$$

where E is the modulus of elasticity and I is the moment of the inertia of beam cross section, and M is the bending moment. Because the bending moment must equal to $-Py$, the foregoing equation becomes

$$y''(x) = \frac{-P}{EI(x)} y(x), \quad y(0) = y(H) = 0$$

This equation is an eigenvalue problem, because solutions exist only when P takes certain discrete values. The constants are given by

$$I(x) = 6 \times 10^{-5}(2 - 0.1x) \text{ m}^4$$
$$E = 200 \times 10^9 \text{ Pa}$$
$$H = 1 \text{ m}$$

Calculate the fundamental eigenvalue P, which corresponds to the deflection illustrated in Figure P10.23a, and the next eigenvalue corresponding to the deflection illustrated in Figure P10.23b. (Usually, the smallest eigenvalue is of the primary interest because the configuration for any higher eigenvalue such as Figure 10.23b is unstable and not experimentally obtained unless additional special devices are used.) The unit of P is 1000 Newton

A   B

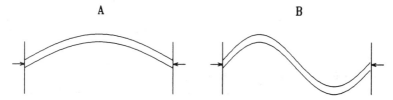

Figure P10.23 Deflection of a beam

(10.24) The eigenvalue problem of the set of difference equations

$$-\phi_{i-1} + 2\phi_i - \phi_{i+1} = \lambda\phi_i$$

with boundary conditions

$$\phi_0 = \phi_{N+1} = 0$$

has N eigenvalues and corresponding eigenfunctions. Show that they can be expressed analytically as follows:

$$\phi_i = \sin(\alpha i), \quad \alpha = (\pi m)/(N - 1)$$
$$\lambda = 2(1 - \cos(\alpha))$$

where $m = 1, 2, \ldots, N$.

Hint: Introduce the eigenvalue and eigenfunction into the difference equation and use the addition theorem of the sine function:

$$\sin (a + b) = \sin (a) \cos (b) + \cos (a) \sin (b)$$
$$\sin (a - b) = \sin (a) \cos (b) - \cos (a) \sin (b)$$

(10.25) The dominance ratio of the inverse power iteration given by Eq. (10.8.5) may be estimated by calculating, in each iteration cycle, the quantity given by

$$r = \sqrt{\frac{\lambda^{(t)} - \lambda^{(t-1)}}{\lambda^{(t-1)} - \lambda^{(t-2)}}}$$

Prove that the r just derived converges to the dominance ratio.

(10.26) The inverse power method given by Eqs. (10.7.10) and (10.7.11) may be modified to

$$A_i f_{i-1}^{(t)} + B_i f_i^{(t)} + C_i f_{i+1}^{(t)} = y_i^{(t-1)}$$

$$\lambda^{(t)} = \frac{\sum_i y_i^{(t-1)} f_i^{(t)}}{\sum_i G_i [f_i^{(t)}]^2}$$

where

$$y_i^{(t-1)} = G_i [\theta \lambda^{(t-1)} f_i^{(t-1)} + (1 - \theta) \lambda^{(t-2)} f_i^{(t-2)}]$$

and θ is an extrapolation parameter satisfying $1 < \theta < 2$. Prove that for any θ between 1 and 2, the foregoing scheme converges faster than does the inverse power method.

(10.27) Flow boundary layer equations for a flow over a flat plate is given (See Example 10.3) by

$$f''' + \frac{1}{2} ff'' = 0 \qquad\qquad\qquad (A)$$

with the boundary conditions,

$$f(0) = 0, \quad f'(0) = 0, \quad f'(\infty) = 1$$

where f is a function of η, and f', f'' and f''' are first, second and third derivatives respectively. If the temperature of the surface is different from that of the fluid, the thermal boundary layer equation is given by

$$T^{*''} + \frac{Pr}{2} f T^{*'} = 0 \qquad\qquad\qquad (B)$$

with boundary conditions

$$T^*(0) = 0, \quad T^*(\infty) = 1$$

where T^* is the nondimensionalized temperature distribution and a function of η. Develop a code to solve both equations (A) and (B), and determine the solution of (B). Assume $Pr = 0.7$. What is the value of $T^{*'}(0)$?

(10.28) A frequently solved boundary layer equation in natural convection is

$$f''' + 3ff'' - 2(f')^2 + T = 0$$
$$T^{*''} + 3Pr f T^{*'} = 0$$

where f and T are function of h and the double and triple primes refer to second and third derivatives respectively. The boundary conditions are

$$f(0) = f'(0) = 0, \quad T^*(0) = 1$$
$$f'(\infty) = 0, \quad T^*(\infty) = 0$$

[Incropera/DeWitt]. Develop an algorithm to solve the boundary layer equations as a boundary value problem. Develop a program and find the solution for $Pr = 0.7$. The last two boundary conditions may be approximated by

$$f'(10) = 0, \quad T(10) = 0$$

for $Pr = 0.7$.

REFERENCES

Arpaci, V. S., and P. S. Larsen, *Convection Heat Transfer*, Prentice-Hall, 1984.

Duderstadt, J. J., and L. J. Hamilton, *Nuclear Reactor Analysis*, Wiley, 1976.

Dusinberre, G. M., *Heat Transfer Calculation by Finite Differences*, International Textbook, 1961.

Eckert, E. R. G., *Heat and Mass Transfer*, McGraw-Hill, 1959.

Habib, I. S., *Engineering Analysis Methods*, Lexington Books, 1975.

Hall G., and J. M. Watt, ed., *Modern Numerical Methods for Ordinary Differential Equations*, Clarendon Press, 1976.

Incropera, F. P., and D. P. Dewitt, *Fundamentals of Heat Transfer*, Wiley, 1981.

Kreith, F., and W. Z. Black, *Basic Heat Transfer*, Harper & Row, 1980.

Kubicek, M., and V. Hlavacek, *Numerical Solution of Nonlinear Boundary Value Problems with Applications*, Prentice-Hall, 1983.

Nakamura, S., *Computational Methods in Engineering and Science with Applications to Fluid Dynamics and Nuclear Systems*, Krieger, 1986.

Nakamura, S., *Applied Numerical Methods with Software*, Prentice-Hall, 1991

Nishida, T., M. Miura, and H. Fujii, eds., *Patterns and Waves*, North-Holland, 1986.

Smith, G. D., *Numerical Solution of Partial Differential Equations*, Oxford University Press, 1978.

Thomson, W. T., *Theory of Vibrations*, Prentice-Hall, 1981.

11

Elliptic Partial Differential Equations

11.1 INTRODUCTION

Second-order partial differential equations (PDE) may be classified into three types: (1) parabolic, (2) elliptic, and (3) hyperbolic.

To distinguish elliptic partial differential equations from the other two kinds, let us consider the following second-order PDE with two independent variables in a general form:

$$A \frac{\partial^2 \phi}{\partial x^2} + B \frac{\partial^2 \phi}{\partial x \, \partial y} + C \frac{\partial^2 \phi}{\partial y^2} + D \frac{\partial \phi}{\partial x} + E \frac{\partial \phi}{\partial y} + F\phi = S \qquad (11.1.1)$$

where x and y are independent variables, and A, B, C, D, E, F, and S are all known functions of x and y. Equation (11.1.1) becomes one of the three types depending on the following conditions:

$$
\begin{array}{lll}
\text{Parabolic} & \text{if } B^2 - 4AC = 0 & \\
\text{Elliptic} & \text{if } B^2 - 4AC < 0 & (11.1.1a) \\
\text{Hyperbolic} & \text{if } B^2 - 4AC > 0 &
\end{array}
$$

Elliptic PDEs appear in many two- and three-dimensional stationary problems. Typical problems of elliptic PDEs include heat conduction in solids, particle diffusion, and vibration of a membrane, among many others. Elliptic PDEs have close relations to parabolic PDEs. In solving a parabolic PDE, for example, numerical methods for an elliptic PDE are often used as a part of the solution scheme. Elliptic PDEs can be viewed as steady-state counterparts of parabolic PDEs. The Poisson and Laplace equations are special cases of elliptic PDEs.

The primary objective of this chapter is to study finite difference methods to solve elliptic PDEs that can be written in a general form as

$$-\nabla p(x, y)\nabla \phi(x, y) + q(x, y)\phi(x, y) = S(x, y) \tag{11.1.2}$$

where p, q, and S are known functions and we assume $q \geqslant 0$. (If q in the given problem has a negative value, the solution methods written in this chapter may not be applicable.) When $p = 1$ and $q = 0$, Eq. (11.1.2) becomes a Poisson or Laplace equation:

$$\text{Poisson equation} \quad -\nabla^2 \phi(x, y) = S(x, y)$$

$$\text{Laplace equation} \quad -\nabla^2 \phi(x, y) = 0$$

An elliptic PDE can include first derivative terms such that

$$-\nabla p \,\nabla \phi + u(x, y)\frac{\partial}{\partial x}\phi + v(x, y)\frac{\partial}{\partial y}\phi + q\phi = S \tag{11.1.3}$$

where u and v are known functions. In fluid dynamics, the second and third terms are called *advective terms*. If the advective terms dominate the first term, the equation behaves more like a hyperbolic PDE. Therefore, the solution methods for hyperbolic PDEs must be applied.

Numerical solution methods for elliptic PDEs are classified largely into two categories: (a) finite difference methods and (b) finite element methods. The finite difference methods, which are the main subjects of this chapter, are derived on a rectangular grid and have a major advantage in that numerous solution methods are available. The advantage of the finite element methods is that the discrete equations can be derived for almost any arbitrary geometry. Therefore, the finite element method is often selected when a complicated geometry is considered. However, in recent years, geometrically complicated problems have also come to be solved by finite difference method with a coordinate transformation [Thompson; Thompson/Wasri/Mastin]. A coordinate transformation is to mathematically transform a given nonrectangular

Table 11.1 A brief comparison between finite difference methods and finite element methods

	Advantages	Disadvantages
Finite Difference Methods	Numerous efficient solution methods available. Easy to vectorize.	Less adaptive to curved geometry than finite elements.
Finite Element Methods	Easy to adapt to curved geometry.	Solution algorithms limited and less efficient than finite difference methods.

geometry to a rectangular computational coordinate. With the transformation, a finite difference method can be used on a rectangular grid on the computational coordinates.

In the remainder of this chapter, however, we focus on finite difference methods for elliptic PDEs given by Eq. (11.1.2).

11.2 DIFFERENCE EQUATIONS

This section consists of four subsections. The first two subsections discuss difference approximations for rectangular geometries and difference approximations for curved geometries, respectively. The third subsection describes derivation of difference equations by the method of integration. The fourth subsection summarizes the properties of the difference approximations.

11.2.1 Difference Approximations for Rectangular Geometries

In this section, we derive finite difference equations for the Poisson equation on the two-dimensional Cartesian coordinates:

$$-\nabla^2 \phi(x, y) = S(x, y) \tag{11.2.1}$$

or equivalently

$$-\frac{\partial^2 \phi(x, y)}{\partial x^2} - \frac{\partial^2 \phi(x, y)}{\partial y^2} = S(x, y) \tag{11.2.2}$$

where $S(x, y)$ is a prescribed function called the inhomogeneous term (or source term).

For simplicity of explanation, we consider the domain [see Figure 11.1(a)] defined by

$$0 \leqslant x \leqslant x_{max}, \quad 0 \leqslant y \leqslant y_{max}$$

Boundary conditions are assumed to be as follows:

$$
\begin{aligned}
&\text{left boundary} &&\frac{\partial \phi}{\partial x} = 0 \quad \text{(Neumann type)} \\
&\text{right boundary} &&\phi = 0 \quad \text{(Dirichlet type)} \\
&\text{bottom boundary} &&\frac{\partial \phi}{\partial y} = 0 \\
&\text{top boundary} &&\phi = 0
\end{aligned}
\tag{11.2.3}
$$

To derive finite difference equations, a grid system with equispaced intervals is imposed on the rectangular domain as shown in Figure 11.1(b). The grid spacings in

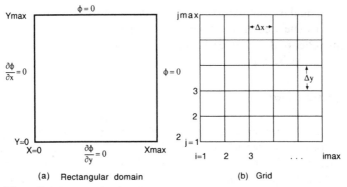

(a) Rectangular domain   (b) Grid

Figure 11.1 Rectangular domain and a grid

the x and y directions are denoted by Δx and Δy respectively. The grid points are numbered by i and j, where i is the grid index in the x direction and j the same in the y direction.

The difference equation for a grid point (i, j) located inside the boundary is derived by considering the grid point (i, j) and four surrounding grid points as shown in Figure 11.2(a). By applying the central difference approximation, the first term of Eq. (11.2.2) is approximated by

$$\frac{\partial^2 \phi}{\partial x^2} = \frac{\phi_{i-1,j} - 2\phi_{i,j} + \phi_{i+1,j}}{\Delta x^2} \tag{11.2.4}$$

Similarly, the difference approximation for the second term is

$$\frac{\partial^2 \phi}{\partial y^2} = \frac{\phi_{i,j-1} - 2\phi_{i,j} + \phi_{i,j+1}}{\Delta y^2} \tag{11.2.5}$$

Introducing Eqs. (11.2.4) and (11.2.5) into Eq. (11.2.2) yields

$$\frac{-\phi_{i-1,j} + 2\phi_{i,j} - \phi_{i+1,j}}{\Delta x^2} + \frac{-\phi_{i,j-1} + 2\phi_{i,j} - \phi_{i,j+1}}{\Delta y^2} = S_{i,j} \tag{11.2.6}$$

where $S_{i,j} = S(x_i, y_j)$. Equation (11.2.6) applies to all the grid points except those on the boundary.

Difference equations for the grid points on the boundary need special treatments because (a) the number of neighboring points is less than four and (b) the boundary conditions must be taken into consideration. For the present geometry, however, the difference equations for the points along the right and top boundaries are not necessary because ϕ values are known ($\phi = 0$) from the boundary conditions given in Eq. (11.2.3).

Considering the bottom boundary [see Figure 11.2(b)], the difference equation for a point, $1 < i < i_{max}$ and $j = 1$, is derived as follows: The first term of Eq. (11.2.2)

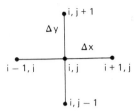

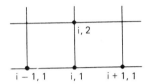

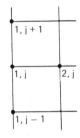

(a) Five grid points used in the difference equation for an internal point.

(b) Four grid points used in the difference equation for point (i, 1) which is along the bottom boundary.

(c) Four grid points used in the difference equation for point (1, j) along the left boundary.

Figure 11.2 Grid points used in difference equations

is approximated by Eq. (11.2.4). The second term of Eq. (11.2.2) is approximated by

$$\left(\frac{\partial^2 \phi}{\partial y^2}\right)_{i,1} \simeq \frac{\left(\frac{\partial \phi}{\partial y}\right)_{i,1+\frac{1}{2}} - \left(\frac{\partial \phi}{\partial y}\right)_{i,1}}{\frac{\Delta y}{2}} \qquad (11.2.7)$$

The first term in the numerator of Eq. (11.2.7) is approximated by the central difference approximation:

$$\left(\frac{\partial \phi}{\partial y}\right)_{i,1+\frac{1}{2}} \simeq \frac{\phi_{i,2} - \phi_{i,1}}{\Delta y} \qquad (11.2.8)$$

The bottom boundary condition given in Eq. (11.2.3) shows that the second term in the numerator on the right side of Eq. (11.2.7) is zero. Therefore, Eq. (11.2.7) becomes

$$\left(\frac{\partial^2 \phi}{\partial y^2}\right)_{i,1} \simeq \frac{2\phi_{i,2} - 2\phi_{i,1}}{\Delta y^2} \tag{11.2.9}$$

Thus, by introducing Eq. (11.2.4) and (11.2.9) into Eq. (11.2.2), the difference equation for a point along the bottom boundary becomes

$$\frac{-\phi_{i-1,1} + 2\phi_{i,1} - \phi_{i+1,1}}{\Delta x^2} + \frac{-2\phi_{i,2} + 2\phi_{i,1}}{\Delta y^2} = S_{i,1} \tag{11.2.10}$$

For a point $i = 1$ and $1 < j < j_{max}$ along the left boundary [see Figure 11.2(c)], the first term of Eq. (11.2.2) is approximated by

$$\frac{\partial^2 \phi}{\partial x^2} \simeq \frac{\left(\dfrac{\partial \phi}{\partial x}\right)_{1+\frac{1}{2},j} - \left(\dfrac{\partial \phi}{\partial x}\right)_{1,j}}{\dfrac{\Delta x}{2}}$$

$$= \frac{2\phi_{2,j} - 2\phi_{1,j}}{\Delta x^2} \tag{11.2.11}$$

where the left boundary condition given in Eq. (11.2.3) is used to eliminate $(\partial\phi/\partial x)_{1,j}$, and the central difference approximation is used for $(\partial\phi/\partial x)_{1+\frac{1}{2},j}$. By introducing Eq. (11.2.11) and Eq. (11.2.5) into Eq. (11.2.2), the difference equation becomes

$$\frac{2\phi_{1,j} - 2\phi_{2,j}}{\Delta x^2} + \frac{-\phi_{1,j-1} + 2\phi_{1,j} - \phi_{1,j+1}}{\Delta y^2} = S_{1,j} \tag{11.2.12}$$

For the corner point at $i = j = 1$, each term on the left side of Eq. (11.2.2) is approximated respectively by Eq. (11.2.9) and Eq. (11.2.11). Hence, the difference equation becomes

$$\frac{2\phi_{1,1} - 2\phi_{2,1}}{\Delta x^2} + \frac{2\phi_{1,1} - 2\phi_{1,2}}{\Delta y^2} = S_{1,1} \tag{11.2.13}$$

Example 11.1

(a) Write the difference approximation of the Poisson equation for the grid shown in Figure E11.1.
(b) Express the difference equations in the matrix and vector form.
(c) Show that the coefficient matrix obtained in (b) can be transformed to a symmetric form by dividing or multiplying each row by a constant.

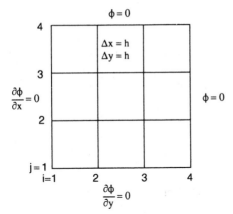

Figure E11.1

⟨Solution⟩

(a) Because of the equispaced grid in both directions, the difference equations may be first written as

$$4\phi_{1,1} - 2\phi_{2,1} - 2\phi_{1,2} = h^2 S_{1,1}$$
$$4\phi_{2,1} - \phi_{1,1} - \phi_{3,1} - 2\phi_{2,2} = h^2 S_{2,1}$$
$$4\phi_{3,1} - \phi_{2,1} - 2\phi_{3,2} = h^2 S_{3,1}$$
$$4\phi_{1,2} - 2\phi_{2,2} - \phi_{1,1} - \phi_{1,3} = h^2 S_{1,2}$$
$$4\phi_{2,2} - \phi_{1,2} - \phi_{3,2} - \phi_{2,1} - \phi_{2,3} = h^2 S_{2,2} \qquad \text{(A)}$$
$$4\phi_{3,2} - \phi_{2,2} - \phi_{3,1} - \phi_{3,3} = h^2 S_{3,2}$$
$$4\phi_{1,3} - 2\phi_{2,3} - \phi_{1,2} = h^2 S_{1,3}$$
$$4\phi_{2,3} - \phi_{1,3} - \phi_{3,3} - \phi_{2,2} = h^3 S_{2,3}$$
$$4\phi_{3,3} - \phi_{3,2} - \phi_{2,3} = h^2 S_{3,3}$$

where $\phi_{4,1} = \phi_{4,2} = \phi_{4,3} = \phi_{4,4} = \phi_{1,4} = \phi_{2,4} = \phi_{3,4} = 0$ have been used.

(b) In matrix notations, the above equations given in (A) are written as follows:

$$
\begin{bmatrix}
4 & -2 & 0 & -2 & 0 & 0 & 0 & 0 & 0 \\
-1 & 4 & -1 & 0 & -2 & 0 & 0 & 0 & 0 \\
0 & -1 & 4 & 0 & 0 & -2 & 0 & 0 & 0 \\
-1 & 0 & 0 & 4 & -2 & 0 & -1 & 0 & 0 \\
0 & -1 & 0 & -1 & 4 & -1 & 0 & -1 & 0 \\
0 & 0 & -1 & 0 & -1 & 4 & 0 & 0 & -1 \\
0 & 0 & 0 & -1 & 0 & 0 & 4 & -2 & 0 \\
0 & 0 & 0 & 0 & -1 & 0 & -1 & 4 & -1 \\
0 & 0 & 0 & 0 & 0 & -1 & 0 & -1 & 4
\end{bmatrix}
\begin{bmatrix}
\phi_{1,1} \\ \phi_{2,1} \\ \phi_{3,1} \\ \phi_{1,2} \\ \phi_{2,2} \\ \phi_{3,2} \\ \phi_{1,3} \\ \phi_{2,3} \\ \phi_{3,3}
\end{bmatrix}
=
\begin{bmatrix}
h^2 S_{1,1} \\ h^2 S_{2,1} \\ h^2 S_{3,1} \\ h^2 S_{1,2} \\ h^2 S_{2,2} \\ h^2 S_{3,2} \\ h^2 S_{1,3} \\ h^2 S_{2,3} \\ h^2 S_{3,3}
\end{bmatrix}
\qquad \text{(B)}
$$

(c) The coefficient matrix can be transformed to a symmetric form by dividing the first equation by 4, and also dividing the second, third, fourth, and seventh equations by 2:

$$
\begin{bmatrix}
1 & -\tfrac{1}{2} & 0 & -\tfrac{1}{2} & 0 & 0 & 0 & 0 & 0 \\
-\tfrac{1}{2} & 2 & -\tfrac{1}{2} & 0 & -1 & 0 & 0 & 0 & 0 \\
0 & -\tfrac{1}{2} & 2 & 0 & 0 & -1 & 0 & 0 & 0 \\
-\tfrac{1}{2} & 0 & 0 & 2 & -1 & 0 & -\tfrac{1}{2} & 0 & 0 \\
0 & -1 & 0 & -1 & 4 & -1 & 0 & -1 & 0 \\
0 & 0 & -1 & 0 & -1 & 4 & 0 & 0 & -1 \\
0 & 0 & 0 & -\tfrac{1}{2} & 0 & 0 & 2 & -1 & 0 \\
0 & 0 & 0 & 0 & -1 & 0 & -1 & 4 & -1 \\
0 & 0 & 0 & 0 & 0 & -1 & 0 & -1 & 4
\end{bmatrix}
\begin{bmatrix}
\phi_{1,1} \\ \phi_{2,1} \\ \phi_{3,1} \\ \phi_{1,2} \\ \phi_{2,2} \\ \phi_{3,2} \\ \phi_{1,3} \\ \phi_{2,3} \\ \phi_{3,3}
\end{bmatrix}
=
\begin{bmatrix}
\tfrac{1}{4}h^2 S_{1,1} \\ \tfrac{1}{2}h^2 S_{2,1} \\ \tfrac{1}{2}h^2 S_{3,1} \\ \tfrac{1}{2}h^2 S_{1,2} \\ h^2 S_{2,2} \\ h^2 S_{3,2} \\ \tfrac{1}{2}h^2 S_{1,3} \\ h^2 S_{2,3} \\ h^2 S_{3,3}
\end{bmatrix}
\quad \text{(C)}
$$

Note: As Eq. (C) illustrates, the coefficient matrix of the difference equations for the Poisson equation on a rectangular grid has the following properties:

(a) The matrix is a block-tridiagonal matrix.

(b) The diagonal blocks are tridiagonal submatrices.

(c) The off-diagonal blocks adjacent to the diagonal blocks are diagonal submatrices with negative diagonal elements.

(d) Other blocks are all null submatrices.

(e) The entire matrix is symmetric.

The number of null elements rapidly increases with any increase of the total number of grid points. If the computer memory space is limited, iterative solution methods are useful for elliptic partial differential equations because they store only nonzero elements of the coefficient matrix so that far less memory space is required than in the direct solution method.

The boundary conditions are often given in a more general form as

$$
\frac{\partial \phi}{\partial n} + \alpha \phi = \beta \quad \text{(mixed type)} \tag{11.2.14}
$$

where α and β are constants, and $\partial/\partial n$ is the derivative outward normal to the boundary. In a rectangular domain, $\partial/\partial n$ has the following interpretation for each of the left, right, top, and bottom boundaries:

$$
\frac{\partial}{\partial n} = -\frac{\partial}{\partial x} \text{ for the left boundary}
$$

$$
\frac{\partial}{\partial n} = \frac{\partial}{\partial y} \text{ for the top boundary}
$$

$$
\frac{\partial}{\partial n} = \frac{\partial}{\partial x} \text{ for the right boundary} \tag{11.2.15}
$$

$$
\frac{\partial}{\partial n} = -\frac{\partial}{\partial y} \text{ for the bottom boundary}
$$

Implementation of the boundary conditions given in the form of Eq. (11.2.14) is quite similar to that for $\partial\phi/\partial x = 0$ or $\partial\phi/\partial y = 0$. For example, if the top boundary condition is given in the form of Eq. (11.2.14), the second term in Eq. (11.2.2) is approximated as follows:

$$\left(\frac{\partial^2\phi}{\partial y^2}\right)_{i,J} \simeq \frac{(\partial\phi/\partial y)_{i,J} - (\partial\phi/\partial y)_{i,J-1/2}}{\Delta y/2}$$

$$\simeq \frac{(-\alpha\phi_{i,J} + \beta) - (\phi_{i,J} - \phi_{i,J-1})/\Delta y}{\Delta y/2}$$

$$= \frac{-2\alpha\Delta y\phi_{i,J} + 2\beta\Delta y - 2\phi_{i,J} + 2\phi_{i,J-1}}{\Delta y^2} \qquad (11.2.16)$$

Introducing Eq. (11.2.16) and Eq. (11.2.4) into Eq. (11.2.2) yields

$$\frac{-\phi_{i-1,J} + 2\phi_{i,J} - \phi_{i+1,J}}{\Delta x^2} + \frac{(2\alpha\Delta y + 2)\phi_{i,J} - 2\phi_{i,J-1}}{\Delta y^2} = S_{i,J} + \frac{2\beta}{\Delta y} \qquad (11.2.17)$$

Equation (11.2.14) is a universal form of boundary conditions because all three types of boundary conditions—the Dirichlet, Neumann, and mixed types—can be represented in this form. If $\alpha = 0$, it reduces to the Neumann (derivative) boundary condition, $\partial\phi/\partial n = \beta$. On the other hand, if β is substituted by $\gamma\alpha$ and α is increased to infinity, then it reduces to the Dirichlet (fixed value) boundary condition, $\phi = \gamma$ (constant).

Although infinity is not allowed in a computer program, practically the same effect is achieved by setting α to a very large number such as 10^{10}. The advantage of using this form is that once a program is written, the type of boundary condition may be changed easily only by revising the parameters α and β for each boundary.

Example 11.2

Considering the geometry and the grid shown in Figure E11.2, derive the difference equations for the Poisson equation,

$$-\nabla^2\phi = S \qquad (A)$$

The boundary conditions are

$$\frac{\partial\phi}{\partial x} = \phi \qquad \text{for the left boundary}$$

$$\frac{\partial\phi}{\partial y} = \phi - 2 \quad \text{for the bottom boundary}$$

$$\phi = 5 \qquad \text{for the right boundary}$$

$$\phi = 7 \qquad \text{for the top boundary}$$

The grid intervals are assumed to be unity in both directions.

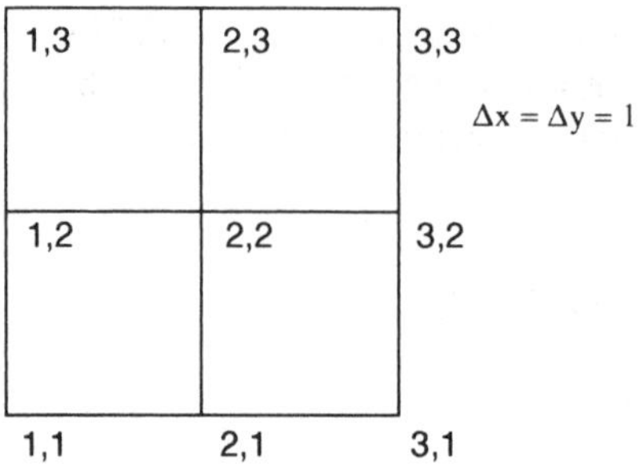

1,3	2,3	3,3
1,2	2,2	3,2
1,1	2,1	3,1

$\Delta x = \Delta y = 1$

Figure E11.2

⟨**Solution**⟩

Because the boundary conditions for the top and right boundaries are the fixed value type, we derive difference equations for only the four grid points: (1, 1), (2, 1), (1, 2), and (2, 2).

Point (1, 1). The partial derivative with respect to x is approximated by

$$\left(\frac{\partial^2 \phi}{\partial x^2}\right)_{1,1} \simeq \frac{\left(\frac{\partial \phi}{\partial x}\right)_{1+\frac{1}{2},1} - \left(\frac{\partial \phi}{\partial x}\right)_{1,1}}{\frac{1}{2}}$$

$$\simeq \frac{(\phi_{2,1} - \phi_{1,1}) - \phi_{1,1}}{\frac{1}{2}}$$

$$= -4\phi_{1,1} + 2\phi_{2,1} \tag{B}$$

where the left boundary condition is used to eliminate $(\partial\phi/\partial x)_{1,1}$. The partial derivative with respect to y is approximated by

$$\left(\frac{\partial^2 \phi}{\partial y^2}\right)_{1,1} \simeq \frac{\left(\frac{\partial \phi}{\partial y}\right)_{1,1+\frac{1}{2}} - \left(\frac{\partial \phi}{\partial y}\right)_{1,1}}{\frac{1}{2}}$$

$$\simeq \frac{(\phi_{1,2} - \phi_{1,1}) - (\phi_{1,1} - 2)}{\frac{1}{2}}$$

$$= -4\phi_{1,1} + 2\phi_{1,2} + 4 \tag{C}$$

where the bottom boundary condition is used to eliminate $(\partial\phi/\partial y)_{1,1}$. Introducing Eq. (B) and (C) into Eq. (11.2.2) yields

$$8\phi_{1,1} - 2\phi_{1,2} - 2\phi_{2,1} = 4 + S \tag{D}$$

Point (2, 1). The partial derivatives are approximated by

$$\frac{\partial^2 \phi}{\partial x^2} = \phi_{1,1} - 2\phi_{2,1} + \phi_{3,1} \tag{E}$$

$$\frac{\partial^2 \phi}{\partial y^2} \simeq \frac{\left(\dfrac{\partial \phi}{\partial y}\right)_{2,1+\frac{1}{2}} - \left(\dfrac{\partial \phi}{\partial x}\right)_{2,1}}{\dfrac{1}{2}}$$

$$= 2\phi_{2,2} - 4\phi_{2,1} + 4 \tag{F}$$

Introducing the above two equations into Eq. (11.2.2) yields

$$6\phi_{2,1} - \phi_{1,1} - 2\phi_{2,2} - \phi_{3,1} = 4 + S \tag{G}$$

Point (1, 2). The partial derivatives are approximated by

$$\left(\frac{\partial^2 \phi}{\partial x^2}\right)_{1,2} \simeq \frac{\left(\dfrac{\partial \phi}{\partial x}\right)_{1+\frac{1}{2},2} - \left(\dfrac{\partial \phi}{\partial x}\right)_{1,2}}{\dfrac{1}{2}}$$

$$\simeq 2\phi_{2,2} - 4\phi_{1,2} \tag{H}$$

$$\left(\frac{\partial \phi^2}{\partial y^2}\right)_{1,2} \simeq \phi_{1,1} - 2\phi_{1,2} + \phi_{1,3} \tag{I}$$

Introducing the above equations into Eq. (11.2.2) yields

$$6\phi_{1,2} - \phi_{1,1} - 2\phi_{2,2} - \phi_{1,3} = S \tag{J}$$

Point (2, 2). The difference equation is

$$4\phi_{2,2} - \phi_{1,2} - \phi_{3,2} - \phi_{2,1} - \phi_{2,3} = S \tag{K}$$

The whole equation set is summarized as

$$\begin{aligned}
8\phi_{1,1} - 2\phi_{1,2} - 2\phi_{2,1} &= 4 + S \\
6\phi_{2,1} - \phi_{1,1} - 2\phi_{2,2} &= 9 + S \\
6\phi_{1,2} - \phi_{1,1} - 2\phi_{2,2} &= 7 + S \\
4\phi_{2,2} - \phi_{1,2} - \phi_{2,1} &= 12 + S
\end{aligned} \tag{L}$$

11.2.2 Geometries with Curved Boundaries

We have assumed in the preceding section that the domains for elliptic partial differential equations are rectangular. In practice, however, the geometries often have irregular or curved boundaries [Nogotov].

There are three major approaches applicable to nonrectangular geometries:

(a) Use of the rectangular grid with adjustment to the difference equations for the grid points near the boundary.

(b) Finite element method.

(c) Mathematical mapping of the given geometry to a rectangular computational domain (coordinate transformation).

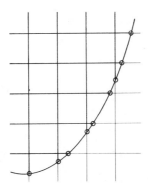

Figure 11.3 A domain with curved boundaries

For simplicity, the remainder of this subsection explains the first approach. Consider a curved boundary as shown in Figure 11.3, on which a rectangular grid is imposed. To conform the curved boundary, we impose special grid points at intersections of regular grid lines and the curved boundary as marked by open circles in Figure 11.3. The difference equation for the grid points adjacent to the curved boundary may be easily written. For example, considering the grid configuration shown in Figure 11.4, the difference equation for Eq. (11.2.2) may be written as

$$
-\frac{\left(\dfrac{\phi_a - \phi_{i,j}}{\alpha\,\Delta x} - \dfrac{\phi_{i,j} - \phi_{i-1,j}}{\Delta x}\right)}{\dfrac{1}{2}(1+\alpha)\,\Delta x} - \frac{\left(\dfrac{\phi_b - \phi_{i,j}}{\beta\,\Delta y} - \dfrac{\phi_{i,j} - \phi_{i,j+1}}{\Delta y}\right)}{\dfrac{1}{2}(1+\beta)\,\Delta y}
$$

$$
= \frac{-\phi_{i-1,j} + \left(1+\dfrac{1}{\alpha}\right)\phi_{i,j} - \dfrac{\phi_a}{\alpha}}{\dfrac{1}{2}(1+\alpha)\,\Delta x^2} + \frac{-\phi_{i,j+1} + \left(1+\dfrac{1}{\beta}\right)\phi_{i,j} - \dfrac{\phi_b}{\beta}}{\dfrac{1}{2}(1+\beta)\,\Delta x^2} = S_{i,j} \qquad (11.2.18)
$$

where ϕ_a and ϕ_b are given by the boundary conditions.

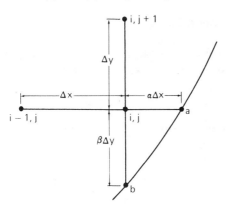

Figure 11.4 A grid point adjacent to the curved boundary

11.2.3 Method of Integration to Derive Difference Equations

In this section, we restrict ourselves to rectangular domains and study a universal method of deriving difference equations for the elliptic partial differential equation, which is based on integrating the elliptic partial differential equation in the volume that belongs to a grid point. By this method, difference equations may be derived in almost any situation including variable coefficients of the elliptic partial differential equation, variable grid spacings, and cylindrical and spherical coordinates.

Consider the equation given by

$$-\nabla p(x, y)\nabla \phi(x, y) + q(x, y)\phi(x, y) = S(x, y) \qquad (11.2.19)$$

In Eq. (11.2.19) the operator ∇ can be in any coordinate system, but it is assumed here to be in the rectangular coordinates. We consider a rectangular grid system in which (1) grid spacings change from an interval to the next, (2) p, q, and S are space-dependent functions but constant in each rectangle cornered by four adjacent grid points—for example, (i, j), $(i - 1, j)$, $(i, j - 1)$, and $(i - 1, j - 1)$.

Now we consider a grid point (i, j) with its adjacent four grid points as shown in Figure 11.5. The rectangular box containing the point (i, j) consists of the four sides each of which passes the midpoint between the grid point (i, j) and an adjacent grid point. The rectangle and its entire boundary are denoted by D and G respectively.

Figure 11.5 Internal grid points with variable spacings (D denotes the rectangular domain including v_1 through v_4, and G is its boundary)

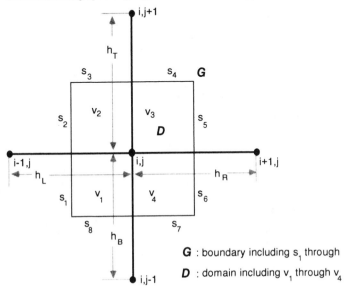

Integrating Eq. (11.2.19) in domain D shown in Figure 11.5 gives

$$-\int_G p(x, y) \frac{\partial}{\partial n} \phi(x, y)\, ds + \iint_D q(x, y)\phi(x, y)\, dx\, dy = \iint_D S(x, y)\, dx\, dy \quad (11.2.20)$$

where the Green's theorem has been used for the first term, the integral of the first term is extended along G; and $\partial/\partial n$ is the derivative outward normal to the boundary [Nakamura].

The first term of Eq. (11.2.20) is partitioned into four parts as

$$-\int_G p(x, y) \frac{\partial}{\partial n} \phi(x, y)\, ds = -\int_{s_1 + s_2} p(x, y) \frac{\partial}{\partial n} \phi(x, y)\, ds$$

$$-\int_{s_3 + s_4} p(x, y) \frac{\partial}{\partial n} \phi(x, y)\, ds$$

$$-\int_{s_5 + s_6} p(x, y) \frac{\partial}{\partial n} \phi(x, y)\, ds$$

$$-\int_{s_7 + s_8} p(x, y) \frac{\partial}{\partial n} \phi(x, y)\, ds \quad (11.2.21)$$

where s_n are to denote the parts of the boundary of D (see Figure 11.5). The partial derivatives in Eq. (11.2.21) may be approximated by the difference approximation at the midpoints between two adjacent grid points. For example, the difference approximation for $\partial\phi/\partial n$ in the first term on the right side of Eq. (11.2.21) is

$$\frac{\partial \phi}{\partial n} = -\frac{\partial \phi}{\partial x} \simeq -\frac{\phi_{i,j} - \phi_{i-1,j}}{h_L} \quad (11.2.22)$$

Thus, the first term after the equality sign in Eq. (11.2.21) can be written as

$$-\int_{s_1 + s_2} p(x, y) \frac{\partial}{\partial n} \phi(x, y)\, ds \simeq (s_1 p_1 + s_2 p_2) \frac{\phi_{i,j} - \phi_{i-1,j}}{h_L} \quad (11.2.23)$$

On the right side of Eq. (11.2.23), s_1 and s_2 are lengths of parts of the left boundary of D. For the plane $x - y$ coordinates, s_k, $k = 1, 2, \ldots 8$, are given by

$$s_1 = \frac{h_B}{2}, \quad s_2 = \frac{h_T}{2}, \quad s_3 = \frac{h_L}{2}, \quad s_4 = \frac{h_R}{2},$$

$$s_5 = \frac{h_T}{2}, \quad s_6 = \frac{h_B}{2}, \quad s_7 = \frac{h_R}{2}, \quad s_8 = \frac{h_L}{2} \quad (11.2.24)$$

Other integrals in Eq. (11.2.21) are approximated similarly. Therefore, Eq. (11.2.21) becomes

$$-\int p(x, y) \frac{\partial}{\partial n} \phi(x, y)\, ds \simeq (s_1 p_1 + s_2 p_2) \frac{\phi_{i,j} - \phi_{i-1,j}}{h_L}$$

$$+ (s_3 p_2 + s_4 p_3) \frac{\phi_{i,j} - \phi_{i,j+1}}{h_T}$$

$$+ (s_5 p_3 + s_6 p_4) \frac{\phi_{i,j} - \phi_{i+1,j}}{h_R}$$

$$+ (s_7 p_4 + s_8 p_1) \frac{\phi_{i,j} - \phi_{i,j-1}}{h_B} \tag{11.2.25}$$

The right side of Eq. (11.2.20) and its second term on the left side are both approximated respectively by

$$\iint_D q(x, y)\phi(x, y)\, dx\, dy \simeq (v_1 q_1 + v_2 q_2 + v_3 q_3 + v_4 q_4)\phi_{i,j} \tag{11.2.26}$$

$$\iint_D S(x, y)\, dx\, dy \simeq v_1 S_1 + v_2 S_2 + v_3 S_3 + v_4 S_4 \tag{11.2.27}$$

where

$$v_1 = \frac{h_L h_B}{4}, \quad v_2 = \frac{h_L h_T}{4}, \quad v_3 = \frac{h_R h_T}{4}, \quad v_4 = \frac{h_R h_B}{4}$$

By collecting all the terms of Eqs. (11.2.25) through (11.2.27), the difference equation involving the five grid points is obtained as

$$a^C \phi_{i,j} + a^L \phi_{i-1,j} + a^R \phi_{i+1,j} + a^B \phi_{i,j-1} + a^T \phi_{i,j+1} = S_{i,j} \tag{11.2.28}$$

$$a_L = -\left[\frac{s_1 p_1 + s_2 p_2}{h_L} \right]_{i,j}$$

$$a_T = -\left[\frac{s_3 p_2 + s_4 p_3}{h_T} \right]_{i,j}$$

$$a_R = -\left[\frac{s_5 p_3 + s_6 p_4}{h_R} \right]_{i,j}$$

$$a_B = -\left[\frac{s_7 p_4 + s_8 p_1}{h_B} \right]_{i,j}$$

$$a_C = [-a_L - a_T - a_R - a_B + v_1 q_1 + v_2 q_2 + v_3 q_3 + v_4 q_4]_{i,j}$$

$$S_{i,j} = [v_1 S_1 + v_2 S_2 + v_3 S_3 + v_4 S_4]_{i,j}$$

where subscripts i and j after the brackets indicate that the values are evaluated for the grid point (i, j).

Derivation of difference equations for the grid points on the boundaries is similar to that for internal grid points. To illustrate this, we consider a corner point as shown in Figure 11.6 and assume that the boundary conditions are given in the form:

$$\frac{\partial}{\partial y} \phi = -\alpha_T \phi + \beta_T \quad \text{(top boundary condition)}$$

$$\frac{\partial}{\partial x} \phi = -\alpha_R \phi + \beta_R \quad \text{(right boundary condition)}$$

(11.2.29)

The difference equation for the corner point can be derived from Eq. (11.2.20) if we denote the rectangular domain belonging to the corner point (i, j) by D and the boundary by G (see Figure 11.6).

The first term of Eq. (11.2.20) is partitioned into four parts as

$$-\int_G p(x, y) \frac{\partial}{\partial n} \phi(x, y)\, ds = -\int_{S_1} p(x, y) \frac{\partial}{\partial n} \phi(x, y)\, ds$$

$$= -\int_{S_3} p(x, y) \frac{\partial}{\partial n} \phi(x, y)\, ds$$

$$= -\int_{S_6} p(x, y) \frac{\partial}{\partial n} \phi(x, y)\, ds$$

$$= -\int_{S_8} p(x, y) \frac{\partial}{\partial n} \phi(x, y)\, ds \qquad (11.2.30)$$

Difference approximations for the first and the fourth terms of Eq. (11.2.30) are similar to that for an internal grid point and given as

$$-\int_{S_1} p(x, y) \frac{\partial}{\partial n} \phi(x, y)\, ds \simeq p_1 s_1 \frac{\phi_{i,j} - \phi_{i-1,j}}{h_L}$$

$$-\int_{S_8} p(x, y) \frac{\partial}{\partial n} \phi(x, y)\, ds \simeq p_1 s_8 \frac{\phi_{i,j} - \phi_{i,j-1}}{h_B}$$

(11.2.31)

The second and the third terms are evaluated by using the boundary conditions:

$$-\int_{S_3} p(x, y) \frac{\partial}{\partial n} \phi(x, y)\, ds \simeq p_1 s_3 (\alpha_T \phi_{i,j} - \beta_T)$$

$$-\int_{S_6} p(x, y) \frac{\partial}{\partial n} \phi(x, y)\, ds \simeq p_1 s_6 (\alpha_R \phi_{i,j} - \beta_R)$$

(11.2.32)

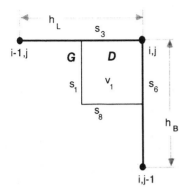

i,j-1   **Figure 11.6** Grid points on a boundary

The second term on the left side of Eq. (11.2.20) is approximated by

$$\iint_D q(x, y)\phi(x, y)\, dx\, dy \simeq (v_1 q_1)\phi_{i,j}$$ (11.2.33)

The right side of Eq. (11.2.20) becomes

$$\iint_D S(x, y)\, dx\, dy \simeq v_1 S_1$$

By collecting all the terms, we get

$$a^C \phi_{i,j} + a^L \phi_{i-1,j} + a^B \phi_{i,j-1} = S_{i,j}$$ (11.2.34)

$$a_L = -\left[\frac{p_1 s_1}{h_L}\right]_{i,j}$$

$$a_B = -\left[\frac{p_1 s_8}{h_B}\right]_{i,j}$$

$$a_C = \left[-a_L - a_B + + v_1 q_1 + p_1 s_3 \alpha_T + p_1 s_6 \alpha_R\right]_{i,j}$$

$$S_{i,j} = \left[v_1 S_1 + p_1 s_3 \beta_T + p_1 s_6 \beta_R\right]_{i,j}$$

Difference equations for other coordinate systems are derived by the same procedure if the volume integral is treated appropriately for the given coordinate system. As an example, let us consider the equation for the *r-z* coordinate system written as

$$-\left[\frac{1}{r}\frac{\partial}{\partial r} pr \frac{\partial}{\partial r}\phi(r,z) + \frac{\partial}{\partial z}\phi(r,z)\right] + q(r, z)\phi(r, z) = S(r, z)$$ (11.2.35)

A grid point (i, j) and four adjacent grid points in the cylindrical coordinate system as shown in Figure 11.7 may be used if x and y are changed to r and z, respectively. In the *r-z* coordinates, however, the domain D represents a donut-shaped space as shown in Figure 11.8.

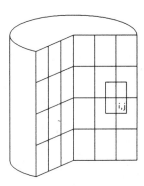

Figure 11.7 Grid on the cylindrical coordinates

Equation (11.2.35) is integrated in domain D. This means that the integration is over a volume rather than on a plane and is given by

$$2\pi \iint_D \left[-\frac{1}{r}\frac{\partial}{\partial r} pr \frac{\partial}{\partial r}\phi(r,z) - \frac{\partial_2}{\partial z^2}\phi(r,z) + q(r,z)\phi(r,z) \right] r\,dr\,dz$$

$$= 2\pi \iint_D S(r,z)r\,dr\,dz \tag{11.2.36}$$

or equivalently, by using the Green's theorem,

$$-2\pi \int_G p \frac{\partial}{\partial n}\phi(r,z)r\,ds + 2\pi \iint_D q\phi(r,z)r\,dr\,dz = 2\pi \iint_D S(r,z)r\,dr\,dz \tag{11.2.37}$$

where the integral in the first term is over the surface (or boundary) of D. The remainder of operation is very similar to the case of the plane geometry described earlier. The resulting difference equation becomes exactly the same form as Eq. (11.2.28) provided that v and s are interpreted in the following manner. That is, v and s in Figure 11.5 are interpreted as partial volumes and surface areas of the domain D. For example, v_1 and s_1 are respectively written as

$$v_1 = \pi(r_i^2 - r_a^2)\frac{h_B}{2}$$

$$s_1 = 2\pi r_a \frac{h_B}{2} = \pi r_a h_B$$

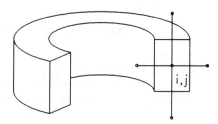

Figure 11.8 Donut-shaped domain

where r_a is the radius at the midpoint between the grid points $(i-1, j)$ and (i, j), and r_i is the radius at (i, j).

11.2.4 Properties of the Difference Equations

All the difference equations derived in the previous subsections can be written in the form

$$a^C \phi_{i,j} + a^L \phi_{i-1,j} + a^R \phi_{i+1,j} + a^B \phi_{i,j-1} + a^T \phi_{i,j+1} = S_{i,j} \qquad (11.2.38)$$

Equation (11.2.38) may be used for all the grid points hereafter with the following interpretations:

(a) Although subscripts i and j for a's are omitted for simplicity, a's are dependent on i and j.

(b) If the point (i, j) is outside the domain, $\phi_{i,j}$ and its coefficients are interpreted as zero.

(c) All unknown terms are on the left side, and all the known terms are combined into $S_{i,j}$. For example, if $\phi_{i,j+1}$ is known from the boundary condition, the term is brought to the right side and added to $S_{i,j}$.

With appropriate boundary conditions, the difference equations Eq. (11.2.38) for the entire domain have the following properties:

(a) The coefficients of Eq. (11.2.38) are all negative or zero except a^C.

(b) The coefficient a^C is positive and called a diagonal coefficient.

(c) Symmetry property.*

$$(a^L)_{i,j} = (a^R)_{i-1,j}$$
$$(a^B)_{i,j} = (a^T)_{i,j-1} \qquad (11.2.39)$$

(d) The coefficient a^C is equal to or greater than the sum of the absolute values of all other coefficients (diagonal dominance)

$$a^C \geq |a^L| + |a^R| + |a^B| + |a^T| \qquad (11.2.40)$$

---

* As illustrated in Example 11.1, the difference equations for an elliptic PDE without first order derivative terms may be written in the symmetric form if the set of equations is in the conservation form. Because of this property of symmetry, only three coefficients for each equation need to be stored in the memory space.

where inequality is satisfied for at least one grid point.** When the inequality is satisfied, the equation is said to have a strong diagonal dominance.

(e) No part of the set of equations can be solved independently of others (irreducibility).

The five properties listed are important because they are sufficient conditions for an iterative scheme to converge [Varga; Wachspress]. When the coefficients of the difference equations having the above properties are written in a matrix form, the matrix is called a *Stieltjes matrix*, or *S*-matrix.

SUMMARY OF THIS SECTION

(a) Difference equations for an elliptic PDE may be derived by both direct application of difference approximation and the method of integration. The former is simpler, but the latter is more powerful if the coefficients of the PDE are space-dependent and if grid spacing is variable.

(b) Boundary conditions are incorporated into the difference equations.

(c) The difference equations in the matrix form are symmetric, pentadiagonal, and have diagonal dominance.

11.3 OVERVIEW OF SOLUTION METHODS FOR ELLIPTIC DIFFERENCE EQUATIONS

Numerical methods to solve a set of difference equations may be classified into two classes: iterative and direct. Iterative methods can be used universally for any size of problem, but direct solution methods are feasible only when one of the following

---

** In the case of the Poisson or Laplace equation, the inequality can be provided only by a boundary condition. If there is a positive removal term q such as in Eq. (11.2.26), it also contributes to enhancing the inequality. The physical meaning of a strong diagonal dominance is interpreted as an exit, or removal, of the physical quantity (e.g., particles or heat) represented by the solution. Without an exit, the physical system has no steady state unless the source term is zero or the total of the source is zero. If an iterative method is applied to a system without an exit or diagonal dominance, the iterative method may not converge. Exceptions often occur in solving a Poisson equation that appears in computational fluid dynamics. The situation is that:

(a) The difference equations have no strong diagonal dominance, that is, no exit.

(b) However, if all the difference equations are added, the total of the inhomogeneous terms vanishes (the inhomogeneous terms have both positive and negative terms).

(c) The total of all the homogeneous terms also vanishes.

The Poisson equation with these properties has a solution, but it is not unique because the solution plus an arbitrary constant is also a solution. If an iterative solution method is applied to such a problem, the method converges, but the final value depends on the initial guess or the iteration parameters used. The solution can be added or subtracted by a constant.

conditions is satisfied:

(a) The number of grid points is very small.

(b) The coefficients of the difference equations have a special and simple form.

(c) A computer with a huge memory space (supercomputer) is available.

To explain the basic difficulties associated with direct solution of difference equations, we must recognize that the elements of the coefficient matrix representing the linear set of difference equations are all zero except along five diagonal lines, as illustrated even in a small set of equations such as those in Example 11.1 (see also Matrix M in Figure 11.11). If the Gauss elimination method is to be used, all the coefficients including zeroes must be stored in the computer memory, so the required memory space can easily exceed the available core memory. For example, the size of the matrix for 20×20 grid points becomes $400^2 = 160,000$.

However, an examination of the matrix reveals that the matrix has a band diagonal form. For example, the coefficient matrix for N times M grid has a band of $2N + 1$ elements wide. In applying the Gauss elimination, a program can be developed in such a way that only $2N + 1$ times M coefficients, including zeroes, are stored. In this case, the necessary amount of memory space is $(2N + 1)(MN)$ rather than $(NM)^2$.

If both the coefficients of the original PDE and the grid have some simple structure, the fast Fourier transform (FFT) [Nussbaumer] or the fast direct solution method (see Section 11.8) may be applicable. Most difference equation sets do not satisfy these conditions, however. On the other hand, iterative methods need to store only nonzero coefficients. Therefore, even a problem on a large grid can be solved iteratively with a minimal requirement of core memory.

Table 11.2 summarizes advantages and disadvantages of both direct solution methods and iterative methods.

SUMMARY OF THIS SECTION

(a) The difference equations for an elliptic PDE may be solved by either an iterative method or a direct method.

(b) Iterative methods are more popularly used than direct methods. They are simple and often the only ways to solve difference equations.

(c) Under certain conditions, direct solution methods can be very efficient.

11.4 SUCCESSIVE RELAXATION METHODS

Iterative methods are far more extensively used to solve elliptic difference equations than direct solution methods because they require much less core memory than the direct solution methods and are applicable to almost any difference equations that arise from elliptic PDEs. Iterative methods are versatile, can be implemented with

Table 11.2 Summary of the finite difference solution methods for elliptic partial differential equations

Solution Methods	Advantages	Disadvantages
Iterative Methods		
Relaxation methods such as Jacobi-iterative method and SOR	Simple programming. Complete theoretical proof. Iteration parameter easily optimized.	Convergence rate becomes slower as the size of the grid increases.
Extrapolated Jacobi-iterative method	Same efficiency as SOR on a scalar computer, but much faster on a supercomputer.	Same as above.
Alternating direction implicit method (ADI) (see Section 11.7)	Faster than SOR for large classes of problems.	Difficult to optimize acceleration parameters. More effort for programming than SOR. For some problems convergence is slower than SOR.
Direct Solutions		
Gauss elimination and LU decomposition	Fast for small and intermediate size problems. Robustness.	Not applicable to large problems because of long computing time and memory requirements.
Fast direct solution (FDS)	Noniterative. Faster than iterative methods.	Limitations in the geometry and the number of grid points.
Fast fourier transform (FFT)	Same as above.	Same as above.

relatively minor programming efforts, and are often the only ways to solve difference equations.

The successive relaxation methods include the Jacobi-iterative method, the Gauss-Seidel method, and the successive-over-relaxation (SOR) methods. Each of Jacobi-iterative, Gauss-Seidel, and SOR methods has two versions—the point relaxation version and the line relaxation version.

In this section, we concentrate our attention on the three iterative methods: the Jacobi-iterative method, the successive-over-relaxation (SOR) method, and an extrapolated Jacobi-iterative method (EJ). Line relaxation methods are not discussed here because of the limited space. Notice also that the gains with line relaxation methods over the point relaxation methods are not substantial except for rectangular geometries with a large aspect ratio.

The SOR methods, particularly point SOR, are well-known iterative methods. Although EJ is less known, it is almost as simple to program as SOR and has exactly the same computational efficiency on the scalar computers. However, on a supercomputer, its efficiency is several times greater than that of SOR.

11.4.1 Jacobi-Iterative (EJ) Method

From the efficiency viewpoint, the Jacobi-iterative method introduced in this section is impractical if used alone. Nonetheless, it is theoretically important when it is necessary to analyze SOR and EJ.

We first rewrite Eq. (11.2.28) to

$$a^C \phi_{i,j} = S_{i,j} - (a^L \phi_{i-1,j} + a^R \phi_{i+1,j} + a^B \phi_{i,j-1} + a^T \phi_{i,j+1}) \qquad (11.4.1)$$

The Jacobi-iterative method is derived from Eq. (11.4.1) by adding the iteration number t or $t - 1$ as superscript to each of ϕ and dividing through by a^C as

$$\phi_{i,j}^{(t)} = [S_{i,j} - (a^L \phi_{i-1,j}^{(t-1)} + a^R \phi_{i+1,j}^{(t-1)} + a^B \phi_{i,j-1}^{(t-1)} + a^T \phi_{i,j+1}^{(t-1)})]/a^C \qquad (11.4.2)$$

For the first iteration cycle $t = 1$, $\phi_{i,j}^{(0)}$ on the right side of Eq. (11.4.2) is an initial guess. In each iteration cycle, Eq. (11.4.2) is evaluated for all the grid points except those where fixed value boundary conditions are given. The scheme converges with any initial guess for the solution, although the closer the initial guess is to the exact solution, the faster is the convergence.

A closer examination of the Jacobi-iterative method reveals, however, that not all the grid points have to be computed in each iteration cycle. Only half the grid points should be swept (2-cyclic property).* Suppose the grid points are colored in an alternating and staggered manner with black and red. For example, if the point (i, j) is painted red, then its neighbors, $(i - 1, j)$, $(i + 1, j)$, $(i, j - 1)$, and $(i, j + 1)$, are painted black. It is easily seen that the computation for the red points for the iteration cycle t needs only the black points from the cycle $t - 1$, and vice versa. If the computations in iteration cycle t is performed for red points only and then the computation for iteration cycle $t + 1$ is done for black points only, all the points are updated in two iteration cycles. The iteration can be continued in this manner until the solution converges. This approach reduces the overall computational work by half.

The number of iteration steps necessary for an iterative method to converge is not known beforehand. Therefore, the iterative method is continued until a convergence test is satisfied. A few convergence tests are described next.

Iteration is terminated if

$$|\phi_{i,j}^{(t)} - \phi_{i,j}^{(t-2)}| < \varepsilon \qquad (11.4.3)$$

is satisfied at all the grid points swept, where ε is a prescribed criterion. In case the solutions for different points vary by orders of magnitude, the test of Eq. (11.4.3)

---

* If all the points are swept in every iteration cycle, two independent series of computations are simultaneously carried on, which converge to the same solution. The two series of computation do not exchange information in each iteration cycle. This property of the dual series of computation in Jacobi-iterative method is called *the 2-cyclic property* [Varga].

should be modified to

$$\left| 1 - \frac{\phi_{i,j}^{(t)}}{\phi_{i,j}^{(t-2)}} \right| < \varepsilon \tag{11.4.4}$$

Thus, very small and very large values of the solution are equally tested on the basis of relative changes per iteration cycle.

   Equation (11.4.3) and (11.4.4) both need a "if" statement in the "for" loop, which may not be desirable for certain computers. An alternative approach is to replace Eq. (11.4.4) by

$$\frac{\sum_{i,j} \left| 1 - \phi_{i,j}^{(t)}/\phi_{i,j}^{(t-2)} \right|}{\text{total number of points}} < \varepsilon$$

11.4.2 Successive-Over-Relaxation (SOR) Method

The SOR method is given by

$$\phi_{i,j}^{(t)} = \omega[S_{i,j} - (a^L \phi_{i-1,j}^{(t)} + a^R \phi_{i+1,j}^{(t-1)} + a^B \phi_{i,j-1}^{(t)}$$
$$+ a^T \phi_{i,j+1}^{(t-1)})]/a_C + (1 - \omega)\phi_{i,j}^{(t-1)} \tag{11.4.5}$$

where ω is the over-relaxation parameter and $1 < \omega < 2$.* As can be seen easily, the SOR method is derived by modifying the Jacobi-iterative method in two respects. First, the superscripts of $\phi_{i-1,j}$ and $\phi_{i,j-1}$ are changed from $t - 1$ to t. To make this possible, the grid points are swept in the increasing order in i and j, because, by doing so, $\phi_{i-1,j}^{(t)}$ and $\phi_{i,j-1}^{(t)}$ of the current iteration step become always available for calculation of $\phi_{i,j}^{(t)}$. Second, an acceleration parameter ω is introduced.

   In SOR all the grid points are swept in each iteration cycle.

   If $\omega = 1$, the method is called the Gauss-Seidel method. The optimum ω is in the low side of $1 < \omega < 2$ if the number of grid points is very small, but it approaches 2 if the number of grid points is increased. For small problems, the convergence rate is intrinsically fast and relatively insensitive to the value of ω, so ω may be set to 1.8 as a rule of thumb. As the number of grid points increases, the convergence rate becomes slow and sensitive to the value of ω selected. The effect of ω is discussed in more detail in Section 11.5. An implementation of the point-SOR is demonstrated with PROGRAM 11–1. If the iteration takes more than a few hundred iteration cycles, the calculation of an optimal ω described in Section 11.6 is strongly recommended.

---

   * Equation (11.4.5) is over-relaxation or under-relaxation depending on whether $1 < \omega < 2$ or $0 < \omega < 1$, respectively. Under-relaxation is useless for linear elliptic PEDs because the convergence rate becomes slower than over-relaxation. However, for a nonlinear equation where the coefficients of the difference equations are revised after each iteration using the previous iteratives, instability may occur unless under-relaxation is used.

Example 11.3

Solve the set of difference equations derived in Example 11.2 by SOR with $\omega = 1.5$.

⟨**Solution**⟩

The solution is obtained through the following steps:
(a) Move all the terms with negative signs to the right side.
(b) Divide each equation by the coefficient of the left side term:

$$\phi_{1,1} = \frac{1}{8}(2\phi_{1,2} + 2\phi_{2,1} + 4 + S)$$

$$\phi_{2,1} = \frac{1}{6}(\phi_{1,1} + 2\phi_{2,2} + 9 + S)$$

$$\phi_{1,2} = \frac{1}{6}(\phi_{1,1} + 2\phi_{2,2} + 7 + S) \qquad \text{(A)}$$

$$\phi_{2,2} = \frac{1}{4}(\phi_{1,2} + \phi_{2,1} + 12 + S)$$

(c) Denoting the iteration number by t, SOR is written as

$$\phi_{1,1}^{(t)} = \frac{1}{8}\omega(2\phi_{1,2}^{(t-1)} + 2\phi_{2,1}^{(t-1)} + 4 + S) + (1 - \omega)\phi_{1,1}^{(t-1)}$$

$$\phi_{2,1}^{(t)} = \frac{1}{6}\omega(\phi_{1,1}^{(t)} + 2\phi_{2,2}^{(t-1)} + 9 + S) + (1 - \omega)\phi_{2,1}^{(t-1)}$$

$$\phi_{1,2}^{(t)} = \frac{1}{6}\omega(\phi_{1,1}^{(t)} + 2\phi_{2,2}^{(t-1)} + 7 + S) + (1 - \omega)\phi_{1,2}^{(t-1)} \qquad \text{(B)}$$

$$\phi_{2,2}^{(t)} = \frac{1}{4}\omega(\phi_{1,2}^{(t)} + \phi_{2,1}^{(t)} + 12 + S) + (1 - \omega)\phi_{2,2}^{(t-1)}$$

Here, the equations are swept in the order of $\phi_{1,1}$, $\phi_{2,1}$, $\phi_{1,2}$ and $\phi_{2,2}$. After the new value $\phi_{i,j}^{(t)}$ is calculated, the old value $\phi_{i,j}^{(t-1)}$ is not necessary any more, so the new value is written over the old value using the same memory space. The initial guesses are all set to zero.
(d) The iteration of Eq. (B) is continued until all $\phi_{i,j}$ converge.

11.4.3 Extrapolated Jacobi-iterative (EJ) method

The point Jacobi-iterative method based on the 2-cyclic property, which allows sweeping of only half the grid points in each iteration step, is accelerated by introducing an extrapolation parameter as

$$\phi_{i,j}^{(t)} = \frac{\theta}{a^C} [S_{i,j} - (a^L \phi_{i-1,j}^{(t-1)} + a^R \phi_{i+1,j}^{(t-1)} + a^B \phi_{i,j-1}^{(t-1)}$$

$$+ a^T \phi_{i,j+1}^{(t-1)})] + (1 - \theta)\phi_{i,j}^{(t-2)} \quad (11.4.6)$$

where θ is the extrapolation parameter satisfying $1 < \theta < 2$. Notice that the second term on the right side causes no difficulty in implementation, because when only half the points are swept, the value of $\phi_{i,j}^{(t-2)}$ rather than $\phi_{i,j}^{(t-1)}$ is found in the memory space for point (i, j). The present method is known also as *red-black SOR* [Hageman/Young].

It is shown in Section 11.6 that the optimum θ is identical to the optimum ω of SOR for the same problem. The convergence rate is half that of SOR in terms of iteration times, but the total amount of computational work does not change because only half the points are swept in an iteration cycle.

On a scalar computer such as IBM 370, VAX, or IBM PC, the choice between SOR and EJ is just a matter of taste because both have the same simplicity and same efficiency. However, the real advantage of EJ is realized on the supercomputer with a vector processor.

In processing a DO loop, the vector processor vectorizes the computations if no result of computation is used within the same loop except in the same cycle. For example, a loop such as

```
for ( i = 1 ; i =< 10 ; i++)
{
    f[i] = 0.5*( f[i-1] + f[i+1] );
}
```

cannot be vectorized because f(i − 1) calculated in the same loop is reused, but the loop

```
for ( i = 1 ; i =< 10 ; i += 2)
{
    f[i] = 0.5*( f[i-1] + f[i+1] );
}
```

can be vectorized. If the loop is vectorized, the computational speed increases significantly (typically, a factor of approximately five). The innermost loop of the point-SOR cannot be vectorized, whereas that of EJ can be vectorized.

The EJ method is a variation of cyclic Chebyshev semi-iterative method [Varga]. In the cyclic Chebyshev semi-iterative method, θ for every iteration step is determined by the Chebyshev polynomial and not a constant. However, as the iteration number increases, it approaches an assymptotic value, which equals θ_{opt}. The overall convergence rate of the cyclic Chebyshev semi-iterative method is only modestly better than EJ—say, a 10% reduction of the total iteration steps.

(a) Among several iterative methods that belong to the relaxation methods, the point-SOR method and the extrapolated Jacobi-iterative (EJ) method are introduced.

(b) The line versions of the relaxation methods are only slightly more efficient than the point versions, except for rectangular geometries with a high aspect ratio.

(c) The Jacobi-iterative method using the 2-cyclic property has the same efficiency as the Gauss-Seidel method. When extrapolated, its efficiency becomes equivalent to that of SOR.

(d) On a supercomputer with a vector processor, EJ becomes significantly faster than SOR.

11.5 ANALYSIS OF CONVERGENCE

The objective of this section is to analyze the convergence rates of the Jacobi-iterative method, SOR, and the extrapolated Jacobi-iterative (EJ) method. Basic analysis of iterative convergence is important in practical application of the methods, particularly when computing efficiency is concerned.

For simplicity, the analysis is performed on a one-dimensional boundary value problem, but the result is universally valid in two and three dimensions. We first study the convergence of the Jacobi-iterative method because it is the basis for the analysis of SOR and EJ.

CONVERGENCE OF JACOBI-ITERATIVE METHOD.   We consider a one-dimensional boundary value problem given by

$$-\frac{d^2\phi}{dx^2} = S \tag{11.5.1}$$

$$\phi(0) = \phi(H) = 0$$

The difference equations for Eq. (11.5.1) are written as

$$-\phi_{i-1} + 2\phi_i - \phi_{i+1} = h^2 S, \quad i = 1, 2, \ldots, I \tag{11.5.2}$$

$$\phi_0 = \phi_{I+1} = 0$$

where $h = H/(I + 1)$ is the grid spacing. The Jacobi-iterative method for Eq. (11.5.2) is

$$\phi_i^{(t)} = \frac{1}{2}\left[h^2 S + \phi_{i-1}^{(t-1)} + \phi_{i+1}^{(t-1)}\right] \tag{11.5.3}$$

Denoting the exact solution of Eq. (11.5.2) by ϕ_i without the superscript, $\phi_i^{(t)}$ may be expressed by

$$\phi_i^{(t)} = \phi_i - e_i^{(t)} \tag{11.5.4}$$

where $e_i^{(t)}$ is the error. Then, Eq. (11.5.3) becomes

$$e_i^{(t)} = \frac{1}{2}[e_{i-1}^{(t-1)} + e_{i+1}^{(t-1)}] \tag{11.5.5}$$

An eigenvalue problem associated with Eq. (11.5.2) is written as

$$\eta_m \psi_{m,i} = \frac{1}{2}(\psi_{m,i-1} + \psi_{m,i+1})$$

$$\psi_{m,0} = \psi_{m,i+1} = 0 \tag{11.5.6}$$

where η_m is the mth eigenvalue, and $\psi_{m,i}$ is the corresponding eigenfunction. It is shown next that eigenvalues and eigenfunctions are

$$\psi_{m,i} = \sin(m\alpha i), \quad m = 1, 2, \ldots, I \tag{11.5.7a}$$

$$\eta_m = \cos(m\alpha), \quad m = 1, 2, \ldots, I \tag{11.5.7b}$$

where subscript m denotes the m-th solution and $\alpha = \pi/(I + 1)$. Introducing Eq. (11.5.7a) into the right side of Eq. (11.5.6) gives

$$\text{RHS} = \frac{1}{2}[\sin(m\alpha(i - 1)) + \sin(m\alpha(i + 1))] = \cos(m\alpha)\sin(m\alpha i) \tag{11.5.8}$$

where the addition theorem of the sine function is used.* The left side of Eq. (11.5.6) becomes the same when Eqs. (11.5.7a) and Eq. (11.5.7b) are both introduced. This proves that Eqs. (11.5.7a) and (11.5.7b) are eigenvalues and eigenfunctions.**

* Addition theorem of the sine function:

$$\sin(njB \pm nB) = \sin(njB)\cos(nB) \pm \cos(njB)\sin(nB)$$

** One may wonder if $\sin(m\alpha i)$ for other m than $1, 2, \ldots, I$ are also eigenfunctions. If $m = 0$ or $I + 1$, $\sin(m\alpha i)$ becomes zero for any i, so $\sin(m\alpha i)$ for these m's are trivial solutions. For all other values of $m < 0$ and $m > I + 1$, we can show that $\sin(m\alpha i)$ becomes equal to a constant times $\sin(m'\alpha i)$ where m' is an integer in $0 < m' < I + 1$. To prove this, we first recognize that any m can be written in the form $m = m' + n(I + 1)$ or $m = -m' + (n + 1)(I + 1)$ where $0 < m' < I + 1$ and n is an integer. We will use the former if n is even and the latter if n is odd. For n even, $\sin(m\alpha i)$ becomes

$$\sin(m\alpha i) = \sin(n\pi i + m'\alpha i) = \sin(m'\alpha i)$$

where $\alpha(I + 1) = \pi$ is used. For n odd, $\sin(m\alpha i)$ can be written as

$$\sin(m\alpha i) = \sin[(n + 1)\pi i - m'\alpha i] = -\sin(m'\alpha i)$$

Thus, $\sin(m\alpha i)$ for $m = 1$ through I are the only independent eigenfunctions.

The eigenfunctions of Eq. (11.5.7a) can be used to expand $e_i^{(t)}$. We first expand the initial error $e_i^{(0)}$ as

$$e_i^{(0)} = \sum_{m=1}^{I} A_m \psi_{m,i} \tag{11.5.9}$$

where A_m is a coefficient.

Setting $t = 1$ in Eq. (11.5.5) and substituting Eq. (11.5.9) to the right side of Eq. (11.5.5) yield

$$e_i^{(1)} = \sum_{m=1}^{I} A_m \eta_m \psi_{m,i} \tag{11.5.10}$$

where Eq. (11.5.6) is used. Equation (11.5.10) is the error of the iterative solution after the first iteration cycle. By repeating the substitution, the error after the tth iteration cycle becomes

$$e_i^{(t)} = \sum_{m=1}^{I} A_m (\eta_m)^t \psi_{m,i} \tag{11.5.11}$$

If $|\eta_m| < 1$ for all m, the error dies away as the iteration number t increases. The overall rate of error decay is governed by

$$\mu_J = \max_m |\eta_m| \tag{11.5.12}$$

which is the spectral radius of the Jacobi-iterative method. Because the maximum of $|\eta_m|$ occurs at both $m = 1$ and $m = I$ [see Eq. (11.5.7b)], the spectral radius equals

$$\mu_J = \cos(\alpha) = -\cos(I\alpha) \tag{11.5.13}$$

and can be approximated by

$$\mu_J \simeq 1 - \frac{1}{2}\alpha^2 = 1 - \frac{1}{2}\left(\frac{\pi}{I+1}\right)^2 \tag{11.5.14}$$

where $\alpha = \pi/(I + 1)$. The convergence rate of an iterative method is defined by

$$R = -\log_{10}\mu$$

For the Jacobi-iterative method of one-dimensional problem, the convergence rate is

$$R = -\log_{10}\mu_J \tag{11.5.15}$$

$$\simeq -\log_{10}\left[1 - \frac{1}{2}\left(\frac{\pi}{I+1}\right)^2\right] \simeq \frac{1}{2}\left(\frac{\pi}{I+1}\right)^2 \bigg/ \ln(10)$$

It is observed that the convergence rate of the Jacobi-iterative method is a function of only the number of grid points. As I increases, the convergence rate R approaches zero.

CONVERGENCE OF SOR.   The convergence rate analysis for SOR is more complicated than for the Jacobi-iterative method because the eigenvalues include complex conjugate pairs.

Considering the same one-dimensional problem as used for the Jacobi-iterative method, SOR is written as

$$\phi_i^{(t)} = \frac{\omega}{2} \left[h^2 S + \phi_{i-1}^{(t)} + \phi_{i+1}^{(t-1)} \right] + (1 - \omega)\phi_i^{(t-1)} \tag{11.5.16}$$

In terms of the error defined by Eq. (11.5.4), Eq. (11.5.16) becomes

$$e_i^{(t)} = \frac{\omega}{2} \left[e_{i-1}^{(t)} + e_{i+1}^{(t-1)} \right] + (1 - \omega)e_i^{(t-1)} \tag{11.5.17}$$

The eigenvalue problem associated with Eq. (11.5.17) is

$$\xi_m \left[v_{m,i} - \frac{\omega}{2} v_{m,i-1} \right] = \frac{\omega}{2} v_{m,i+1} + (1 - \omega)v_{m,i} \tag{11.5.18}$$

where ξ_m is the m-th eigenvalue, $v_{m,i}$ is the m-th eigenfunction and the boundary conditions are

$$v_{m,0} = v_{m,I+1} = 0$$

The error can be expanded in terms of the eigenfunctions of Eq. (11.5.18) as

$$e_i^{(t)} = \sum_{m=1}^{I} A_m (\xi_m)^t v_{m,i} \tag{11.5.19}$$

where A_m are expansion coefficients that depend on the initial error, $e_i^{(0)}$. Because the decay rate of error is determined by the largest value of $|\xi_m|$, we must evaluate ξ_m next.

We show first that the eigenfunctions of Eq. (11.5.18) are given by

$$v_{m,i} = \xi_m^{i/2} \psi_{m,i} \tag{11.5.20}$$

where $\psi_{m,i}$ is an eigenfunction of the Jacobi-iterative method. We rewrite, Eq. (11.5.18) as

$$\xi_m v_{m,i} = \frac{\omega}{2} (\xi_m v_{m,i-1} + v_{m,i+1}) + (1 - \omega)v_{m,i} \tag{11.5.21}$$

Substituting Eq. (11.5.20) and using Eq. (11.5.6), Eq. (11.5.21) becomes

$$\xi_m^{(i/2)+1}\,\psi_{m,i} = \omega\xi_m^{(i+1)/2}\eta_m\psi_{m,i} + (1-\omega)\xi_m^{i/2}\psi_{m,i} \tag{11.5.22}$$

Then, dividing Eq. (11.5.22) through by $\xi_m^{i/2}\psi_{m,i}$ yields

$$\xi_m = \omega\xi_m^{1/2}\eta_m + (1-\omega) \tag{11.5.23}$$

where $\eta_m = \cos(m\alpha)$ is the Jacobi eigenvalue given by Eq. (11.5.7b). Equation (11.5.23) is the characteristic equation that ξ_m must satisfy. Then, $v_{m,i}$ satisfies Eq. (11.5.18), that is, it is an eigenfunction.

Because Eq. (11.5.23) is a quadratic equation of $\xi_m^{1/2}$, its roots are

$$\xi_m^{1/2} = \frac{\omega\eta_m}{2} \pm \sqrt{\frac{\omega^2\eta_m^2}{4} + 1 - \omega} \tag{11.5.24}$$

By taking square, we obtain

$$\xi_m = \frac{\omega^2\eta_m^2}{2} + 1 - \omega \pm \omega\eta_m\sqrt{\frac{\omega^2\eta_m^2}{4} + 1 - \omega} \tag{11.5.25}$$

This equation relates the SOR eigenvalues to the corresponding Jacobi eigenvalues.

Equation (11.5.23) can be used to derive the relation between the spectral radius of the Jacobi-iterative method, $\mu_J = \max|\eta_m|$, and the spectral radius of SOR, $\mu_\omega = \max|\xi_m|$. As seen from Eq. (11.5.7b), the maximum value of η_m is real and positive. It can be shown that the maximum value of ξ_m is real and positive if $\omega \leqslant \omega_{\text{opt}}$, where ω_{opt} is the optimum value of ω (which is explained later). When both μ_J and μ_ω are real and positive, Eq. (11.5.23) can be written for spectral radii as

$$\mu_\omega = \omega\mu_\omega^{1/2}\mu_J + (1-\omega) \tag{11.5.25a}$$

On the other hand, if $\omega > \omega_{\text{opt}}$, Eq. (11.5.25a) does not hold because the SOR eigenvalue corresponding to $\mu_J = \max|\eta_m|$ becomes complex. Including both cases, however, the spectral radius of SOR is related to the Jacobi spectral radius by

$$\mu_\omega = \left|\frac{\omega^2\mu_J^2}{2} + 1 - \omega + \omega\mu_J\sqrt{\frac{\omega^2\mu_J^2}{4} + 1 - \omega}\right| \tag{11.5.26}$$

If $\omega > \omega_{\text{opt}}$, the inside of the absolute sign is a complex root of Eq. (11.5.25a). The absolute sign is necessary to compute the spectral radius. If $\omega \leqslant \omega_{\text{opt}}$, μ_ω equals the largest root of Eq. (11.5.25a), that is positive. Therefore, the absolute sign on the right side of Eq. (11.5.26) is not necessary.

To analyze the distribution of ξ_m given by Eq. (11.5.25) and the influence of ω on μ_ω, we first recall that η_m are all real and $-\mu_J \leqslant \eta_m \leqslant \mu_J$. We define $\overline{\omega}$ as the larger root of

$$\frac{1}{4}\overline{\omega}^2\mu_J^2 + 1 - \overline{\omega} = 0$$

Then, for any η_m satisfying $|\eta_m| < \mu_J$, we have

$$\frac{1}{4}\overline{\omega}^2\eta_m^2 + 1 - \overline{\omega} < 0$$

If $\omega = \overline{\omega}$, the square root term in Eq. (11.5.25) becomes imaginary except for $\eta_m = \pm\mu_J$, and it can be shown easily that $|\xi_m| = \overline{\omega} - 1$ for every m. Thus, if $\omega = \overline{\omega}$, then $\mu_\omega = \overline{\omega} - 1 \equiv \mu_{\overline{\omega}}$.

If $\omega > \overline{\omega}$ in Eq. (11.5.25), the square root term always becomes imaginary, and $|\xi_m| = \omega - 1$ for every m, so $\mu_\omega = \omega - 1$ becomes greater than $\overline{\omega} - 1$.

If $\omega < \overline{\omega}$, the square root term becomes imaginary only for those η_m that satisfy $\omega^2\eta_m^2/4 + 1 - \omega < 0$, but it becomes real for any other η_m including $\eta_m = \pm\mu_J$. The largest real value of Eq. (11.5.25), which equals μ_ω, occurs for $\eta_m = \mu_J$. This μ_ω becomes greater than $\mu_{\overline{\omega}}$.

Thus, the minimum of μ_ω possible equals $\mu_{\overline{\omega}}$, so $\overline{\omega}$ is called the optimum ω and denoted by ω_{opt}. These three cases of ω are pictorially illustrated in Figure 11.9.

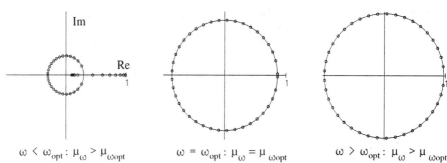

Figure 11.9 Distribution of SOR eigenvalues for three values of ω

The ω_{opt} satisfies, as already explained,

$$\frac{1}{4}(\omega_{opt})^2\mu_J^2 + 1 - \omega_{opt} = 0$$

or by solving it as a quadratic equation

$$\omega_{opt} = \frac{2}{1 + \sqrt{1 - \mu_J^2}} \tag{11.5.27}$$

With this optimum ω_{opt}, μ_ω becomes [see Eq. (11.5.26)]

$$\mu_{\omega\,opt} = \omega_{opt} - 1 = \frac{2}{1 + \sqrt{1 - \mu_j^2}} - 1$$

$$= \frac{1 - \sqrt{1 - \mu_j^2}}{1 + \sqrt{1 - \mu_j^2}} \tag{11.5.28}$$

Equations (11.5.26) through (11.5.28) are valid not only for the one-dimensional model but for two- and three-dimensional problems. They apply to the relation between the line Jacobi-iterative method and the line SOR also. The results of Example 11.4 show an important trend of the eigenvalues as well as the effect of ω on the convergence rate of SOR.

Example 11.4

Assuming $I = 20$ in Eqs. (11.5.2) and (11.5.3), evaluate all the eigenvalues of the Jacobi-iterative method. Next, evaluate ω_{opt} for SOR applied to the same problem using Eq. (11.5.28). Calculate all the SOR eigenvalues for these four values of ω: $\omega = 1.2$, 1.5, ω_{opt}, and 1.8. Plot them on the complex plane.

⟨Solution⟩

The Jacobi eigenvalues are given by Eq. (11.5.7b). The Jacobi eigenvalues η_M calculated for $\alpha = \pi/21 = 0.149559$ are listed in the second column of Table E11.4, which shows $\mu_J = 0.98883$. All the eigenvalues of the Jacobi-iterative method appear in pairs, each consisting of one positive eigenvalue and one negative eigenvalue with the same magnitude.

The ω_{opt} given by Eq. (11.5.28) becomes

$$\omega_{opt} = \frac{2}{1 + \sqrt{1 - 0.98883^2}} = 1.74057$$

Table E11.4 SOR Eigenvalues

m	η_m	$\omega = 1.2$	$\omega = 1.5$	$\omega = \omega_{opt} = 1.741$	$\omega = 1.8$
1	$+0.988$	0.966, 0.041	0.931, 0.268	0.741, 0.741	$0.784 \pm 0.159j$
2	$+0.956$	0.869, 0.046	0.695, 0.360	$0.642 \pm 0.368j$	$0.679 \pm 0.423j$
3	$+0.901$	0.713, 0.056	$0.413 \pm 0.281j$	$0.489 \pm 0.556j$	$0.515 \pm 0.612j$
4	$+0.826$	0.504, 0.079	$0.268 \pm 0.422j$	$0.293 \pm 0.680j$	$0.305 \pm 0.739j$
5	$+0.733$	$0.187 \pm 0.071j$	$0.105 \pm 0.489j$	$0.073 \pm 0.736j$	$0.071 \pm 0.797j$
6	$+0.623$	$0.080 \pm 0.183j$	$-0.063 \pm 0.496j$	$-0.152 \pm 0.725j$	$-0.170 \pm 0.782j$
7	$+0.500$	$-0.020 \pm 0.199j$	$-0.219 \pm 0.450j$	$-0.362 \pm 0.646j$	$-0.395 \pm 0.696j$
8	$+0.365$	$-0.104 \pm 0.171j$	$-0.350 \pm 0.357j$	$-0.538 \pm 0.508j$	$-0.584 \pm 0.547j$
9	$+0.223$	$-0.164 \pm 0.114j$	$-0.444 \pm 0.229j$	$-0.665 \pm 0.325j$	$-0.720 \pm 0.349j$
10	$+0.075$	$-0.196 \pm 0.040j$	$-0.494 \pm 0.079j$	$-0.732 \pm 0.112j$	$-0.791 \pm 0.120j$
	$\mu_J = 0.988$	$\mu_\omega = 0.966$	$\mu_\omega = 0.931$	$\mu_\omega = 0.741$	$\mu_\omega = 0.8$

The SOR eigenvalues for each of $\omega = 1.2, 1.5, \omega_{opt} = 1.74057$, and $\omega = 1.8$ are also shown in Table E11.4. Although each Jacobi eigenvalue yields two SOR eigenvalues, an examination of the table reveals that a pair of positive and negative eigenvalues of the Jacobi-iterative method produces the same pairs of SOR eigenvalues. So, the total number of SOR eigenvalues is identical to that of the Jacobi eigenvalues.

It is seen that the SOR eigenvalues for $\omega = 1.5$ include both complex and real eigenvalues. For $\omega = \omega_{opt}$ all the eigenvalues become complex except the two eigenvalues (corresponding to μ_J) that are real and double roots. For $\omega > \omega_{opt}$, all eigenvalues become complex.

A graph of the eigenvalues for three values of ω on the complex plane, similar to Figure 11.9, is shown in Figure E11.4. The spectral radius (SR) of each case is given beneath each graph (see also the underlined eigenvalues in Table 11.4 that correspond to the spectral radii). Notice that the complex values for each case lie on a circle of radius $\omega - 1$. The largest distances between an eigenvalue and the origin is the spectral radius. For $\omega = 1.5$, the largest real eigenvalue constitutes the spectral radius. On the other hand, the eigenvalues for $\omega > \omega_{opt}$ are on the circle of radius of $\omega - 1$, so that the spectral radius is equal to $\omega - 1$ and is independent of the Jacobi eigenvalues. When $\omega = \omega_{opt}$, a pair of double eigenvalues is on the real axis. The spectral radius for ω_{opt} equals $\omega_{opt} - 1$, which is smaller than for two other values of ω.

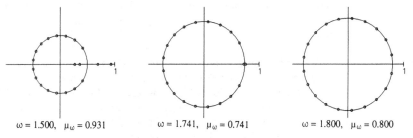

$\omega = 1.500, \quad \mu_\omega = 0.931$ $\omega = 1.741, \quad \mu_\omega = 0.741$ $\omega = 1.800, \quad \mu_\omega = 0.800$

Figure E11.4 Distribution of SOR eigenvalues for three values of omega

CONVERGENCE OF THE EXTRAPOLATED JACOBI-ITERATIVE METHOD. The EJ for the one-dimensional model is

$$\phi_i^{(t)} = \frac{\theta}{2}[h^2 S + \phi_{i-1}^{(t-1)} + \phi_{i+1}^{(t-1)}] + (1-\theta)\phi_i^{(t-2)} \tag{11.5.29}$$

In terms of the error defined by Eq. (11.5.4), Eq. (11.5.29) becomes

$$e_i^{(t)} = \frac{\theta}{2}[e_{i-1}^{(t-1)} + e_{i+1}^{(t-1)}] + (1-\theta)e_i^{(t-2)} \tag{11.5.30}$$

The error can be expanded in the eigenfunctions of the Jacobi-iterative method as

$$e_i^{(t)} = \sum_{m=1}^{I} A_m \zeta_m^t \psi_{m,i} \tag{11.5.31}$$

where $\psi_{m,i}$ is the eigenfunction of the Jacobi-iterative method given by Eq. (11.5.7) and ζ_m is the eigenvalue of the present method for the mth eigenfunction. (The EJ shares the same eigenfunctions, but its eigenvalues are different.)

By introducing Eq. (11.5.31) into Eq. (11.5.30), we find

$$\zeta_m^t = \theta \eta_m \zeta_m^{t-1} + (1 - \theta)\zeta_m^{t-2} \tag{11.5.32}$$

where η_m is the mth Jacobi eigenvalue and Eq. (11.5.6) is used to get the first term on the right side. Dividing by ζ_m^{t-2} further yields

$$\zeta_m^2 = \theta \eta_m \zeta_m + (1 - \theta) \tag{11.5.33}$$

Equation (11.5.33) is the characteristic equation that ζ_m has to satisfy, and it has exactly the same form as Eq. (11.5.23) except that $\xi_m^{1/2}$ is replaced by ζ_m. Solving Eq. (11.5.33) for ζ_m yields

$$\zeta_m = \frac{\theta \eta_m}{2} \pm \sqrt{\frac{\theta^2 \eta_m^2}{4} + 1 - \theta} \tag{11.5.34}$$

The objective now is to find θ that minimizes the spectral radius, which is defined by

$$\mu_\theta = \max_m |\zeta_m| \tag{11.5.35}$$

It can be shown by an analysis similar to the one for Eq. (11.5.26) that θ is optimum when the square root term in Eq. (11.5.34) becomes zero for $\eta_m = \mu_J$. Thus, θ_{opt} becomes

$$\theta_{opt} = \frac{2}{1 + \sqrt{1 - \mu_J^2}} \tag{11.5.36}$$

Notice that θ_{opt} is identical to ω_{opt} for SOR. With this θ_{opt}, the spectral radius of EJ becomes

$$\mu_{\theta_{opt}} = \frac{\mu_J}{1 + \sqrt{1 - \mu_J^2}} \tag{11.5.37}$$

We take square of Eq. (11.5.37) to get

$$(\mu_{\theta_{opt}})^2 = \frac{(\mu_J)^2}{(1 + \sqrt{1 - \mu_J^2})^2}$$

$$= \frac{1 - \sqrt{1 - \mu_J^2}}{1 + \sqrt{1 - \mu_J^2}} \qquad (11.5.38)$$

which becomes equal to $\mu_{\omega_{opt}}$ given by Eq. (11.5.28). Thus,

$$(\mu_{\theta_{opt}})^2 = \mu_{\omega_{opt}} \qquad (11.5.39)$$

This result indicates that the convergence rate of EJ with θ_{opt} is just half of that of SOR with ω_{opt}. However, the computational effort per iteration cycle of the former is half of the latter. Therefore, overall computational time for EJ is the same as that of SOR.

SUMMARY OF THIS SECTION

(a) The convergence rate of the Jacobi-iterative method is analyzed by using the eigenfunctions of the Jacobi-iterative method.

(b) The optimum SOR parameter is expressed in terms of the spectral radius of the Jacobi-iterative method.

(c) The optimal extrapolation parameter for EJ, θ_{opt}, equals that of the optimum SOR parameter, ω_{opt}.

(d) With the optimum parameter, the convergence rate of the EJ is half that of SOR in terms of iteration times. However, the amount of computational work of the former per iteration step is half of the latter. Therefore, the overall computational efficiency of EJ is identical to that of SOR.

(e) As explained earlier, however, EJ is several times faster on a supercomputer with a vector processor.

(f) In spite of a simple model used in this subsection, the equations for optimum parameters and the spectral radii of the three methods are unchanged for more general problems in two and three dimensions.

11.6 HOW TO OPTIMIZE ITERATION PARAMETERS

Despite the simple model used in the previous section, the equations derived there apply to point- and line-SOR in both two and three dimensions. The present section explains how to find the μ_J for a given problem and to optimize the convergence rates of SOR and EJ.

For any iterative scheme, the error of iterative solution after each iteration cycle may be expressed by a superposition of spatial modes having different decay rates:

$$e_{i,j}^{(t)} = \sum_{m=0}^{M} a_m (\gamma_m)^t u_{i,j}^{(m)} \tag{11.6.1}$$

where t is the iteration count, $u_{i,j}^{(m)}$ is the mth spatial mode (eigenfunction), γ_m is its amplitude factor (eigenvalue), and $M + 1$ is the total number of eigenvectors. The coefficients a_m depend on the initial guess. Equation (11.6.1), indeed, corresponds to Eq. (11.4.11) and Eq. (11.4.19) in one-dimensional cases. Here, we assume

$$|\gamma_0| = \max_m |\gamma_m| = \mu \tag{11.6.2}$$

and γ_0 is called the *dominant eigenvalue*. The $u_{i,j}^{(0)}$ is called the *dominant eigenfunction*.

All the terms will vanish as t increases if every γ_m satisfies $|\gamma_m| < 1$. Among the terms on the right side of Eq. (11.6.1), the smaller the value of $|\gamma_m|$, the faster the decay to zero. The term with the largest value of $|\gamma_m|$ is the slowest to vanish. In using SOR, we often find that the error of iterative solution decays very quickly in the earlier iteration cycles but gradually slows down, and finally the decay rate of error reaches a constant.* This phenomenon occurs when ω is less than optimum and is exactly what Eq. (11.6.1) indicates. The fast rate of error decay in the beginning is the contribution of small $|\gamma_m|$'s. As those terms die away, the error decay rate is finally governed by the dominant eigenvalue, γ_0. When ω is optimum, all $|\gamma_m|$'s of SOR become equal.

We should note a peculiar phenomenon that may occur when ω_{opt} is used, however. That is, when $\omega = \omega_{opt}$, the error may increase until a certain number of iteration cycles is reached. After then, the error will decrease at the optimum speed. The eigenvector deficiency, which occurs with $\omega = \omega_{opt}$, is responsible for this temporary increase of errors [Wachspress].

The convergence rate of an iterative scheme is given by

$$R = -\log_{10}(\mu) \tag{11.6.3}$$

The error decreases through N iteration steps by a factor,

$$\beta = (\mu)^N \tag{11.6.4}$$

If β is to be reduced to 10^{-5}, for example, the number of necessary iteration cycles becomes $N = 5/R$. Table 11.3 illustrates the number of iteration cycles required for $\beta = 10^{-5}$.

The Jacobi spectral radius μ_J increases as the number of grid points increases and is also influenced by the boundary conditions. For a given geometry with a fixed

---

* It is important to use a semi-log scale to plot iterative errors versus iteration times.

Table 11.3 Effect of μ on the
number of iteration steps to
reduce error by a factor 10^{-5}

μ	R	N
0.5	0.3	17
0.7	0.15	32
0.9	0.045	109
0.95	0.022	224
0.99	0.0043	1145
0.999	0.00043	11507

number of grid points, μ_J is smallest when all the boundary conditions are of the fixed-value type. As the portion of the boundaries with the derivative boundary conditions increases, μ_J increases also. The effect of the mixed boundary conditions is somewhere between the effects of both the fixed value and the derivative boundary conditions.

The effects of the number of grid points and the types of boundary conditions on the spectral radius of SOR are similar to those of the Jacobi-iterative method. However, the spectral radius of SOR is also significantly dependent on the iteration parameter ω. It is related to μ_J and ω as

$$\mu_\omega = \left| \frac{1}{2}\omega^2\mu_J^2 - \omega + 1 + \omega\mu_J\sqrt{\frac{1}{4}\omega^2\mu_J^2 - \omega + 1} \right| \tag{11.6.5}$$

In plotting μ_ω versus ω for Eq. (11.6.5), the following relations are helpful. For $\omega < \omega_{opt}$, the square root term in Eq. (11.6.5) is real and the whole inside the absolute sign becomes real and positive, so we have

$$\mu_\omega = \frac{1}{2}\omega^2\mu_J^2 + 1 - \omega + \omega\mu_J\sqrt{\frac{1}{4}\omega^2\mu_J^2 - \omega + 1} \tag{11.6.6}$$

For $\omega > \omega_{opt}$, the square root term in Eq. (11.6.6) becomes imaginary, and the spectral radius becomes

$$\mu_\omega = \omega - 1 \tag{11.6.7}$$

Figure 11.10 shows plots of Eq. (11.6.5) versus ω. It is observed that μ_ω is very sensitive to ω about $\omega = \omega_{opt}$. It decreases sharply when ω approaches from a lower side to ω_{opt}. This is why finding ω_{opt} is so important in maximizing the convergence rate of SOR.

OPTIMIZATION OF ω. The optimum value of ω for a given problem can be found relatively easily by a pilot run of SOR with a lower estimate of ω. When $\omega < \omega_{opt}$,

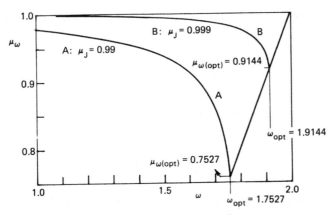

Figure 11.10 Effect of ω on the SOR spectral radius

the dominant eigenvalue of SOR becomes real and positive, (and equals μ_ω), as illustrated in Example 11.4. All other eigenvalues are smaller than the dominant eigenvalue in magnitude. Therefore, as the number of iteration cycles increases, Eq. (11.6.1) approaches

$$e_{i,j}^{(t)} = \sum_{m=0}^{M} a_m \gamma_m^t u_{i,j}^{(m)} \longrightarrow a_0 \mu_\omega^t u_{i,j}^{(0)} \qquad (11.6.8)$$

where $u_{i,j}^{(0)}$ and γ_0 are assumed to be the dominant eigenfunction and the corresponding eigenvalue.

The quantity defined by

$$N^{(t)} = \sum_{i,j} \left[\phi_{i,j}^{(t)} - \phi_{i,j}^{(t-1)} \right]^2 \qquad (11.6.9)$$

can be easily computed in each iteration sweep. Introducing

$$e_{i,j}^{(t)} = (\phi_{i,j})_{\text{exact}} - \phi_{i,j}^{(t)} \qquad (11.6.10)$$

into Eq. (11.6.9) yields

$$N^{(t)} = \sum_{i,j} \left[(e_{i,j}^{(t)} - e_{i,j}^{(t-1)}) \right]^2$$

$$\simeq \sum_{i,j} \left[a_0 \mu_\omega^{t-1} (\mu_\omega - 1) u_{i,j}^{(0)} \right]^2 \qquad (11.6.11)$$

where Eq. (11.6.8) is also used. The ratio of $N^{(t)}$ in two consecutive iteration cycles becomes

$$\frac{N^{(t)}}{N^{(t-1)}} \simeq \mu_\omega^2 \tag{11.6.12}$$

Thus, an estimate for μ_ω is given in every iteration by

$$\mu_\omega^{(t)} = \sqrt{\frac{N^{(t)}}{N^{(t-1)}}} \tag{11.6.13}$$

Once μ_ω is obtained, an estimate for μ_J can be computed by Eq. (11.5.25a), as follows:

$$\mu_J^{(t)} = \frac{\mu_\omega^{(t)} + \omega - 1}{\omega\sqrt{\mu_\omega^{(t)}}} \tag{11.6.14}$$

Furthermore, using Eq. (11.5.27), ω_{opt} is estimated by

$$\omega_{\text{opt}}^{(t)} = \frac{2}{1 + \sqrt{1 - (\mu_J^{(t)})^2}} \tag{11.6.15}$$

Accuracy of these estimates is poor in the beginning of iteration but improves as the number of iteration cycles increases. The algorithm of Eq. (11.6.9) through Eq. (11.6.15) may be incorporated in any program of the point or line SOR. The ω_{opt} may be estimated with a reasonable accuracy during the first small number of iteration steps (for example, at most 20). Then the iteration is continued with $\omega = \omega_{\text{opt}}$ until convergence. If the same equation is solved repeatedly with different source terms, the ω_{opt} found for first problem remains the same for all other problems (ω_{opt} is not affected by the source term). Such repetitive solution of the same elliptic problem is often encountered in nuclear reactor as well as in fluid dynamics calculations.

Implementation of the optimization algorithm is illustrated in PROGRAM 11–3.

ESTIMATION OF θ_{opt} FOR EXTRAPOLATED JACOBI-ITERATIVE METHOD (EJ). Since $\theta_{\text{opt}} = \omega_{\text{opt}}$, the algorithm to find ω_{opt} may be used. However, the algorithm described previously cannot be vectorized. So it is desirable to use an algorithm based on EJ itself.

The summation

$$N^{(t)} = \sum_{i,j} [(\phi_{i,j}^{(t)} - \phi_{i,j}^{(t-2)}]^2 \tag{11.6.16}$$

is computed for each odd iteration cycle t, where the summation is over only the half of the grid points for which actual calculations are performed. The ratio, $N^{(t)}/N^{(t-2)}$

converges to

$$\frac{N^{(t)}}{N^{(t-2)}} \longrightarrow \mu_\theta^4$$

Thus, an estimate for μ_θ is given in every iteration by

$$\mu_\theta^{(t)} = \left(\frac{N^{(t)}}{N^{(t-2)}}\right)^{1/4} \tag{11.6.17}$$

Using μ_θ, an estimate for μ_J is computed by

$$\mu_J^{(t)} = \frac{(\mu_\theta^{(t)})^2 + \theta - 1}{\theta \mu_\theta^{(t)}} \tag{11.6.18}$$

Then, θ_{opt} is estimated by

$$\theta_{opt}^{(t)} = \frac{2}{1 + \sqrt{1 - (\mu_J^{(t)})^2}} \tag{11.6.19}$$

which is exactly the same as Eq. (11.6.15).

An implementation of the algorithm just described is demonstrated in PROGRAM 11–4.

SUMMARY OF THIS SECTION

(a) An optimal SOR parameter, as well as that of EJ, is estimated by running both methods with a low estimate for the parameter.

(b) The optimization of ω or θ may be done in the beginning of the iterative solution, so that the remainder of the iteration is performed with the optimized parameter.

11.7 ALTERNATING DIRECTION IMPLICIT METHOD (ADI)

The ADI is an iterative solution method to solve a large set of difference equations for an elliptic partial differential equation. Although ADI requires more effort in developing a program, it is popular because the convergence rate is generally faster than SOR, particularly for large problems. In recent years, interest in ADI has been renewed because ADI is suitable for supercomputers with a vector processor. The advantages of ADI follow:

(a) Computational efficiency is much higher than SOR for a large class of problems.

(b) Boundary conditions are not so restrictive as for FDS or FFT.

(c) More suitable for vector processors of a supercomputer.

On the other hand, the disadvantages are as follows:

(a) Programming is more cumbersome than for SOR.

(b) Optimizing the iteration parameters is much more difficult than for SOR.

(c) Geometry is restricted to a rectangular grid.

The ADI explained here is based on the approximate factorization method and is closely related to the ADI explained in Chapter 12.

For simplicity of explanations, we consider the Poisson equation

$$(L_x + L_y)\phi(x, y) = S \tag{11.7.1}$$

where L_x and L_y are differential operators,

$$L_x = -\frac{\partial^2}{\partial x^2}, \quad L_y = -\frac{\partial^2}{\partial y^2} \tag{11.7.2}$$

The essence of ADI is separation of variables in the iterative operator, which can be explained before introducing difference approximations. Notice, however, that each of L_x and L_y will be replaced by the three-point central difference operators.

An iterative method for Eq. (11.7.1) may be written in the form

$$M\phi^{(t+1)} = M\phi^{(t)} + \theta[S - (L_x + L_y)\phi^{(t)}(x, y)] \tag{11.7.3}$$

where M is an operator that approximates $L_x + L_y$ but makes solution easy, $\phi^{(t)}$ is the t-th iterative, and θ is an extrapolation parameter that will be set to 2 in the remainder.

If we define

$$\delta\phi(x, y) = \phi^{(t+1)} - \phi^{(t)} \tag{11.7.4}$$

Eq. (11.7.3) can be rewritten as

$$M\delta\phi(x, y) = 2R^{(t)} \tag{11.7.5}$$

where R is the t-th residual defined by

$$R^{(t)} = [S - (L_x + L_y)\phi^{(t)}(x, y)]$$

The convergence rate and efficiency of an iterative method depends on the selection of the iterative operator M. We desire M to be easy to solve as well as to be a good approximation to $L_x + L_y$ as mentioned earlier. The M for the ADI is written as

$$M = \frac{1}{\omega}(\omega + L_x)(\omega + L_y) \tag{11.7.6}$$

where ω is an acceleration parameter and is varied in each iteration to optimize the convergence rate. With this form, the operators in x and in y are separated. By rewriting Eq. (11.7.6) to

$$M = \omega + L_x + L_y + \frac{1}{\omega} L_x L_y \qquad (11.7.7)$$

we can see that M consists of the true differential operator $(L_x + L_y)$ plus two additional terms. Therefore, M may be considered as an approximation to the true differential operator.

Introducing Eq. (11.7.6) into Eq. (11.7.5) yields

$$(\omega + L_x)(\omega + L_y)\delta\phi(x, y) = 2\omega R^{(t)} \qquad (11.7.8)$$

Because the differential operators in x and y are in the factorized form, Eq. (11.7.8) may be solved in two steps. The first step is to solve

$$(\omega + L_x)\psi(x, y) = 2\omega R^{(t)} \qquad (11.7.9)$$

and the second step is to solve

$$(\omega + L_y)\delta\phi(x, y) = \psi(x, y) \qquad (11.7.10)$$

Equation (11.7.9) for any y is a one-dimensional boundary value problem, so its difference approximation is solved by using the tridiagonal solution along each grid line in the x direction. Equation (11.7.10) is a one-dimensional boundary value problem in the y-direction.

The selection of the acceleration parameters ω is important for fast convergence. Finding the optimum acceleration parameters theoretically is possible only for simple ideal problems—that is, rectangular domains with constant coefficients, uniform grid spacings, and fixed value boundary conditions. For any of more complicated problems, one has to guess an optimal value from one's knowledge of the ideal cases. This is a drawback of ADI when compared to SOR. Nevertheless, ADI is useful as long as it converges faster than SOR, even with crude estimates of optimum acceleration parameters. For practical problems, we adopt the following approach.

For a maximal convergence rate, a set of K values for ω are selected, which are used sequentially in a cycle of K iteration steps. The value of K is typically 4 to 8. The set of K values are cyclically used until the iteration converges. They are as follows:

$$\omega_k = \alpha_0 \left(\frac{\alpha_1}{\alpha_0}\right)^{(2k-1)/2K}, \qquad k = 1, 2, 3, \ldots, K \qquad (11.7.3)$$

where

$$\alpha_0 = \frac{p\pi^2}{[\text{the larger of } V \text{ and } H]^2} \tag{11.7.4}$$

$$\alpha_1 = \frac{4p}{\left[\text{the smaller of } \dfrac{\Delta x}{H} \text{ and } \dfrac{\Delta y}{V}\right]^2}$$

Here, H and V are the horizontal and vertical lengths of the domain, assuming the boundary conditions are of the fixed value type. If the left or right boundary condition is of the derivative type, then H in Eq. (11.7.4) should be set to 2 times the actual length. If the boundary conditions for the left and right boundaries are both of the derivative type, we suggest setting the H to 4 times the actual size. The value of V is treated in a similar way.

SUMMARY OF THIS SECTION

(a) In ADI, solving the set of difference equations is reduced to two sets of one-dimensional boundary value problems.

(b) The difference equations along a grid line are solved by the tridiagonal solution scheme.

(c) ADI can be vectorized in two ways: first, by using a vectorized tridiagonal solution scheme [Kershaw]; second, by performing the tridiagonal solutions for all the lines in parallel (without vectorizing each tridiagonal scheme). The latter is more efficient than the former although it needs more memory space in the core.

11.8 DIRECT SOLUTION METHODS

11.8.1 Gauss Elimination Based on Band Structure

The coefficient matrix of difference approximation for an elliptic PDE has a penta-diagonal structure such as Matrix A illustrated in Figure 11.11. Both Gauss elimination and LU decomposition can be used to solve elliptic difference equations. However, the LU decomposition tends to be more efficient than Gauss elimination if the solution is to be repeated with the same coefficients but different inhomogeneous terms.

In applying the LU decomposition for such a matrix, it is important to recognize that the matrix is a band-diagonal matrix of half-width w, where w equals the number of grid points in the x direction plus one (assuming the same numbering system as in Example 11.1 is used). When decomposed, the L and U matrices also have band-diagonal structures as shown in Figure 11.11. Matrix L will be a lower triangular matrix that has nonzero elements only in the band structure. Matrix U will be an

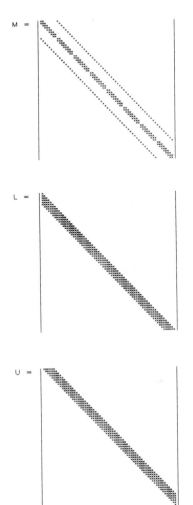

Figure 11.11

upper triangular matrix with nonzero elements only in the band. The LU decomposition and Gauss elimination for elliptic difference equations are essentially same as for ordinary matrices except that only the elements in the band are saved in the computer memory.

For elliptic difference equations, no pivoting is necessary because difference equations for PDE have diagonal dominances.

A band-diagonal matrix A can be stored in the memory space of $(2w - 1)N$, which is much smaller than N^2.

If the coefficient matrix is symmetric, then L and U can be chosen to satisfy $L = U^t$, or equivalently, $A = LL^t$. The Cholesky method [Strang] is based on this property and uses only approximately half the core space that is required to store both L and U. See Swarztrauber, Boisvert/Sweet for more details.

11.8.2 Fast Fourier Transform

The fast Fourier transform (FFT) is applicable to elliptic partial differential equations if the following conditions are satisfied:

(a) The coefficients of Eq. (11.3.1) do not change with j (or in the vertical direction). This is the case if the coefficients of the elliptic partial differential equation and the grid spacing are both constant in the vertical direction.

(b) The top and bottom boundary conditions are cyclic.

(c) The number of grid points in the vertical directions excluding the top and bottom boundaries is given by $J = 2^l$, where l is an integer.

The computational time for the FFT is shorter than for any iterative method, for the direct solution methods based on the Gauss elimination, and for LU decomposition for a large set of difference equations.

As stated earlier, there are freedoms in the grid spacing and variable coefficients in the horizontal direction. The boundary conditions for both left and right boundaries can take any form. Cylindrical coordinates may be considered also in the horizontal direction (or i direction). For more details of FFT, see Nussbaumer.

11.8.3 Fast Direct Solution (FDS) by Cyclic Reduction

The fast direct solution method, which is a variant of fast Fourier transform, is applicable to elliptic partial differential equations on the same conditions as the FFT except:

(a) The top and bottom boundary conditions are the fixed value type.

(b) The number of grid points in the vertical directions, excluding the top and bottom boundaries, is equal to $J = 2^l - 1$, where l is an integer. The admissible numbers for J are in the following table:

l	J
3	7
4	15
5	31
6	63
7	127
8	255
...	...

The computational time of FDS is comparable to that of FFT. It has the same freedom on the grid spacing in the i direction and left and the right boundary conditions. For more details of FDS, see Nakamura.

SUMMARY OF THIS SECTION. When certain conditions are satisfied, the Gauss elimination, the fast Fourier, and fast direct solution methods can provide very fast, robust, and efficient means of solving difference equations for an elliptic PDE.

PROGRAMS

PROGRAM 11–1 Successive-over-relaxation (SOR) Method

(A) Explanations

An elliptic partial differential equation is given by

$$-\nabla p(x, y)\nabla \phi(x, y) + q(x, y)\phi(x, y) = S(x, y) \tag{A}$$

for a rectangular geometry

$$0 < x < H, \quad 0 < y < V$$

with the boundary conditions

$$\frac{\partial \phi}{\partial n} = -\beta_{1,k}\phi + \beta_{2,k}$$

where $\beta_{1,k}$ and $\beta_{2,k}$ are constants; and k is 1, 2, 3, and 4 respectively for left, top, right, and bottom boundaries.

The difference approximation for Eq. (A) can be written as

$$a^C\phi_{i,j} + a^L\phi_{i-1,j} + a^R\phi_{i+1,j} + a^B\phi_{i,j-1} + a^T\phi_{i,j+1} = S_{i,j}$$

The SOR scheme is written as

$$\phi_{i,j}^{(t)} = \frac{\omega}{a^C}(S_{i,j} - a^L\phi_{i-1,j}^{(t)} - a^R\phi_{i+1,j}^{(t-1)} - a^B\phi_{i,j-1}^{(t)} - a^T\phi_{i,j+1}^{(t-1)}) + (1-\omega)\phi_{i,j}^{(t-1)} \tag{B}$$

where t is the iteration times and ω is a SOR parameter satisfying $1 < \omega < 2$.

In this particular program, the Poisson equation

$$-\nabla^2\phi(x, y) = e^{-0.05\sqrt{x^2+y^2}} \quad 0 < x < 5, \quad 0 < y < 5$$

is solved by SOR. The grid spacings are assumed to be variable.

The number of grid points in two directions, the iteration parameter, the total iteration steps allowed, and the convergence test criterion are specified in the beginning of the program. Mesh intervals are specified in the declaration statements for hx[] and hy[]. The boundary conditions specified in the declaration statements for b1[] and b2[] represent

$$\partial\phi/\partial n = -\phi + 1 \quad \text{for the left and bottom boundaries}$$

$$\partial\phi/\partial n = 0 \qquad\qquad \text{for the top and right boundaries}$$

(B) List

```
/*  CSL/c11-1.c       Successive-Over-Relaxation (SOR) */
#include <stdio.h>
#include <stdlib.h>
#include <math.h>
#define j_dim 20
/*          hx[i], hy[j]: grid spacings
            ni,nj: number of grid points in x and y directions
            s[i][j]: source term
            f[i][j]: solution, Greek"phi"
            w: SOR parameter
            ep: convergence criterion
            em: maximum error in each iteration
            x, y: coordinates
            ar[i][j]: aR for point i,j
            at[i][j]: aT for point i,j
            ac[i][j]: aC for point i,j
            itmax: iteration limit
            b1, b2: boundary conditions                        */
main()
{
long int i, itmax, j, k, ni, nj, _i, _r;
float ac[j_dim][j_dim], ar[j_dim][j_dim], at[j_dim][j_dim];
float bb, ec, em, ep, er, f[j_dim][j_dim], hb, hl, hr, ht, q, qq;
float s[j_dim][j_dim], sf[j_dim], w, wb, x, y;
void prnt();
static char car[9] = "AR       ";
static char cat[9] = "AT       ";
static char cac[9] = "AC       ";
static char soc[9] = "Source   ";
static char sol[9] = "Solution";
static char cfinal[9] = "         ";
static int _aini = 1;
static float hx[j_dim] = {0,1,1,1,1,1,0};
static float hy[j_dim] = {0,1,1,1,1,1,0};
static float b1[4]    = {1000,0,0,1000};
static float b2[4]    = {1000,0,0,1000};
    printf( "CSL/C11-1      Successive-Over-Relaxation (SOR)\n" );
    ni = 6;          /* Max. of i */
    nj = 6;          /* Max. of j */
    w = 1.7;         /* Omega, SOR parameter */
    wb = 1.0 - w;
    itmax = 200;     /* Max. iteration no. */
```

```
ep = .00001;        /* Convergence criterion */
y=0.0;
for( j = 0; j <= nj+1; j++ ){
   for( i = 0; i <= ni+1; i++ ){
      f[i][j] = 0; ac[i][j]=0; ar[i][j]=0; at[i][j]; s[i][j]=0;
   }
}
for( j = 1; j <= nj; j++ ){          /* Preparation of coefficients
   ht = hy[j];
   hb = hy[j - 1];
   y = y + hb;
   x = 0;
   for( i = 1; i <= ni; i++ ){
      ec = 0;
      q = 0;
      hl = hx[i - 1];
      hr = hx[i];
      x = x + hx[i - 1];
      s[i][j] = exp(-0.05*sqrt(x*x+y*y))*(hl+hr)*(ht+hb)/4;
      ar[i][j] = 0;
      if( hr > 0 )  ar[i][j] = (ht + hb)/hr/2;
      at[i][j] = 0;
      if( ht > 0 )  at[i][j] = (hl + hr)/ht/2;
      if( ((i == 1 || i == ni) || j == 1) || j == nj ){
         if( i == 1 ){  /* Left boundary condition */
            ec = ec + (hb + ht)*b1[0]/2;
            q = q + (hb + ht)*b2[0]/2;
         }
         if( i == ni ){  /*  Right boundary condition */
            ec = ec + (hb + ht)*b1[2]/2;
            q = q + (hb + ht)*b2[2]/2;
         }
         if( j == 1 ){     /* Bottom boundary condition */
            ec = ec + (hl + hr)*b1[3]/2;
            q = q + (hl + hr)*b2[3]/2;
         }
         if( j == nj ){  /* Top boundary condition */
            ec = ec + (hl + hr)*b1[1]/2;
            q = q + (hl + hr)*b2[1]/2;
         }
         s[i][j] = s[i][j] + q;
         f[i][j] = 0;
      }
      ac[i][j] = ar[i-1][j] + ar[i][j] + at[i][j-1] + at[i][j] + ec;
   }
}
prnt( car,ar, ni, nj);
prnt( cat,at, ni, nj);
prnt( cac,ac, ni, nj);
prnt( soc,s , ni, nj);
   /*---------------------------------------------* SOR starts * */
for( k = 1; k <= itmax; k++ ){
   em = 0;
   for( j = 1; j <= nj; j++ ) {     /*   Point-wise sweep starts */
      for( i = 1; i <= ni; i++ ){
         qq = s[i][j] + ar[i - 1][j]*f[i-1][j] + ar[i][j]*f[i+1][j];
         qq = qq + at[i][j]*f[i][j+1] + at[i][j - 1]*f[i][j-1];
         bb = f[i][j];
```

```
                f[i][j] = w*qq/ac[i][j] + wb*f[i][j];
                er = fabs( bb - f[i][j] );
                if( er > em )   em = er;  /* end of sweep */
            }
        }
        printf ("It. =%ld Error= %10.2e Omega= %10.2e \n", k, em, w );
        if( em < ep )  break;
        if(k == itmax)  printf( "Iteratioh limit exceeded.\n" );
    }                                   /* End of SOR. */
    printf( "\n" );
    printf( " Final solution (in order of increasing i for each j)\n");
    prnt( cfinal,f, ni, nj);
    exit(0);
}

void prnt(capt, f, ni, nj)
char *capt;
float f[][j_dim];
long int ni, nj;
{
long int i, j;
    printf (  "%s \n", capt );
    for( j = nj; j >= 1; j-- ){
        for( i = 1; i <= ni; i++ ){
            printf (  "%12.4e ", f[i][j] );
        }
        printf ( "\n" );
    }
    return;
}
```

(C) Sample Output

```
CSL/C11-1        Successive-Over-Relaxation (SOR)

AR
   5.0000e-01    5.0000e-01    5.0000e-01    5.0000e-01    5.0000e-01    0.0000e+00
   1.0000e+00    1.0000e+00    1.0000e+00    1.0000e+00    1.0000e+00    0.0000e+00
   1.0000e+00    1.0000e+00    1.0000e+00    1.0000e+00    1.0000e+00    0.0000e+00
   1.0000e+00    1.0000e+00    1.0000e+00    1.0000e+00    1.0000e+00    0.0000e+00
   1.0000e+00    1.0000e+00    1.0000e+00    1.0000e+00    1.0000e+00    0.0000e+00
   5.0000e-01    5.0000e-01    5.0000e-01    5.0000e-01    5.0000e-01    0.0000e+00
AT
   0.0000e+00    0.0000e+00    0.0000e+00    0.0000e+00    0.0000e+00    0.0000e+00
   5.0000e-01    1.0000e+00    1.0000e+00    1.0000e+00    1.0000e+00    5.0000e-01
   5.0000e-01    1.0000e+00    1.0000e+00    1.0000e+00    1.0000e+00    5.0000e-01
   5.0000e-01    1.0000e+00    1.0000e+00    1.0000e+00    1.0000e+00    5.0000e-01
   5.0000e-01    1.0000e+00    1.0000e+00    1.0000e+00    1.0000e+00    5.0000e-01
   5.0000e-01    1.0000e+00    1.0000e+00    1.0000e+00    1.0000e+00    5.0000e-01
AC
   5.0100e+02    2.0000e+00    2.0000e+00    2.0000e+00    2.0000e+00    1.0000e+00
   1.0020e+03    4.0000e+00    4.0000e+00    4.0000e+00    4.0000e+00    2.0000e+00
   1.0020e+03    4.0000e+00    4.0000e+00    4.0000e+00    4.0000e+00    2.0000e+00
   1.0020e+03    4.0000e+00    4.0000e+00    4.0000e+00    4.0000e+00    2.0000e+00
   1.0020e+03    4.0000e+00    4.0000e+00    4.0000e+00    4.0000e+00    2.0000e+00
   1.0010e+03    1.0020e+03    1.0020e+03    1.0020e+03    1.0020e+03    5.0100e+02
```

```
Source
   5.0019e+02    3.8748e-01    3.8197e-01    3.7355e-01    3.6302e-01    1.7555e-01
   1.0004e+03    8.1371e-01    7.9963e-01    7.7880e-01    7.5364e-01    3.6302e-01
   1.0004e+03    8.5375e-01    8.3504e-01    8.0886e-01    7.7880e-01    3.7355e-01
   1.0005e+03    8.9422e-01    8.6812e-01    8.3504e-01    7.9963e-01    3.8197e-01
   1.0005e+03    9.3173e-01    8.9422e-01    8.5375e-01    8.1371e-01    3.8748e-01
   1.0003e+03    1.0005e+03    1.0005e+03    1.0004e+03    1.0004e+03    5.0019e+02


It.  =  1 Error=    3.21e+00 Omega=    1.50e+00
It.  =  2 Error=    1.74e+00 Omega=    1.50e+00
It.  =  3 Error=    1.30e+00 Omega=    1.50e+00
It.  =  4 Error=    8.71e-01 Omega=    1.50e+00
It.  =  5 Error=    7.33e-01 Omega=    1.50e+00
It.  =  6 Error=    3.99e-01 Omega=    1.50e+00
It.  =  7 Error=    2.92e-01 Omega=    1.50e+00
It.  =  8 Error=    1.63e-01 Omega=    1.50e+00
It.  =  9 Error=    1.16e-01 Omega=    1.50e+00
It.  =10 Error=    6.86e-02 Omega=    1.50e+00
It.  =11 Error=    5.21e-02 Omega=    1.50e+00
It.  =12 Error=    2.92e-02 Omega=    1.50e+00
It.  =13 Error=    2.11e-02 Omega=    1.50e+00
It.  =14 Error=    1.28e-02 Omega=    1.50e+00
It.  =15 Error=    8.82e-03 Omega=    1.50e+00
It.  =16 Error=    5.51e-03 Omega=    1.50e+00
It.  =17 Error=    3.74e-03 Omega=    1.50e+00
It.  =18 Error=    2.34e-03 Omega=    1.50e+00
It.  =19 Error=    1.57e-03 Omega=    1.50e+00
It.  =20 Error=    1.01e-03 Omega=    1.50e+00
It.  =21 Error=    6.64e-04 Omega=    1.50e+00


Final solution   (in order of increasing i for each j)

   1.0027e+00 3.2888e+00 4.8938e+00 5.9312e+00 6.5006e+00 6.6763e+00
   1.0026e+00 3.2418e+00 4.7957e+00 5.7914e+00 6.3344e+00 6.5006e+00
   1.0025e+00 3.0669e+00 4.4563e+00 5.3260e+00 5.7914e+00 5.9312e+00
   1.0022e+00 2.7138e+00 3.8020e+00 4.4563e+00 4.7957e+00 4.8938e+00
   1.0016e+00 2.0905e+00 2.7138e+00 3.0669e+00 3.2418e+00 3.2888e+00
   1.0003e+00 1.0016e+00 1.0022e+00 1.0025e+00 1.0026e+00 1.0027e+00
```

(D) Discussions

The number of grid points for the problem is 6 × 6. The SOR parameter ω is set to 1.5 in the beginning of the program. The program first prints out a^T, a^R, a^C, and S for all the points; a^B and a^L are not printed out because these are the same as the a^T and a^R for neighbor points. The convergence criterion is set to 0.001 in the program, which is satisfied after 21 iteration steps.

QUIZ: If the source term of the Poisson equation is zero, the equation is called the Laplace equation. Among many other solution methods for the Laplace equations other than the finite difference method, an exotic approach is to use random walks [Nakamura 1986]. Suppose the boundary values for the Laplace equation are all specified (Dirichlet boundary condition). Then the solution for one particular point, say P, can be found as follows. A random walker (a particle moving randomly on the domain) is released from P. The random walks may be simulated on a computer. As the walker moves around the domain, it ultimately arrives at the boundary. When this occurs, find the value of the given boundary condition at that point and score it. If this is repeated many times, the average of the scores will converge to the solution at P. If the domain is circular and the values of the solution are all known along the circular boundary, what is the solution of the Laplace equation at the center? No computer is necessary to answer this question.

PROGRAM 11–2 Extrapolated Jacobi-iterative (EJ) Method

(A) Explanations

Consider Eqs. (A) and (B) written for PROGRAM 11–1. The extrapolated Jacobi scheme is written as

$$\phi_{i,j}^{(t)} = \frac{\theta}{a^C}\left(S_{i,j} - a^L\phi_{i-1,j}^{(t-1)} - a^R\phi_{i+1,j}^{(t-1)} - a^B\phi_{i,j-1}^{(t-1)} - a^T\phi_{i,j+1}^{(t-1)}\right) + (1-\theta)\phi_{i,j}^{(t-2)} \quad \text{(B)}$$

where t is the iteration counter and θ is an extrapolation parameter satisfying $1 < \theta < 2$. In the extrapolated Jacobi scheme, it is important to sweep only half of the grid points in each iteration cycle in accordance with the rules:

For odd t: Sweep only for odd i with odd j, and even i with even j.

For even t: Sweep only for odd i with even j, and even i with odd j.

Thus, painting the points swept in odd t and those in even t in different colors would make a checkerboard pattern.

For the same value of θ and ω, the theoretical convergence rate of EJ is exactly half that for SOR in terms of the iteration numbers. However, because the EJ skips half of the points per iteration cycle, the total CPU requirements for computations are the same. A program written for SOR can be changed to EJ with minor modifications.

However, the remarkable advantage of EJ is realized if it is used on a super-computer with a vector processor because EJ is vectorizable while SOR is not.

The program is essentially identical to PROGRAM 11–1. In the following list only a part of the program that has significant changes is shown. The differences are highlighted.

(B) List

```
/*  CSL/c11-2.c       Extrapolated-Jacobi (EJ) */
#include <stdio.h>
#include <stdlib.h>
#include <math.h>
#define j_dim 20
```

```
/*              hx[i], hy[j]: grid spacings
                ni,nj: number of grid points in x and y directions
                s[i][j]: source term
                f[i][j]: solution, Greek"phi"
                w: SOR parameter
                ep: convergence criterion
                em: maximum error in each iteration
                x, y: coordinates
                ar[i][j]: aR for point i,j
                at[i][j]: aT for point i,j
                ac[i][j]: aC for point i,j
                itmax: iteration limit                              */
                b1, b2: boundary conditions
main()
{
long int i, i1, itmax, j, k, ni, nj, _i, _r;
float ac[j_dim][j_dim], ar[j_dim][j_dim], at[j_dim][j_dim];
float bb, ec, em, ep, er, f[j_dim][j_dim], hb, hl, hr, ht, q, qq;
float  s[j_dim][j_dim], sf[j_dim], w, wb, x, y;
void prnt();
static char car[9] = "AR       ";
static char cat[9] = "AT       ";
static char cac[9] = "AC       ";
static char soc[9] = "Source   ";
static char sol[9] = "Solution";
static char cfinal[9] = "        ";
static int _aini = 1;
static float hx[j_dim] = {0,1,1,1,1,1,0};
static float hy[j_dim] = {0,1,1,1,1,1,0};
static float b1[4]  = {1000,0,0,1000};
static float b2[4]  = {1000,0,0,1000};
    printf( "CSL/C11-2       Extrapolated Jacobi Iteration\n" );
    ni = 6;           /* Max. of i */
    nj = 6;           /* Max. of j */
    w = 1.7;          /* Omega, SOR parameter */
    wb = 1.0 - w;
    itmax = 200;    /* Max. iteration no. */
    ep = .00001;      /* Convergence criterion */
    y=0.0;
    for( j = 0; j <= nj+1; j++ ){
       for( i = 0; i <= ni+1; i++ ){
          f[i][j]=0;
       }
    }
    for( j = 1; j <= nj; j++ ){                /* Preparation of coefficients   */
       ht = hy[j];
       hb = hy[j - 1];
       y = y + hb;
       x = 0;
          for( i = 1; i <= ni; i++ ){
             ec = 0;
             q = 0;
             hl = hx[i - 1];
             hr = hx[i];
             x = x + hx[i - 1];
             s[i][j] = exp(-0.05*sqrt(x*x+y*y))*(hl+hr)*(ht+hb)/4;
             ar[i][j] = 0;
             if( hr > 0 )  ar[i][j] = (ht + hb)/hr/2;
```

```
            at[i][j] = 0;
            if( ht > 0 )   at[i][j] = (hl + hr)/ht/2;
            if( ((i == 1 || i == ni) || j == 1) || j == nj ){
                if( i == 1 ){   /* Left boundary condition */
                    ec = ec + (hb + ht)*b1[0]/2;
                    q = q + (hb + ht)*b2[0]/2;
                }
                if( i == ni ){   /*  Right boundary condition */
                    ec = ec + (hb + ht)*b1[2]/2;
                    q = q + (hb + ht)*b2[2]/2;
                }
                if( j == 1 ){      /* Bottom boundary condition */
                    ec = ec + (hl + hr)*b1[3]/2;
                    q = q + (hl + hr)*b2[3]/2;
                }
                if( j == nj ){   /* Top boundary condition */
                    ec = ec + (hl + hr)*b1[1]/2;
                    q = q + (hl + hr)*b2[1]/2;
                }
                s[i][j] = s[i][j] + q;
                f[i][j] = 0;
            }
            ac[i][j] = ar[i-1][j] + ar[i][j] + at[i][j-1] + at[i][j] + ec;
        }
    }
    prnt( car,ar, ni, nj);
    prnt( cat,at, ni, nj);
    prnt( cac,ac, ni, nj);
    prnt( soc,s , ni, nj);
        /*-------------------------------* Extraporated Jacobi starts*/
    for( k = 1; k <= itmax; k++ ){
        em = 0;
        for( j = 1; j <= nj; j++ ) {      /*   Point-wise sweep starts */
            i1=1;
            if ( (k/2)*2 == k) i1=2;
            for( i = i1; i <= ni; i=i+2){
                qq = s[i][j] + ar[i-1][j]*f[i-1][j] + ar[i][j]*f[i+1][j];
                qq = qq + at[i][j]*f[i][j+1] + at[i][j-1]*f[i][j-1];
                bb = f[i][j];
                f[i][j] = w*qq/ac[i][j] + wb*f[i][j];
                er = fabs( bb - f[i][j] );
                if( er > em )   em = er;  /* end of sweep */
            }
        }
        if( (k/2)*2 == k)
                printf("It. =%ld Error= %10.2e Omega= %10.2e \n",k,em,w);
        if( em < ep )  break;
        if(k == itmax)  printf(  "Iteratioh limit exceeded.\n" );
    }                                    /* End of Iteration */
    printf( "\n" );
    printf( "Final solution (in order of increasing i for each j)\n");
    prnt( cfinal,f, ni, nj);
    exit(0);
}

void prnt(capt, f, ni, nj)
char *capt;
float f[][j_dim];
```

```
long int ni, nj;
{
long int i, j;
    printf (  "%s \n", capt );
    for( j = nj; j >= 1; j-- ){
        for( i = 1; i <= ni; i++ ){
            printf (  "%12.4e ", f[i][j] );
        }
        printf ( "\n" );
    }
    return;
}
```

PROGRAM 11-3 Optimization of SOR Parameter

(A) Explanations

This program demonstrates the algorithm to find the optimum SOR parameter. As a sample problem, the Poisson equation

$$-\nabla^2 \phi(x, y) = 1$$

is solved with 22×22 equispaced grid points including the points along the boundaries. The boundary conditions are assumed to be $\phi = 0$ along all sides.

(B) List

```
/* CSL/c11-3.c  Demonstration of Optimum Omega for SOR */
#include <stdio.h>
#include <stdlib.h>
#include <math.h>
#define TRUE 1
/*              mj: Jacobi eigenvalue
                ws: SOR eigenvalue
                w : SOR parameter used currently                    */
                wo: optimum omega
main()
{
int i, j, k, ni, nj;
float bb, bl, bs, f[41][41], fb, fs, s, sb, w, wb, wo;
double mj, ms;
    printf("\nCSL/C11-3   Demonstration of Optimum Omega for SOR\n" );
    ni = 20;     nj = 20;     s = 1;
    w = 1.3; /* 1.3 is selected as a low of omega. */
    wb = 1 - w;     k = 0;     bs = 1;
    printf( " Grid for this run is %d x %d \n", ni+2 , nj+2 );
    printf( "                           (including boundaries)\n" );
    printf( " A low estimate of Omega used = %g \n", w );
    for( i = 0; i <= (ni + 1); i++ ){
        for( j = 0; j <= (nj + 1); j++ ) f[j][i] = 0;
    }
    printf( "            Spectral Radii\n" );
    printf( " Itr.No.  Jacobi      SOR   Optimum-Omega\n" );
```

```
k = 0;                    /* Iteration count initialization */
while( TRUE ){
  k = k + 1;
  fs = 0;
  for( j = 1; j <= nj; j++ ){
    for( i = 1; i <= ni; i++ ){
      bb = f[j - 1][i];
      if( j == nj ) bb = bb*2;          /* BC for j=NJ */
      bl = f[j][i - 1];
      if( i == ni ) bl = bl*2;          /* BC for i=NI */
      fb = f[j][i];
      f[j][i] = w*(s +  bl + f[j][i+1] + bb
        + f[j+1][i])/4.0  + wb*f[j][i];   /* SOR: Eq.(11.4.5) */
      fs = fs + pow(f[j][i] - fb, 2.0);
    }
  }
  sb = ms;        ms = sqrt( fs/bs );        bs = fs;
  if( ms <= 1.0 ){
    mj = (ms + w - 1)/(w*sqrt( ms ));             /* Eq.(11.6.14) */
    if( mj < 1 ) wo = 2/(1 + sqrt( 1 - mj*mj ));{ /* Eq.(11.6.15) */
      printf( "%7d %10.6f %10.6f %10.6f \n", k, mj, ms, wo );
      if( fabs( sb - ms ) < 0.00001 ) exit(0);
    }
  }
}
}
```

(C) Sample Output

```
CSL/C11-3  Demonstration of Optimum Omega for SOR

Grid for this run is 22 x 22
                        (including boundaries)
A low estimate of Omega used = 1.3
          Spectral Radii
Itr.No.  Jacobi      SOR      Optimum-Omega
      2  0.988341  0.956754   1.735723
      3  0.991948  0.970122   1.775180
      4  0.993335  0.975265   1.793299
      5  0.994116  0.978161   1.804532
      6  0.994630  0.980069   1.812430
      7  0.995001  0.981446   1.818412
      8  0.995284  0.982493   1.823142
      9  0.995506  0.983317   1.826986
     10  0.995686  0.983984   1.830178
     11  0.995834  0.984533   1.832863
     12  0.995957  0.984992   1.835152
     13  0.996062  0.985379   1.837116
     14  0.996151  0.985711   1.838820
     15  0.996228  0.985996   1.840305
     16  0.996294  0.986243   1.841606
     17  0.996352  0.986458   1.842752
     18  0.996404  0.986648   1.843770
     19  0.996449  0.986815   1.844675
     20  0.996489  0.986963   1.845480
     21  0.996524  0.987096   1.846206
     22  0.996556  0.987214   1.846856
```

23	0.996585	0.987321	1.847446
24	0.996611	0.987416	1.847976
25	0.996634	0.987502	1.848458
26	0.996655	0.987581	1.848899
27	0.996674	0.987652	1.849297
28	0.996691	0.987716	1.849662
29	0.996707	0.987775	1.849995
30	0.996722	0.987830	1.850306

(D) Discussions

The foregoing output illustrates that the optimum ω can be found by running SOR with a low estimate for ω. The optimum ω can be estimated by this algorithm with a relatively small number of iterations. Therefore, the present method is useful if a very slowly convergent problem is to be run, or if the same problem has to be solved many times with different inhomogeneous terms. The author's study shows that if the estimate for ω is too low, the rate of the convergence for the optimum ω becomes slow.

PROGRAM 11-4 Optimization of EJ Parameter

(A) Explanations

The program is very similar to the preceding one except that it optimizes the iteration parameter for the extrapolated Jacobi-iterative method. See Section 11.6 for details. Note that the optimal parameters of SOR and EJ are identical for the same problem. Therefore, PROGRAMS 11-3 and 11-4 are interchangeable.

(B) List

```
/*  CSL/c11-4.c   Demonstration of Optimum Theta */
#include <stdio.h>
#include <stdlib.h>
#include <math.h>
#define TRUE 1
/*              mj: Jacobi eigenvalue
                mth2: square of EJ eigenvalue
                th : EJ parameter used currently                        */
                thopt: optimum theta
main()
{
int i, i1, j, k, ni, nj;
float bb, bl, bs, f[41][41], fb, fs, s, sb, th, thb, thopt;
double mj, mth2;
    printf( "\nCSL/C11-4  Demonstration of Optimum Theta \n" );
    ni = 20;     nj = 20;     s = 1;
    th = 1.3;                  /* 1.3 is an low estimate of theta. */
    thb = 1 - th;
    k = 0;      bs = 1;
    printf( " Grid for this run is %d x %d \n", ni + 2, nj + 2 );
    printf( "                          (Including boundaries)\n" );
```

```
printf( " Theta used = %g \n", th );
for( i = 0; i <= (ni + 1); i++ ){
   for( j = 0; j <= nj+1; j++) f[j][i] = 0; /*Iteratives initialized */
}
printf( "             Spectral radii\n" );
printf( " Itr.No.      Jacobi       EJ**2    Optimum-Theta\n" );
k = 0; /* Iteration count initialization */
while( TRUE ){
   k = k + 1;
   fs = 0;
   for( j = 1; j <= nj; j++ ){
      i1 = 1;
      if( (k + j)/2*2 == k + j )   i1 = 2;      /* I1=2 if K+J is even */
      for( i = i1; i <= ni; i += 2 ){
         bb = f[j - 1][i];
         if( j == nj )    bb = bb*2;                     /* BC for j=NJ */
         bl = f[j][i - 1];
         if( i == ni )    bl = bl*2;                     /* BC for i=NI */
         fb = f[j][i];
         f[j][i] = th*(s +  bl + f[j][i + 1] + bb + f[j + 1][i])/4
                       + thb*f[j][i];          /* EJ: Eq.(11.4.6) */
         fs = fs + pow(f[j][i] - fb, 2.0);
      }
   }
   if( k/2*2 == k ){                            /* Skip the next if k is odd */
      sb = mth2;
      mth2 = sqrt( fs/bs );
      bs = fs;
      if( mth2 <= 1.0 ){
         mj = (mth2 + th - 1)/(th*sqrt( mth2 ));     /*  Eq.(11.6.18) */
         if( mj < 1 ) thopt = 2/(1 + sqrt(1-mj*mj));;/*  Eq.(11.6.19) */
         if( k != 1 ){
            printf("%7d %10.6f %10.6f %10.6f \n", k,mj,mth2,thopt);
            if( fabs( sb - mth2 ) < 0.00001 ) exit(0);
                                        /*Convergence test passed.*/
         }
      }
   }
}
}
```

(C) Sample Output

```
CSL/C11-4  Demonstration of Optimum Theta

Grid for this run is 22 x 22
                        (Including boundaries)
   Theta used = 1.3
            Spectral radii
    Itr.No.    Jacobi      EJ**2     Optimum-Theta
        6    0.995352    0.982748    1.824319
        8    0.993104    0.974410    1.790137
       10    0.993751    0.976807    1.799173
       12    0.994359    0.979062    1.808205
       14    0.994799    0.980696    1.815121
       16    0.995129    0.981920    1.820534
```

18	0.995386	0.982873	1.824902
20	0.995593	0.983641	1.828527
22	0.995764	0.984275	1.831594
24	0.995908	0.984808	1.834230
26	0.996030	0.985263	1.836522
28	0.996136	0.985655	1.838531
30	0.996228	0.985997	1.840310
32	0.996309	0.986295	1.841884
34	0.996380	0.986559	1.843292
36	0.996442	0.986792	1.844545
38	0.996498	0.986997	1.845665
40	0.996547	0.987180	1.846668
42	0.996591	0.987343	1.847571
44	0.996630	0.987488	1.848377
46	0.996664	0.987616	1.849098
48	0.996696	0.987732	1.849750
50	0.996723	0.987834	1.850329
52	0.996748	0.987926	1.850852
54	0.996770	0.988007	1.851317
56	0.996789	0.988080	1.851734
58	0.996807	0.988145	1.852111
60	0.996822	0.988202	1.852440

(D) Discussions

The foregoing output illustrates that the optimum θ can be found by running EJ with a low estimate for θ. The present program is run for 22×22 points including the points along the boundaries. All the comments given in the discussion on PROGRAM 11–3 apply to this program. By setting θ to 1.75, the convergence of θ to the optimum parameter will become faster.

PROBLEMS

(11.1) The Laplace equation

$$\nabla^2 \phi(x, y) = 0$$

is given for the geometry shown in Figure P11.1. Using the boundary conditions specified there, derive difference equations and solve them by SOR with $\omega = 1.3$.

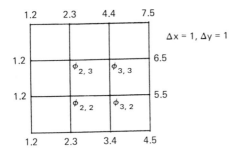

Figure P11.1

(11.2) Consider the Laplace equation for a plane geometry:

$$\nabla^2 \phi(x, y) = 0$$

(a) Write difference equations for the geometry shown in Figure P11.2. The boundary condition for the right side is given by $\partial\phi/\partial x = 0$. The boundary conditions for the other sides are of the fixed type, and the values are shown in Figure P11.2.

(b) Solve the difference equations by SOR with $\omega = 1.3$.

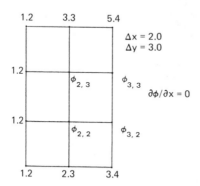

Figure. P11.2

(11.3) Rewrite Eq. (L) in Example 11.2 in a matrix form and show that it has the properties explained in the note of Example 11.1. (*Hint:* Multiply the first equation by 1/4, and the second and third equations by 1/2.)

(11.4) In a two-dimensional domain shown in Figure P11.4, the heat conduction equation is given by

$$-k\nabla^2 \phi = 0$$

The temperatures of the top, right, and bottom boundaries are printed in Figure P11.4. The left boundary is subject to convection heat transfer and the boundary condition is given by

$$k\frac{\partial T}{\partial x} = h_c(T(x, y) - T_\infty)$$

where k is the thermal conductivity of the medium (25 W/mK), h_c is the heat transfer coefficient (250 W/m²K), and T_∞ is the temperature of the fluid left to the boundary (10° C).

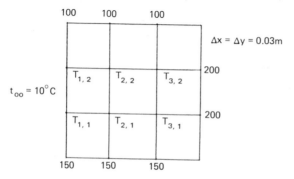

Figure P11.4

(a) Derive the difference equations for the six unknown temperatures.

(b) Solve the difference equations by SOR.

(11.5) An elliptic partial differential equation is given by

$$-\left(\frac{\partial^2}{\partial x^2} + \frac{\partial^2}{\partial y^2}\right)\phi + 0.1\phi = 0$$

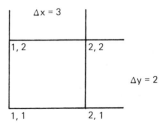

Figure P11.5

with the boundary conditions

$$\frac{\partial\phi}{\partial n} = -\phi \qquad \text{(left)}$$

$$\frac{\partial\phi}{\partial n} = -2\phi + 0.5 \quad \text{(bottom)}$$

Derive the difference equation for the corner point shown in Figure P11.5.

(11.6) Derive the difference equation for the Laplace equation in the geometry shown in Figure P11.6. Solve the difference equations by SOR with $\omega = 1.3$.

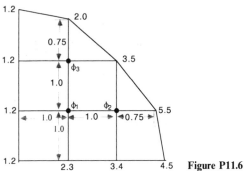

Figure P11.6

(11.7) Using PROGRAM 11–1, determine the solution of the Laplace equation $\nabla^2\phi(x, y) = 0$ for the geometry shown in Figure P11.7. Use 12 grid intervals in the x direction and 18 grid intervals in the y direction.

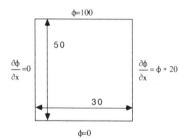

Figure P11.7

(11.8) A rectangular sheet of carbon film has the cathode and anode as shown in Fig. P11.8. The voltage of the cathode is fixed at 5 v and that of the anode is 0 v. However, because the film is not uniform, the voltage distribution between the cathode and anode is not linear. The electric potential (voltage) distribution is the solution of

$$\nabla k(x, y)\nabla E(x, y) = 0$$

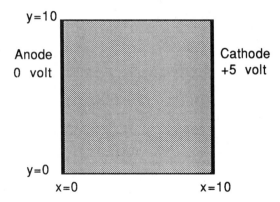

Figure P11.8

where $k(x, y)$ is the film conductivity, ohm^{-1}; E is the electric potential; and x and y are in cm. The boundary conditions are

$$\frac{\partial E}{\partial y} = 0 \qquad \text{at top and bottom sides}$$

$$E = 0 \text{ v} \qquad \text{along the left side}$$

$$E = 5 \text{ v} \qquad \text{along the right side}$$

The electric conductivity k is given by

$$k(x, y) = 1 + \frac{1}{20}[(x - 3)^2 + (y - 3)^2] \, ohm^{-1}$$

Derive difference equations using $\Delta x = \Delta y = 1$ cm, and solve the difference equations by SOR with $\omega = 1.7$. What is the value of $E(5, 5)$?

(11.9) Consider the Laplace equation on the r-z coordinates:

$$\nabla^2 \phi(r, z) = 0$$

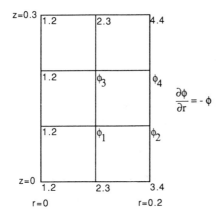

Figure P11.9

(a) Write difference equations for the Laplace equation for the geometry shown in Figure P11.9. The boundary condition for the right side is given by

$$\frac{\partial \phi}{\partial r} = -\phi$$

The values of ϕ for other boundaries are shown in the figure.

(b) Solve the difference equations by the Gauss-Seidel method (or equivalently, SOR with $\omega = 1$).

(11.10) Rewrite the difference equations in Example 11.1 assuming that the coordinates are $r - z$ coordinates. Assume that the left side is the center line of the cylinder. Show also that the properties of the difference equations described in the notes of Example 11.1 apply to the $r - z$ coordinates also.

(11.11) Modify the iterative method in PROGRAM 11–1 to the extrapolated Jacobi-iterative method, and find how many iteration steps are necessary with $\theta = 1.5$ to satisfy the same convergence criterion.

(11.12) Write a program to solve

$$\phi_i = \frac{1}{2}(\phi_{i-1} + \phi_{i+1}) + 1, \quad i = 1, 2, \ldots, N$$

$$\phi_0 = \phi_{N+1} = 0$$

by SOR. Set the initial guess $\phi = 0$ for all the grid points. Stop the iteration when $|\phi_i^{(t)} - \phi_i^{(t-1)}| < 0.001$ is satisfied at all the grid points. When the program is completed, run the program for $N = 30$ with various values of ω in $1 < \omega < 2$ and find the total iteration steps to satisfy this test. Plot the number of iteration steps versus ω.

(11.13) A one-dimensional difference equation is given by

$$-\phi_{i-1} + 2\phi_i - \phi_{i+1} = 1$$

with the boundary conditions, $\phi_0 = \phi_{100} = 0$. If SOR is used to solve the foregoing equation with $\omega = 1.3$ that is less than ω_{opt}, what would be the spatial distribution of the error after a sufficiently larger number of iteration steps? At what rate does this error decay through one iteration cycle?

(11.14) Consider the difference equations given by

$$-\phi_{i-1} + 2\phi_i - \phi_{i+1} = 1$$
$$i = 1, 2, \ldots, 20$$

with boundary conditions, $\phi_0 = \phi_{21} = 0$.

(a) Show that the ω_{opt} for the SOR applied to the equation above is 1.7406.

(b) Plot all the eigenvalues of the SOR on the complex plane for the following three cases: $\omega = 1.7$, $\omega = \omega_{opt}$, and $\omega = 1.8$.

(c) Calculate the spectral radius for each ω in (b).

(11.15) Repeat the previous problem with the extrapolated Jacobi-iterative method.

(11.16) In a typical boiling water reactor (BWR), the power distribution is controlled by cruciform control rods that absorb neutrons. The neutron flux distribution in the neighborhood of a control rod may be calculated by solving the equation

$$-D\nabla^2\psi(x, y) + \Sigma_a\psi(x, y) = S$$

in the geometry shown in Figure P11.16. In the foregoing equation, D and Σ_a and S are neutron flux, absorption cross section and neutron source, respectively. Assuming $D = 0.2 \text{ cm}^2$, $\Sigma_a = 0.1 \text{ cm}^{-1}$, and $S = 1 \text{ neutron sec}^{-1} \text{ cm}^{-3}$, solve the difference equation for the foregoing equation with the boundary conditions given by

$$\frac{\partial\psi}{\partial n} = 0 \quad \text{along all the boundaries except on the control rod}$$

$$\psi = 0 \quad \text{along the surface of the control rod}$$

Assume that the thickness of the control rod is zero and the pitch of the control rods is 24 cm.

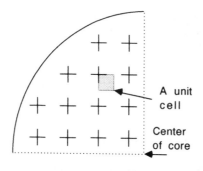

Array of control rods in a
boiling water nuclear reactor
(a quadrant of a core is shown)

Figure P11.16

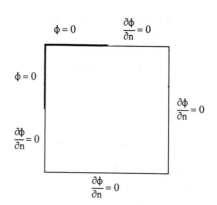

A unit cell around a control rod

(**11.17**) Repeat the previous problem using the boundary condition along the control rod given by

$$\frac{\partial \psi}{\partial n} = -\psi$$

(**11.18**) Equation (11.5.16) becomes a Gauss-Seidel method if $\omega = 1$. Show that half of the eigenvalues of the Gauss-Seidel method are zero, while all other eigenvalues are real and positive.

(**11.19**) Incorporate the automatic algorithm to find the optimum ω into PROGRAM 11–1.

(**11.20**) Incorporate the automatic algorithm to find the optimum ω into PROGRAM 11–2, and run the program for

$$\left(\frac{\delta^2}{\delta x^2} + \frac{\delta^2}{\delta y^2}\right)\phi_{i,j} = 0$$

with boundary conditions,

$$\phi_{0,j} = \phi_{i,0} = 1$$
$$\phi_{30,j} = \phi_{i,30} = 0$$

(**a**) What is the solution for $\phi_{15,15}$?

(**b**) What is θ_{opt}?

(**11.21**) The eigenvalue problem associated with the point Jacobi-iterative method is written as

$$4\eta u_{i,j} = u_{i-1,j} + u_{i+1,j} + u_{i,j-1} + u_{i,j+1} \qquad \text{(A)}$$

where η is an eigenvalue. If the boundary conditions are

$$u_{0,j} = u_{I+1,j} = u_{i,0} = u_{i,J+1} = 0$$

then eigenfunctions are given by

$$u_{i,j}^{(m,n)} = \sin(miA)\sin(njB)$$

where m and n are integers satisfying $0 < m < I + 1$, and $0 < n < J + 1$; $A = \pi/(I + 1)$ and $B = \pi/(J + 1)$.

(**a**) Show that the eigenvalues are

$$\eta^{(m,n)} = \frac{1}{2}(\cos(mA) + \cos(nB))$$

(**b**) Show that the maximum eigenvalue occurs when $m = n = 1$, and that the minimum occurs at $m = I$ and $n = J$.

(**c**) Show that the absolute value of the minimum η is equal to the maximum η.

(**11.22**) Show that $\eta^{(1,1)}$ of the previous problem may be approximated by

$$\eta^{(1,1)} = 1 - \frac{1}{4}(A^2 + B^2)$$

(11.23) Consider the difference equation

$$\frac{-\phi_{i-1,j} + 2\phi_{i,j} - \phi_{i+1,j}}{h_x^2} + \frac{-\phi_{i,j-1} + 2\phi_{i,j} - \phi_{i,j+1}}{h_y^2} = S$$

with boundary conditions

$$\phi_{0,j} = \phi_{I+1,j} = 0$$
$$\phi_{i,0} = \phi_{i,J+1} = 0$$

(a) Write the eigenvalue problem associated with the Jacobi-iterative method.

(b) Write the eigenfunctions and corresponding eigenvalues.

(c) Show that the spectral radius of the Jacobi-iterative method is

$$\mu = \frac{1}{2}[\cos(\alpha) + \cos(\beta)]$$

where $\alpha = \pi/(I+1)$ and $\beta = \pi/(J+1)$

(11.24) The eigenvalue problem associated with SOR is

$$\xi[a^C v_{i,j} + \omega(a^L v_{i-1,j} + a^B v_{i,j-1})] = -[a^R v_{i+1,j} + a^T v_{i,j+1}] + (1-\omega)a^C v_{i,j} \qquad \text{(A)}$$

The SOR eigenfunction is related to the associated Jacobi eigenfunction by

$$v_{i,j} = \xi^{(i+j)/2} u_{i,j} \qquad \text{(B)}$$

where $u_{i,j}$ is the Jacobi eigenfunction. Introducing Eq. (B) into Eq. (A) yields

$$\xi a^C u_{i,j} = -\omega\xi^{1/2}[a^L u_{i-1,j} + a^R u_{i+1,j}$$
$$+ a^B u_{i,j-1} + a^T u_{i,j+1}] + (1-\omega)a^C u_{i,j}$$

Show that the Jacobi eigenvalue and the SOR eigenvalue are related by

$$\xi - \omega\xi^{1/2}\eta + (\omega - 1) = 0$$

where η is the Jacobi eigenvalue.

(11.25) Prove that the spectral radius of SOR becomes Eq. (11.6.6) if $\omega < \omega_{opt}$, and becomes Eq. (11.6.7) if $\omega > \omega_{opt}$.

(11.26) After a certain number of SOR iteration cycles with $\omega = 1.3$, the following statistics of the iteratives was found:

$$N^{(t)} = 50.32$$
$$N^{(t+1)} = 49.31$$
$$N^{(t+2)} = 48.33$$

where

$$N^{(t)} = \sum_{i,j}[\phi_{i,j}^{(t)} - \phi_{i,j}^{(t-1)}]^2$$

Estimate:

(a) SOR spectral radius with $\omega = 1.3$, μ_ω.

(b) Jacobi spectral radius, μ_J.

(c) Optimum ω, ω_{opt}.

REFERENCES

Becker, E. B., G. F. Carey, and J. T. Oden, *Finite Elements: an Introduction*, Prentice-Hall, 1981.

Boisvert, R. F., and R. A. Sweet, "*Mathematical Software for Elliptic Boundary Value Problems*," in *Sources and Development of Mathematical Software* (Colwel, ed.), Prentice-Hall, 1984.

Hageman, L. A. and D. M. Young, *Applied Iterative Methods*, Academic Press, 1981.

Kershaw, D., "Solution of single tridiagonal linear systems and vectorization of the ICCG algorithm on the Cray-1," in *Parallel Computations* (G. Rodrigue, ed.), Academic Press, 1982.

Myint-U, U., and L. Debnath, *Partial Differential Equations for Scientists and Engineers*, North-Holland, 1987.

Nakamura, S, *Computational Methods in Engineering and Science with Application to Fluid Dynamics and Nuclear Systems*, Krieger, 1986.

Nogotov, E. F., *Applications of Numerical Heat Transfer*, Hemisphere, 1978.

Norrie, D. H., and G. Devies, *An Introduction to Finite Element Analysis*, Academic Press, 1978.

Nussbaumer, H. J., *Fast Fourier Transform and Convolution Algorithms*, 2nd ed., Springer-Verlag, 1982.

Segerlind, L. J., *Applied Finite Element Analysis*, Wiley, 1976.

Smith, G. D., *Numerical Solution of Partial Differential Equations: Finite Difference Methods*, 2nd ed., Oxford University Press, 1978.

Strang, G. W., *Linear Algebra and Its Application*, 2nd ed., Academic Press, 1980.

Swarztrauber, P. N., "The methods of cyclic reduction, Fourier analysis and FACR algorithm for discrete solution of Poisson's equation on a rectangle," *SIAM Rev.*, Vol 19, 490–501, 1977.

Sweet, R. A., "A cyclic reduction algorith for solving block tridiagonal systems of arbitrary dimension," SIAM *J. Numer. Anal.*, Vol 14, 706–720, 1977.

Thompson, J. F., ed., *Numerical Grid Generation*, North-Holland, 1982.

Thompson, J. F., Z. U. A. Wasri, and C. W. Mastin, *Numerical Grid Generation: Foundation and Applications*, North-Holland, 1985.

Varga, R. S., *Matrix iterative analysis*, Prentice-Hall, 1962.

Wachspress, E. L., *Iterative Solution of Elliptic Systems*, Prentice-Hall, 1966.

12

Parabolic Partial Differential Equations

12.1 INTRODUCTION

Governing equations for transient particle diffusion, or heat conduction, are partial differential equations (PDE) of the parabolic type. Therefore, the numerical solution methods for the parabolic PDEs are important in such fields as molecular diffusion, heat transfer, nuclear reactor analysis, and fluid flow. Because the parabolic PDEs represent time-dependent diffusion processes, we usually use t and x as independent variables where t is time and x is the one-dimensional space coordinate. Parabolic PDEs in two-space dimensions will be written using x and y as space coordinates and t as time.

Following are some examples of parabolic PDEs:

(a) Transient heat conduction in one-space dimension [Incropera/DeWitt]:

$$\rho c \frac{\partial T}{\partial t} = k \frac{\partial^2 T(x, t)}{\partial x^2} + Q(x) \qquad (12.1.1)$$

(b) Transient neutron diffusion equation in one-space dimension [Hetric]:

$$\frac{1}{v} \frac{\partial}{\partial t} \phi(x, t) = D \frac{\partial^2 \phi}{\partial x^2} - \Sigma_a \phi + v \Sigma_f \phi + S \qquad (12.1.2)$$

where ϕ is the neutron flux.

(c) Convective transport of a chemical specie with diffusion [Brodkey/Hershey]:

$$\frac{\partial}{\partial t}\,\phi = -\frac{\partial}{\partial x}\,u(x)\phi + D\,\frac{\partial^2}{\partial x^2}\,\phi \qquad (12.1.3)$$

where ϕ is the density of the chemical species and $u(x)$ is the flow velocity, and D is the diffusion constant.

The parabolic PDE for two- and three-space dimensions can be written by extending the space variable to two- and three-space dimensions. For example, the transient heat conduction equation in two-space dimensions is

$$\rho c\,\frac{\partial \phi}{\partial t} = k\!\left(\frac{\partial^2 \phi}{\partial x^2} + \frac{\partial^2 \phi}{\partial y^2}\right) + Q(x,\,y) \qquad (12.1.4)$$

In the remainder of this chapter, finite difference methods for partial differential equations of the parabolic type in one-space dimensions as well as in two-space dimensions are described. The advantages and disadvantages of the methods are summarized in Table 12.1.

Table 12.1 Finite difference methods for parabolic PDE

Method	Advantages	Disadvantages
One-space Dimension		
Forward Euler (explicit)	Simplicity.	Δt must be smaller than a stability limit.
Backward Euler (implicit)	Unconditionally stable.	Needs the tridiagonal solution scheme.
Crank-Nicolson	Unconditionally stable; more accurate than backward Euler.	Same as above.
Two-space Dimension		
Forward Euler (explicit)	Simplicity.	For stability, Δt must be less than a certain criterion.
Backward Euler (implicit)	Unconditionally stable.	Needs simultaneous solution in each time step.
Crank-Nicolson	Better accuracy than the former method. Unconditionally stable.	Same disadvantage as the implicit method.
ADI	Unconditionally stable. Uses tridiagonal solution.	Requires more programming effort than the two methods above.

12.2 DIFFERENCE EQUATIONS

As explained in the preceding section, a parabolic PDE in one-space dimension is a transient counterpart of a boundary value problem of second order ODE (see Chapter 10). It is an initial value problem, also. Indeed, if a parabolic PDE is discretized in space first, it becomes an initial value problem of coupled ordinary differential equations (see Example 9.16). Therefore, numerical methods for a parabolic PDE include both (1) a boundary value problem and (2) an initial value problem.

For these reasons, numerical methods for a parabolic PDE can be developed by combining a numerical method for the initial value problems of ODEs and a numerical method for the boundary value problems. Although most of the numerical methods for initial value problems discussed in Chapter 9 may be used in principle, using higher order Runge-Kutta methods or the predictor-corrector methods can result in very complicated or, at least, inefficient methods. This limitation leads us to consideration of the simplest group of numerical methods for initial value problems—that is, the Euler methods.

12.2.1 Difference Approximations on the Time Domain

As an example of parabolic PDE, we consider the equation:

$$\frac{\partial \phi}{\partial t} = \alpha \frac{\partial^2 \phi(x, t)}{\partial x^2} + S(x, t), \quad 0 \leqslant x \leqslant H \tag{12.2.1}$$

where α is a constant. The boundary and initial conditions are

$$\phi(0, t) = \phi_L, \qquad \phi(H, t) = \phi_R \quad \text{(boundary conditions)}$$

$$\phi(x, 0) = \phi_{\text{ini}}(x) \qquad\qquad \text{(initial condition)}$$

The boundary conditions may be given in a more general form similar to Eqs. (10.2.8) and (10.2.9).

To derive finite difference approximations, we consider the grid shown in Figure 12.1. The spatial grid points are numbered here by i, whereas the grid points on the

Figure 12.1 Grid system for a one-dimensional parabolic partial differential equation

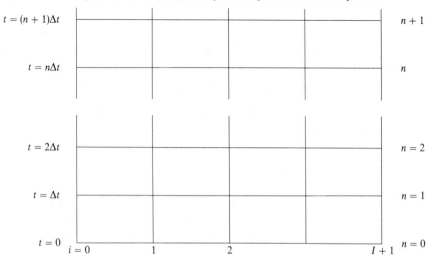

time coordinate are numbered by n. The solution at x_i and t_n will be denoted by

$$\phi_i^{(n)} \equiv \phi(x_i, t_n), \quad i = 0, 1, 2, \ldots, I + 1 \tag{12.2.2}$$

We first apply the difference approximation to the time derivative of Eq. (12.2.1). If we rewrite Eq. (12.2.1) for simplicity as

$$\frac{\partial}{\partial t} \phi = RHS(x, t) \tag{12.2.3}$$

where $RHS(x, t)$ represents the right side of Eq. (12.2.1), then the forward and backward Euler methods on the time domain can be written, respectively, as follows:

$$\frac{\phi(x, t_{n+1}) - \phi(x, t_n)}{\Delta t} = RHS(x, t_n) \quad \text{(Forward Euler: explicit)} \tag{12.2.4}$$

$$\frac{\phi(x, t_{n+1}) - \phi(x, t_n)}{\Delta t} = RHS(x, t_{n+1}) \quad \text{(Backward Euler: implicit)} \tag{12.2.5}$$

Furthermore, using the modified Euler method yields

$$\frac{\phi(x, t_{n+1}) - \phi(x, t_n)}{\Delta t}$$

$$= \frac{1}{2}[RHS(x, t_n) + RHS(x, t_{n+1})] \quad \text{(Modified Euler: Crank-Nicolson)} \tag{12.2.6}$$

Equation (12.2.4) is called the explicit method, and Eqs. (12.2.5) and (12.2.6) are the implicit and Crank-Nicolson methods, respectively [Richtmyer/Morton].

12.2.2 Forward Euler Method

The second derivative term of Eq. (12.2.1) may be discretized by the central difference approximation. Then, the forward Euler method given by Eq. (12.2.4) becomes

$$\frac{\phi_i^{(n+1)} - \phi_i^{(n)}}{\Delta t} = \alpha \frac{\phi_{i-1}^{(n)} - 2\phi_i^{(n)} + \phi_{i+1}^{(n)}}{(\Delta x)^2} + S_i \tag{12.2.7}$$

with

$$\phi_0 = \phi_L, \quad \phi_{I+1} = \phi_R$$

Rewriting Eq. (12.2.7) yields

$$\phi_i^{(n+1)} = \phi_i^{(n)} + \gamma(\phi_{i-1}^{(n)} - 2\phi_i^{(n)} + \phi_{i+1}^{(n)}) + \Delta t S_i \tag{12.2.8}$$

where

$$\gamma = \frac{\alpha \Delta t}{\Delta x^2} \tag{12.2.9}$$

The present scheme is called an explicit method because, if $\phi_i^{(n)}$ are given for t_n at all the grid points, $\phi_i^{(n+1)}$ for the new time t_{n+1} are calculated without solving any simultaneous equations.

The time step Δt of the explicit method must satisfy

$$\Delta t \leqslant \frac{0.5\Delta x^2}{\alpha}$$

or equivalently,

$$\gamma \leqslant 0.5 \tag{12.2.10}$$

(Derivation of this criterion will be discussed in Section 12.3.) Otherwise, instability of solution will occur. No matter how slow the physical change of the system is, Δt must be equal to or smaller than $0.5\Delta x^2/\alpha$. Equation (12.2.10) also indicates that Δt must become increasingly smaller if Δx is reduced.

PROGRAM 12–1 includes solution of a heat condition problem by the explicit method. Programs of explicit method are also found in Ferziger, Incropera/Dewitt, and Rieder/Busby.

Example 12.1

Using the explicit method, solve the heat conduction equation

$$\frac{\partial}{\partial t} T(x, t) = \alpha \frac{\partial^2}{\partial x^2} T(x, t), \quad 0 < x < 10, \quad t > 0 \tag{A}$$

where boundary and initial conditions are given by

$$T(0, t) = 0, \quad T(10, t) = 100 \quad \text{(boundary conditions)}$$
$$T(x, 0) = 0 \quad\quad\quad\quad\quad\quad \text{(initial condition)}$$

and symbols are

  x space coordinate, cm
  α thermal diffusivity, cm²/sec
  T temperature, °C
  t time, sec

Assume $\alpha = 10$ and use 10 grid intervals on the x coordinate. Use two values of Δt: $\Delta t = 0.02$ and $\Delta t = 0.055$.

⟨**Solution**⟩

Because the length of the domain is 10 cm and the number of grid intervals is 10, the grid spacing is $\Delta x = 1$ cm. The total number of grid points is 11, among which two are boundary points with fixed temperatures, so calculations of temperature need to be done for only nine points. The finite difference approximation of the explicit method for Eq. (A) is

$$T_i^{(n+1)} = T_i^{(n)} + \gamma(T_{i-1}^{(n)} - 2T_i^{(n)} + T_{i+1}^{(n)}) \tag{B}$$

where γ is defined by Eq. (12.2.9). The stability criterion for Δt to satisfy $\gamma \leqslant 0.5$ is $\Delta t \leqslant 0.5(\Delta x)^2/\alpha = 0.05$ (sec).

The computational results using $\Delta t = 0.02$ ($\gamma = 0.2$) are shown in Table 12.2.

Table 12.2 Computed temperature distribution

n	t(s)	i = 0	1	2	3	4	5	6	7	8	9	10
1	0.02	0.0	0.0	0.0	0.0	0.0	0.0	0.0	0.0	0.0	20.0	100.0
2	0.04	0.0	0.0	0.0	0.0	0.0	0.0	0.0	0.0	4.0	32.0	100.0
3	0.06	0.0	0.0	0.0	0.0	0.0	0.0	0.0	0.8	8.8	40.0	100.0
4	0.08	0.0	0.0	0.0	0.0	0.0	0.0	0.2	2.2	13.4	45.8	100.0
5	0.10	0.0	0.0	0.0	0.0	0.0	0.0	0.5	4.1	17.7	50.1	100.0
6	0.12	0.0	0.0	0.0	0.0	0.0	0.1	1.1	6.1	21.4	53.6	100.0
7	0.14	0.0	0.0	0.0	0.0	0.0	0.3	1.9	8.2	24.8	56.5	100.0
8	0.16	0.0	0.0	0.0	0.0	0.1	0.6	2.9	10.2	27.8	58.8	100.0
9	0.18	0.0	0.0	0.0	0.0	0.2	0.9	3.9	12.3	30.5	60.9	100.0
10	0.20	0.0	0.0	0.0	0.0	0.3	1.4	5.0	14.2	32.9	62.6	100.0
20	0.40	0.0	0.1	0.5	1.4	3.5	8.0	16.2	29.5	48.5	72.7	100.0
30	0.60	0.0	0.8	2.1	4.4	8.5	15.2	25.2	39.0	56.7	77.5	100.0
40	0.80	0.0	1.9	4.4	8.1	13.5	21.4	32.0	45.6	61.9	80.4	100.0
50	1.00	0.0	3.1	6.7	11.5	18.0	26.5	37.3	50.4	65.6	82.4	100.0
60	1.20	0.0	4.2	8.9	14.7	21.8	30.7	41.5	54.2	68.4	83.9	100.0
70	1.40	0.0	5.2	10.8	17.3	25.0	34.2	44.9	57.1	70.6	85.0	100.0
80	1.60	0.0	6.0	12.4	19.5	27.7	37.0	47.6	59.4	72.3	86.0	100.0
90	1.80	0.0	6.7	13.8	21.4	29.9	39.3	49.9	61.4	73.7	86.7	100.0
100	2.00	0.0	7.3	14.9	22.9	31.7	41.3	51.7	62.9	74.9	87.3	100.0

The results of using $\Delta t = 0.055$ ($\gamma = 0.55$), which is slightly higher than the stability limit, is also shown in Table 12.3. Oscillations in both space and time with an increasing magnitude are typical symptoms of instability.

Table 12.3 Computed temperature distribution

n	t(s)	i = 0	1	2	3	4	5	6	7	8	9	10
1	0.05	0.0	0.0	0.0	0.0	0.0	0.0	0.0	0.0	0.0	55.0	100.0
2	0.11	0.0	0.0	0.0	0.0	0.0	0.0	0.0	0.0	30.3	49.5	100.0
3	0.17	0.0	0.0	0.0	0.0	0.0	0.0	0.0	16.6	24.2	66.7	100.0
4	0.22	0.0	0.0	0.0	0.0	0.0	0.0	9.2	11.6	43.4	61.6	100.0
5	0.28	0.0	0.0	0.0	0.0	0.0	5.0	5.5	27.7	36.0	72.7	100.0
6	0.33	0.0	0.0	0.0	0.0	2.8	2.5	17.5	20.0	51.7	67.5	100.0
7	0.38	0.0	0.0	0.0	1.5	1.1	10.9	10.7	36.0	43.0	76.7	100.0
8	0.44	0.0	0.0	0.8	0.5	6.7	5.4	24.7	25.9	57.7	71.0	100.0
9	0.50	0.0	0.5	0.2	4.1	2.5	16.8	14.7	42.7	47.5	79.6	100.0
10	0.55	0.0	0.0	2.5	1.1	11.2	7.8	31.2	30.0	62.5	73.2	100.0
15	0.82	0.0	4.2	0.1	15.9	4.4	34.7	20.0	59.5	53.0	86.4	100.0
20	1.10	0.0	-2.2	19.7	-3.7	41.8	4.9	64.6	30.7	84.3	74.3	100.0
25	1.38	0.0	18.5	-15.0	53.7	-19.1	82.1	-2.2	98.3	40.0	101.4	100.0
30	1.65	0.0	-21.6	66.4	-54.2	117.4	-56.9	140.2	-18.8	131.0	55.6	100.0
35	1.92	0.0	63.8	-93.5	172.3	-145.8	228.5	-128.5	216.6	-37.8	146.5	100.0
40	2.20	0.0	-105.6	232.0	-273.8	385.2	-328.1	407.7	-237.9	296.2	-28.1	100.0
45	2.47	0.0	234.0	-412.6	617.8	-662.0	778.9	-644.4	661.7	-356.6	316.5	100.0

12.2.3 Implicit Method

The implicit method is based on the backward Euler method in the time domain as already shown by Eq. (12.2.5). Discretizing Eq. (12.2.5) on the space coordinate yields

$$\frac{\phi_i^{(n+1)} - \phi_i^{(n)}}{\Delta t} = \alpha \frac{\phi_{i-1}^{(n+1)} - 2\phi_i^{(n+1)} + \phi_{i+1}^{(n+1)}}{\Delta x^2} + S_i \qquad (12.2.11)$$

Using the definition of γ given by Eq. (12.2.9), Eq. (12.2.11) can be rewritten as

$$-\gamma\phi_{i-1}^{(n+1)} + (1 + 2\gamma)\phi_i^{(n+1)} - \gamma\phi_{i+1}^{(n+1)} = \phi_i^{(n)} + \Delta t S_i \qquad (12.2.12)$$

Equation (12.2.12) cannot be solved individually for each grid point i. The equations for all the grid points must be solved simultaneously. The set of equations for $i = 1, 2, \ldots, N$ form a tridiagonal equation. PROGRAM 12–1 includes the implicit method. See also Ferziger for a program.

The implicit method is always stable regardless of the size of the time step.

Example 12.2

Repeat the problem of Example 12.1 by using the implicit method with the same grid points but $\gamma = 2$.

⟨Solution⟩

The difference equations are

$$-\gamma T_{i-1}^{(n+1)} + (2\gamma + 1)T_i^{(n+1)} - \gamma T_{i+1}^{(n+1)} = T_i^{(n)} \quad i = 1, 2, \ldots, I-1$$

where $I = 10$, $T_0 = 0$ and $T_I = 100$. The foregoing equation may be written in a matrix form as

$$\begin{bmatrix} 2\gamma + 1 & -\gamma & & & \\ -\gamma & 2\gamma + 1 & -\gamma & & \\ & -\gamma & 2\gamma + 1 & -\gamma & \\ & & & \ddots & \\ & & & -\gamma & 2\gamma + 1 \end{bmatrix} \begin{bmatrix} T_1^{(n+1)} \\ T_2^{(n+1)} \\ T_3^{(n+1)} \\ \vdots \\ T_{I-1}^{(n+1)} \end{bmatrix} = \begin{bmatrix} T_1^{(n)} \\ T_2^{(n)} \\ T_3^{(n)} \\ \vdots \\ T_{I-1}^{(n)} + 100\gamma \end{bmatrix}$$

which is solved by the tridiagonal algorithm described in Section 8.3. Computational results obtained by using PROGRAM 12–1 are shown in Table 12.4.

Table 12.4 Computed temperature distribution

n	t(s)	i = 0	1	2	3	4	5	6	7	8	9	10
0	.00	0.0	0.0	0.0	0.0	0.0	0.0	0.0	0.0	0.0	0.0	100.0
1	.2	0.0	0.2	0.4	0.8	1.6	3.1	6.3	12.5	25.0	50.0	100.0
2	.4	0.0	0.6	1.3	2.5	4.7	8.3	14.6	25.0	41.7	66.7	100.0
3	.6	0.0	1.2	2.8	5.0	8.6	14.1	22.4	34.7	51.8	74.1	100.0
4	.8	0.0	2.1	4.5	7.9	12.7	19.5	29.1	41.9	58.4	78.2	100.0
5	1	0.0	3.0	6.4	10.8	16.6	24.3	34.5	47.4	63.0	80.8	100.0
6	1.2	0.0	3.9	8.2	13.5	20.1	28.5	38.9	51.5	66.3	82.7	100.0
7	1.4	0.0	4.7	9.9	15.9	23.2	32.0	42.5	54.8	68.8	84.1	100.0
8	1.6	0.0	5.5	11.4	18.1	25.8	34.9	45.5	57.5	70.8	85.1	100.0
9	1.8	0.0	6.2	12.8	20.0	28.1	37.4	47.9	59.6	72.4	86.0	100.0
10	2.0	0.0	6.8	13.9	21.6	30.0	39.4	49.9	61.4	73.7	86.7	100.0

12.2.4 Crank-Nicolson Method

The numerical method based on the modified Euler method given by Eq. (12.2.6) is named the *Crank-Nicolson method* and is written as follows:

$$\frac{\phi_i^{(n+1)} - \phi_i^{(n)}}{\Delta t} = \frac{\alpha}{2(\Delta x)^2} [(\phi_{i-1}^{(n+1)} - 2\phi_i^{(n+1)} + \phi_{i+1}^{(n+1)})$$

$$+ (\phi_{i-1}^{(n)} - 2\phi_i^{(n)} + \phi_{i+1}^{(n)})] + S_i \qquad (12.2.13)$$

Equation (12.2.13) may be equivalently written as follows:

$$-\frac{\gamma}{2} \phi_{i-1}^{(n+1)} + (1 + \gamma)\phi_i^{(n+1)} - \frac{\gamma}{2} \phi_{i+1}^{(n+1)}$$

$$= \phi_i^{(n)} + \frac{\gamma}{2} (\phi_{i-1}^{(n)} - 2\phi_i^{(n)} + \phi_{i+1}^{(n)}) + \Delta t S_i \qquad (12.2.14)$$

This method requires tridiagonal solution for each time step.

The Crank-Nicolson method is unconditionally stable (regardless of the value of Δt). Although computational time is not much different from the backward Euler implicit method, its accuracy is higher than that of the latter because the Crank-Nicolson method is based on the modified Euler method, which is second-order accurate in time as discussed in Section 9.2. PROGRAM 12–1 may be run for the Crank-Nicolson scheme.

SUMMARY OF THIS SECTION

(a) The forward Euler method is an explicit method. It is simple to program, but the solution becomes unstable if Δt is greater than the stability criterion. This method is recommended when easy programming is desired, or when a very fast and short transient is computed.

(b) The backward Euler implicit and Crank-Nicolson methods need tridiagonal solution in each time step, but they are unconditionally stable. These methods are recommended when long and slow transients are computed.

(c) Accuracy of the Crank-Nicolson method is one order higher than that of the other two methods.

12.3 STABILITY ANALYSIS

As already discussed in the preceding section, one major concern with numerical methods for parabolic PDEs is stability. If unstable, the numerical results will behave erratically and diverge with oscillatory behavior in both time and space. In this section, therefore, we study basic methods to analyze stability of a method and find the stability criteria.

There are at least four mathematical methods for analyzing the stability of difference equations [Godunov/Ryabenkii; Smith]:

(a) Eigenfunction method

(b) Fourier expansion method

(c) Matrix method

(d) Modified equation method

These methods are not only applicable to PDEs of the parabolic type but also the hyperbolic type. In all these methods, the difference equations for the homogeneous part of the PDE is considered. This is because without the inhomogeneous (or source) term, the original PDE has no solution that unboundedly increases in time. Therefore, if the numerical solution of the homogeneous part of the PDE increases unboundedly, it is attributed to instability of the numerical approximation used.

In the eigenfunction method, the numerical solution is expanded into eigenfunctions of the matrix representing the difference equations for each time step. Then, the change in the amplitude of each eigenfunction with advance of time steps is found. In this approach, the effects of boundary conditions are incorporated. If eigenfunctions are known for the set of given finite difference equations, this approach is useful.

In the Fourier expansion method, stability of the method on an infinite domain is investigated by expanding the solution into a Fourier series. This approach is most widely used.

If eigenfunctions cannot be found or written easily, but yet the effect of boundary conditions and other detail of the difference equations has to be considered, the only way is to investigate eigenvalues of the matrix representing the set of difference equations. The modified equation method is described in Section 13.4.

In the remainder of this section, we study the first two methods.

12.3.1 Stability Analysis with Eigenfunctions

When we consider a difference approximation for a linear parabolic PDE with no advective term, its solution may be expressed in terms of eigenfunctions that are sine functions for one-dimensional slab geometries. Each sine function has its own time-dependent coefficient. We will show that if the method is unstable, the time-dependent coefficients for high-frequency modes will show abnormal time behaviors.

For simplicity of explanation, let us consider Eq. (12.2.1)* where $S = 0$, or $\partial\phi/\partial t = \beta(\partial^2\phi/\partial x^2)$, with the fixed boundary conditions:

$$\phi(0, t) = \phi(H, t) = 0 \tag{12.3.1}$$

and the initial condition:

$$\phi(x, 0) = \phi_0(x) \tag{12.3.2}$$

It can be shown easily that an analytical solution of Eq. (12.2.1) satisfying the boundary conditions can be written by

$$\phi(x, t) = f_k(t) \sin(\eta_k x) \tag{12.3.3}$$

where k is a positive integer and

$$\eta_k = \frac{k\pi}{H} \tag{12.3.4}$$

$$f_k(t) = \exp\left[-\beta\eta_k^2 t\right] \tag{12.3.5}$$

Therefore, the general solution is written as a summation of all possible solutions, namely,

$$\phi(x, t) = \sum_{k=0}^{\infty} a_k \exp\left[-\beta\eta_k^2 t\right] \sin(\eta_k x) \tag{12.3.6}$$

where a_k is a coefficient determined by the initial condition. An important feature in this analytical solution is that all the terms vanish as time increases because the sign in the exponential function, Eq. (12.3.5), is negative.

Now we analyze stability of the explicit difference scheme,

$$\phi_i^{(n+1)} = \phi_i^{(n)} + \gamma(\phi_{i-1}^{(n)} - 2\phi_i^{(n)} + \phi_{i+1}^{(n)}), \quad 0 < i < I + 1$$

$$\gamma = \frac{\beta\Delta t}{\Delta x^2} \tag{12.3.7}$$

---

* In this section, thermal diffusivity α in Eq. (12.2.1) will be denoted by β.

with the boundary conditions

$$\phi_0^{(n)} = \phi_{I+1}^{(n)} = 0$$

and an initial condition for $\phi_i^{(0)}$.

The solution of Eq. (12.3.7) may be found analytically in the form

$$\phi_i^{(n)} = (\lambda_k)^n \sin(i\alpha_k) \tag{12.3.8}$$

where α_k is given by

$$\alpha_k = \frac{k\pi}{I+1}, \quad k = 1, 2, \ldots, I$$

and λ_k is a constant called the *amplitude factor*. To find the value of λ_k we introduce Eq. (12.3.8) into Eq. (12.3.7). Left and right sides of Eq. (12.3.7) respectively become

$$LHS = (\lambda_k)^{n+1} \sin(i\alpha_k) \tag{12.3.9}$$

$$RHS = (\lambda_k)^n\{\sin(i\alpha_k) + \gamma[\sin(i\alpha_k - \alpha_k) - 2\sin(i\alpha_k) + \sin(i\alpha_k + \alpha_k)]\}$$

$$= (\lambda_k)^n\{\sin(i\alpha_k) + 2\gamma[\cos(\alpha_k) - 1]\sin(i\alpha_k)\}$$

$$= (\lambda_k)^n\{1 + 2\gamma[\cos(\alpha_k) - 1]\}\sin(i\alpha_k) \tag{12.3.10}$$

Thus, equating Eq. (12.3.9) to Eq. (12.3.10) yields

$$\lambda_k = 1 + 2\gamma[\cos(\alpha_k) - 1] \tag{12.3.11}$$

Thus λ_k and $\phi_i = \sin(i\alpha_k)$ are eigenvalue and eigenfunction, respectively, of

$$\lambda\phi_i = \phi_i + \gamma(\phi_{i-1} - 2\phi_i + \phi_{i+1})$$

Since Eq. (12.3.8) for each k of $k = 1, 2, \ldots, I$ satisfies Eq. (12.3.7), the general solution is a linear combination of all the possible solutions and is written as

$$\phi_i^{(n)} = \sum_{k=1}^{I} a_k(\lambda_k)^n \sin(i\alpha_k) \tag{12.3.12}$$

where a_k is a coefficient determined by the initial condition.

The analytical solution of the explicit method is very similar to that of Eq. (12.3.6). But one major difference is that Eq. (12.3.12) has $(\lambda_k)^n$ instead of $\exp(-\beta\eta_k^2 t)$. It is shown next that λ_k for a small k is an approximation for $\exp(-\beta\eta_k^2 t)$. If

$\alpha_k \ll 1$, λ_k becomes

$$\lambda_k = 1 + 2\gamma[\cos(\alpha_k) - 1] \simeq 1 - \gamma\alpha_k^2$$

$$= 1 - \beta\frac{\Delta t}{\Delta x^2}\left(\frac{k\pi}{I+1}\right)^2 = 1 - \beta\Delta t\left(\frac{k\pi}{H}\right)^2$$

$$= 1 - \beta\Delta t\eta_k^2 \simeq \exp(-\beta\eta_k^2\Delta t) \tag{12.3.13}$$

where Taylor expansions of $\cos(\alpha_k)$ and $\exp(-\beta\eta_k^2\Delta t)$ are used. However, as k increases, λ_k no longer approximates $\exp(-\beta\eta_k^2\Delta t)$ but $(\lambda_k)^n$ can behave erratically. To ensure stability, λ_k must satisfy

$$-1 \leqslant \lambda_k \leqslant 1 \tag{12.3.14}$$

for all k. Introducing Eq. (12.3.11), Eq. (12.3.14) is written as

$$-1 \leqslant 1 + 2\gamma[\cos(\alpha_k) - 1] \leqslant 1 \tag{12.3.15}$$

The second inequality is always satisfied because $\cos(\alpha_k) \leqslant 1$. To examine the implication of the first inequality, we first recognize that the minimum of $\cos(\alpha_k)$ equals $\cos(\alpha_I) = \cos(\pi I/(I+1))$ that approaches -1 as I becomes large. Therefore, the necessary condition to satisfy the first inequality can be written as

$$\gamma \leqslant \frac{1}{2}$$

or equivalently

$$\Delta t \leqslant \frac{(\Delta x)^2}{2\beta} \tag{12.3.16}$$

Stability of the implicit and Crank-Nicolson methods may be studied in a similar manner. The amplitude factor for the implicit method becomes

$$\lambda_k = \frac{1}{1 + 2\gamma[1 - \cos(\alpha_k)]} \tag{12.3.17}$$

Since $1 - \cos(\alpha_k) \geqslant 0$, λ_k given in Eq. (12.3.17) is always positive, thus the first inequality of the stability condition, $-1 \leqslant \lambda_k \leqslant 1$, is always satisfied. The second inequality is always satisfied since the denominator of (12.3.17) is greater than 1. Therefore, the implicit method is stable regardless of the value of γ or Δt.

The amplitude factor of the Crank-Nicolson method is

$$\lambda_k = \frac{1 - \gamma[1 - \cos(\alpha_k)]}{1 + \gamma[1 - \cos(\alpha_k)]} \tag{12.3.18}$$

Equation (12.3.18) is shown to satisfy $|\lambda_k| \leqslant 1$ for all α_k's. Thus, the Crank-Nicolson method is also unconditionally stable.

12.3.2 Fourier (Von Neumann) Stability Analysis

One disadvantage of the method discussed in the preceding section is that it works if eigenfunctions are known, but this is not the case for many problems. The Fourier stability analysis introduced here is more universal and applicable to any types of difference equations for space-time problems [Mitchell/Griffiths; Richtmyer/Morton].

The Fourier stability analysis examines the stability of a given method for a linear PDE under the following conditions:

(a) The PDE is a linear PDE.
(b) The domain of interest is infinite.
(c) The grid spacing is constant.
(d) The coefficients of the PDE are constant.

The effects of the actual boundary conditions are ignored. Sometimes, a numerical scheme becomes stable with certain boundary conditions even if it is judged to be unstable by the Fourier stability analysis. Nonetheless, the test of Fourier stability analysis is considered to be an important criterion to assure stability of a scheme.

The source term of the PDE is ignored for the following reason. If no source term exists, the solution should not increase in time. So, if the numerical solution increases, it must be due to the instability of the scheme.

The Fourier stability analysis can be applied to any difference approximation for a PDE of parabolic or hyperbolic type under the conditions stated earlier. Let us consider the explicit method:

$$\phi_i^{(n+1)} = \phi_i^{(n)} + \gamma(\phi_{i-1}^{(n)} - 2\phi_i^{(n)} + \phi_{i+1}^{(n)}) \tag{12.3.19}$$

in the infinite domain.

Suppose the initial condition for the problem is given in a Fourier function:

$$\phi_i^0 = \exp(ij\pi/k)$$

or equivalently

$$\phi_i^0 = \exp(ij\theta) \tag{12.3.20}$$

where $j = \sqrt{-1}$, k is an integer, $-\infty < k < \infty$, excepting $k = 0$, and $\theta = \pi/k$, $-\pi \leqslant \theta \leqslant \pi$. The k is half of the wave length in terms of number of grid intervals.

Then the solution of the numerical method can be written in the form:

$$\phi_i^{(n)} = (G_\theta)^n \exp{(ij\theta)} \qquad (12.3.21)$$

where G_θ is the amplitude factor (generally a complex value). G_θ is found by introducing Eq. (12.3.21) into Eq. (12.3.19), as follows:

$$G_\theta = 1 + \gamma[\exp{(-j\theta)} - 2 + \exp{(+j\theta)}]$$
$$= 1 + 2\gamma[\cos{(\theta)} - 1] \qquad (12.3.22)$$

If $|G_\theta| \leqslant 1$ for $-\pi \leqslant \theta \leqslant \pi$, the method is stable because $\phi_i^{(n)}$ does not increase in time. Since $-1 \leqslant \cos{(\theta)} \leqslant 1$, the condition $0 \leqslant |G_\theta| \leqslant 1$ requires

$$\gamma \leqslant 0.5 \qquad (12.3.23)$$

Thus, the method is stable when Eq. (12.3.23) is satisfied. This criterion is exactly the same as Eq. (12.3.16) obtained by using the eigenfunctions.

One may ask why the analysis in the previous subsection, which includes the effect of boundary conditions, gives the same results as the Fourier stability analysis for the infinite domain. The answer is that stability of the difference schemes for parabolic PDE is determined by the Fourier mode of the shortest wave length ($k = \pm 1$, $\theta = \pm\pi$) that is independent of the boundary conditions but dependent on only the grid spacing.

Similar analysis shows that the amplitude factors for the backward Euler (implicit) method becomes

$$G_\theta = \frac{1}{1 + 2\gamma(1 - \cos{(\theta)})} \qquad (12.3.24)$$

The denominator is greater or equal to unity, so $G_\theta \leqslant 1$. Thus the method is unconditionally stable.

The amplitude factor for the Crank-Nicolson method is

$$G_\theta = \frac{1 - \gamma[1 - \cos{(\theta)}]}{1 + \gamma[1 - \cos{(\theta)}]} \qquad (12.3.25)$$

which is identical to Eq. (12.3.18).

SUMMARY OF THIS SECTION

(a) In the stability analysis using eigenfunctions, the numerical solution is expanded into the eigenfunction of the difference operator. If the numerical method's eigenfunctions are in an analytical form, the analysis includes the effect of boundary conditions.

(b) The analysis with the eigenfunctions shows that (i) the forward Euler method is stable only if $\gamma \leqslant 0.5$ is satisfied; (ii) the backward and modified Euler methods are unconditionally stable.

(c) Although the effect of boundary conditions is included in the stability analysis using eigenfunctions, the results show that stability is unrelated to the boundary conditions. The same stability criterion is obtained by the Fourier (Von Neumann) stability analysis, which considers an infinite domain and ignores the effect of boundary conditions.

(d) In the Fourier stability analysis, an infinite domain is considered, and the solution is expanded into the Fourier series. This approach is based on the fact that, whether the space is continuous or discrete, any function may be expanded into the Fourier integral. However, because of the discrete space on the grid, the integral is reduced to summation of the Fourier components of discrete frequencies. A numerical scheme tested is considered to be stable if the amplitude factors for all the wavelengths are less than or equal to unity in magnitude.

12.4 NUMERICAL METHODS FOR TWO-DIMENSIONAL PARABOLIC PROBLEMS

The three methods explained in Section 12.2 for one-space-dimensional parabolic PDEs may be extended to two-space-dimensional parabolic PDEs, but each has the following disadvantage: The forward Euler method is simple to implement in a program, but time step is limited because of the stability criterion. The backward Euler implicit method and the Crank-Nicolson method are unconditionally stable, but both need simultaneous solution of the difference equations for all the grid points in the whole domain in every time step. The simultaneous solution is performed by either a direct solution or an iterative scheme, both of which are very time consuming because solution is necessary in every time step.

The method based on an approximate factorization, which is explained in the remainder of this section, is unconditionally stable, and yet solution in each time step needs only tridiagonal solution along each grid line. The scheme is known also as the alternating direction implicit (ADI) method. Approximate factorization is applied extensively to other types of PDEs under the name of splitting methods [Mitchell/Griffiths; Steger/Warming].

Let us consider

$$\frac{\partial \phi}{\partial t} = \alpha \nabla^2 \phi + S(x, y) \tag{12.4.1}$$

Discretizing the time variable with the modified Euler method yields

$$\frac{\phi(x, y, t_{n+1}) - \phi(x, y, t_n)}{\Delta t} = \frac{1}{2}[\alpha \nabla^2 \phi(x, y, t_{n+1}) + \alpha \nabla^2 \phi(x, y, t_n)] + S(x, y) \tag{12.4.2}$$

If we define

$$\delta\phi(x, y) = \phi(x, y, t_{n+1}) - \phi(x, y, t_n)$$

then Eq. (12.4.2) may be rewritten as

$$\left[1 - \frac{\Delta t\alpha}{2}\left(\frac{\partial^2}{\partial x^2} + \frac{\partial^2}{\partial y^2}\right)\right]\delta\phi = \frac{\Delta t\alpha}{2}\left[\frac{\partial^2}{\partial x^2} + \frac{\partial^2}{\partial y^2}\right]\phi(x, y, t_n) + \Delta t S \quad (12.4.3)$$

Equation (12.4.3) is an elliptic PDE, which can be approximated by a factorized form:

$$\left[1 - \frac{\Delta t\alpha}{2}\frac{\partial^2}{\partial x^2}\right]\left[1 - \frac{\Delta t\alpha}{2}\frac{\partial^2}{\partial y^2}\right]\delta\phi = RHS \quad (12.4.4)$$

where RHS is the right-hand side of Eq. (12.4.3) and a fourth-order cross derivative term has been added to make the factorization possible. Equation (12.4.4) may be solved in two steps as follows:

$$\left[1 - \frac{\Delta t\alpha}{2}\frac{\partial^2}{\partial x^2}\right]\psi(x, y) = RHS$$

$$\left[1 - \frac{\Delta t\alpha}{2}\frac{\partial^2}{\partial y^2}\right]\delta\phi = \psi(x, y) \quad (12.4.5)$$

Both equations are discretized on the x-y domain:

$$-\frac{1}{2}\gamma_x\psi_{i-1,j} + (1 + \gamma_x)\psi_{i,j} - \frac{1}{2}\gamma_x\psi_{i+1,j} = RHS_{i,j}$$

$$-\frac{1}{2}\gamma_y\delta\phi_{i,j-1} + (1 + \gamma_y)\delta\phi_{i,j} - \frac{1}{2}\gamma_y\delta\phi_{i,j+1} = \psi_{i,j} \quad (12.4.6)$$

where $\gamma_x = \Delta t\alpha/\Delta x^2$ and $\gamma_y = \Delta t\alpha/\Delta y^2$. The first equation comprises a set of tridiagonal equations for each j (horizontal grid line), and the second equation comprises a set of tridiagonal equations for each i (a vertical grid line). Thus, the solution for ϕ at the new time point is obtained using only the tridiagonal scheme without any iterative solution.

SUMMARY OF THIS SECTION

(a) The numerical method introduced in this section is known as the alternating direction implicit (ADI) method, or the approximate factorization method. It requires only tridiagonal solution for each vertical and horizontal grid lines.

(b) The ADI method is unconditionally stable.

PROGRAMS

PROGRAM 12–1   Parabolic PDE (Heat Conduction Equation)

(A) Explanations

The one-dimensional heat conduction equation

$$\frac{\partial T(x, t)}{\partial t} = \alpha \frac{\partial^2 T(x, t)}{\partial x^2} \quad 0 < x < H \tag{A}$$

is considered where T is the temperature and α is the thermal diffusivity. Assume boundary conditions are

$$T(0, t) = T_L, \quad T(H, t) = T_R$$

The finite difference approximation for (A) may be written as

$$\frac{T_i^n - T_i^{n-1}}{\Delta t} = \alpha \left(\theta \frac{T_{i-1}^n - 2T_i^n + T_{i+1}^n}{\Delta x^2} + (\theta - 1) \frac{T_{i-1}^{n-1} - 2T_i^{n-1} + T_{i+1}^{n-1}}{\Delta x^2} \right) \tag{B}$$

with

$$T_0^n = T_L, \quad \text{and} \quad T_I^n = T_R$$

Here subscript i denotes the spacial grid point, superscript n the time step, I-1 is the number of grid points excluding the points on the boundaries, and θ specifies:

$$\text{Forward Euler scheme} \quad \text{if} \quad \theta = 0$$
$$\text{Backward Euler scheme} \quad \text{if} \quad \theta = 1$$
$$\text{Crank-Nicolson scheme} \quad \text{if} \quad \theta = 0.5$$

Equation (B) may be equivalently written as

$$-\theta\gamma T_{i-1}^n + (2\theta\gamma + 1)T_i^n - \theta\gamma T_{i+1}^n = T_i^{n-1} + (\theta - 1)\gamma(T_{i-1}^{n-1} - 2T_i^{n-1} + T_{i+1}^{n-1}) \tag{C}$$

with

$$\gamma = \frac{\alpha\Delta t}{\Delta x^2}$$

If $\theta = 0$, solution of Eq. (C) is explicit. If $\theta > 0$, it is solved by using the tridiagonal scheme.

The boundary conditions are set to $T_L = 0$ and $T_R = 100$. The initial condition is $T_i = 0$. (See Example 12.2.)

(B) List

```
/* CSL/c12-1.c    Implicit/Explicit Scheme for Parabolic PDE*/
#include <stdio.h>
#include <stdlib.h>
#include <math.h>
/*        mi : grid point next to the right boundary
          nt : time step number
     nmax : maximum time step number
     time : t
     ga : gamma
     tb[] : temperature
     s[] : heat source term            */
main()
{
int i, mi, nmax, nt=0;
float al, a[101], b[101], c[101], d[101];
float dt, dummy, dx, et, ga, s[101], t[10], tb[11], th, time, ze;
void trid();
    printf("\n\nCSL/C12-1  Implicit/Explicit Scheme for Parabolic PDE\n");
    printf( "Type alpha\n" );                    scanf( "%f", &th );
    if( th == 0 ) printf( "Explicit method is selected.\n" );
    if( th > 0.5 ) printf( "Implicit method is selected\n" );
    if( th > 0.4999 && th < 0.5001 )
        printf( "Crank-Nicolson method is selected.\n" );
    printf( "Type dt (time step size)\n" );      scanf( "%f", &dt );
    printf( "Type maximum number of time steps\n" );scanf( "%d", &nmax );
    al = 10;     dx = 1;     mi = 9;
    ga = al*dt/dx/dx;   ze = th*ga; et=(1-th)*ga; printf("Gamma=%g\n",ga );
    for( i = 0; i <= mi; i++ ) tb[i] = 0;
    tb[0] = 0;    tb[mi + 1] = 100;        /* Boundary conditions*/
    nt=0; time=0.0;
    printf( "Time step=%d  time=%g \n", nt, time );
    for( i = 0; i <= (mi + 1); i++) printf(" %6.2f",tb[i]); printf("\n");
    for( i = 1; i <= mi; i++ )    s[i] = 0;        /* Heat source is zero*/
    do {
        nt = nt + 1; time = nt*dt;
        printf( "Time step=%d  time=%g \n", nt, time );
        for( i = 1; i <= mi; i++ )
                   d[i] = tb[i] + et*(tb[i - 1] - 2*tb[i] + tb[i + 1]);
        d[mi] = d[mi] + ze*tb[mi + 1];
        if( th != 0 ){
            for( i = 1; i <= mi; i++ ){
               a[i] = -ze;  b[i] = 1 + 2*ze;   c[i] = -ze;
            }
            trid( a, b, c, d, &mi );
        }
        for( i = 1; i <= mi; i++ )    tb[i] = d[i];
        for( i = 0; i <= (mi + 1); i++) printf(" %6.2f",tb[i]);
        printf("\n");
    } while (nt<=nmax);
    printf("\n");
    exit(0);
}
```

```
void trid(a, b, c, d, n)
float a[], b[], c[], d[]; int *n;
{
int i;  float r;
   for( i = 2; i <= *n; i++ ){
      r = a[i]/b[i-1]; b[i] = b[i] - r*c[i-1]; d[i] = d[i] - r*d[i-1];
   }
   d[*n] = d[*n]/b[*n];
   for( i = *n - 1; i >= 1; i-- ) d[i] = (d[i] - c[i]*d[i + 1])/b[i];
   return;
}
```

(C) Sample Output

Please see Examples 12.1 and 12.2.

PROBLEMS

(12.1) To investigate the effect of changing Δt on the results of the explicit scheme in Example 12.1, calculate the solution of the problem in Example 12.1 with each of $\Delta t = 0.001$, 0.01, 0.1, 0.5, and compare the results for $x = 0.5$ at $t = 1, 2, 5, 10$.

(12.2) Solve the problem of Example 9.16 by the explicit method.

(12.3) Repeat the computation of Example 12.1 by the Crank-Nicolson method using PROGRAM 12–1.

(12.4) Verify Eqs. (12.3.22), (12.2.24), and (12.3.25) by Fourier stability analysis.

(12.5) The two-step method called ADEP (alternating direction explicit) method for the heat conduction equation is given by

$$T_i^{(n+1)} - T_j^{(n)} = \gamma(T_{i+1}^{(n)} - T_i^{(n)} + T_i^{(n+1)} - T_{i-1}^{(n+1)})$$
$$T_i^{(n+2)} - T_j^{(n+1)} = \gamma(T_{i+1}^{(n+2)} - T_i^{(n+2)} + T_i^{(n+1)} - T_{i-1}^{(n+1)})$$

where $\gamma = \alpha \Delta t / \Delta x^2$

Show by Fourier stability analysis that the method is unconditionally stable.

(12.6) (a) By Fourier stability analysis, show that the following method is unconditionally stable:

$$3u_i^{(n+1)} - 4u_i^{(n)} + u_i^{(n-1)} - 2\gamma(u_{i-1}^{(n+1)} - 2u_i^{(n+1)} + u_{i+1}^{(n+1)}) = 0$$

where $\gamma = \beta \Delta t / (\Delta x)^2$

(b) Write the PDE that is approximated by the foregoing difference approximation.

REFERENCES

Brodkey, R. S., and H. C. Hershey, *Transport Phenomena*, McGraw-Hill, 1988.

Ferziger, J. H., *Numerical Methods for Engineering Applications*, Wiley-Interscience, 1988.

Godunov, S. K., and V. S. Ryabenkii, *Difference Schemes*, North-Holland, 1987.

Hetric, L. H., *Dynamics of Nuclear Reactors*, Chicago University Press, 1971.

Incropera, F. P., and D. P. DeWitt, *Introduction to Heat Transfer*, Wiley, 1985.

Mitchell, A. R., and D. F. Griffiths, *The Finite Difference Method in Partial Differential Equations*, Wiley-Interscience, 1980.

Oran, E. S., and J. P. Boris, *Numerical Simulation of Reactive Flow*, Elsevier, 1987.

Richtmyer, R. D., and K. W. Morton, *Difference Methods for Initial-Value Problems*, Wiley-Interscience, 1957.

Rieder, W. G., and H. R. Busby, *Introductory Engineering Modeling*, Wiley, 1986.

Smith, G. D., *Numerical Solution of Partial Differential Equations*, Oxford University Press, 1978.

Steger, J., and R. F. Warming, "Flux Vector Splitting of the Inviscid Gas Dynamic Equations with Application to Finite Difference Methods," *J. Comp. Phys.*, Vol. 40, 1981.

13

Hyperbolic Partial Differential Equations

13.1 INTRODUCTION

The governing equations for convective transport of matters and their physical quantities, as well as for elastic, acoustic, and electomagnetic waves, are hyperbolic PDEs. The remarkable progress in numerical schemes for hyperbolic PDEs in the recent past has been closely related to the progress in computational fluid dynamics, however. The basic equations of inviscid fluid flow consist of hyperbolic PDEs. Even those equations for viscous flows must be treated as hyperbolic when the viscous effect is weak. The success in a computational simulation of fluid flow depends on the accuracy and efficiency of solving hyperbolic PDEs. This is why development of numerical schemes for hyperbolic PDEs has been an urgent research topic in computational fluid dynamics.

A hyperbolic PDE can be written in both the first-order and second-order forms. The former can be easily rewritten in the latter form and shown to satisfy the criterion in Eq.(11.1.1a). Hyperbolic PDEs for transport of matters and their properties are mostly in the first-order form, whereas those for elastic, acoustic and electromagnetic waves are in the second-order form. However, most numerical schemes for hyperbolic PDEs are based on the first-order form. In this chapter, therefore, we consider the following three first-order forms of hyperbolic PDEs:

(a) Linear hyperbolic PDEs in the conservative form:

$$\frac{\partial}{\partial t} u(x, t) + \frac{\partial}{\partial x} f(x, t) = s(x, t) \tag{13.1.1}$$

where f is a linear function of u, for example—$f = a(x)u(x, t)$—and where $a(x)$ is a known function and $s(x, t)$ is a source term.

(b) Linear hyperbolic PDEs in the nonconservative form:

$$\frac{\partial}{\partial t} u(x, t) + a(x, t) \frac{\partial}{\partial x} u(x, t) = s(x, t) \tag{13.1.2}$$

(c) Nonlinear hyperbolic PDEs in the conservative form:

$$\frac{\partial}{\partial t} u(x, t) + \frac{\partial}{\partial x} f(u(x, t)) = s(x, t) \tag{13.1.3}$$

where f is a nonlinear function of u. The physical interpretations of Eqs. (13.1.1) through (13.1.3) are given in Appendix D. Numerical schemes for nonlinear hyperbolic PDEs are discussed in Section 13.7.

Difference approximations for the hyperbolic equations work best when the solution is smooth. However, the solution of hyperbolic PDEs can include discrete jumps. (This is contrary to the solution of elliptic or parabolic PDEs, which are always continuous in space and time.) For example, in a pipe flow of some chemicals, the concentration of a chemical species can have a sudden jump. As another example, the spatial distribution of momentum becomes discontinuous at a shock in a compressible fluid flow.

To illustrate the behavior of the discontinuous solution of a hyperbolic PDE, consider

$$u_t + au_x = 0, \quad x \geqslant 0, \quad t \geqslant 0 \tag{13.1.4}$$

which is identical to Eq. (13.1.2) except a is constant and $s = 0$. The initial condition is

$$\begin{aligned} u(x, 0) &= 1 \quad \text{for } x \leqslant 1 \\ &= 0 \quad \text{for } x > 1 \end{aligned} \tag{13.1.5}$$

and boundary condition is

$$u(0, t) = 0 \quad \text{for } t > 0 \tag{13.1.6}$$

The analytical solution for this problem is

$$u(x, t) = u(x - at, 0) \tag{13.1.7}$$

(See Example 13.1 for derivation.) This solution represents the wave of a rectangular shape traveling at the speed a, as depicted in Figure 13.1.

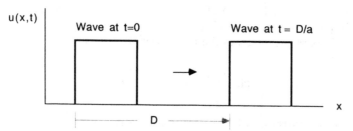

Figure 13.1 A square-shaped wave traveling at speed *a*

The traveling wave of the rectangular shape presents some basic difficulties associated with the finite difference approximation. First, the discontinuity of the solution cannot be exactly represented on the grid. Second, an accurate approximation of the derivatives with respect to time and space about a discrete jump is not easy. Many researchers have investigated the numerical schemes to increase their accuracy in simulating discrete jumps.

In the remainder of this chapter, we focus on numerical schemes for the simple hyperbolic PDEs in one dimension. For those who have been little exposed to hyperbolic PDEs, the concept of characteristics is introduced. Then, the first-order accurate and second-order accurate numerical schemes are explained including the theoretical and computational aspects of stability, diffusion, and aliasing errors. Discussions of the errors of difference methods follow. Finally, the solution schemes for nonlinear hyperbolic PDEs and flux-corrected methods are discussed.

13.2 METHOD OF CHARACTERISTICS

This section is concerned with characteristics of a hyperbolic PDE, which is important in understanding both analytical solution and numerical schemes [Courant/Hilbert; Garabedian; Mitchell/Griffiths; Smith].

Suppose we try to find an analytical solution of

$$u_t + a(x)u_x = s(x, t) \tag{13.2.1}$$

along an arbitrary curve on which the points P and Q are located with an infinitesimal distance as shown in Figure. 13.2. The change of u from P to Q is denoted by du, which may be written by

$$du = u_t dt + u_x dx$$

Dividing the foregoing equation through by dt yields

$$\frac{du}{dt} = u_t + u_x \frac{dx}{dt} \tag{13.2.2}$$

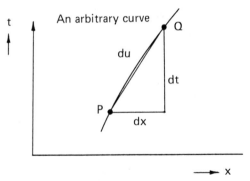

t

An arbitrary curve

Q

du

dt

P

dx

x

Figure 13.2 Solution along a curve

where dx/dt is the gradient of the curve PQ on the x-t plane. If the curve is chosen to satisfy

$$\frac{dx}{dt} = a(x) \tag{13.2.3}$$

the right side of Eq. (13.2.2) becomes equal to the left side of Eq. (13.2.1). Hence, we get

$$\frac{du}{dt} = s \tag{13.2.4}$$

The hyperbolic PDE is now represented by a pair of ODEs, namely Eqs. (13.2.3) and (13.2.4). Equation (13.2.3) represents a curve (or line) on the x-t plane, called the *characteristic curve* (or *line*), and Eq. (13.2.4) is an ODE along the curve.

If we determine the curve $x = x(t)$ by integrating Eq. (13.2.3), then the solution of Eq. (13.2.4) is obtained by integrating Eq. (13.2.4) along the curve.

Example 13.1

By using the method of characteristics, solve Eq. (13.1.4) and show that the analytical solution is given by Eq.(13.1.7). Assume that $a =$ constant, $s = 0$, and the initial and boundary conditions are given by Eqs. (13.1.5) and (13.1.6), respectively.

⟨**Solution**⟩

The equation for the characteristic lines is given by

$$\frac{dx}{dt} = a \quad \text{(constant)} \tag{A}$$

Integrating Eq. (A) yields a characteristic line:

$$x = at + b \tag{B}$$

where b is a constant. If we consider the characteristic line passing through the point $x = x_0$ at $t = 0$, then Eq. (B) becomes

$$x = at + x_0 \tag{C}$$

Along this line, the solution of Eq. (13.1.1) is obtained by integrating Eq. (13.2.4), or $du/dt = 0$ since $s = 0$.

Integrating $du = 0$ yields

$$u(x, t) = k \quad \text{along } x = at + x_0 \tag{D}$$

where k is a constant determined by the initial condition. At $t = 0$ the initial condition must be satisfied. Therefore, Eq. (D) becomes

$$u(x, t) = \begin{cases} u(x_0, 0) & \text{along } x = at + x_0, \ x_0 \geqslant 0 \\ 0 & x_0 < 0 \end{cases}$$

or equivalently

$$u(x, t) = \begin{cases} u(x - at, 0) & \text{for } x \geqslant at \\ 0 & \text{for } x < at \end{cases} \tag{E}$$

where x_0 is eliminated by using $x_0 = x - at$. Thus, we have proved Eq. (13.1.7).

Example 13.2

A hyperbolic PDE is given by

$$u_t + a(x, t)u_x = s(x, t) \tag{A}$$

where

$$a(x, t) = 3x + 0.1 \tag{B}$$
$$s(x, t) = 1 - x^2 + 0.1t \tag{C}$$

Assuming the initial condition is given by $u(x, 0) = 1$ for $t = 0$, calculate the solution along the characteristic curve that passes through $x = 0.2$ at $t = 0$.

⟨**Solution**⟩

Using Eq. (B), Eq. (13.2.3) for the characteristic line becomes

$$\frac{dx}{dt} = 3x + 0.1, \quad x(0) = 0.2 \tag{D}$$

where the second equation is the initial condition. Its analytical solution satisfying the initial condition is

$$x(t) = \frac{1}{3}(0.7e^{3t} - 0.1) \tag{E}$$

With Eq. (C), the characteristic equation becomes

$$\frac{du}{dt} = s(x, t) = 1 - x^2 + 0.1t$$

$$= 1 - \frac{1}{9}(0.49e^{6t} - 0.14e^{3t} + 0.01) + 0.1t \qquad \text{(F)}$$

where Eq. (E) is used and the initial condition is $u(0) = 1$.

Since Eq. (F) is dependent on only t, it can be now integrated as

$$u(x, t) = \int_0^t s(x, t')\, dt'$$

$$= t - \frac{1}{9}\left[\frac{0.49}{6} e^{6t} - \frac{0.14}{3} e^{3t} + 0.01t\right] + \frac{0.1}{2} t^2$$

The method of characteristics may be implemented with finite difference approximation on a $x\text{-}t$ grid as shown in Figure 13.3.

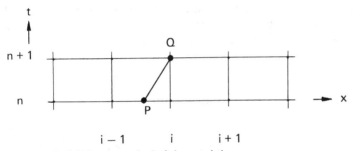

Figure 13.3 Grid for the method of characteristics

A characteristic line passing through Q located at $(i, n + 1)$ is drawn. The intersection of the characteristic line and $t = t_n$ is denoted by P. The two ODEs, namely, $dx = a\,dt$ and $du = s\,dt$, may be approximated along a finite length of the characteristic line by

$$\delta x = a\delta t \quad \text{and} \quad \delta u = s\delta t$$

Applying these relations to line PQ (see Figure 13.3) yields

$$\delta x = x_Q - x_P = a\Delta t, \qquad \delta u = u_Q - u_P = s\Delta t \qquad (13.2.5)$$

If $u_i^{(n)}$ for all the grid points are known, u_P may be calculated by a linear interpolation written as

$$u_P = \frac{\Delta x - a\Delta t}{\Delta x} u_i^{(n)} + \frac{a\Delta t}{\Delta x} u_{i-1}^{(n)}$$

$$= (1 - \gamma)u_i^{(n)} + \gamma u_{i-1}^{(n)} \qquad (13.2.6)$$

where

$$\gamma = \frac{a\Delta t}{\Delta x}$$

is the Courant number. Thus, u_Q that equals $u_i^{(n)}$ is calculated by introducing Eq. (13.2.6) into the second equation of Eq. (13.2.5):

$$u_i^{(n+1)} = (1 - \gamma)u_i^{(n)} + \gamma u_{i-1}^{(n)} + s\Delta t \tag{13.2.7}$$

The values of $u_i^{(n+1)}$ for all the points are obtained by repeating the same calculations for every point. The scheme is called the *explicit characteristic method*.

If $\gamma = 1$, Eq. (13.2.7) reduces to

$$u_i^{(n+1)} = u_{i-1}^{(n)} + s\Delta t \tag{13.2.8}$$

which is exact when a and s are both constant.

The method of characteristic on a grid involves two kinds of errors. The first is the numerical diffusion effect, which smears sharp changes of the solution. The second is instability. Numerical diffusion is caused by the interpolation to compute u_P when $\gamma \neq 1$.

The scheme is stable if

$$\gamma \leqslant 1 \quad \text{(stability criterion)}$$

but unstable for $\gamma > 1$.

The method of characteristics on a grid is identical with the FTBS method derived in Section 13.3. More details of instability and numerical diffusion are discussed in Sections 13.3 and 13.4.

SUMMARY OF THIS SECTION

(a) A first-order hyperbolic PDE may be reduced to an ODE along a characteristic curve (or line). Therefore, it can be solved by integrating the ODE along the characteristic line. The solution scheme based on this principle is named the *method of characteristics*.

(b) The method of characteristics on a grid is developed by using an interpolation method to compute u at the previous step.

13.3 FIRST-ORDER DIFFERENCE SCHEMES

Most numerical schemes for hyperbolic PDEs are based on finite difference approximations. In this section, basic difference schemes that are first-order-accurate are derived. The stability of each scheme is examined by Fourier stability analysis.

Throughout this section we will consider

$$u_t + au_x = 0 \qquad\qquad (13.3.1)$$

with

$$u(x, 0) = u_0(x) \quad \text{(initial condition)} \qquad\qquad (13.3.2)$$

$$u(0, t) = u_L(t) \quad \text{(boundary condition)} \qquad\qquad (13.3.3)$$

where a is a constant and $a > 0$. An inhomogeneous (source) term may be added to the right side without any essential changes to the discussions that follow.

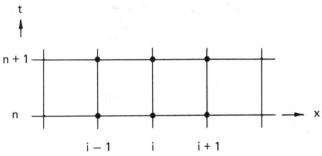

Figure 13.4 A grid on the x-t plane

Considering the grid as illustrated in Figure 13.4, a number of different numerical schemes can be derived depending on the difference approximation selected for each of u_t and u_x. Because u_x and u_t are first-order partial derivatives, candidates include backward difference, forward difference, and central difference approximations in both x and t. When $a > 0$ (flow is in the positive direction), the backward difference approximation

$$u_x \simeq \frac{u_i - u_{i-1}}{\Delta x} \qquad\qquad (13.3.4)$$

is called the *first-order upwind difference approximation*, where Δx is the grid interval in space, because the difference approximation is based on the information from the upwind domain [Anderson/Tannehill/Pletcher; Patanka]. If $a < 0$ on the other hand, the first-order upwind difference approximation equals the forward difference approximation:

$$u_x \simeq \frac{u_{i+1} - u_i}{\Delta x} \qquad\qquad (13.3.5)$$

Considering the time interval between t_n and t_{n+1}, the simplest difference approximation for u_t is

$$u_t \simeq \frac{u_i^{(n+1)} - u_i^{(n)}}{\Delta t} \tag{13.3.6}$$

where $\Delta t = t_{n+1} - t_n$. The whole scheme becomes the backward Euler or forward Euler or Modified Euler (Crank-Nicolson) schemes depending on for which of t_n and t_{n+1} the spatial derivatives are evaluated.

Forward Euler in Time and Backward Difference in Space (FTBS)

The "FTBS" in the title stands "forward Euler in time and backward difference in space." The phrase "forward Euler" indicates that u_x is evaluated at t_n, so the scheme becomes an explicit scheme. When $a > 0$, backward difference is an "upwind scheme" as explained earlier. Evaluating u_x in Eq. (13.3.1) by the backward difference approximation at time point n yields

$$\frac{u_i^{(n+1)} - u_i^{(n)}}{\Delta t} + a\,\frac{u_i^{(n)} - u_{i-1}^{(n)}}{\Delta x} = 0 \tag{13.3.7}$$

Solving for $u_i^{(n+1)}$ and rewriting yield

$$\begin{aligned}
u_i^{(n+1)} &= u_i^{(n)} - \gamma(u_i^{(n)} - u_{i-1}^{(n)}) \\
&= (1 - \gamma)u_i^{(n)} + \gamma u_{i-1}^{(n)}
\end{aligned} \tag{13.3.8}$$

where $\gamma = a\Delta t/\Delta x$ is the Courant number. When $n = 0$, $u_0^{(1)}$ is given by the boundary condition and all the values of $u_i^{(0)}$ are given by the initial condition, so Eq. (13.3.8) can be evaluated for all the grid points. The same is true for any time step because $u_0^{(n+1)}$ is always given by the boundary condition and $u_i^{(n)}$ are all known from the previous step. The FTBS scheme happens to be identical to the method of characteristics on a grid.

The FTBS scheme introduced so far is based on the assumption that $a > 0$. Therefore, for $a \leqslant 0$, the difference scheme has to be switched to the forward difference scheme so the scheme remains as an "upwind scheme." If the sign of $a(x, t)$ changes in the middle of the domain, the scheme is switched from one to the other. However, both cases can be written in a single equation as

$$\frac{u_i^{(n+1)} - u_i^{(n)}}{\Delta t} + a\,\frac{u_{i+1}^{(n)} - u_{i-1}^{(n)}}{2\Delta x} - |a|\Delta x\,\frac{u_{i+1}^{(n)} - 2u_i^{(n)} + u_{i-1}^{(n)}}{2\Delta x^2} = 0 \tag{13.3.9}$$

The second term is the central difference approximation for u_x. The third term is a central difference approximation for $-|a|\Delta x u_{xx}/2$. We can interpret the FTBS scheme as using the central difference approximation for u_x and artificially adding the central difference approximation for $-|a|\Delta x u_{xx}/2$, which is named the *numerical viscosity term*.

When the initial and boundary conditions are all nonnegative, the solution of the present scheme never becomes negative. This property is important for the obvious reason that, if transport of a real matter is represented by the equation, the solution should never become negative.

Stability of the present scheme is investigated next by applying the Fourier stability analysis introduced in Section 12.3. In the Fourier stability analysis, the inhomogeneous term is set to zero. In an infinite domain, the solution is expanded in a Fourier series. Considering one Fourier component at a time, the solution of the equation being investigated, for example Eq. (13.3.8), is written as

$$u_i^{(n)} = G^n \exp(ij\pi/k)$$

or equivalently

$$u_i^{(n)} = G^n \exp(ij\theta) \qquad (13.3.10)$$

with

$$\theta = \pi/k, \quad k = \pm 1, \pm 2, \pm 3, \ldots, \pm\infty$$

$G = G(\theta)$ is the amplification factor (generally a complex function of θ), $j = \sqrt{-1}$, k is the wave length in terms of the number of grid intervals. Since the smallest magnitude of k is 1, the largest magnitude of θ is π, so θ satisfies $-\pi \leqslant \theta \leqslant \pi$. Indeed, introducing Eq. (13.3.10) into Eq. (13.3.8) and then dividing by $\exp(ij\theta)$ yield

$$G = 1 - \gamma(1 - e^{-j\theta}) \qquad (13.3.11)$$

Figure 13.5 Effect of γ on the amplitude factor of FTBS

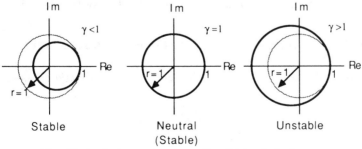

Stable   Neutral   Unstable
                (Stable)
(Amplitude factors are along the thick circles)

The dependence of $|G(\theta)|$ on γ may be examined by plotting it on the complex plane as shown in Figure 13.5. It shows that, if $\gamma \leqslant 1$, the curve representing G is on or inside the unit circle, which indicates that the amplification factor never exceeds unity for any value of θ. Thus, the scheme is stable if $\gamma \leqslant 1$. However, if $\gamma > 1$, it goes outside the unit circle, so the scheme is unstable.

Forward Euler in Time and Central Difference in Space (FTCS)

Here, the central difference approximation in space is used to approximate u_x:

$$u_x \simeq \frac{u_{i+1}^{(n)} - u_{i-1}^{(n)}}{2\Delta x} \tag{13.3.12}$$

Thus, the difference equation becomes

$$u_i^{(n+1)} = u_i^{(n)} - \frac{\gamma}{2}(u_{i+1}^{(n)} - u_{i-1}^{(n)}) \tag{13.3.13}$$

The amplification factor G for this equation becomes

$$G = 1 - \frac{\gamma}{2}(e^{j\theta} - e^{-j\theta})$$

$$= 1 - \gamma j \sin(\theta) \tag{13.3.14}$$

where $j = \sqrt{-1}$. Its magnitude is

$$|G| = \sqrt{G\overline{G}} = \sqrt{1 + \gamma^2 \sin^2 \theta} \geqslant 1 \quad \text{for all } \theta \tag{13.3.15}$$

Therefore, this scheme is unstable always.

Equation (13.3.13) is identical to Eq. (13.3.9) except that the third term of the latter does not exist in the former. This observation indicates that the third term of Eq. (13.3.9) plays an important role of stabilizing the first-order upwind scheme.

Backward Euler in Time and Central Difference in Space (BTCS)

With the backward Euler approximation in time, the central difference approximation in space at time $n + 1$ is used:

$$u_x \simeq \frac{u_{i+1}^{(n+1)} - u_{i-1}^{(n+1)}}{2\Delta x} \tag{13.3.16}$$

The difference approximation for Eq. (13.3.1) is written as

$$-\frac{\gamma}{2} u_{i-1}^{(n+1)} + u_i^{(n+1)} + \frac{\gamma}{2} u_{i+1}^{(n+1)} = u_i^n \qquad (13.3.17)$$

The left side of Eq. (13.3.17) has three unknowns. Using the left boundary condition, Eq. (13.3.17) for $i = 1$ becomes

$$u_1^{(n+1)} + \frac{\gamma}{2} u_2^{(n+1)} = u_1^{(n)} + \frac{\gamma}{2} u_0 \qquad (13.3.18)$$

When $a > 0$, the term u_{I+1} (for the equation for $i = I$) is not given a priori. Therefore, an artificial boundary condition for u_{I+1} becomes necessary. Although there are several alternative schemes for artificial boundary conditions, a frequently used approach is to extrapolate u_{I+1} from inside [Yee/Beam/Warming] as

$$u_{I+1} = 2u_I - u_{I-1} \qquad (13.3.19)$$

By using this equation, Eq. (13.3.17) for $i = I$ becomes

$$-\gamma u_{I-1}^{(n+1)} + (\gamma + 1)u_I^{(n+1)} = u_I^{(n)} \qquad (13.3.20)$$

The set of equations for $i = 1, 2, \ldots, I$ forms a tridiagonal set of simultaneous equations.

The amplification factor becomes

$$G = \frac{1}{1 + \frac{\gamma}{2}(e^{j\theta} - e^{-j\theta})}$$

$$= \frac{1}{1 + j\gamma \sin(\theta)} \qquad (13.3.21)$$

The absolute value of G becomes

$$|G| = \frac{1}{\sqrt{1 + \gamma^2 \sin^2 \theta}} \leqslant 1 \qquad (13.3.22)$$

Therefore, the scheme is unconditionally stable.

SUMMARY OF THIS SECTION

(a) The FTBS scheme, which is also called *first-order upwind explicit scheme*, is stable if γ (Courant number) $\leqslant 1$.

(b) The FTCS scheme is always unstable.

(c) The BTCS scheme is stable unconditionally.

13.4 TRUNCATION ERROR ANALYSIS

Numerical schemes for hyperbolic PDEs have errors originated from truncation errors of difference approximations. Truncation errors give certain artificial nature to the solution of a numerical scheme. To investigate the effect of truncation errors, the modified equations are used, which are differential equations similar to the original hyperbolic PDEs. The modified equations include all the effects of the truncation errors. They are derived by introducing Taylor expansions into the difference approximations. The solutions of both original and modified equations can be obtained in an analytical form on an equispaced grid in an infinite domain. So, by comparison of the two analytical solutions, the effects of the truncation errors can be analyzed.

We first examine the FTBS scheme derived in the preceding section:

$$u_i^{(n+1)} - u_i^{(n)} + \gamma[u_i^{(n)} - u_{i-1}^{(n)}] = 0 \tag{13.4.1}$$

Taylor expansions of $u_i^{(n+1)} = u(x_i, t_{n+1})$ and $u_{i-1}^{(n)} = u(x_{i-1}, t_n)$ about $x = x_i$ and $t = t_n$ are, respectively,

$$u_i^{(n+1)} = u + \Delta t u_t + \frac{\Delta t^2}{2} u_{tt} + \cdots \tag{13.4.2}$$

$$u_{i-1}^{(n)} = u - \Delta x u_x + \frac{\Delta x^2}{2} u_{xx} - \cdots \tag{13.4.3}$$

where u without superscript expresses $u = u(x_i, t_n)$, and u_t and u_x are partial derivatives of u at (x_i, t_n). Introducing the Taylor expansions into Eq. (13.4.1) yields

$$u_t + \frac{\Delta t}{2} u_{tt} + \frac{\Delta t^2}{6} u_{ttt} + \cdots + \left[au_x - \frac{a\Delta x}{2} u_{xx} + \frac{a\Delta x^2}{6} u_{xxx} - \cdots \right] = 0 \tag{13.4.4}$$

Equation (13.4.4) is impossible to analyze in its original form because it involves higher derivatives in both t and x. So, it is transformed to a first-order PDE in t by eliminating all the second and higher order partial derivatives with respect to t.

In general, an equation given by

$$u_t + A_2 u_{tt} + A_3 u_{ttt} + \cdots + B_1 u_x + B_2 u_{xx} + B_3 u_{xxx} + B_4 u_{xxxx} + \cdots = 0 \tag{13.4.5}$$

with $B_1 = a$ may be transformed to

$$u_t + c_1 u_x + c_2 u_{xx} + c_3 u_{xxx} + c_4 u_{xxxx} + \cdots = 0 \tag{13.4.6}$$

where

$$c_1 = B_1 = a$$

$$c_2 = B_2 + a^2 A_2$$

$$c_3 = B_3 + 2aA_2B_2 + a^3(2A_2^2 - A_3) \qquad (13.4.7)$$

$$c_4 = B_4 + A_2B_2^2 + 2aA_2B_3 + 6a^2A_2^2B_2 - 3a^2A_3B_2$$

$$\qquad + a^4(5A_2^3 - 5A_2A_3 + A_4)$$

The foregoing transformation is explained in Appendix F. Difference equations such as Eq. (13.4.1) can be transformed to the form of Eq. (13.4.6), which is named the *modified equation*.

Applying the relation between Eq. (13.4.5) and Eq. (13.4.6) to Eq. (13.4.4), the modified equation becomes

$$u_t + au_x - \frac{a\Delta x}{2}(1 - \gamma)u_{xx} + \frac{a\Delta x^2}{6}(2\gamma^2 - 3\gamma + 1)u_{xxx} + \cdots = 0 \qquad (13.4.8)$$

where

$$c_2 = -\frac{a\Delta x}{2}(1 - \gamma)$$

$$c_3 = \frac{a\Delta x^2}{6}(2\gamma^2 - 3\gamma + 1)$$

are introduced. The modified equation given by Eq. (13.4.8) is a differential equation representing the difference equation that is the FTBS scheme in the present analysis. By comparing Eq. (13.4.8) to Eq. (13.3.1), we find that all the terms other than the first two are the effects of truncation errors.

The analytical solution of Eq. (13.4.8) in an infinite domain is now sought in the form

$$u(x, t) = f_\theta(t) \exp(jx\theta) \qquad (13.4.9)$$

where $j = \sqrt{-1}$ and θ has the same meaning as defined after Eq. (13.3.10) and is a constant related to the frequency of a Fourier component in space. Introducing Eq. (13.4.9) into Eq. (13.4.6) yields

$$\frac{d}{dt}f_\theta(t) + [ja\theta - c_2\theta^2 - jc_3\theta^3 + c_4\theta^4 + \cdots]f_\theta(t) = 0 \qquad (13.4.10)$$

where $c_1 = a$ is used. An analytical solution of Eq. (13.4.10) is

$$f_\theta(t) = f_\theta(0) \exp(-ja\theta t + c_2\theta^2 t + jc_3\theta^3 t - c_4\theta^4 t + \cdots) \qquad (13.4.11)$$

where $f_\theta(0)$ is determined by an initial condition. The general solution for Eq. (13.4.8) is the summation of Eq. (13.4.11) for all possible θ (or more precisely it is an integral over θ from $-\pi$ to π), but we are interested in Eq. (13.4.11) with only one value of θ at a time.

Equation (13.4.11) can be expressed as

$$f_\theta(t) = f_\theta(0) \exp\left(-ja\theta t\right) \exp\left(c_2\theta^2 t\right) \exp\left(jc_3\theta^3 t\right) \exp\left(-c_4\theta^4 t\right) \cdots \quad (13.4.12)$$

[Anderson et al.]. On the other hand, the exact solution of Eq. (13.3.1) is

$$f_\theta(t) = f_\theta(0) \exp\left(-ja\theta t\right) \quad (13.4.13)$$

Thus, all the exponential terms except the first one in Eq. (13.4.12) are the effects of truncation errors.

Now we examine exponential terms in Eq. (13.4.12). If $c_2 > 0$, the second exponential term, namely, $\exp\left(c_2\theta^2 t\right)$, increases in time, that is, the scheme becomes unstable. The absolute value of the third exponential term, $\exp\left(jc_3\theta^3 t\right)$, does not change in time because its argument is imaginary. However, if the third exponential term is combined with the first term, the product becomes

$$\exp\left(-aj\theta t\right)\exp\left(jc_3\theta^3 t\right) = \exp\left[-j(a - c_3\theta^2)\theta t\right] \quad (13.4.14)$$

which means that the speed of the wave in the numerical solution is $a - c_3\theta^2$ rather than a. In the exact solution, the wave speed is a and independent of θ. The dependence of the wave speed on θ in the numerical solution is the effect of the third derivative in the truncation error. The higher the value of θ becomes, the more retarded or advanced is the wave speed in the numerical solution. The effect of variable wave speed is called *aliasing error*, and it causes oscillation of the numerical solution particularly where the solution has a sharp spatial change such as a shock.

The fourth-order term $\exp\left(-c_4\theta^4 t\right)$ is increasing or decreasing in time depending on whether $c_4 < 0$ or $c_4 > 0$, respectively. The combined effect of the c_2 and c_4 terms is expressed by

$$\exp\left[(c_2 - c_4\theta^2)\theta^2 t\right] \quad (13.4.15)$$

If $c_2 = 0$, then c_4 determines stability: The scheme is stable if $c_4 > 0$, or unstable if $c_4 < 0$. If $c_2 < 0$ and $c_4 \geqslant 0$, the scheme is stable. If $c_2 < 0$ but $c_4 < 0$, the scheme is stable when $|c_2| > |c_4|\pi^2$, where the fact that the maximum of θ equals π, because θ is bounded by $-\pi \leqslant \theta \leqslant \pi$, is used. This stability condition becomes always satisfied as Δt is decreased. If $c_2 > 0$, the scheme becomes unstable even when $c_4 > 0$, because $c_2 - c_4\theta^2$ becomes positive for small values of θ^2.

We conclude that the FTBS scheme is unstable for $\gamma > 1$ because c_2 given after Eq. (13.4.8) becomes positive for $\gamma > 1$. However, when $\gamma = 1$, c_3 vanishes so there is no aliasing error. For $\gamma < 1$, the scheme is stable but with a significant disadvantage.

That is, the second exponential term of Eq. (13.4.12) damps the solution, or the solution decays to zero as time increases. The same effect occurs when the travel distance of a wave increases. Since the exact solution given by Eq. (13.4.3) does not have such a term, the damping effect of a negative c_2 is the effect of the truncation error of the difference equation, which is named *numerical damping* or *second-order numerical viscosity effect*. A positive c_4 also has the damping effect named *fourth-order numerical viscosity effect*. Both the second-order numerical viscosity and fourth-order numerical viscosity effects damp more strongly the waves of higher spatial frequency than lower frequency waves. As the spatial frequency of the wave increases, the damping effect of the fourth-order damping effect increases more rapidly than the second-order effect because the former is proportional to θ^4 in the exponential term while the latter is proportional to θ^2.

A similar analysis for BTCS gives the following modified equation:

$$u_t + au_x - \frac{1}{2}a^2 \Delta t u_{xx} + \left[\frac{1}{6}a(\Delta x)^2 + \frac{1}{3}a^3 \Delta t^2\right] u_{xxx} + \cdots \qquad (13.4.16)$$

Equation (13.4.16) indicates that, because $c_2 = -\frac{1}{2}a^2\Delta t < 0$, the scheme is unconditionally stable, but it has aliasing errors because $c_3 > 0$ all the time.

SUMMARY OF THIS SECTION

(a) The effects of truncation errors are analyzed by transforming a difference equation to the modified equation.

(b) The scheme is stable if $c_2 \leqslant 0$. If $c_2 = 0$, then $c_4 > 0$ is necessary for stability. The value c_3 has no effect on stability, but it causes aliasing errors.

(c) A positive c_4 has strong damping effects on the waves of high spatial frequencies.

13.5 HIGHER-ORDER SCHEMES

Lax-Wendroff Scheme

Here, we consider

$$u_t + au_x = 0 \quad \text{(We set } s = 0 \text{ for simplicity.)} \qquad (13.5.1)$$

The Taylor expansion of $u_i^{(n+1)}$ about x_i and t_n may be written as

$$u_i^{(n+1)} = u_i^{(n)} + \Delta t(u_t)_i^n + \frac{1}{2}\Delta t^2(u_{tt})_i^n + \cdots$$

$$= u_i^{(n)} - a\Delta t(u_x)_i^n + \frac{1}{2}a^2\Delta t^2(u_{xx})_i^n + \cdots \qquad (13.5.2)$$

where u_t is eliminated using Eq. (13.5.1) and u_{tt} is eliminated by

$$u_{tt} = -au_{xt} = a^2 u_{xx}$$

Truncating after the second-order term of Eq. (13.5.2) and applying central difference approximations for u_x and u_{xx} yield

$$u_i^{(n+1)} = u_i^{(n)} - \frac{\gamma}{2}(u_{i+1}^{(n)} - u_{i-1}^{(n)}) + \frac{\gamma^2}{2}(u_{i-1}^{(n)} - 2u_i^{(n)} + u_{i+1}^{(n)}) \qquad (13.5.3)$$

where

$$\gamma = a\Delta t/\Delta x$$

Equation (13.5.3) is an explicit scheme named the *Lax-Wendroff scheme*.

The truncation error of the Lax-Wendroff scheme comes from two sources: (1) the truncation of the Taylor expansion after the second derivative, and (2) the central difference approximations for u_x and u_{xx}. The order of truncation error from the Taylor expansion of $u_i^{(n+1)}$ is Δt^3, the order of error of the central difference approximation for u_x is $\Delta t\Delta x^2$, and that for u_{xx} is $\Delta t^2\Delta x^2$.

The amplification factor of the Lax-Wendroff scheme is

$$G = 1 - \gamma^2[1 - \cos(\theta)] - j\gamma \sin(\theta) \qquad (13.5.4)$$

The Lax-Wendroff scheme is stable if $0 \leqslant |\gamma| \leqslant 1$. When $\gamma = 1$, the scheme reduces to $u_i^{(n+1)} = u_{i-1}^{(n)}$ and is exact.

The modified equation is

$$u_t + au_x + \frac{1}{6}a\Delta x^2(1 - \gamma^2)u_{xxx} + \frac{1}{8}a\Delta x^3\gamma(1 - \gamma^2)u_{xxxx} + \cdots = 0 \qquad (13.5.5)$$

Equation (13.5.5) indicates that the truncation error of the Lax-Wendroff scheme vanishes when $\gamma = 1$. If $\gamma < 1$, the leading truncation error is the third derivative with a positive coefficient. Thus, the scheme is second-order accurate. The magnitude of each error term increases when Δt is decreased, but Δx is fixed (γ approaches 0). The scheme is stable for $\gamma < 1$ because c_4, which is the coefficient of the fourth derivative term, satisfies $c_4 > 0$, although the second-order derivative term is zero.

MacCormack Scheme

The MacCormack scheme is

$$\bar{u}_i^{(n+1)} = u_i^{(n)} - \gamma(u_{i+1}^{(n)} - u_i^{(n)})$$

$$u_i^{(n+1)} = \frac{1}{2}[u_i^{(n)} + \bar{u}_i^{(n+1)} - \gamma(\bar{u}_i^{(n+1)} - \bar{u}_{i-1}^{(n+1)})] \qquad (13.5.6)$$

The first equation is a predictor, and the second is the corrector. For the linear problems as considered here, the predictor can be eliminated by introducing the first equation into the second, so that the MacCormack scheme becomes identical to the Lax-Wendroff scheme. The modified equation and the stability criterion are the same as for the Lax-Wendroff scheme.

Third-order Upwind Scheme

The spatial derivative can be approximated by the third-order accurate difference approximation given by

$$a(u_x)_i \simeq \begin{cases} a \dfrac{2u_{i+1} + 3u_i - 6u_{i-1} + u_{i-2}}{6\Delta x} & \text{for } a > 0 \\[4mm] a \dfrac{-u_{i+2} + 6u_{i+1} - 3u_i - 2u_{i-1}}{6\Delta x} & \text{for } a < 0 \end{cases}$$

which can be written in a single form as

$$a(u_x)_i \simeq a \frac{-u_{i+2} + 8u_{i+1} - 8u_{i-1} + u_{i-2}}{12\Delta x}$$

$$+ |a| \frac{u_{i+2} - 4u_{i+1} + 6u_i - 4u_{i-1} + u_{i-2}}{12\Delta x} \tag{13.5.7}$$

[Kawamura; Kawamura/Kuwahara; Leonard]. It can be shown that, by expanding each term into Taylor series, the right side of Eq. (13.5.7) becomes

$$a(u_x)_i + \frac{1}{12}|a|\Delta x^3(u_{xxxx})_i + \cdots \tag{13.5.8}$$

Here, the first term equals the left side of Eq. (13.5.7), and the second term is the truncation error, which is proportional to Δx^3 and to the fourth derivative of u.

Using Eq. (13.5.7), a semidifference approximation for Eq. (13.5.1) is written as

$$u_t + a \frac{-u_{i+2} + 8u_{i+1} - 8u_{i-1} + u_{i-2}}{12\Delta x}$$

$$+ |a| \frac{u_{i+2} - 4u_{i+1} + 6u_i - 4u_{i-1} + u_{i-2}}{12\Delta x} = 0 \tag{13.5.9}$$

Equation (13.5.9) has no aliasing error due to a third derivative term.

Equation (13.5.9) can be fully discretized by introducing the forward or backward Euler schemes or the Crank-Nicolson scheme for u_t.

With the forward Euler scheme in time, it becomes an explicit scheme and written as

$$u_i^{(n+1)} = u_i^{(n)} + \Delta t \left[-a \frac{-u_{i+2}^{(n)} + 8u_{i+1}^{(n)} - 8u_{i-1}^{(n)} + u_{i-2}^{(n)}}{12\Delta x} \right.$$

$$\left. - |a| \frac{u_{i+2}^{(n)} - 4u_{i+1}^{(n)} + 6u_i^{(n)} - 4u_{i-1}^{(n)} + u_{i-2}^{(n)}}{12\Delta x} \right] \qquad (13.5.10)$$

To study the effects of truncation errors in the explicit scheme, we derive the modified equation. Using the Taylor expansions in Eq. (13.5.10) yields

$$u_t + \frac{\Delta t}{2} u_{tt} + \frac{\Delta t^2}{6} u_{ttt} + \cdots + au_x + \frac{\Delta x^3}{12} |a| u_{xxxx} + \cdots = 0 \qquad (13.5.11)$$

where Eq. (13.5.8) has been also used. By eliminating the second-order and higher-order derivatives with respect to t in accordance with the algorithm written in the preceding section, we get

$$u_t + au_x + a^2 \Delta t u_{xx} + \frac{5a^3 \Delta t^2}{6} u_{xxx} + \left(\frac{|a| \Delta x^3}{12} + \frac{a^4 \Delta t^3}{4} \right) u_{xxxx} + \cdots = 0 \quad (13.5.12)$$

The leading error term of truncation is second order (derivative) and has a positive sign, so the whole scheme based on the forward Euler scheme reduces to a first-order accurate scheme. The second-order term with the positive sign has an antidiffusive effect that makes the scheme unstable unless the antidiffusive effect is made smaller than the effect of the fourth-order truncation error. However, when Δt is decreased, the second- and third-order errors both approach zero, so they can be made as small as desired at the expense of using a very small Δt.

If the backward Euler differencing in time is used, the scheme becomes an implicit scheme:

$$u_i^{(n+1)} + a \frac{-u_{i+2}^{(n+1)} + 8u_{i+1}^{(n+1)} - 8u_{i-1}^{(n+1)} + u_{i-2}^{(n+1)}}{12\Delta x}$$

$$+ |a| \frac{u_{i+2}^{(n+1)} - 4u_{i+1}^{(n+1)} + 6u_i^{(n+1)} - 4u_{i-1}^{(n+1)} + u_{i-2}^{(n+1)}}{12\Delta x} = u_i^{(n)} \qquad (13.5.13)$$

The set of simultaneous equations must be solved by a pentadiagonal scheme in each time step. The implicit scheme is unconditionally stable. A modified equation analysis shows that the leading error term of truncation is second order and the same as for the forward Euler explicit version (but its sign becomes opposite). The second-order and third-order error terms can be decreased by using a small Δt, but the benefit of using the implicit scheme then disappears.

An explicit scheme that is second-order accurate in time based on the Adams-Bashforth predictor is written [Kawamura] as

$$\frac{u_i^{(n+1)} - u_i^{(n)}}{\Delta t} + \frac{1}{2}[3F^{(n)} - F^{(n-1)}] = 0 \qquad (13.5.14)$$

where F_n is the third-order difference approximation for au_x. Because of the second-order accuracy in time, the modified equation for Eq. (13.5.14) does not include the u_{xx} term (the cause of the antidiffusive effect in the forward Euler scheme in time). Therefore, Eq. (13.5.14) is more stable than Eq. (13.5.10).

SUMMARY OF THIS SECTION

(a) The Lax-Wendroff and MacCormack schemes are second-order accurate. The accuracy of the schemes are best when $\gamma = 1$. Therefore, accuracy decreases even when Δt is decreased unless Δx is also decreased.

(b) The third-order upwind scheme has an accuracy of the third order in space, but forward or backward Euler differencing in time introduces errors of the second order. However, the magnitude of the second-order errors can be decreased by using a small Δt independently of Δx.

13.6 DIFFERENCE SCHEMES IN THE CONSERVATIVE FORM

In the preceding sections, difference equations in the nonconservative forms are discussed. With the nonconservative form, the total amount of the quantity in the whole domain may be lost or gained in each time step because of numerical errors, and such effects may accumulate as time progresses. If a difference equation is written in the conservative form, the summation of the equations in space satisfies the conservation of the quantity in the whole domain.

To discuss the concept of the conservative form for a hyperbolic PDE, consider a compressible fluid flow in a straight tube of a constant cross-sectional area. The continuity equation is given

$$\frac{\partial}{\partial t}\rho(x, t) + \frac{\partial}{\partial x}f(x, t) = 0 \qquad (13.6.1)$$

where $\rho(x, t)$ is density of the fluid, $f(x, t) = u(x, t)\rho(x, t)$ is the mass flow rate per unit cross-sectional area, and $u(x, t)$ is the velocity of the fluid. Equation (13.6.1) represents conservation of mass. If we integrate the equation in space from $x = a$ to $x = b$, we get

$$\frac{d}{dt}\left[\int_a^b \rho(x,t)\,dx\right] = f(a,t) - f(b,t) \tag{13.6.2}$$

On the basis of unit cross-sectional area, the left side is the rate of change in the total mass in $a < x < b$, the first term on the right side is the rate of mass flow entering through $x = a$, and the second term is the same exiting through $x = b$. Thus, Eq. (13.6.2) represents the mass conservation in $a < x < b$.

Equation (13.6.1) may be alternatively written, after introducing $f = u(x,t)\rho(x,t)$ and performing differentiation of the second term, as

$$\frac{\partial}{\partial t}\rho(x,t) + \rho(x,t)\frac{\partial}{\partial x}u(x,t) + u(x,t)\frac{\partial}{\partial x}\rho(x,t) = 0 \tag{13.6.3}$$

Equation (13.6.3) is mathematically equivalent to Eq. (13.6.1), but mass conservation cannot be explained immediately in this form, so the physical meaning of conservation is lost. Thus, we call Eq. (13.6.3) a nonconservative form and Eq. (13.6.1) a conservative form.

The differences between the conservative form and the nonconservative form can cause profound effects for difference equations for PDEs. The conservative form of a difference equation may always be written as

$$\frac{\rho_i^{(n+1)} - \rho_i^{(n)}}{\Delta t} + \frac{g_{i+\frac{1}{2}} - g_{i-\frac{1}{2}}}{\Delta x} = 0 \tag{13.6.4}$$

where $g_{i+(1/2)}$ is a numerical approximation for f at $x_{i+(1/2)}$, and usually a function of f_{i+1} and f_i. There is a freedom in choosing the particular form for $g_{i+(1/2)}$. For example, if

$$g_{i+\frac{1}{2}} = \frac{f_{i+1} + f_i}{2} \tag{13.6.5}$$

then the second term of Eq. (13.6.4) reduces to

$$\frac{\rho_i^{(n+1)} - \rho_i^{(n)}}{\Delta t} + \frac{f_{i+1} - f_{i-1}}{2\Delta x} = 0 \tag{13.6.6}$$

which is a central difference approximation and is unstable, however.

As another example, $g_{i+\frac{1}{2}}$ may be written as

$$g_{i+\frac{1}{2}} = \frac{1}{2}(f_{i+1} + f_i) - |a_{i+\frac{1}{2}}|(u_{i+1} - u_i) \tag{13.6.7}$$

where

$$a_{i+\frac{1}{2}} = \frac{f_{i+1} - f_i}{u_{i+1} - u_i}$$

With this choice, Eq. (13.6.4) becomes

$$\frac{\rho_i^{(n+1)} - \rho_i^{(n)}}{\Delta t} + \frac{f_{i+1} - f_{i-1}}{2\Delta x}$$

$$+ \frac{-\left|a_{i+\frac{1}{2}}\right|u_{i+1} + \left(\left|a_{i+\frac{1}{2}}\right| + \left|a_{i-\frac{1}{2}}\right|\right)u_i - \left|a_{i-\frac{1}{2}}\right|u_{i-1}}{2\Delta x} = 0 \qquad (13.6.8)$$

The last term can be interpreted as a numerical viscosity term.

In Eq. (13.6.4), g is called numerical flux because, as Eq. (13.6.7) shows, it consists of mass flux and an artificial term that yields a viscosity effect.

The reason Eq. (13.6.6) is the conservative form is obvious: Adding Eq. (13.6.4) for $i = j, j + 1, \ldots, k$, and reorganizing yields

$$\Delta x \sum_{i=j}^{k} \rho_i^{(n+1)} - \Delta x \sum_{i=j}^{k} \rho_i^{(n)} = -\Delta t(g_{k+\frac{1}{2}} - g_{j-\frac{1}{2}}) \qquad (13.6.9)$$

The left side is the change of the total mass in $\left[x_{j-(1/2)}, x_{k+(1/2)}\right]$ between time t_n and time t_{n+1}. The first term on the right side is the total flow of the numerical quantity at $x_{k+(1/2)}$ in Δt and the second term on the right side is the same at $x_{j-(1/2)}$. Equation (13.6.9) indicates that the total mass in the portion of the pipe considered is governed by numerical flux at two end points. If the boundary conditions for the numerical fluxes at the end points are set exactly to the mass flux, Eq. (13.6.9) maintains the balance of mass in the pipe. It is important to recognize that if the numerical scheme is in the conservative form, the particular choice of a numerical approximation does not affect the total mass.

Example 13.3

Write the difference approximation in the conservation form, which is first-order explicit in time and the third-order upwind difference approximation in space, for

$$u_t(x, t) + [a(x)u(x, t)]_x = 0$$

⟨Solution⟩

The explicit difference approximation based on the third-order upwind differencing can be written as

$$\frac{u_i^{(n+1)} - u_i^{(n)}}{\Delta t} + \frac{G_{i+\frac{1}{2}} - G_{i-\frac{1}{2}}}{\Delta x} = 0 \qquad (A)$$

where G is a flux given by

$$G_{i+\frac{1}{2}} = \frac{-(au)_{i+2}^{(n)} + 7(au)_{i+1}^{(n)} + 7(au)_i^{(n)} - (au)_{i-1}^{(n)}}{12}$$

$$+ \frac{(|a|u)_{i+2}^{(n)} - 3(|a|u)_{i+1}^{(n)} + 3(|a|u)_i^{(n)} - (|a|u)_{i-1}^{(n)}}{12} \tag{B}$$

Equation (A) reduces to Eq. (13.5.9) if a = constant.

Difference equations can become nonconservative for different reasons. One of the major causes for the nonconservative form is to derive difference equations from a nonconservative form of the PDE. The difference equations derived from the nonconservative form of a PDE cannot be written in the form of Eq. (13.6.4). An example of nonconservative form is the upwind difference approximation (FTBS when $u > 0$). If u changes its sign from positive to negative between two consecutive grid points, say i and $i + 1$, the difference equations are

$$\frac{\rho_i^{(n+1)} - \rho_i^{(n)}}{\Delta t} + \frac{u_i \rho_i^{(n)} - u_{i-1} \rho_{i-1}^{(n)}}{\Delta x} = 0, \quad u_i > 0 \tag{13.6.10}$$

and

$$\frac{\rho_{i+1}^{(n+1)} - \rho_{i+1}^{(n)}}{\Delta t} + \frac{u_{i+2} \rho_{i+2}^{(n)} - u_{i+1} \rho_{i+1}^{(n)}}{\Delta x} = 0, \quad u_{i+1} < 0 \tag{13.6.11}$$

When Eqs. (13.6.10) and (13.6.11) are added, no cancellation of the flux at the interface between grid points i and $i + 1$ occurs. Thus, conservation is not satisfied.

A nonconservative form may also result when the geometrical consideration is not appropriate. Consider the equation

$$\frac{\partial}{\partial t} A(x)\rho(x, t) + \frac{\partial}{\partial x} A(x)u(x, t)\rho(x, t) = 0 \tag{13.6.12}$$

which is a conservation law for one-dimensional flow with variable cross-sectional areas, where $A(x)$ is the cross-sectional area at x. Performing differentiation of the second term yields

$$A(x)\frac{\partial}{\partial t}\rho(x, t) + A(x)\frac{\partial}{\partial x}u(x, t)\rho(x, t) + A_x(x)u(x, t)\rho(x, t) = 0 \tag{13.6.13}$$

A difference approximation for Eq. (13.6.13) may then be written, for example, as

$$A(x_i)\frac{\rho_i^{(n+1)} - \rho_i^{(n)}}{\Delta t} + A(x_i)\frac{u_{i+(1/2)}^{(n)}\rho_{i+(1/2)}^{(n)} - u_{i-(1/2)}^{(n)}\rho_{i-(1/2)}^{(n)}}{\Delta x} + (A_x)_i u_i^{(n)}\rho_i^{(n)} = 0 \tag{13.6.14}$$

When the equations for consecutive grid points are added, no cancellation of the flux terms occurs, so Eq. (13.6.14) does not satisfy conservation.

In actual solution of a hyperbolic PDE, both the conservative form and the nonconservative form are used. Often, a nonconservative form is used for simplicity of the solution algorithm. However, the conservative form is preferred if possible.

SUMMARY OF THIS SECTION

(a) Both the nonconservative and conservative forms of difference equations are used, but the latter is more desirable.

(b) Conservative difference equations are derived from a hyperbolic PDE in the conservative form.

(c) The difference equations in the conservative form can always be written in the form of Eq. (13.6.4).

13.7 COMPARISON OF SCHEMES THROUGH WAVE TESTS

An effective method to investigate the performance of a scheme is to run test problems. Here, we solve the equation

$$u_t + u_x = 0$$

with initial conditions of a square wave. The exact solution of the equation at any time has the same square shape as the initial distribution, but the location of the wave is continuously advanced by the unit velocity.

Throughout the tests, grid spacing is $\Delta x = 0.1$, and the time step $\Delta t = \rho \Delta x$ is set to $\Delta t = \rho = 0.01$. Figure 13.6 shows the computed results for the square wave of the following three numerical schemes:

FTBS

Lax-Wendroff

Third-order-upwind explicit

The FTBS scheme never gets negative values of numerical solution, nor does it have any oscillatory behavior. However, the wave tends to spread and be smeared as it travels. The height of the wave becomes lower and its width increases.

The Lax-Wendroff scheme has much less smearing effects of waves than the FTBS scheme, but it has a significant oscillatory behavior, which is the aliasing error associated with the third-order truncation error term. This trend increases as γ approaches 0.

With the third-order upwind scheme, the height and width of the square wave are both better maintained. However, heights of the wave become negative in the front and rear of the wave. The trend of oscillation around a sharp change of the solution is due to aliasing error. The flux-corrected scheme in Section 13.9 approximates the traveling wave better than the examples given in this section.

13.8 NUMERICAL SCHEMES FOR NONLINEAR HYPERBOLIC PDEs

The nonlinear hyperbolic PDEs may be written in the conservative form

$$u_t + F_x = 0 \tag{13.8.1}$$

1st-order upwind, explicit (FTBS)

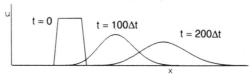

Lax-Wendroff

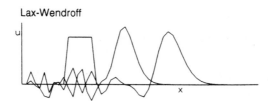

3rd-order upwind, explicit

Figure 13.6 Numerical simulation of a square wave moving at a constant speed a

where $F = F(u)$ is a nonlinear function of u. Equation (13.8.1) may also be written in a nonconservative form as

$$u_t + \frac{\partial F}{\partial u}\, u_x = 0 \tag{13.8.2}$$

Courant-Isaacson-Rees Scheme

For a nonlinear hyperbolic PDE, the FTBS does not work in its original form. However, schemes similar to FTBS for linear hyperbolic PDEs have been developed to meet the following criteria:

(a) F_x is approximated by the central difference approximation.

(b) A numerical diffusion term is added.

(c) Difference equations are in the conservative form.

The Courant-Isaacson-Rees scheme and the Lax-Friedrich scheme fall into this category. The former is given by

$$u_i^{(n+1)} = u_i^{(n)} - k(f_{i+(1/2)}^{(n)} - f_{i-(1/2)}^{(n)}) \tag{13.8.3}$$

where $k = \Delta t/\Delta x$, and with omissions of superscript (n) to F and u,

$$f_{i+\frac{1}{2}} = \frac{1}{2}\left[F_i + F_{i+1} - \left|a_{i+\frac{1}{2}}\right|(u_{i+1} - u_i)\right]$$

$$a_{i+\frac{1}{2}} = \begin{cases} \dfrac{F_{i+1} - F_i}{u_{i+1} - u_i}, & \text{if } u_{i+1} - u_i \neq 0 \\ 0, & \text{if } u_{i+1} - u_i = 0 \end{cases} \tag{13.8.4}$$

By eliminating $f_{i+\frac{1}{2}}$, Eq. (13.8.3) becomes

$$u_i^{n+1} = u_i^n - \frac{\Delta t}{2\Delta x}(F_{i+1} - F_{i-1})$$

$$- \frac{\Delta t}{2\Delta x}\left[-\left|F_{i+1} - F_i\right| \text{sign}\,(u_{i+1} - u_i)\right.$$

$$\left. + \left|F_i - F_{i-1}\right| \text{sign}\,(u_i - u_{i-1})\right] \tag{13.8.5}$$

The last term of Eq. (13.8.5) is the numerical diffusion term.

Lax-Wendroff Scheme

Derivation of the Lax-Wendroff scheme for the nonlinear hyperbolic PDE is essentially the same as for the linear version and starts with the Taylor expansion of $u_i^{(n+1)}$ as

$$u_i^{(n+1)} = u_i^{(n)} + \Delta t(u_t)_i^{(n)} + \frac{\Delta t^2}{2}(u_{tt})_i^{(n)} + \cdots \tag{13.8.6}$$

The u_t in the equation is eliminated by using Eq. (13.8.1), and u_{tt} is eliminated in the following manner.

Differentiating F with respect to t gives

$$F_t = F_u u_t = A u_t \tag{13.8.7}$$

where $A = F_u$. By eliminating u_t in Eq. (13.8.7), using Eq. (13.8.1), we get

$$F_t = -AF_x \tag{13.8.8}$$

Differentiating Eq. (13.8.1) with respect to t gives

$$u_{tt} = -F_{xt} = -\frac{\partial}{\partial x}F_t = \frac{\partial}{\partial x}AF_x \tag{13.8.9}$$

Thus, elimination of u_t and u_{tt} in Eq. (13.8.6) yields the Lax-Wendroff scheme:

$$u_i^{(n+1)} = u_i^{(n)} - \frac{\Delta t}{2\Delta x}(F_{i+1}^{(n)} - F_{i-1}^{(n)})$$

$$+ \frac{1}{2}\left(\frac{\Delta t}{\Delta x}\right)^2 [A_{i+(1/2)}^{(n)}(F_{i+1}^{(n)} - F_i^{(n)}) - A_{i-(1/2)}^{(n)}(F_i^{(n)} - F_{i-1}^{(n)})] \quad (13.8.10)$$

The Lax-Wendroff scheme may also be written in the form

$$u_i^{(n+1)} = u_i^{(n)} - k(f_{i+(1/2)}^{(n)} - f_{i-(1/2)}^{(n)}) \quad (13.8.11)$$

where f is called *numerical flux* defined by

$$f_{i+\frac{1}{2}} = \frac{1}{2}[F_{i+1}^{(n)} + F_i^{(n)} - kA_{i+(1/2)}^{(n)}(F_{i+1}^{(n)} - F_i^{(n)})]$$

$$f_{i-\frac{1}{2}} = \frac{1}{2}[F_i^{(n)} + F_{i-1}^{(n)} - kA_{i-(1/2)}^{(n)}(F_i^n - F_{i-1}^n)] \quad (13.8.12)$$

where $k = \Delta t / \Delta x$.

MacCormack Scheme

The MacCormack scheme [von Lavante/Thompkins] for the nonlinear PDEs is given by

$$\bar{u}_i^{(n+1)} = u_i^{(n)} - \frac{\Delta t}{\Delta x}(F_{i+1}^{(n)} - F_i^{(n)})$$

$$u_i^{(n+1)} = \frac{1}{2}\left[u_i^{(n)} + \bar{u}_i^{(n+1)} - \frac{\Delta t}{\Delta x}(\bar{F}_i - \bar{F}_{i-1})\right] \quad (13.8.13)$$

where the first equation is the predictor, the second is the corrector, and

$$\bar{F}_i = F(\bar{u}_i^{(n+1)})$$

As discussed earlier, the MacCormack scheme has the same order of accuracy as the Lax-Wendroff scheme, but it is easier to use because F values are directly evaluated.

Beam-Warming Implicit Scheme

The Beam-Warming scheme for Eq. (13.8.1) starts with the modified Euler scheme in time and central difference approximation in space:

$$u_i^{(n+1)} - u_i^{(n)} = \frac{\Delta t}{4\Delta x}[F_{i+1}^{(n+1)} - F_{i-1}^{(n+1)} + F_{i+1}^{(n)} - F_{i-1}^{(n)}] \quad (13.8.14)$$

In Eq. (13.8.14), $F_{i\pm1}^{(n+1)}$ are the nonlinear function of unknowns $u_{i\pm1}^{(n+1)}$. So, they are expanded into a Taylor series

$$F_i^{(n+1)} = F_i^{(n)} + \Delta t A_i^{(n)} \delta u_i + \cdots$$

where

$$\delta u_i = u_i^{(n+1)} - u_i^{(n)} \tag{13.8.15}$$

$$A_i^{(n)} = \left(\frac{\partial F}{\partial u}\right)_i^{(n)} \tag{13.8.16}$$

Introducing Eqs. (13.8.15) and (13.8.16) into Eq. (13.8.14) yields

$$-\Delta t A_{i-1}\delta u_{i-1} + \delta u_i + \Delta t A_{i+1}\delta u_{i+1} = \frac{\Delta t}{2\Delta x}\left[F_{i+1}^{(n)} - F_{i-1}^{(n)}\right] \tag{13.8.17}$$

where the superscripts of A_is are omitted for simplicity, but they are evaluated by $u_i^{(n)}$. The set of Eq. (13.8.17) for all the points are simultaneously solved by the tri-diagonal scheme. The requirement of the numerical boundary condition for the right end is fulfilled in the same way as described for the implicit scheme in Section 13.3.

The linear stability for this scheme is neutral. However, nonlinearity of the equation often induces instability, so a usual practice is to add a fourth-order numerical viscosity term:

$$-\Delta t A_{i-1}\delta u_{i-1} + \delta u_i + \Delta t A_{i+1}\delta u_{i+1} = \frac{\Delta t}{2\Delta x}\left[F_{i+1}^{(n)} - F_{i-1}^{(n)}\right]$$

$$-\frac{\varepsilon}{\Delta x^4}\left(u_{i-2}^{(n)} - 4u_{i-1}^{(n)} + 6u_i^{(n)} - 4u_{i+1}^{(n)} + u_{i+2}^{(n)}\right)$$

where the last term is the fourth-order artificial viscosity term and ε is the artificial viscosity coefficient.

13.9 FLUX CORRECTED SCHEMES

The tests of Section 13.7 show that numerical diffusion and aliasing error are major problems of the numerical schemes for hyperbolic PDEs. Positivity of the solution is violated by the aliasing effects of the truncated terms. In general, if a higher-order scheme is used, the numerical diffusion is reduced but aliasing error appears. On the other hand, suppression of the aliasing error means an increase of the numerical viscosity (or numerical diffusion) effect. Thus, suppressing numerical diffusion and eliminating aliasing errors simultaneously in a single difference scheme is impossible [Oran/Boris].

A number of approaches to improve the solution have been proposed under different names, among which are the *flux corrected scheme* and the total variation diminishing (TVD) scheme [Yee; Yee/Warming/Harten; Nittmann; Book/Boris/Zalesak]. The basic principle of these schemes is to use a low-order scheme where there is a threat of oscillation but to use a higher-order scheme wherever aliasing error effect is not large enough to cause oscillation. In each time step, both a low-order solution (with the second-order numerical viscosity) and a higher-order solution (without the second-order numerical viscosity effect) are computed, and the solution for that time step is finalized by blending the two computational results.

We explain the flux-corrected scheme for a hyperbolic PDE in the conservative form given by

$$u_t + F_x = 0 \tag{13.9.1}$$

The scheme consists of two steps. The first step is a first-order numerical scheme written as

$$\bar{u}_i^{(n+1)} = u_i^{(n)} - \frac{\Delta t}{\Delta x}\left(g_{i+(1/2)}^{(n)} - g_{i-(1/2)}^{(n)}\right) \tag{13.9.2}$$

where g is a numerical flux. If the Courant-Issacson-Rees scheme is used in this step, $g_{i+(1/2)}^{(n)}$ is written as

$$g_{i+(1/2)}^{(n)} = \frac{1}{2}\left[F_i^{(n)} + F_{i+1}^{(n)} - \left|a_{i+\frac{1}{2}}\right|\left(u_{i+1}^{(n)} - u_i^{(n)}\right)\right]$$

$$a_{i+\frac{1}{2}} = \begin{cases} \dfrac{F_{i+1} - F_i}{u_{i+1}^{(n)} - u_i^{(n)}}, & \text{if } u_{i+1}^{(n)} - u_i^{(n)} \neq 0 \\ 0, & \text{if } u_{i+1}^{(n)} - u_i^{(n)} = 0 \end{cases} \tag{13.9.3}$$

The second step is called an *antidiffusion process* and is written as

$$u_i^{(n+1)} = \bar{u}_i^{(n+1)} - \frac{\Delta t}{\Delta x}\left(\delta f_{i+\frac{1}{2}} - \delta f_{i-\frac{1}{2}}\right) \tag{13.9.4}$$

where δf corrects flux, and its purpose is to cancel the numerical viscosity (or diffusion) effect in the first step.

To analytically find the amount of diffusion effect to be canceled is difficult because, in real problems, grid spacing and the coefficients (such as fluid velocity and pipe cross section) change in space. This is the reason a higher-order scheme, which has no second-order diffusion effect, is also used.

A higher-order scheme is written here as

$$u_i^{(n+1)} = u_i^{(n)} - \frac{\Delta t}{\Delta x}\left(G_{i+(1/2)}^{(n)} - G_{i-(1/2)}^{(n)}\right) \tag{13.9.5}$$

The difference between $u_i^{(n+1)}$ calculated by the first-order scheme and $u_i^{(n+1)}$ by the higher-order accurate scheme is primarily the second-order numerical diffusion effect. Therefore, to remove the numerical diffusion effect, one can set $\delta f_{i+\frac{1}{2}}$ to

$$\delta f_{i+\frac{1}{2}} = G_{i+(1/2)}^{(n)} - g_{i+(1/2)}^{(n)} \tag{13.9.6}$$

If $\delta f_{i+(1/2)}$ is set as Eq. (13.9.6) for every point, then the second step simply becomes the higher-order scheme. Therefore, some adjustment of δf is neccessary from point to point. The adjustment is such that $\delta f_{i+(1/2)}$ remains the same whenever there is no threat to spurious oscillation. However, if there is a threat of oscillation, then δf for that interval must be reduced or even set to zero. The algorithm based on this concept is given by

$$\delta f_{i+\frac{1}{2}} = S \max\{0, \min [S(\bar{u}_{i+2}^{(n+1)} - \bar{u}_{i+1}^{(n+1)}), |\delta \hat{f}_{i+\frac{1}{2}}|, S(\bar{u}_i^{(n+1)} - \bar{u}_{i-1}^{(n+1)})]\} \tag{13.9.7}$$

with

$$\delta \hat{f}_{i+\frac{1}{2}} = G_{i+(1/2)}^{(n)} - g_{i+(1/2)}^{(n)}$$
$$S = \text{sign} (\bar{u}_{i+1}^{(n+1)} - \bar{u}_i^{(n+1)}) \text{ with } |S| = 1.$$

To explain the meaning of Eq. (13.9.7), let us assume that $\bar{u}_{i+1}^{(n+1)} - \bar{u}_i^{(n+1)} > 0$ or equivalently, the $\bar{u}_i^{(n+1)}$ is increasing from point i to point $i+1$, so $S = 1$. Then Eq. (13.9.7) takes

$$\delta f_{i+\frac{1}{2}} = \min [S(\bar{u}_{i+2}^{(n+1)} - \bar{u}_{i+1}^{(n+1)}), |\delta f_{i+\frac{1}{2}}|, S(\bar{u}_i^{(n+1)} - \bar{u}_{i-1}^{(n+1)})] \tag{13.9.8}$$

or

$$\delta f_{i+\frac{1}{2}} = 0 \tag{13.9.9}$$

whichever is the larger. Now, if

$$\bar{u}_{i+2}^{(n+1)} - \bar{u}_{i+1}^{(n+1)} < 0 \tag{13.9.10}$$

or

$$\bar{u}_i^{(n+1)} - \bar{u}_{i-1}^{(n+1)} < 0 \tag{13.9.11}$$

which means that an oscillation exists, Eq. (13.9.8) takes the smaller of these, but then Eq. (13.9.8) sets $\delta f_{i+(1/2)} = 0$. In other words, if Eq. (13.9.10) or Eq. (13.9.11) is satisfied, u_i is oscillating, so $\delta f_{i+(1/2)}$ is set to zero, and the flux computed by the lower-order scheme remains effective.

Example 13.4

(a) Develop a flux corrected scheme for

$$u_t(x, t) + au_x(x, t) = 0 \qquad (A)$$

where a = constant. Use the first-order upwind scheme as the low-order scheme and the third-order upwind scheme developed in Example 13.3 as the high-order scheme.

(b) Numerically compute the solution for Eq. (A) with $a = 1$, $\Delta x = 1$, $\Delta t = \gamma = 0.01$, and the initial and boundary conditions as follows:

$$u(0, t) = 0$$

$$u(x, 0) = 1 \quad \text{for } 10\Delta x_i \leqslant x \leqslant 15\Delta x_i$$

but

$$u(x, 0) = 0 \quad \text{for } x < 10\Delta x_i \text{ and } 15\Delta x_i < x$$

Plot u_i for $t = 0$, $t = 100\Delta t$ and $t = 200\Delta T$. Compare the results to those in Figure 13.5 because the results in Figure 13.5 are solutions for the same problem.

⟨Solution⟩

(a) The low-order scheme is written as

$$\frac{\bar{u}_i^{(n+1)} - u_i^{(n)}}{\Delta t} + \frac{g_{i+\frac{1}{2}} - g_{i-\frac{1}{2}}}{\Delta x} = 0 \qquad (B)$$

with

$$g_{i+\frac{1}{2}} = au_i^{(n)}$$

The flux based on the third-order upwind scheme is given by

$$G_{i+\frac{1}{2}} = a\frac{-u_{i+2}^{(n)} + 7u_{i+1}^{(n)} + 7u_i^{(n)} - u_{i-1}^{(n)}}{12} + |a|\frac{u_{i+2}^{(n)} - 3u_{i+1}^{(n)} + 3u_i^{(n)} - u_{i-1}^{(n)}}{12} \qquad (C)$$

If g's in Eq. (B) were substituted by Eq. (C), Eq. (B) would become the third-order upwind scheme (see Example 13.3).

The flux corrected scheme is

$$u_i^{(n+1)} = \bar{u}_i^{(n+1)} - \frac{\Delta t}{\Delta x}(\delta f_{i+\frac{1}{2}} - \delta f_{i-\frac{1}{2}})$$

where

$$\delta f_{i+\frac{1}{2}} = S \max \{0, \min [S(\bar{u}_{i+2} - \bar{u}_{i+1}), |\delta \hat{f}_{i+\frac{1}{2}}|, S(\bar{u}_i - \bar{u}_{i-1})]\}$$

and

$$\delta \hat{f}_{i+\frac{1}{2}} = G_{i+\frac{1}{2}} - g_{i+\frac{1}{2}}$$

$$S = \begin{cases} 1 & \text{if } \bar{u}_{i+1}^{(n+1)} > \bar{u}_i^{(n+1)} \\ -1 & \text{if } \bar{u}_{i+1}^{(n+1)} < \bar{u}_i^{(n+1)} \end{cases}$$

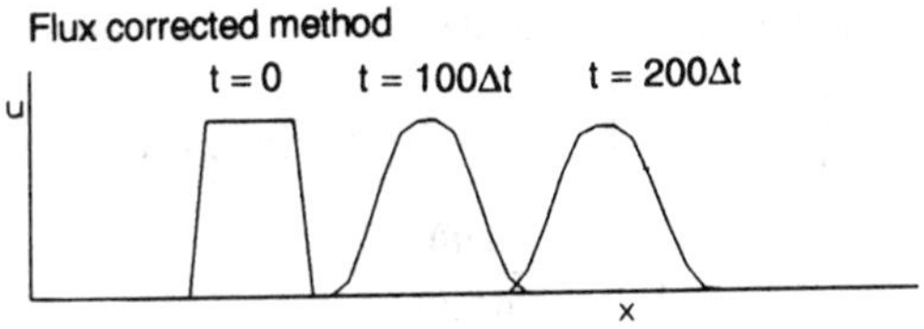

Figure 13.7 Numerical simulation of a square wave moving at a constant speed using the fulx corrected method

(b) The computed results are shown in Figure 13.7. Figure 13.7 can be compared with the results of the first-order upwind, Lax-Wendroff, and third-order upwind method in Figure 13.6 because all the solutions are for the same problem. Figure 13.7 shows that the flux corrected method better simulates the traveling wave than the third-order upwind method. In Figure 13.6, the waves calculated by the third-order upwind method oscillates and has negative values before and after the wave, but the oscillation does not appear in the result of the flux corrected method.

PROBLEMS

(13.1) Derive the modified equation for the FTCS scheme given by Eq. (13.3.13).

(13.2) By eliminating the predictor, show that the MacCormack scheme given by Eq. (13.8.13) becomes identical to the Lax-Wendroff scheme.

(13.3) The Lax scheme is written as

$$u_i^{(n+1)} = \frac{1}{2}(u_{i+1}^{(n)} + u_{i-1}^{(n)}) - \frac{\gamma}{2}\left[u_{i+1}^{(n)} - u_{i-1}^{(n)}\right]$$

where $\gamma \equiv a\Delta t/\Delta x$.

(a) Show that its amplification factor is

$$G = \cos(\theta) - j\gamma \sin(\theta), \quad j = \sqrt{-1}$$

(b) Plot G on the complex plane.

(c) Show that the scheme is stable if

$$0 < \gamma < 1$$

(13.4) Show that the modified equation for the Lax scheme introduced in Problem (13.3) is

$$u_t + au_x - \frac{a\Delta x}{2}(1/\gamma - \gamma - 1)u_{xx} - \frac{a\Delta x^2}{3}\left(1 - \gamma^2 - \frac{3}{2}\gamma\right)u_{xxx} + \cdots = 0$$

(13.5) Show that the modified equation for the BTCS scheme (the backward Euler in time and central difference in space) and its amplitude factor become, respectively,

$$u_t + au_x - \frac{1}{2}a^2\Delta t u_{xx} + \left[\frac{a}{6}\Delta x^2 + \frac{1}{3}a^3\Delta t^2\right]u_{xxx} = 0$$

$$G = \frac{1 - j\gamma \sin(\theta)}{1 + \gamma^2 \sin^2(\theta)}$$

(13.6) The scheme that follows is named the *two-step Lax-Wendroff* scheme. Show that it is identical with the original Lax-Wendroff scheme for a linear hyperbolic PDE such as Eq. (13.3.1):

Step 1

$$\frac{1}{2}\Delta t[2u_{i+(1/2)}^{n+(1/2)} - (u_{i+1}^{(n)} + u_i^{(n)})] + \frac{a}{\Delta x}(u_{i+1}^{(n)} - u_i^{(n)}) = 0$$

Step 2

$$\frac{1}{\Delta t}(u_i^{(n+1)} - u_i^{(n)}) + \frac{a}{\Delta x}\left[u_{i+(1/2)}^{n+(1/2)} - u_{i-(1/2)}^{n+(1/2)}\right] = 0$$

(13.7) Prove that, if the boundary value and the initial values of u are nonnegative, then the solution of the FTBS scheme is nonnegative for $\gamma < 1$.

(13.8) The difference equation for

$$u_t + au_x = bu_{xx}, \quad a > 0 \text{ and } b > 0$$

may be written using the forward Euler scheme in time, and the central difference approximation for u_x as well as for u_{xx}. Find under what condition the difference scheme becomes stable.

(13.9) By using the Fourier analysis, determine the stability of the difference approximation for

$$u_t + au_x = 0, \quad a > 0$$

given by

$$\frac{1}{\Delta t}\left[u_i^{(n+1)} - u_i^{(n)}\right] + a\frac{3u_i^{(n)} - 4u_{i-1}^{(n)} + u_{i-2}^{(n)}}{2\Delta x} = 0$$

(13.10) Derive the modified equation for the difference scheme in Problem (13.9).

(13.11) Show that the following scheme is a conservative difference approximation for Eq. (13.6.1) and that it is based on the third-order upwind scheme:

$$u_i^{(n+1)} = u_i^{(n)} - k(f_{i+(1/2)}^{(n)} - f_{i-(1/2)}^{(n)})$$

where $k = \Delta t/\Delta x$

$$f_{i+\frac{1}{2}} = \frac{1}{12}[-F_{i+2} + 7F_{i+1} + 7F_i - F_{i-1}$$

$$+ |a_{i+\frac{1}{2}}|(u_{i+2} - 3u_{i+1} + 3u_i - u_{i-1})]$$

$$a_{i+\frac{1}{2}} = \begin{cases} \dfrac{F_{i+1} - F_i}{u_{i+1} - u_i}, & \text{if } u_{i+1} - u_i \neq 0 \\ 0, & \text{if } u_{i+1} - u_i = 0 \end{cases}$$

(13.8.4)

REFERENCES

Anderson, D. A., J. C. Tannehill, and R. H. Pletcher, *Computational Fluid Mechanics and Heat Transfer*, Hemisphere, 1984.

Book, D. L., J. P. Boris, and S. T. Zalesak, "Flux-Corrected Transport," *Finite-Difference Techniques for Vectorized Fluid Dynamics Calculations*, D. L. Book, ed., Springer-Verlag, 1981.

Courant, R., and D. Hilbert, *Methods of Mathematical Physics*, Wiley-Interscience, 1962.

Ferziger, J. H., *Numerical Methods for Engineering Application*, Wiley-Interscience, 1981.

Garabedian, P. R., *Partial Differential Equations*, Wiley, 1965.

Jameson, A., "Numerical Solution of the Euler Equation for Compressible Inviscid Fluids," *Numerical Methods for Euler Equations of Fluid Dynamics*, SIAM, 1985.

Kawamura, T., "Computation of Turbulent Pipe and Duct Flow Using Third Order Upwind Scheme," AIAA–86–1042, AIAA/ASME 4th Fluid Mechanics, Plasma Dynamics and Lasers Conference, May 12–14, 1986/Atlanta, Ga.

Kawamura, T., and K. Kuwahara, "Computation of High Reynolds Number Flow around a Circular Cylinder with Surface Roughness," AIAA 22nd Aerospace Science Meeting, Jan 9–12, Reno, Nevada, 1984: AIAA–84–0340.

Leonard, B. P., "Third-order Upwind as a Rational Basis for Computational Fluid Dynamics," in *Computational Techniques and Applications: CTA–83*, Noye & Fletcher, eds., Elsevier, 1984.

Mitchell, A. R., and Griffiths, D. F., *The Finite Difference Methods in Partial Differential Equations*, Wiley-Interscience, 1980.

Nittmann, J., "Doner Cell, FCT-Shasta and Flux Splitting Method: Three Finite Difference Equations Applied to Astrophysical Schock-Cloud Interactions," *Numerical Methods for Fluid Dynamics*, Morton and Baines, eds., Academic Press, 1982.

Oran, E. S., and J. P. Boris, *Numerical Simulation of Reactive Flow*, Elsevier, 1987.

Patankar, S. V., *Numerical Heat Transfer and Fluid Flow*, Hemisphere, 1980.

Roe, P. L., *Upwind Schemes Using Various Formulations of the Euler Equations, Numerical Methods for Euler Equations of Fluid Dynamics*, SIAM, 1985.

Smith, D. G., *Numerical Solution of Partial Differential Equations*, Oxford University Press, 1978.

von Lavante, E., and W. T. Thoompkins, *An Implicit, Bi-Diagonal Numerical Method for Solving the Navier-Stokes Equations*, AIAA–82–0063.

Yee, H. C., On Symmetric and Upwind TVD Schemes, Proceedings of the Sixth GAMM-Conference on Numerical Methods in Fluid Mechanics, Notes on Numerical Fluid Mechanics, Vol 13, Friedr, Vieweg & Sohn, 1986.

Yee, H. C., R. M. Beam, and R. F. Warming, Stable Boundary Approximations for a Class of Implicit Schemes for the One-Dimensional Inviscid Equations of Gas Dynamics, AIAA Computational Fluid Dynamics Conference, Palo Alto, CA., June 22–23, 1981.

Yee, H. C., and J. L. Shinn, Semi-Implicit and Fully Implicit Shock-Capturing Methods for Hyperbolic Conservation Laws with Stiff Source Terms, NASA TM–89415, 1986.

Yee, H. C., R. F. Warming, and A. Harten, Application of TVD for the Euler Equations of Gas Dynamics, Lecture in Applied Mathematics, Vol 22, 1985.

APPENDIX A

Error of Polynomial Interpolations

A.1 ERROR OF A LINEAR INTERPOLATION

The error of a linear interpolation given by Eq. (2.2.1) is defined by

$$e(x) = f(x) - g(x) \tag{A.1}$$

where $f(x)$ is the exact function. Because the error becomes zero at $x = a$ and $x = b$, $e(x)$ may be expressed in the form

$$e(x) = (x - a)(x - b)s(x) \tag{A.2}$$

where $s(x)$ is a function.

Let us select a fixed value η satisfying $a < \eta < b$ and define a new function by

$$p(x) = f(x) - g(x) - (x - a)(x - b)s(\eta) \tag{A.3}$$

or equivalently,

$$p(x) = (x - a)(x - b)[s(x) - s(\eta)] \tag{A.4}$$

Obviously, the function $p(x)$ becomes zero at three points, $x = a, b$, and η, as depicted in Figure A–1. We can observe in Figure A–1 that $p'(x)$ has two zeroes, one left to η and another right to η. Furthermore, $p''(x)$ has one zero denoted here by ξ in between the two roots of $p'(x)$.

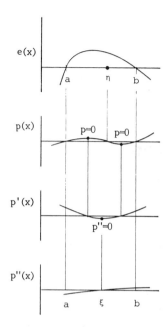

Figure A1 Zeroes of $p(x)$, $p'(x)$, and $p''(x)$

Differentiating Eq. (A.3) twice yields

$$p''(x) = f''(x) - 0 - 2s(\eta) \tag{A.5}$$

where $g''(x)$ becomes zero because $g(x)$ is a linear function by definition. At $x = \xi$, which is the root of $p''(x) = 0$, Eq. (A.5) becomes

$$0 = f''(\xi) - 2s(\eta)$$

or equivalently

$$s(\eta) = \frac{1}{2}f''(\xi) \tag{A.7}$$

This equation indicates that $s(\eta)$ for a given value of η satisfying $a < \eta < b$ is given by $\frac{1}{2}f''(\xi)$ where ξ is in $a < \xi < b$. Thus, by changing the symbol η to x, we can write

$$s(x) = \frac{1}{2}f''(\xi), \quad a < \xi < b, \quad a < x < b$$

where ξ depends on x but is always in $[a, b]$. So, the error expressed by Eq. (A.2) becomes

$$e(x) = \frac{1}{2}(x - a)(x - b)f''(\xi), \quad a < \xi < b, \quad a < x < b \tag{A.8}$$

Assuming that the change of $f''(x)$ in the interval $[a, b]$ is small, $f''(\xi)$ may be approximated by $f''(a)$ or $f''(b)$ or $f''(x_m)$ where $x_m = (a + b)/2$. Thus, the error is now written approximately as

$$e(x) \simeq \frac{1}{2}(x - a)(x - b)f''(x_m) \tag{A.9}$$

This equation indicates that the maximum of $|e(x)|$ occurs approximately at the midpoint $[a, b]$, and

$$\max_{a < x < b} |e(x)| \simeq \frac{h^2}{8} |f''(x_m)| \tag{A.10}$$

A.2 ERROR OF A LINEAR EXTRAPOLATION

Linear extrapolation is the use of linear interpolation outside the two data points. Error of a linear extrapolation may be expressed in a similar form as an extension of the previous analysis.

Error of a linear extrapolation is given also by Eq. (A.2). It is shown here that $s(x)$ in Eq. (A.2) for extrapolation is given by

$$s(x) = f''(\xi) \tag{A.11a}$$

where

$$x \leqslant \xi \leqslant b \quad \text{if } x < a < b \tag{A.11b}$$
$$a \leqslant \xi \leqslant x \quad \text{if } a < b < x \tag{A.11c}$$

We consider an extrapolation extended to the left of the interpolation range. In Eq. (A.3), or equivalently Eq. (A.4), we fix η to a value satisfying $\eta < a$. Notice that $p(x)$ thus defined is a new function of x. It is seen that $p(x)$ has three zeroes, $x = \eta$, a, and b, and that the zero of $p''(\xi) = 0$ is in $\eta \leqslant \xi \leqslant b$:

$$0 = f''(\xi) = 2s(\eta), \quad \eta < a, \quad \eta \leqslant \xi \leqslant b$$

By changing notations, the foregoing equation is written as

$$s(x) = \frac{1}{2}f''(\xi), \quad x < a, \quad x \leqslant \xi \leqslant b \tag{A.12a}$$

Similarly, if the extrapolation is extended to the right of the interpolation range, $b < x$, then

$$s(x) = \frac{1}{2}f''(\xi), \quad b < x, \quad a \leqslant \xi \leqslant x \tag{A.12b}$$

A.3 ERROR OF A POLYNOMIAL INTERPOLATION

The expression of error of the linear interpolation discussed in the previous section can be extended to higher-order polynomial interpolations, including Lagrange and Newton interpolations.

We define the error of a polynomial interpolation by

$$e(x) = f(x) - g(x) \tag{A.13}$$

where $f(x)$ is the exact function from which f_i are sampled and $g(x)$ is polynomial of order N. It is assumed that $g(x)$ and $f(x)$ do not intersect at any other point than $x = x_i$, where $x_0 \leqslant x \leqslant x_N$. Since $g(x)$ becomes exact at the grid points x_i, $e(x)$ may be written in the form

$$e(x) = (x - x_0)(x - x_1) \cdots (x - x_N)s(x) \tag{A.14}$$

We will now show that, for a value x in $[x_0, x_N]$, $s(x)$ is given by

$$s(x) = \frac{1}{N!} f^{(N+1)}(\xi), \quad x_0 < \xi < x_N \tag{A.15}$$

where ξ depends on x but is always in $[x_0, x_N]$.

Let us select a fixed value η satisfying $x_0 < \eta < x_N$ and define a new function by

$$p(x) = f(x) - g(x) - (x - x_0)(x - x_1) \cdots (x - x_N)s(\eta) \tag{A.16}$$

or equivalently

$$p(x) = (x - x_0)(x - x_1) \cdots (x - x_N)[s(x) - s(\eta)] \tag{A.17}$$

Obviously, the function $p(x)$ becomes zero at $N + 1$ grid points x_i, $i = 0, 1, 2, \ldots, N$ and also at $x = \eta$. It has no other zeroes in $[x_0, x_N]$. Following a similar argument as in the preceding section, we can say that all the roots of $p'(x)$ are in between the two extreme roots of $p(x)$, and that all the roots of $p''(x)$ are in between two extreme roots of $p'(x)$, and so on. Thus, the root of $p^{(N+1)}$ must be also in between x_0 and x_N. The $(N + 1)$th derivative of Eq. (A.17) is

$$p^{(N+1)}(x) = f^{(N+1)}(x) - 0 - s(\eta)N! \tag{A.18}$$

which becomes

$$p^{(N+1)}(\xi) = f^{(N+1)}(\xi) - s(\eta)N! = 0 \tag{A.19}$$

where ξ denotes the root of $p^{(N+1)} = 0$ satisfying $x_0 < \xi < x_N$. Thus,

$$s(\eta) = \frac{1}{N!} f^{(N+1)}(\xi) \tag{A.20}$$

where both η and ξ are in $a < \eta < b$ and $a < \xi < b$, respectively.

By changing the notations η and ξ to x and ξ, respectively, we get

$$s(x) = \frac{1}{N!} f^{(N+1)}(\xi) \tag{A.21}$$

So, the error of an interpolation of order N is expressed by

$$e(x) = \frac{1}{N!} (x - x_0)(x - x_N) \ldots (x - x_N) f^{(N+1)}(\xi), \quad x_0 < \xi < x_N \tag{A.22}$$

If $g(x)$ is used as an extrapolation with $x < a$, or $x > b$, the error is given by Eq. (A.22) except ξ is a value satisfying $x < \xi < b$, or $a < \xi < x$, respectively.

APPENDIX B

Legendre Polynomials

The Legendre polynomials, on which the Gauss-Legendre quadratures are based, are written as

$$P_0(x) = 1$$

$$P_1(x) = x$$

$$P_2(x) = \frac{1}{2}[3x^2 - 1]$$

$$P_3(x) = \frac{1}{2}[5x^3 - 3x]$$

$$P_4(x) = \frac{1}{8}[35x^4 - 30x^2 + 3]$$

$$\vdots$$

Any higher-order Legendre polynomial may be derived using the recursion formula

$$nP_n(x) - (2n - 1)xP_{n-1}(x) + (n - 1)P_{n-2}(x) = 0$$

A Legendre polynomial of order n may be also alternatively expressed by

$$P_n(x) = \frac{1}{2^n n!} \frac{d^n(x^2 - 1)^n}{dx^n}$$

One important property of the Legendre polynomials is orthogonality:

$$\int_{-1}^{1} P_m(x)P_n(x)\,dx = 0, \qquad \text{for } n \neq m$$

$$= \frac{2}{2n+1}, \quad \text{for } m = n$$

The foregoing equation indicates that the integral of two different Legendre polynomials in $[-1, 1]$ is zero.

Every polynomial of order $N - 1$ or less is orthogonal to the Legendre polynomial of order N. This can be proven easily because a polynomial of order $N - 1$ or less can be expressed as a linear combination of Legendre polynomials of order at most $N - 1$.

APPENDIX C

Calculation of Higher-order Differences by Using the Shift Operator

The shift operator is defined by

$$Ef_i = f_{i+1} \tag{C.1}$$

Here the operator, E, shifts the index of f_i by one in the positive direction. Its inverse, E^{-1}, shifts the index in the negative direction, namely

$$E^{-1}f_i = f_{i-1} \tag{C.2}$$

A multiple application of E is

$$E^n f_i = f_{i+n} \tag{C.3}$$

where n can be any negative or positive integer.

With E, the forward and backward difference operators can be written respectively as

$$\Delta = E - 1 \tag{C.4}$$

$$\nabla = 1 - E^{-1} \tag{C.5}$$

Therefore, the nth-order forward difference can be derived by

$$\Delta^n f_i = (E - 1)^n f_i$$

$$= \left[E^n - nE^{n-1} + \binom{n}{2}E^{n-2} - \binom{n}{3}E^{n-3} + \cdots + (-1)^n \binom{n}{n}E^0 \right] f_i$$

$$= f_{i+n} - nf_{i+n-1} + \frac{1}{2}n(n-1)f_{i+n-2} - \frac{1}{6}n(n-1)(n-2)f_{i+n-3}$$

$$+ \cdots + (-1)^{n-1}nf_{i+1} + (-1)^n f_i \qquad (C.6)$$

where $\binom{n}{k}$ is a binomial coefficient that equals $n!/(n-k)!k!$

The nth-order backward difference is derived similarly as

$$\nabla^n f_i = (1 - E^{-1})^n f_i$$

$$= \left[1 - nE^{-1} + \binom{n}{2}E^{-2} - \binom{n}{3}E^{-3} + \cdots + (-1)^n \binom{n}{n}E^{-n} \right] f_i$$

$$= f_i - nf_{i-1} + \frac{1}{2}n(n-1)f_{i-2} - \cdots + (-1)^{n-1}nf_{i-n+1} + (-1)^n f_{i-n} \qquad (C.7)$$

APPENDIX D

Derivation of One-dimensional Hyperbolic PDEs for Flow Problems

Suppose an incompressible fluid flow in a perfectly insulated pipe with variable cross-sectional area $A(x)$, and the temperature of the fluid changes in the direction of the flow. We ignore the heat conduction in the axial direction of the pipe and also assume the temperature is constant across the cross sectional-plane perpendicular to the axial direction. Then, the heat balance in a small control volume located at x as shown in Figure A2 is as follows:

$$c_p \rho A(x)\, dx\, dT = \left[c_p \rho A(x)v(x)T(x, t) - c_p \rho A(x + dx)v(x + dx)T(x + dx, t) \right] dt \quad \text{(D.1)}$$

where c_p is the specific heat and ρ is the density of the fluid. The left side is the rate of increase of internal energy during dt, the first term on the right side is the inflow of heat through the left boundary of the control volume during dt, and the second term is the outflow of heat through the right boundary during dt. If we assume that both c_p and ρ are constant, dividing Eq. (D.1) by dt yields

$$A(x)\, dx\, \frac{\partial T}{\partial t} = A(x)v(x)T(x, t) - A(x + dx)v(x + dx)T(x + dx, t) \quad \text{(D.2)}$$

By dividing by dx and expressing the right side in a derivative form we get

$$A(x)\frac{\partial T}{\partial t} = -\frac{\partial}{\partial x}\left[A(x)v(x)T(x) \right] \quad \text{(D.3a)}$$

or, equivalently,

$$T_t + v(x)[T(x)]_x = 0 \tag{D.3b}$$

where $A(x)v(x) = $ constant is used. Equation (D.3a) is in the conservation form, whereas Eq. (D.3b) is in nonconservation form. If a heat source exists, the heat source term is added to the right side as

$$T_t + v(x)[T(x)]_x = q(x)/c_p\rho \tag{D.4}$$

A(x) : cross-sectional area

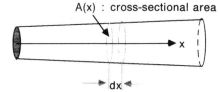

x

dx   **Figure A2** A nonuniform pipe

In case the flow in the pipe is of a chemical species that is nonuniformly distributed, the equation for the chemical species distribution is

$$c_t(x, t) + v(x)[c(x, t)]_x = s(x, t) \tag{D.5}$$

where $c(x, t)$ is concentration of the chemical species and s is the volume source of the species. Equation (D.5) is in the nonconservation form.

The mass conservation of a compressible fluid flowing in a pipe is

$$A(x)\rho_t(x, t) + [A(x)v(x)\rho(x, t)]_x = 0 \tag{D.6}$$

If the cross-sectional area $A(x)$ is constant, the equations for temperature, the chemical species, and the density of the compressible fluid become, respectively

$$T_t + [v(x)T(x)]_x = q(x, t)/c_p\rho \tag{D.7}$$

$$c_t(x, t) + [v(x)c(x, t)]_x = s(x, t) \tag{D.8}$$

$$\rho_t(x, t) + [v(x)\rho(x, t)]_x = 0 \tag{D.9}$$

APPENDIX E

Total Variation Diminishing (TVD)

The total variation (TV) of a wave function $u(x)$ is defined by

$$TV = \int \left| \frac{\partial u}{\partial x} \right| dx \qquad \text{(E.1)}$$

[Jameson]. The TV of the numerical solution is defined accordingly by

$$TV = \sum_{i=-\infty}^{i=+\infty} |u_{i+1} - u_i| \qquad \text{(E.2)}$$

The analysis of the entropy of fluid shows that the TV can never increase. Therefore, we desire that TV of the solution of the numerical scheme will not increase.

If the difference equation can be written in the form

$$\frac{d}{dt} u_i(t) = c^+_{i+(1/2)}(u_{i+1} - u_i) - c^-_{i-(1/2)}(u_i - u_{i-1}) \qquad \text{(E.3)}$$

where c^- and c^+ are both nonnegative, then the method is a TVD method.

Equation (E.2) may be written as

$$TV = \sum_{i=-\infty}^{i=+\infty} s_{i+\frac{1}{2}}(u_{i+1} - u_i) \qquad \text{(E.4)}$$

where the summation is over i from $-\infty$ to $+\infty$ and

$$s_{i+\frac{1}{2}} = \begin{cases} 1 & \text{if } u_{i+1} - u_i \geqslant 0 \\ -1 & \text{if } u_{i+1} - u_i < 0 \end{cases} \tag{E.5}$$

The time derivative of TVD becomes

$$\frac{d}{dt} TVD = \sum_{i=-\infty}^{i=+\infty} s_{i+(1/2)} \frac{d}{dt}(u_{i+1} - u_i)$$

$$= \sum_{i=-\infty}^{i=+\infty} s_{i+(1/2)} [c_{i+(3/2)}^+ (u_{i+2} - u_{i+1}) - c_{i+(1/2)}^- (u_{i+1} - u_i)$$

$$- c_{i+(1/2)}^+ (u_{i+1} - u_i) + c_{i-(1/2)}^- (u_i - u_{i-1})]$$

$$= \sum_{i=-\infty}^{i=+\infty} v_{i+(1/2)}(u_{i+1} - u_i) \tag{E.6}$$

where Eq. (E.3) is used and

$$v_{i+(1/2)} = c_{i+(1/2)}^+ (s_{i+(1/2)} - s_{i-(1/2)}) + c_{i+(1/2)}^- (s_{i+(1/2)} - s_{i+(3/2)}) \tag{E.7}$$

The second term on the right side of Eq. (E.6) has the same sign as $u_{i+1} - u_i$ or else zero, and the same for the second term. Therefore, the term after the summation sign is nonnegative. Thus,

$$\frac{d}{dt} TV \leqslant 0 \tag{E.8}$$

This completes the proof of Eq. (E.3).

REFERENCE

Jameson, A., "Numerical Solution of the Euler Equation for Compressible Inviscid Fluids," *Numerical Methods for Euler Equations of Fluid Dynamics* (F. Angrand, A. Dervieux, J. A. Desideri and R. Glowinski, eds.) SIAM, 1985.

APPENDIX F

Derivation of the Modified Equations

Suppose an equation is given by

$$u_t + A_2 u_{tt} + A_3 u_{ttt} + \cdots + B_1 u_x + B_2 u_{xx} + B_3 u_{xxx} + B_4 u_{xxxx} + \cdots = 0 \quad \text{(F.1)}$$

We desire to transform Eq. (F.1) to the form

$$u_t + c_1 u_x + c_2 u_{xx} + c_3 u_{xxx} + c_4 u_{xxxx} + \cdots \quad \text{(F.2)}$$

by eliminating u_{tt}, u_{ttt}, and all higher derivative terms with respect to t.

Let us write Eq. (F.1) as

$$F(x, t) = 0 \quad \text{(F.1a)}$$

where F is exactly the left side of Eq. (F.1). By taking partial derivatives of Eq. (F.1a), we can write

$$F_t = 0 \quad \text{(F.3a)}$$

$$F_x = 0 \quad \text{(F.3b)}$$

$$F_{tt} = 0 \quad \text{(F.3c)}$$

$$F_{tx} = 0 \quad \text{(F.3d)}$$

$$F_{xx} = 0 \quad \text{(F.3e)}$$

$$\vdots$$

Obviously, the lowest derivative terms of Eq. (F.3a) are u_{tt} and u_{xt}; the lowest terms of Eq. (F.3b) are u_{tx} and u_{xx}; and the lowest terms of Eq. (F.3c) are u_{ttt}, u_{xtt}, and so on.

We write a linear combination of all the equations as

$$F + P_1 F_t + P_2 F_x + P_3 F_{tt} + P_4 F_{tx} + P_5 F_{xx} + P_6 F_{ttt}$$
$$+ P_7 F_{ttx} + P_8 F_{txx} + P_9 F_{xxx} + \cdots = 0 \tag{F.4}$$

where $P_1, P_2, \ldots$ are undetermined constants. Then it should be possible to determine the coefficients P_n's so the time derivatives such as $u_{tt}, u_{tx}, u_{ttt}, u_{ttx}, u_{txx}$, and so on are all eliminated except u_t.

To implement this algorithm, we express all the Eqs. (F.1a) and Eq. (F.3a), Eq. (F.3b), ... in a tabular form. In Table F.1, the leftmost column shows derivatives of u in increasing order in t and x. The second column is for the coefficients of the derivatives in Eq. (F.1a), $F = 0$. The remaining columns are for Eqs. (F.3a), (F.3b), (F.3c), and so on.

By setting the coefficients of undesired derivatives, namely $u_{tt}, u_{tx}, u_{ttt}, u_{ttx}, u_{txx}$, $u_{tttt}, u_{tttx}, u_{ttxx}, u_{txxx}, \ldots$, in Eq. (F.4) to zero, we obtain the following equations:

$$A_2 + P_1 = 0$$
$$P_1 B_1 + P_2 = 0$$
$$A_3 + P_1 A_2 + P_3 = 0$$
$$P_2 A_2 + P_3 B_1 + P_4 = 0$$
$$P_1 B_2 + P_4 B_1 + P_5 = 0 \tag{F.5}$$
$$A_4 + P_1 A_3 + P_3 A_2 + P_6 = 0$$
$$P_2 A_3 + P_4 A_2 + P_6 B_1 + P_7 = 0$$
$$P_3 B_2 + P_5 A_2 + P_7 B_1 + P_8 = 0$$
$$P_1 B_3 + P_4 B_2 + P_8 B_1 + P_9 = 0$$

The coefficients P_n's may be determined by solving Eqs. (F.5) sequentially from the top. Setting $B_1 = c_1 = a$ in accordance with Eq. (13.4.5), the solutions are as follows:

$$P_1 = -A_2$$
$$P_2 = aA_2$$
$$P_3 = -A_3 + A_2^2$$
$$P_4 = a(-2A_2^2 + A_3)$$
$$P_5 = A_2 B_2 + a^2(2A_2^2 - A_3) \tag{F.6}$$
$$P_6 = -A_4 + 2A_2 A_3 - A_2^3$$
$$P_7 = a(-4A_2 A_3 + 3A_2^3 + A_4)$$
$$P_8 = (A_3 - 2A_2^2)B_2 + a^2(-5A_2^3 + 5A_2 A_3 - A_4)$$
$$P_9 = A_2 B_3 + a(4A_2^2 - 2A_3)B_2 + a^3(5A_2^3 - 5A_2 A_3 + A_4)$$

From Table F.1, the coefficients of u_x, u_{xx}, u_{xxx}, and u_{xxxx} of Eq. (F.1) are as follows:

$$c_1 = B_1$$
$$c_2 = B_2 + P_2 B_1$$
$$c_3 = B_3 + P_2 B_2 + P_5 B_1 \tag{F.7}$$
$$c_4 = B_4 + P_2 B_3 + P_5 B_2 + P_9 B_1$$

By introducing Eq. (F.6), Eq. (F.7) becomes

$$c_1 = B_1 = a$$
$$c_2 = B_2 + a^2 A_2$$
$$c_3 = B_3 + 2a A_2 B_2 + a^3(2A_2^2 - A_3) \tag{F.8}$$
$$c_4 = B_4 + A_2 B_2^2 + 2a A_2 B_3 + 6a^2 A_2^2 B_2 - 3a^2 A_3 B_2 + a^4(5A_2^3 - 5A_2 A_3 + A_4)$$

Thus, using Eq. (F.8), the coefficients of Eq. (F.2) can be easily computed in terms of the coefficients of the original equation, Eq. (F.1).

Table F.1 Derivatives of the modified equation

	1 F	P_1 F_t	P_2 F_x	P_3 F_{tt}	P_4 F_{tx}	P_5 F_{xx}	P_6 F_{ttt}	P_7 F_{ttx}	P_8 F_{txx}	P_9 F_{xxx}
u_t	1									
u_x	B1									
u_{tt}	A2	1								
u_{tx}		B1	1							
u_{xx}	B2		B1							
u_{ttt}	A3	A2		1						
u_{ttx}			A2	B1	1					
u_{txx}		B2			B1	1				
u_{xxx}	B3		B2			B1				
u_{tttt}	A4	A3		A2						
u_{tttx}			A3		A2		1			
u_{ttxx}			B2			A2		B1	1	
u_{txxx}		B3			B2				B1	1
u_{xxxx}	B4		B3			B2				B1
	F	F_t	F_x	F_{tt}	F_{tx}	F_{xx}	F_{ttt}	F_{ttx}	F_{txx}	F_{xxx}

APPENDIX G

Cubic Spline Interpolation

Often a large number of data points have to be fitted by a single smooth curve, but the Lagrange interpolation or Newton interpolation polynomial of a high order is not suitable for this purpose, because the errors of a single polynomial tend to increase drastically as its order becomes large. Cubic spline interpolation is designed to suit this purpose.

In cubic spline interpolation, a cubic polynomial is used in each interval between two consecutive data points. One cubic polynomial has four free coefficients, so it needs four conditions. Two of them come from the requirements that the polynomial must pass through the data points at the two end points of the interval. The other two are the requirements that the first and second derivatives of the polynomial become continuous across each data point.

Cubic splines are piecewise cubic polynomials and similar to cubic Hermite polynomials to some extent. Although cubic Hermite interpolation is more accurate than cubic spline interpolation with respect to functional values, cubic spline interpolation is smoother than cubic Hermite interpolation because the interpolating function is required to have continuity in the functional value, the first derivative and the second derivative.

Consider one interval, $x_i \leqslant x \leqslant x_{i+1}$, of length $h_i = x_{i+1} - x_i$ in the interpolating range shown in Figure G.1. Using the local coordinate $s = x - x_i$, a cubic polynomial for one interval may be written as

$$g(s) = a + bs + cs^2 + es^3 \tag{G.1}$$

$$x_i \leqslant x \leqslant x_{i+1} \quad \text{or equivalently} \quad 0 \leqslant s \leqslant h_i$$

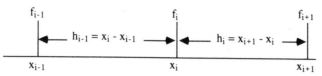

Figure G.1 Notations in cubic spline interpolations

We first require $g(s)$ to equal the known value of the function $f(s)$ at $s = 0$ and $s = h_i$, namely

$$f_i = a \qquad (G.2)$$

$$f_{i+1} = a + bh_i + ch_i^2 + eh_i^3 \qquad (G.3)$$

where f_i and f_{i+1} are known values at $s = 0$ and $s = h_i$, respectively. In addition, g' and g'' are required to be continuous at i and $i + 1$ with the cubic polynomial for the adjacent intervals. We denote g' and g'' at grid point i by g_i' and g_i''.

The second derivative of Eq. (G.1), namely

$$g''(s) = 2c + 6es \qquad (G.4)$$

is equated to g_i'' and g_{i+1}'' (which are still unknowns) at i and $i + 1$, respectively:

$$g_i'' = 2c \qquad (G.5)$$

$$g_{i+1}'' = 2c + 6eh_i \qquad (G.6)$$

By solving the preceding two equations, c and e are expressed in terms of g_i'' and g_{i+1}'' as

$$c = \frac{g_i''}{2} \qquad (G.7)$$

$$e = \frac{g_{i+1}'' - g_i''}{6h_i} \qquad (G.8)$$

The coefficient a is already given by Eq. (G.2). The coefficient b is determined by eliminating a, c, and e in Eq. (G.3) using Eqs. (G.2), (G.7), and (G.8), and then solving:

$$b = \frac{f_{i+1} - f_i}{h_i} - \frac{g_{i+1}'' + 2g_i''}{6} h_i \qquad (G.9)$$

Thus, the cubic polynomial Eq. (G.1) may be written as

$$g(s) = f_i + \left[\frac{f_{i+1} - f_i}{h_i} - \frac{g_{i+1}'' + 2g_i''}{6} h_i \right] s + \frac{g_i''}{2} s^2 + \frac{g_{i+1}'' - g_i''}{6h_i} s^3 \qquad (G.10)$$

The first derivative of Eq. (G.10) at $s = 0$ and $s = h$ becomes respectively

$$g'_i = -\frac{h_i}{6}[g''_{i+1} + 2g''_i] + \frac{1}{h_i}[f_{i+1} - f_i] \tag{G.11}$$

$$g'_{i+1} = \frac{h_i}{6}[2g''_{i+1} + g''_i] + \frac{1}{h_i}[f_{i+1} - f_i] \tag{G.12}$$

where $h = x_{i+1} - x_i$. For another interval of $x_{i-1} < x < x_i$, (G.12) becomes

$$g'_i = \frac{h_{i-1}}{6}[2g''_i + g''_{i-1}] + \frac{1}{h_{i-1}}[f_i - f_{i-1}] \tag{G.13}$$

where $h_{i-1} = x_i - x_{i-1}$. Since g'_i of Eq. (G.14) must be equal to that of g'_i of Eq. (G.11) for continuity of the first derivative, eliminating g'_i between those equations should yield

$$h_{i-1}g''_{i-1} + (2h_{i-1} + 2h_i)g''_i + h_ig''_{i+1} = 6\left[\frac{1}{h_{i-1}}f_{i-1}\right.$$
$$\left. -\left(\frac{1}{h_{i-1}} + \frac{1}{h_i}\right)f_i + \frac{1}{h_i}f_{i+1}\right] \tag{G.15}$$

Notice that Eq. (G.15) applies to each grid point except the two end points. If we prescribe the value of g'' at the two end points, or estimate g'' at the end points by extrapolation from two inner grid points, then the number of undetermined constants g''_i and the number of equations match. Assuming the data points are denoted by $i = 0, 1, \ldots, N$, the set of equations becomes

$$(2h_0 + 2h_1)g''_1 + h_ig''_2 = 6\left[\frac{1}{h_0}f_0 - \left(\frac{1}{h_0} + \frac{1}{h_1}\right)f_1\right.$$
$$\left. + \frac{1}{h_1}f_2\right] - h_0g''_0$$

$$\vdots$$

$$h_{i-1}g''_{i-1} + (2h_{i-1} + 2h_i)g''_i + h_ig''_{i+1} = 6\left[\frac{1}{h_{i-1}}f_{i-1}\right.$$
$$\left. -\left(\frac{1}{h_{i-1}} + \frac{1}{h_i}\right)f_i + \frac{1}{h_i}f_{i+1}\right] \tag{G.16}$$

$$\vdots$$

$$h_{N-1}g''_{N-2} + (2h_{N-2} + 2h_{N-1})g''_{N-1} = 6\left[\frac{1}{h_{N-2}}f_{N-2} - \left(\frac{1}{h_{N-2}} + \frac{1}{h_{N-1}}\right)f_{N-1}\right.$$
$$\left. + \frac{1}{h_{N-1}}f_N\right] - h_{N-1}g''_N$$

Therefore, we can solve the set of equations to determine g_i''. The set of equations is a tridiagonal equation set, for which the solution algorithm is described in Section 10.3.

There are three ways of specifying boundary conditions as follows:

(a) Specify g'' at the boundary (as explained already)
(b) Extrapolation from inside
(c) Cyclic boundary condition

If the second derivative of the function is known at the end points, (a) above can be used. In most cases, however, the second derivative is not known. One approach in specifying the boundary conditions is to assume $g'' = 0$ at the end points. Another approach is (b), namely, to extrapolate from inside. Considering the point, $i = 0$, extrapolating for g_0'' can be written as

$$g_0'' = -\frac{h_0}{h_1} g_2'' + \left(1 + \frac{h_0}{h_1}\right) g_1''$$

Eliminating g_0'' in Eq. (G.15) with $i = 1$ yields

$$\left(3h_0 + 2h_1 + \frac{h_0^2}{h_1}\right) g_1'' + \left(h_1 - \frac{h_0^2}{h_1}\right) g_2'' = 6\left[\frac{1}{h_0} f_0 - \left(\frac{1}{h_0} + \frac{1}{h_1}\right) f_1 + \frac{1}{h_1} f_2\right] \quad \text{(G.17)}$$

The cyclic boundary condition is applied when the first data and the last data are identical and the second derivatives at these data points are required to be identical also. This occurs if the whole data set represents points on a closed loop of a contour, for example.

Assuming there are four equally spaced grid intervals for simplicity of explanation, applications of two types of boundary conditions are illustrated next:

PRESCRIBING *g* AT THE END POINTS. The system of equations becomes

$$4g_1'' + g_2'' \quad\quad\quad = \frac{6}{h^2} [f_0 - 2f_1 + f_2] - g_0''$$

$$g_1'' + 4g_2'' + g_3'' = \frac{6}{h^2} [f_1 - 2f_2 + f_3] \quad\quad\quad\quad\quad \text{(G.18)}$$

$$g_2'' + 4g_3'' = \frac{6}{h^2} [f_2 - 2f_3 + f_4] - g_4''$$

where g_0'' and g_4'' are prescribed values.

EXTRAPOLATING g'' AT THE END POINTS FROM INTERNAL GRID POINTS. The g_0'' may be extrapolated by

$$g_0'' = 2g_1'' - g_2'' \tag{G.19}$$

Similarly, g'' at the right boundary may be extrapolated by

$$g_N'' = 2g_{N-1}'' - g_{N-2}'' \tag{G.20}$$

Using those extrapolations, the set of equations becomes

$$6g_1'' = \frac{6}{h^2}[f_0 - 2f_1 + f_2]$$

$$g_1'' + 4g_2'' + g_3'' = \frac{6}{h^2}[f_1 - 2f_2 + f_3] \tag{G.21}$$

$$6g_3'' = \frac{6}{h^2}[f_2 - 2f_3 + f_4]$$

Solution of the preceding equations is almost trivial: g_1'' and g_3'' are immediately obtained from the first and third equations respectively, and g_2'' is then determined by the second equation.

If the number of grid points is larger, the system of equations needs the Gauss elimination (see Chapter 6) or the tridiagonal solution scheme (Section 10.3).

One of the disadvantages of the cubic spline interpolation is that the interpolating polynomials may develop an oscillatory behavior of errors. A technique to suppress the oscillation of cubic spline interpolation has been proposed as the *tension spline method* [Barsky].

Example G.1

A data table is given as follows:

x	$f(x)$
0	0.000000
0.1	0.099833
$\pi/4$	0.707106
$\pi/2$	1.000000
$3\pi/4$	0.707106
π	0.000000
$5\pi/4$	-0.707106
$3\pi/2$	-1.000000
$7\pi/4$	-0.707106
$2\pi - 0.1$	-0.099833
2π	0.000000

By an cubic spline interpolation, estimate the values of $f(x)$ for $x = 0.1, 0.2, \ldots, 1.0$. The foregoing data were sampled from a test function $f(x) = \sin(x)$. Knowing this, evaluate the error of the $f(x)$ estimated by spline interpolation.

⟨**Solution**⟩

In the data table given, no boundary values for the second derivative are given. Therefore we use extrapolation. The result of interpolation and error evaluation are shown in Table G.1.

Table G.1

x	$g(x)$	$f(x)$	error
0.000000	0.000000	0.000000	0.000000
0.698131	0.643082	0.642787	−0.000295
1.396262	0.984166	0.984808	0.000642
2.094393	0.865193	0.866026	0.000833
2.792525	0.341525	0.342022	0.000497
3.490656	−0.341526	−0.342017	−0.000492
4.188787	−0.865193	−0.866024	−0.000831
4.886918	−0.984167	−0.984808	−0.000641
5.585049	−0.643088	−0.642791	0.000297
6.283180	0.000000	−0.000005	−0.000005

$g(x)$: spline interpolation
$f(x)$: exact function
error $= f - g$

PROGRAM

PROGRAM G-1 Cubic Spline Interpolation

(A) Explanations

This program performs spline interpolation for the given function table specified in the declaration statements. The abscissas of the functional values to be found by interpolation are also specified in a similar manner within the code. The user can choose one of the three types of boundary conditions, (1) specifying f'' at the boundaries, (2) extrapolation, and (3) cyclic. In case of (3) the last and first data points must have identical functional value.

(B) List

```
/* CSL/spline.c          C-Spline Interpolation Code   */
#include <stdio.h>
#include <stdlib.h>
#include <math.h>
```

```
/*                    x[i]:   x values of data points
                      f[i]:   Functional values of the data table.
                      h[i]:   grid point interval between points i and i+1
                      ni :    the number of points in x[] less one
                      jan :   number of x values in xa[]
                      kbc :   boundary condition type
                      xa[i]:  x values for which functional values are to be found
*/
main()
{
int i, i_, _i, _r;
float a[51], b[51], c[51], dd[51], fa[99], h[51], s[51];
void spl1();
int ni = 10;
int jan = 10;
float x[21] = {0.0,0.1,0.78539,1.57079,2.35619,3.14159,
                  3.92699,4.71239,5.4978,6.18318,6.28318};
float f[21] = {0.0,0.099833,0.707106,1.0,0.707106,
                  0.0,-0.707106,-1.0,-0.707106,-0.099833,0.0};
float xa[21] = {0.0,0.698131,1.396262,2.094393,2.792525,
                  3.490656,4.188787,4.886918,5.585049,6.28318};
    printf( "\n\nCSL/SPLINE   C-Spline Interpolation Code \n" );
    spl1( &ni, x, f, &jan, xa, fa, dd );
    printf( "\n      i           f[i]            f''[i] (solution)\n" );
    for( i = 0; i <= ni; i++ )
       printf( "%3d      %12.5e      %12.5e \n", i, f[i], dd[i] );
    printf( "\n      x          g(x): result of interpolation \n" );
    for( i = 1; i <= jan; i++ )
       printf( "%12.5e      %12.5e \n", xa[i-1], fa[i-1] );
    printf( "\n\n" );
    exit(0);
}


void spl1(ni, x, f, jan, xa, fa, dd)
int *ni, *jan;
float x[], f[], xa[], fa[], dd[];
{
int _10, i, i_, im, j, k, k_, kbc;
float a[101], b[101], c[101], h[101], s[101],z;
void trid(), tridcy();
    /*-------------------------- Input boundary conditions */
    printf("Type 0 to specify second derivatives at the end points, or\n");
    printf( "     1 to exprapolate, or \n" );
    printf( "     2 fo cyclic boundary conditions. \n" );
    scanf( "%d", &kbc );
    if( kbc == 0 ){
       printf( "Type the value of second derivative for the left end.\n");
       scanf( "%f", &dd[0] );
       printf( "Type the value of second derivative of the right end.\n");
       scanf( "%f", &dd[*ni] );
}
/*---------------------------------> Determining spline   */
im = *ni - 1;
for( i = 0; i <= (im); i++ )    h[i] = x[i + 1] - x[i];
for( i = 1; i <= (im); i++ ){
   a[i] = h[i - 1];
   c[i] = h[i];
   b[i] = 2*(a[i] + c[i]);
```

```
        s[i] = 6*((f[i - 1] - f[i])/h[i - 1] + (f[i + 1] - f[i])/h[i]);
    }
    if( kbc < 2 ){
        if( kbc == 0 ){
            s[1] = s[1] - a[1]*dd[0]; s[im] = s[im] - c[im]*dd[*ni];
        }
        b[1] = b[1] + 2*a[1];
        c[1] = c[1] - a[1];
        b[im] = b[im] + 2*c[im];
        a[im] = a[im] - c[im];
        trid( a, b, c, s, dd, &im );
        if( kbc == 1 ){
            dd[0] = 2*dd[1] - dd[2];
            dd[*ni] = 2*dd[im] - dd[*ni - 2];
        }
    }
    else{
        a[*ni] = h[im];
        c[*ni] = h[0];
        b[*ni] = 2*(a[*ni] + c[*ni]);
        s[*ni] = 6*((f[im] - f[*ni])/h[im] + (f[1] - f[0])/h[0]);
        tridcy( a, b, c, s, dd, ni );
        dd[0] = dd[*ni];
    }
    /* -------------------------------> Interpolation */
    j = 0;
    for( k = 1; k <= *jan; k++ ){
        k_ = k - 1;
        if( xa[k_] < x[0] || xa[k_] > x[*ni] ) {
            printf( " XA(K) = %g : out of range%d\n", xa[k - 1], k );
            return;
        }
        while( !(xa[k_] >= x[j] && xa[k_] <= x[j + 1]) ){
            if( j > *ni ) {
                printf( " J = %d: out of range\n", j );
                return;
            }
            j = j + 1;
        }
        z = xa[k_] - x[j];
        fa[k_] = f[j] + (-(2*dd[j] + dd[j + 1])/6*h[j] + (f[j+1] -
            f[j])/h[j])*z + (dd[j+1] - dd[j])/6/h[j]*pow(z,3) +
            dd[j]*pow(z,2)/2;
    }
    return;
}


void trid(a, b, c, s, dd, im)        /* Tridiagonal solution */
float a[], b[], c[], s[], dd[];  int *im;
{
int i, i_;    float r;
    for( i = 2; i <= *im; i++ ){
        i_ = i - 1;
        r = a[i]/b[i - 1];
        b[i] = b[i] - r*c[i - 1];
        s[i] = s[i] - r*s[i - 1];
    }
```

```
      dd[*im] = s[*im]/b[*im];
      for( i = *im - 1; i>= 1;i-- ) dd[i] = (s[i] - c[i]*dd[i + 1])/b[i];
      return;
}

void tridcy(a, b, c, s, dd, n)  /* tridiagonal solution with cyclic b.c.*/
float a[], b[], c[], s[], dd[];
int *n;
{
int i, i_, im;
float h[51], p, r, t, v[51];
   im = *n - 1;   v[1] = a[1];   h[1] = c[*n];   h[im] = a[*n];
   h[*n] = b[*n]; v[im] = c[im];
   for( i = 2; i <= im; i++ ){
      i_ = i - 1;
      r = a[i]/b[i_];
      b[i] = b[i] - r*c[i_];
      s[i] = s[i] - r*s[i_];
      v[i] = v[i] - r*v[i_];
      p = h[i_]/b[i_];
      h[i] = h[i] - p*c[i_];
      h[*n] = h[*n] - p*v[i_];
      s[*n] = s[*n] - p*s[i_];
   }
   t = h[im]/b[im];
   h[*n] = h[*n] - t*v[im];
   dd[*n] = (s[*n] - t*s[im])/h[*n];
   dd[im] = (s[im] - v[im]*dd[*n])/b[im];
   for( i = *n - 2; i >= 1; i-- )
      dd[i] = (s[i] - v[i]*dd[*n] - c[i]*dd[i + 1])/b[i];
   return;
}
```

REFERENCES

Barsky, B. A., *Computer Graphics and Geometric Modeling Using Beta-Splines*, Springer-Verlag, 1988.

Gerald, C. F., and P. O. Wheatley, *Applied Numerical Analysis*, 4th ed., Addison-Wesley, 1989.

Rogers, D. F., and J. A. Adams, *Mathematical Elements for Computer Graphics*, McGraw-Hill, 1976.

Transfinite Interpolation in Two Dimensions

Transfinite interpolation is an interpolation method for a two-dimensional space where the functional values along the external boundaries, as well as along the vertical and horizontal lines inside the boundaries, are known. The double interpolations discussed in Section 4.9 apply when functional values are known only at the intersections of vertical and horizontal lines. In contrast to the double interpolations, the transfinite interpolation fits to continuous functions specified along the horizontal and vertical lines.

To illustrate an application of transfinite interpolation, imagine an architect who is designing a curved roof on a rectangular building, the top view of which satisfies

$$x_0 \leqslant x \leqslant x_1, \quad y_0 \leqslant y \leqslant y_1$$

The client has specified the shape of the roof line along the four edges, which are four analytical functions expressing the height of the roof along the edges. These four functions are continuous across the corners so no sudden change of height happens at any corner of the roof. Now, he wants to create a smoothly curved surface that fits to the edge heights of the client.

This question can be restated as follows: Determine a smooth function $F(x, y)$ that satisfies the boundary conditions given by

$$F(x_0, y) = f_W(y)$$
$$F(x_1, y) = f_E(y)$$
$$F(x, y_0) = f_S(x)$$
$$F(x, y_1) = f_N(x)$$

$$(H.1)$$

where the right side of each equation is the analytical function that the client has given to the architect. We recognize that no unique solution exists to this problem, just as there is no unique way of interpolating given data. However, there are several possible ways to find such a function, among which are (1) to solve a Laplace equation

$$\nabla^2 F(x, y) = 0$$

with the boundary conditions, and (2) the transfinite interpolation.

The transfinite interpolation for the present problem can be written as

$$F(x, y) = \sum_{m=0}^{1} \phi_m(x)F(x_m, y) + \sum_{n=0}^{1} \psi_n(y)F(x, y_n)$$

$$- \sum_{m=0}^{1} \sum_{n=0}^{1} \phi_m(x)\psi_n(y)F(x_m, y_n) \tag{H.2}$$

where

$$\phi_0(x) = \frac{x_1 - x}{x_1 - x_0}$$

$$\phi_1(x) = \frac{x_0 - x}{x_0 - x_1}$$

$$\psi_0(y) = \frac{y_1 - y}{y_1 - y_0} \tag{H.3}$$

$$\psi_1(y) = \frac{y_0 - y}{y_0 - y_1}$$

The transfinite interpolation thus obtained is a smooth function and satisfies the boundary conditions. In the more general discussion that follow, the Lagrange interpolation is used instead of linear interpolation.

The transfinite interpolation above can be more generalized to include the functions specified along multiple lines. Consider a rectangular domain divided by vertical and horizontal lines as shown in Figure H.1. The leftmost vertical line is the left boundary and the right-most vertical lines is the right boundary. The vertical lines are identified by index m, where the left-most vertical lines is indexed by $m = 0$ while the last one is indexed by $m = M$. Likewise, the horizontal lines are indexed by n, where $n = 0$ is the bottom boundary and $n = N$ is the top boundary. Suppose that the values of $F(x, y)$ are known along all the horizontal and vertical lines. The function given along the mth vertical line will be denoted by $f_{v,m}(y)$, and that along the nth horizontal line is $f_{h,n}(x)$. Then, finding $F(x, y)$ that becomes equal to the known functions along the vertical and horizontal lines is the task. This problem can be

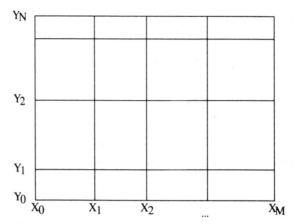

Figure H.1 Rectangular domain with horizontal and vertical lines

restated as: Find a smooth function $F(x, y)$ that satisfies the conditions given by

$$F(x, y) = F(x_m, y) = f_{v, m}(y), \quad \text{along } x = x_m \text{ (mth vertical line)}$$
$$F(x, y) = F(x, y_n) = f_{h, n}(x), \quad \text{along } y = y_n \text{ (nth horizontal line)} \tag{H.4}$$

The transfinite interpolation satisfying the given conditions is

$$F(x, y) = \sum_{m=0}^{M} \phi_m(x) F(x_m, y) + \sum_{n=0}^{N} \psi_n(y) F(x, y_n)$$
$$- \sum_{m=0}^{M} \sum_{n=0}^{N} \phi_m(x) \psi_n(y) F(x_m, y_n) \tag{H.5}$$

where

$$\phi_m(x) = \prod_{k=0, k \neq m}^{M} \frac{x - x_k}{x_m - x_k}$$

$$\psi_n(y) = \prod_{k=0, k \neq n}^{N} \frac{y - y_k}{y_n - y_k}$$

One can see that the first term of Eq. (H.5) is the Lagrange interpolation on the x-coordinate of the functions $F(x_m, y)$, while the second term is the Lagrange interpolation on the y-coordinate of the functions given along the horizontal lines. The third term is a double Lagrange interpolation of the data points given at the intersections of the vertical and horizontal lines. The transfinite interpolation satisfies all the boundary conditions at the external boundaries as well as along the internal boundaries.

Although we assumed that the functions along the vertical and horizontal lines are analytical functions, this can be applied to a discretely defined function as illustrated next.

Example H.1

In the following function table, the functional values are given along certain columns and rows.

Table H.1 A data table given for $F(i, j)$

j\i	1	2	3	4	5	6	7	8	9	10	11
1	0.2955	0.3894	0.4794	0.5646	0.6442	0.7174	0.7833	0.8415	0.8912	0.9320	0.9636
2	0.4794					0.8415					0.9975
3	0.6442					0.9320					0.9917
4	0.7833	0.8415	0.8912	0.9320	0.9636	0.9854	0.9975	0.9996	0.9917	0.9738	0.9463
5	0.8912					0.9996					0.8632
6	0.9636					0.9738					0.7457
7	0.9975	0.9996	0.9917	0.9738	0.9463	0.9093	0.8632	0.8085	0.7457	0.6755	0.5985

Fill the blank by the transfinite interpolation.

⟨**Solution**⟩

The table filled with the transfinite interpolation is shown in Table H.2. The error of interpolation is evaluated and shown in Table H.3.

Table H.2 Results of transfinite interpolation

1	0.2955	0.3894	0.4794	0.5646	0.6442	0.7174	0.7833	0.8415	0.8912	0.9320	0.9636
2	0.4794	0.5647	0.6443	0.7174	0.7834	0.8415	0.8912	0.9320	0.9635	0.9854	0.9975
3	0.6442	0.7174	0.7834	0.8415	0.8912	0.9320	0.9635	0.9854	0.9974	0.9995	0.9917
4	0.7833	0.8415	0.8912	0.9320	0.9636	0.9854	0.9975	0.9996	0.9917	0.9738	0.9463
5	0.8912	0.9320	0.9635	0.9854	0.9975	0.9996	0.9917	0.9739	0.9464	0.9094	0.8632
6	0.9636	0.9854	0.9974	0.9995	0.9916	0.9738	0.9464	0.9094	0.8633	0.8086	0.7457
7	0.9975	0.9996	0.9917	0.9738	0.9463	0.9093	0.8632	0.8085	0.7457	0.6755	0.5985

Table H.3 Error of transfinite interpolation

1	0.0000	0.0000	0.0000	0.0000	0.0000	0.0000	0.0000	0.0000	0.0000	0.0000	0.0000
2	0.0000	-0.0001	-0.0001	-0.0001	0.0000	0.0000	0.0001	0.0001	0.0001	0.0001	0.0000
3	0.0000	-0.0001	-0.0001	-0.0001	0.0000	0.0000	0.0000	0.0001	0.0001	0.0001	0.0000
4	0.0000	0.0000	0.0000	0.0000	0.0000	0.0000	0.0000	0.0000	0.0000	0.0000	0.0000
5	0.0000	0.0001	0.0001	0.0001	0.0000	0.0000	0.0000	-0.0001	-0.0001	-0.0001	0.0000
6	0.0000	0.0001	0.0001	0.0001	0.0001	0.0000	-0.0001	-0.0001	-0.0001	-0.0001	0.0000
7	0.0000	0.0000	0.0000	0.0000	0.0000	0.0000	0.0000	0.0000	0.0000	0.0000	0.0000

PROGRAM

PROGRAM H–1 Transfinite Interpolation Program

(A) Explanations

This program estimates the values of a table for $f(i, j)$ when the values of certain columns i and rows j are all known. The x value corresponding to each i should be equispaced, and the same for the y values corresponding to each j.

(B) List

```
/* CSL/transf.c    Transfinite Interpolation */
#include <stdio.h>
#include <stdlib.h>
#include <math.h>
/*      in, jn: number of points in x and y directions, respectively
        ix(m): i value of the mth column where functional values are known
        je(n) : j value of the nth column where functional values are known
        mmx : number of columns where the functional values are known
        nny : number of rows where functional values are known
                lines separating the regions
        ph[][]: greek "phi" in the text (see Eq.(H.3))
        ps[][]: greek "psi" in the text (see Eq.(H.3))
         f[][]: function F(i,j)
*/
main()
{
int i, i_, im, imd, j, j_, jn, jnd, m, m_, md, md_;
int n, n_, nd, nd_, _i, _r;
static int ix[6], je[6];
float a[20][20], dummy, fi, fim, fimd, fj, fjn, fjnd, g, ph[100][20],
    ps[100][20], z;
static int mmx = 3, nny = 3, ni = 11, nj = 7;
static _itmp0[] = {1,6,11};
static _itmp1[] = {1,4,7};
float f[20][20] =
        {{0.2955,0.3894,0.4794,0.5646,0.6442,
          0.7174,0.7833,0.8415,0.8912,0.9320,0.9636},
         {0.4794,0.,0.,0.,0.8415,0.,0.,0.,    0.,0.9975},
         {0.6442,0.,0.,0.,0.9320,0.,0.,0.,0.,0.9917},
         {0.7833,0.8415,0.8912,0.9320,0.9636,
          0.9854,0.9975,0.9996,0.9917,0.9738,0.9463},
         {0.8912,0.,0.,0.,0.9996,0.,0.,0.,    0.,0.8632},
         {0.9636,0.,0.,0.,0.9738,0.,0.,0.,    0.,0.7457},
         {0.9975,0.9996,0.9917,0.9738,0.9463,
          0.9093,0.8632,0.8085,0.7457,0.6755,0.5985}}};
    for( n = 1, _r = 0; n <= 3; n++ ) ix[n - 1] = _itmp0[_r++];
    for( m = 1, _r = 0; m <= 3; m++ ) je[m - 1] = _itmp1[_r++];
    printf( "\n\n CSL/TRANSF  Trnasfinite Interpolation \n" );
    for( j = 1; j <= nj; j++ ){
        printf( "%d ", j );
        for( i = 1; i <= ni; i++ ) printf( "%6.4f ", f[j-1][i - 1] );
        printf( "\n" );
    }
    for( i = 1; i <= ni; i++ ){
        i_ = i - 1;
        fi = i;
        for( m = 1; m <= mmx; m++ ){
            m_ = m - 1;
            im = ix[m_];
            fim = im;
            z = 1.0;
            for( md = 1; md <= mmx; md++ ){
                imd = ix[md-1];
                 if( im != imd ){
```

```
                    fimd = imd;
                    z = z*(fi - fimd)/(fim - fimd);
                }
            }
            ph[i_][m_] = z;
        }
    }
    for( j = 1; j <= nj; j++ ){
        j_ = j - 1;
        fj = j;
        for( n = 1; n <= nny; n++ ){
            n_ = n - 1;
            jn = je[n_];
            fjn = jn;
            z = 1.0;
            for( nd = 1; nd <= nny; nd++ ){
                nd_ = nd - 1;
                jnd = je[nd_];
                if( jn != jnd ){
                    fjnd = jnd;
                    z = z*(fj - fjnd)/(fjn - fjnd);
                }
            }
            ps[j_][n_] = z;
        }
    }
    for( i = 1; i <= ni; i++ ) {
        for( j = 1; j <= nj; j++ ) a[j-1][i-1] = 0.0;
    }
    for( j = 1; j <= nj; j++ ){
        for( i = 1; i <= ni; i++ ){
            i_ = i - 1;
            g = 0.0;
            for( m = 1; m <= mmx; m++ ){
                m_ = m - 1;
                im = ix[m_];
                g = g + ph[i_][m_]*f[j-1][im - 1];
            }
            for( n = 1; n <= nny; n++ ){
                n_ = n - 1;
                jn = je[n_];
                g = g + ps[j-1][n_]*f[jn - 1][i_];
            }
            for( m = 1; m <= mmx; m++ ){
                m_ = m - 1;
                im = ix[m_];
                for( n = 1; n <= nny; n++ ){
                    n_ = n - 1;
                    jn = je[n_];
                    g = g - ph[i_][m_]*ps[j-1][n_]*f[jn - 1][im - 1];
                }
            }
            a[j-1][i_] = g;
        }
    }

        }
        printf( " \nFinal results with interpolation: \n" );
        for( j = 1; j <= nj; j++ ){
```

```
      printf( "%d ", j );
      for( i = 1; i <= ni; i++ )  printf( "%6.4f ", a[j-1][i - 1] );
      printf( "\n" );
   }
   printf( "\n\n" );        exit(0);
}
```

(C) Sample Output and Discussions

See Example H.1.

REFERENCES

Erikson, L., "Practical Three-dimensional Mesh Generation Using Transfinite Interpolation," *SIAM J. Sci. Stat. Comput.*, Vol. 6, 1985.

Gordan, W. J., and C. A. Hall, "Construction of Curvilinear Coordinate Systems and Application to Mesh Generation," *International Journal for Numerical Methods in Engineering*, Vol. 7, 461–477, 1973.

Smith, R. E., "Algebraic Grid Generation," *Numerical Grid Generation* (J. Thompson, ed.), North-Holland, 1982.

Thompson, J. F., Z. U. A. Wasri, and C. W. Mastin, "Boundary-Fitted Coordinate Systems for Numerical Solution of Partial Differential Equations—A Review," *J. Comp. Physics*, Vol. 47, 1–108, 1982.

APPENDIX I

Linking C with Fortran

Mixing Fortran and C in one program becomes often advantageous or even necessary. Fortran is still a preferred language in developing scientific computations software. Its major advantages are (1) debugging of Fortran programs is far easier than C, (2) scientific subroutines in Fortran are more abundantly available, (3) existing software for scientific and mathematical computations is mostly in Fortran. Nevertheless, C becomes desirable whenever graphics or other special programming that needs access to machine language level is required. Indeed, in a typical situation, most of a program may be written in Fortran but some subroutines are developed in C. Conversely, one may write the main program in C while subroutines may be in Fortran. In this appendix, we consider the former case. Once this case is understood, the opposite case is easy.

In the following sections, we first discuss how to compile and link Fortran and C programs. Then, we discuss how to pass arguments of subroutines between Fortran and C.

COMMANDS FOR COMPILATION AND LINKING

When a subroutine written in C is called from Fortran, or one in Fortran is called from C, the name of each subroutine is followed by "_", like "gauss_(....)".

We assume that the Fortran program is saved as main.f, and the C subroutines are saved as sub.c. The compile-run procedures shown next are based on IRIS of

Silicon Graphics, but they are applicable to other Unix systems with minor modifications. We also consider the case of VAX.

On IRIS, both Fortran and C programs are compiled and linked with libraries by the following command:

```
> cc -tf  main.f sub.c (-lgl_s -lm) -o demo
```

In the preceding command, > is a prompt sign, "-tf" is a compile option specific to IRIS, which orders compilation of a Fortran program from the "cc" command, "-lgl_s" is to link with the graphic library on IRIS, "-lm" is to link with a math library, and "demo" is the name of the execution file to be created. The library specifications in () are optional (do not type parentheses).

Fortran and C programs can be compiled separately also. In the next example, C and Fortran programs are compiled by cc and f77 commands, respectively, and then linked:

```
> cc -c sub.c
> f77 -c main.f
> cc -tf main.o sub.o (-lgl_s -lm)
```

In the preceding commands, the option "-c" specifies that cc or f77 only compile the source but do not link with any other object or library. The second command may be replaced by

```
> cc -c -tf main.f -o test
```

The "cc" command in the third line is used to link the two object files, one from Fortran and another from C, together with libraries and create an execution file "test".

There can be some variations from the preceding example. If sub.o has been created and need not be compiled again, the following command will compile only Fortran and link:

```
> cc -tf main.f sub.o (-lgl_s -lm)
```

It is simpler to compile and link Fortran and C programs on VAX than in Unix. Assume that the names of the Fortran and C programs on VAX are MAIN.FOR and SUB.C, respectively. The commands to compile these are:

```
VAX> FOR MAIN
VAX> CC SUB
```

The preceding procedures create MAIN.OBJ and SUB.OBJ, which can be linked by

```
VAX> LINK MAIN, SUB
```

Then an executable file MAIN.EXE is created. To run it, the command is

```
VAX> RUN MAIN
```

CALLING SUBROUTINES IN C FROM FORTRAN

Arguments in the calling statements are interpreted differently in different languages. Therefore, understanding it is the key in using subroutines written in different languages. We will discuss how arguments are passed from Fortran to C in the following four sections.

Passing Nonarray Numeric Variables as Arguments

Arguments in the calling statement of a subroutine or function in Fortran are viewed as pointers in C. In the following example, nonarray numeric variables are in the arguments of subroutine orange. Notice that "orange" is lower case in C, and is followed by "_".

Fortran

```
    X=3.14159
    CALL ORANGE(X)
    PRINT *,X
    END
```

C

```
#include <stdio.h>
orange_(x)
float *x;
{   int i;
    printf("%12.5e\n", *x);
    *x=2.888;
    return;
}
```

Output

```
3.14159
2.888
```

In the foregoing example, X is a floating variable in Fortran, but when received in C, it is a pointer. In C, therefore, $*x$ represents X in Fortran.

Passing Numeric Array Variables as Arguments

Numeric array variables are viewed as pointers from C just as nonarray variables are. In C subroutines, however, the indirection operator (*) is not necessary in front of the array variables.

Another important aspect is that the base of an array variable in C is 0, while in Fortran it is 1 unless the base is otherwise specified in DIMENSION or COMMON statement. This causes the need of shifting the index of an array variable by one as illustrated in the following example:

Fortran

```
DIMENSION X(10)
IMAX = 3
DO K = 1, IMAX
X(K) = K
END DO
CALL APPLE(X,IMAX)
END
```

C

```
#include <stdio.h>
void apple_(a,imax)
float a[]; int *imax;
{
int k;
        for (k=1; k<=*imax; k++)
        {
                printf("%d, %f\n", k, a[k-1]);
        }
        return;
}
```

Output

```
1       1.0
2       2.0
3       3.0
```

Passing Numeric Two-dimensional Array Variables as Arguments

All the rules for one-dimensional array variables apply to two-dimensional or multi-dimensional array variables, but one more complication is added. This is because there is a difference in the way two-dimensional (as well as multidimensional) array variables are stored in the memory. In Fortran a variable $P(I, J)$ is saved starting with $P(1, 1)$ and then in the sequence, $P(2, 1), P(3, 1), \ldots, P(I_{max}, 1), P(1, 2), P(2, 2), \ldots$ That is, the first index is changed first. In C, on the other hand, the last index is changed first. Therefore, $s[i][j]$ is saved in the sequence of $s[0][0], s[0][1], s[0][2], \ldots$ $s[0][j_{max}], s[1][0], s[1][1] \ldots$ Therefore, if P is passed through an argument of Fortran and it is received as s in C, the contents are not passed correctly unless the order of i and j in C is reversed. The following example illustrates the use of a two-dimensional array variable.

Fortran

```
DIMENSION P(4,3)
IMAX=4
JMAX=3
CALL GRAPE(P,IMAX,JMAX)
DO I=1,IMAX
        PRINT *, (P(I,J), J=1, JMAX)
END DO
END
```

C

```
#include <math.h>
void grape_(s,imax,jmax)
float s[3][4]; int *imax, *jmax;
{
int     i, j;
        printf("----\n");
        printf("    %d     %d    \n",*imax,*jmax);
        for (i=0; i<=*imax-1; i++)
        {
                for (j=0; j<=*jmax-1; j++)
                {
                        s[j][i] = (i+1)*100 + (j+1);
                }
        }
        return;
}
```

Output

```
101.0    102.0    103.0
201.0    202.0    203.0
301.0    302.0    303.0
401.0    402.0    403.0
```

Passing Character Variables as Arguments

Character variables in the arguments are treated differently from numeric variables. That is, when a character variable appears as an argument in the Fortran calling statement, the Fortran compiler places the pointer of the character variable first, and next comes an integer representing the length of the character variable. On the receiving side in C, therefore, two arguments must be received. An illustration is given next:

Fortran

```
CHARACTER *8, A
A = 'ANALYTIC'
CALL SHOWCHAR( A )
END
```

C

```
#include <stdio.h>
void showchar_( a, j )
char a[]; int j;
{
int i;
      printf("%d\n",j);
      for (i=0; i<j; i++)
      {
            printf("%c", a[i]);
      }
      printf("\n");
      return;
}
```

Output

```
8
ANALYTIC
```

Notice in the foregoing example that *j* in C is not a pointer but an integer variable.

If two character variables are placed in the Fortran calling statement, there will be four arguments in the C subroutine, the first two are character variables and next two are integers for the lengths of the two character variables.

Answer Key for Chapters 1–9

Chapter 1

(1.2) (a) -1.

(b) -32768

(1.9) (b) $\dfrac{1 + e^{2x}}{2}$

Chapter 2

(2.3)

(a) $g(x) = \dfrac{(x - 0.5)(x - 0.75)}{(-0.25)(-0.5)} 0.8109$

$\qquad + \dfrac{(x - 0.25)(x - 0.75)}{(0.25)(-0.25)} 0.6931$

$\qquad + \dfrac{(x - 0.25)(x - 0.5)}{(0.50)(0.25)} 0.5596$

(b) Error of the Lagrange interpolation formula given in item (a) is

$\qquad \text{Error} \simeq (x - 0.25)(x - 0.5)(x - 0.75)f'''(0.5)/6$
$\qquad\qquad = 0.00023 \text{ for } x = 0.6$

(2.5) (b) Equation. 92.3.40 becomes

$$e(x) = \frac{(x - 0.0)(x - 0.4)(x - 0.8)(x - 1.2)}{(4)(3)(2)(1)} \exp (0.6)$$

where $f''(x_M) = \exp (0.6)$. Estimates of error for $x = 0.2, 0.6,$ and 1.0 are

$$\begin{aligned} e(0.2) &\simeq -0.0018 \\ e(0.6) &\simeq 0.0011 \\ e(1.0) &\simeq -0.0018 \end{aligned}$$

(c) Actual error evaluated by $e(x) = \exp (x) - g(x)$ are

$$\begin{aligned} e(0.2) &= -0.0017 \\ e(0.6) &= 0.011 \\ e(1.0) &= -0.0020 \end{aligned}$$

(2.16) The Newton forward interpolation polynomial passing through $i = 2, 3,$ and 4 is

$$g(x) = 0.8109 - 0.1178s - 0.0157s(s - 1)/2$$

where $s = (x - 0.25)/0.25$.

(b) The error of the Newton forward polynomial above is the term added if one more data point is used, so

$$\text{Error} = -0.0049s(s - 1)(s - 2)/6$$

The value of s for $x = 0.6$ is $s = (0.6 - 0.25)/0.25 = 1.4$, so

$$\text{Error} = -0.0049(1.4)(1.4 - 1)(1.4 - 2)/6 = 0.00027$$

(2.27) (a) The four Chebyshev points and corresponding functional values are:

i	x_i	f_i
1	1.0381	0.0374
2	1.3087	0.2690
3	1.6913	0.5255
4	1.9619	0.6739

(b)/(c) Estimate for errors by Eq. (2.3.4) is

$$e(x) \simeq \frac{(x - 1.0381)(x - 1.3087)}{(x - 1.6913)(x - 1.9619)}{4!} f''''(x_m)$$

where $f''''(x_M) = -6/(1.5)^4$. The estimated errors are calculated and shown with the actual errors evaluated for (c):

x	Estimated error	Actual error
1	$-3.8E - 4$	$-5.7E - 4$
1.1	$3.2E - 4$	$4.4E - 4$
1.2	$3.2E - 4$	$4.2E - 4$
1.3	$2.9E - 5$	$3.5E - 5$
1.4	$-2.6E - 4$	$-3.0E - 4$
1.5	$-3.8E - 4$	$-4.1E - 4$
1.6	$-2.6E - 4$	$-2.7E - 4$
1.7	$2.9E - 4$	$2.8E - 4$
1.8	$3.2E - 4$	$3.0E - 4$
1.9	$3.2E - 4$	$2.8E - 4$
2.0	$-3.9E - 4$	$-3.2E - 4$

Chapter 3

(3.1) $x = 1.7626$ (Exact $= 1.762441$)

(3.2) Final answer $x = 0.3193$ (Exact $= 0.3189289$)

(3.5)

(a) $[0.6, 0.7]$, $[4.7, 4.8]$: roots are 0.60527, 4.70794

(b) $[1.6, 1.7]$: root is 1.61804

(c) $[4, 4.1]$: root is 4.0

(3.6) $x = 0.7697$

(3.7)

(a) $[3.14, 4.71]$ $\quad x = 4.4283$
 $[6.28, 7.85]$ $\quad x = 7.7056$

(b) $[0, 1]$ $\quad x = 0.5419$
 $[1, 2]$ $\quad x = 1.0765$

(c) $[1, 2]$ $\quad x = 1.3248$

(d) $[0.1, 0.2]$ $\quad x = 0.10214$
 $[0.5, 0.6]$ $\quad x = 0.55635$
 $[0.9, 1.0]$ $\quad x = 0.92595$

(3.8)

(a) $[-5, -4]$ $\quad x = -4.7368$
 $[-2, -1]$ $\quad x = -1.3335$
 $[0, 1]$ $\quad x = -0.8530$
 $[50, 51]$ $\quad x = 50.1831$

(b) $[0.3, 0.4]$ $\quad x = 0.3786$
 $[3.3, 3.4]$ $\quad x = 3.3155$

(c) $[-3.2, -3.1]$ $\quad x = -3.1038$

(3.9) $v = 37.73$ m/sec

(3.13)

(a) $x = 0.6772$, $x = 1.9068$

(b) $x = 0.0$, $x = 0.7469$

(c) $x = -0.3714$, $x = 0.6053$, $x = 4.7079$

(d) $x = 0.4534$

(e) $x = 2$

(3.14)

K	Roots	
1	-1.1183	-35.77
2	-2.5917	-25.42
3	-4.894	-17.81

(3.15) $x = 3.47614$

(3.16)

(b) $f(x) = \sin(x) - 0.3 \exp(x)$
The root found: $x = 1.0764$ (with $x_0 = 5$)
The root found: $x = 0.5419$ (with $x_0 = 0$)

(c) $f(x) = -x^3 + x + 1$
The root found: $x = 1.3247$ (with $x_0 = 5$)

(d) $f(x) = 16x^5 - 20x^3 + 5x$

Initial guess x_0	Roots found
-5.0	-0.9511
1	0.9511
0	0
0.5	0.5877

(3.17) 1.8751, 4.6940, 7.8547

(3.18) First root $\quad$ 3.927
 Second root $\quad$ 7.068
 Third root $\quad$ 10.210

(3.20)

(a) $f(x) = 0.5 \exp(x/3) - \sin(x), \quad x > 0$

Initial guess	x converged
1	0.6772
3	1.9068

(b) $f(x) = \log(1 + x) - x^2$

Initial guess	x converged
1	0.7468
0.1	0

(c) $f(x) = \exp(x) - 5x^2$

Initial guess	x converged
5	4.7079
3	0.6052
0	-0.3715

(d) $f(x) = x^3 + 2x - 1$

Initial guess	x converged
5	0.4533

(e) $f(x) = \sqrt{x + 2} - x$

Initial guess	x converged
4	2

(3.22) $x = 0.1929$

(3.23)

Initial guess	x converged
0.5	0.4717

(3.27)

(a) $f(x) = 0.5 \exp(x/3) - \sin(x) = 0$
Converged solution: $x = 0.6773$

(b) $f(x) = \log(1 + x) - x^2 = 0$
Initial guess $= 0.1$
Converged solution: $x = 0.7469$

(3.28) (a) Using an estimate of $f = 0.01$ on the right side:

Iteration times	f
1	0.05557
2	0.05409
3	0.05411
4	0.05411

(b) With an estimate of $f = 0.01$, the equation converges with two iteration steps to $f = 0.01967$.

(3.29)

(a) $x^2 - 1$ with deflated polynomial of $x^2 - 4$

(b) $2.386 + 1.9676x + x^2$
Deflated polynomial is $-2.934 + 2x$

(c) $0.3733 - 0.325x + x^2$
Deflated polynomial is $-8.034 - 4.325x - x^2$ which is also a quadratic factor.

(d) $2.403 - 2.343x + x^2$
Deflated polynomial is $(0.6657 - x)$

(e) $0.563 - 2.068x + x^2$
Deflated polynomial $= x^2 - 13.931 + 42626$

(f) $x^2 - 2x + 1$
Deflated polynomial $= x^4 - 4x^3 + 5x^2 - 4x + 1$
Applying the Bairstow program once again yields

$$x^2 - x + 1$$

with a deflated polynomial of $x^2 - 3x + 1$
So, the three quadratic factors found are:

$$(x^2 - 2x + 1), \quad (x^2 - x + 1), \quad \text{and} \quad (x^2 - 3x + 1)$$

(3.31)

(a) Quadratic factors found: $x^2 - x + 1, x^2 + x + 2$
Roots: $(1 \pm \sqrt{3}i)/2, (1 \pm \sqrt{7}i)/2$

(b) Quadratic factor found: $x^2 + 2x + 2$
Deflated polynomial: $x + 1$
Roots: $-1 \pm i, -1$

(c) Quadratic factor found: $x^2 + 0.5x - 0.5$
Deflated polynomial: $x - 2.2$
Roots: $0.5, -1, 2.2$

(d) Quadratic factor: $x^2 - 0.4x - 1.65$
Deflated polynomial: $x^2 + 1.5x - 7$
Roots: $1.5, -1.1, -3.5, 2$

(e) Quadratic factor: $x^2 - 2x + 1$
Deflated polynomial: $x^2 - 4x + 4$
Roots: $1, 1, 2, 2$

(3.33) $k = 0 \quad s = 0, -3, 1 \pm i$
$k = 1 \quad s = -0.472, -3.065, -0.731 \pm 0.92i$
$k = 10: \ s = -1.679, -3.570, 0.126 \pm 3.64i$

(3.34)

K	Roots		
0	1,	-2,	-5
1	0.945,	-1.78,	-5.05
10	0,	-0.55,	-5.45
10.392	-0.260,	-0.27,	-5.46
11	-0.256	$\pm 0.68i$,	-5.48
20	-0.109	$\pm 2.62i$,	-5.78
25	-0.039	$\pm 3.18i$,	-5.92
35	0.085	$\pm 4.02i$,	-6.17
50	0.244	$\pm 4.94i$,	-6.48
100	0.642	$\pm 6.91i$,	-7.28

Chapter 4

(4.1) $3x^3 + 5x - 1$ $[0, 1]$

$N = 2$	$I = 2.62500$
$N = 4$	$I = 2.53125$
$N = 8$	$I = 2.50781$
$N = 16$	$I = 2.50195$

$x^3 - 2x^2 + x + 2$ $[0, 3]$

$N = 2$	$I = 15.56250$
$N = 4$	$I = 13.45312$
$N = 8$	$I = 12.92578$
$N = 16$	$I = 12.79394$
$N = 32$	$I = 12.76098$

$x^4 + x^3 - x^2 + x + 3$ $[0, 1]$

$N = 2$	$I = 3.71875$
$N = 4$	$I = 3.64257$
$N = 8$	$I = 3.62316$
$N = 16$	$I = 3.61829$
$N = 32$	$I = 3.61707$

$\tan(x)$ $\left[0, \dfrac{\pi}{4}\right]$

$N = 2$	$I = .35901$
$N = 4$	$I = .34975$
$N = 8$	$I = .34737$
$N = 16$	$I = .34677$
$N = 32$	$I = .34662$

e^x $[0, 1]$

$N = 2$	$I = 1.75393$
$N = 4$	$I = 1.72722$
$N = 8$	$I = 1.72051$
$N = 16$	$I = 1.71884$
$N = 32$	$I = 1.71842$

$1/(2 + x)$ $[0, 1]$

$N = 2$	$I = .40833$
$N = 4$	$I = .40618$
$N = 8$	$I = .40564$
$N = 16$	$I = .40551$
$N = 32$	$I = .40547$

(4.2)

N	I	Error
2	0.94805	0.05195
4	0.98711	0.01289
8	0.99687	0.00313
25	0.99966	0.00034
50	0.99991	0.00009
100	0.99997	0.00003

(4.3) $h = 0.4$ 1.62312
$h = 0.2$ 1.91924
$h = 0.1$ 1.99968

(4.4) $I = 0.1 + \dfrac{1}{3}(I_{0.1} - I_{0.2})$

$$= 1.99968 + \frac{1}{3}(1.99968 - 1.91924) = 2.02649$$

(4.5)
(a) With $h = 0.25$:

$$I_{0.25} = \frac{0.25}{2}[0.9162 + 2(0.8109 + 0.6931$$
$$+ 0.5596) + 0.4055] = 0.68111$$

With $h = 0.5$:

$$I_{0.5} = \frac{0.5}{2}[0.9162 + 2(0.6931) + 0.4055] = 0.67697$$

(b) $I = I_{0.25} + \dfrac{1}{3}[I_{0.25} - I_{0.5}] = 0.68249$

(4.6)

N	I	Error, %
2	1.00228	-0.22
4	1.00013	-0.013
8	1.00001	-0.001
16	1.00000	-0.00005

(4.7) $3x^3 + 5x - 1$ $[0, 1]$

$N = 4$	$I = 2.5000$
$N = 8$	$I = 2.5000$
$N = 16$	$I = 2.5000$

$x^3 - 2x^2 + x + 2$ $[0, 3]$

$N = 4$	$I = 12.7500$
$N = 8$	$I = 12.7500$
$N = 16$	$I = 12.7500$

$x^4 + x^3 - x^2 + x + 3$ $[0, 1]$

$N = 4$	$I = 3.61718$
$N = 8$	$I = 3.61669$
$N = 16$	$I = 3.61666$

$\tan(x)$ $\left[0, \dfrac{\pi}{4}\right]$

$N = 4$	$I = .34667$
$N = 8$	$I = .34657$
$N = 16$	$I = .34657$

e^x $[0, 1]$

$N = 4$	$I = 1.71831$
$N = 8$	$I = 1.71828$
$N = 16$	$I = 1.71828$

$\dfrac{1}{2+x}$ $[0, 1]$

$N = 4$	$I = .40547$
$N = 8$	$I = .40546$
$N = 16$	$I = .40546$

(4.10)

(a) 1.8137

(b) 0.6142

(c) 1.1107

(4.11)

(a) 2.0972

(b) 1.2912

(c) 8.5527

(4.12)

N	I
2	11.7809
4	11.7318
8	11.7288

(4.13) Length = 56.52 m

(4.14) Distance = 291.59m   (Exact = 291.86)

(4.16)

$N = 2$	$I = 18.805$
$N = 4$	$I = 21.526$
$N = 6$	$I = 21.540$

(4.18)

$N = 4$	$I = 0.82224$
$N = 6$	$I = 0.82246$ $(I_{\text{exact}} = 0.82246)$

(4.19) 1.34329

(4.20) **(a)** 0.9063, **(b)** 3.104

(4.22) **(a)** 1.8138, **(b)** 0.6142, **(c)** 2.0972, **(d)** 1.2914

(4.23) $I = \dfrac{0.5}{2}[0.93644 + 2(0.85364) + 0.56184]$

$\qquad = 0.80139$

(4.24) $I = \dfrac{0.5}{3}[0 + 4(0.4309) + 1.2188] = 0.4904$

(4.25)

2×2	2.6666
4×4	2.9760
8×8	3.0835
16×16	3.1211
32×32	3.1343
64×64	3.1390

(4.26) $I = 0.36686$

(4.27) $I \simeq \dfrac{0.5}{3}[1.9682 + 4(1.8154) + 1.5784] = 1.8014$

(4.28) $I = \left[\left(5 - \dfrac{1}{2}\right)\right][(0.55555)(1.29395)$

$\qquad + (0.88888)(0.04884)$

$\qquad + (0.55555)(0.63524)]$

$\qquad = 2.23036$

Chapter 5

(5.13) $\begin{aligned} 27a_3 + 8a_2 + a_1 + 0 &= 0 \\ 9a_3 + 4a_2 + a_1 + 0 &= 1 \\ 3a_3 + 2a_2 + a_1 + 0 &= 0 \\ a_3 + a_2 + a_1 + a_0 &= 0 \end{aligned}$

(5.14)

(a) $[f(0.2) + 3f(0) - 4f(-0.1)]/(6*0.1)$

(b) $0(h^3) = -\dfrac{1}{3}h^2 f'''$

(c) $(4.441 + 3 * 4.020 - 4 * 4.157)/0.6 = -0.2117$

(5.15) $f'_i(x) = \dfrac{-f_{i+3} + 9f_{i+1} - 8f_i}{6h} + 0(n^2)$

$\qquad 0(h^2) = \dfrac{1}{2}h^2 f'''_i$

(5.18) $m = -4$

(5.28) $g'(x_i) = \dfrac{1}{h}\left[\Delta f_i + \dfrac{1}{2}(-1)\Delta^2 f_i + \dfrac{1}{6}(2)\Delta^3 f_i\right]$

$\qquad = \dfrac{1}{h}\left[f_{i+1} - f_i - \dfrac{1}{2}(f_{i+2} - 2f_{i+1} + f_i)\right.$

$\qquad\qquad \left. + \dfrac{1}{3}(f_{i+3} - 3f_{i+2} + 3f_{i+1} - f_i)\right]$

$\qquad = \dfrac{2f_{i+3} - 9f_{i+2} + 18f_{i+1} - 11f_i}{6h}$

$\qquad \text{error} = -\dfrac{1}{4}h^3 f''''$

$\qquad g''(x_i) = \dfrac{1}{h^2}(\Delta^2 f_i - \Delta^3 f_i)$

$\qquad = \dfrac{1}{h^2}[(f_{i+2} - 2f_{i+1} + f_i)$

$\qquad\qquad - (f_{i+3} - 3f_{i+2} + 3f_{i+1} - f_i)]$

$\qquad = \dfrac{1}{h^2}[-f_{i+3} + 4f_{i+2} - 5f_{i+1} + 2f_i]$

$\qquad \text{error} = \dfrac{11}{12}h^2 f''''$

(5.34)

(a) $\dfrac{\partial}{\partial y} f(1, 0) \simeq \dfrac{-f(1, 1) + 4f(1, 0.5) - 3f(1, 0)}{2h}$

$= \dfrac{-0.7002 + 4(0.4767) - 3(0.2412)}{2(0.5)}$

$= 0.483$

(b) $\dfrac{\partial^2}{\partial x^2} f \simeq \dfrac{f(0.5, 1) - 2f(1, 1) + f(1.5, 1)}{(0.5)^2}$

$= \dfrac{0.4547 - 2(0.7002) + 0.9653}{(0.5)^2} = 0.0784$

(c) $\dfrac{\partial^2 f}{\partial x\, \partial y} = \dfrac{\partial}{\partial y} f_x \simeq \dfrac{-f_x(0, 1) + 4f_x(0, 0.5) - 3f_x(0, 0)}{2h}$

$= \dfrac{-0.4481 + 4(0.3065) - 3(0.1555)}{2(0.5)} = 0.3114$

Chapter 6

(6.1)

(a) $x_1 = 1, x_2 = 3, x_3 = 2$

(b) $x_1 = 1, x_2 = 66, x_3 = 23$

(6.2)

(a) $x = 2.3829, y = 1.4893, z = 2.0212$

(b) $x = 11, y = 11, z = 10$

(6.6) $\begin{bmatrix} 1 & 0 & 0 & 0.25 & 2.00 & 0.25 \\ 0 & 1 & 0 & -0.3125 & -3.25 & 1.6875 \\ 0 & 0 & 1 & 1.0625 & -0.75 & 0.0625 \end{bmatrix}$

(a) $x_1 = 0.25, x_2 = -0.3125, x_3 = 1.0625$

(b) $x_1 = 2.00, x_2 = -3.25, x_3 = -0.75$

(c) $x_1 = 0.25, x_2 = -1.6875, x_3 = 0.0625$

(6.10)
$$F = AB = \begin{bmatrix} 13 & 9 & 10 \\ 9 & 8 & 9 \\ 21 & 8 & 10 \end{bmatrix}$$

$$H = BC = \begin{bmatrix} 37 \\ 27 \\ 9 \\ 22 \end{bmatrix}$$

(6.11) $E = \begin{bmatrix} 14 & 6 & 15 \\ -3 & 5 & 3 \\ 4 & 7 & 20 \end{bmatrix}$

(6.12) Inverse is $A^{-1} = \begin{bmatrix} 0.16129, & -0.032258 \\ -0.12903, & 0.22581 \end{bmatrix}$

(6.13)

(a) $A^{-1} = \begin{bmatrix} 4 & 3 & 2 & 1 \\ 3 & 3 & 2 & 1 \\ 2 & 2 & 2 & 1 \\ 1 & 1 & 1 & 1 \end{bmatrix}$, $\det(A) = 1$

(b) $A^{-1} = \begin{bmatrix} -0.04 & 0.04 & 0.12 \\ 0.56 & -1.56 & 0.32 \\ -0.24 & 1.24 & -0.28 \end{bmatrix}$, $\det(B) =$

(6.14) $\begin{bmatrix} \frac{1}{2} & -\frac{1}{2} & \frac{1}{2} \\ -\frac{1}{2} & \frac{3}{2} & -\frac{3}{2} \\ \frac{1}{2} & -\frac{3}{2} & \frac{5}{2} \end{bmatrix}$

(6.16) In compact arrays:

(a) LU: $\begin{matrix} 2 & -0.5 & 0 \\ -1 & 1.5 & -0.6666 \\ 0 & -1 & 1.3333 \\ -3 & -1.3333 & 0.3333 \end{matrix}$

(b) LU: $\begin{matrix} 1 & 0.3333 & 5 \\ 2 & 1.6666 & -8 \end{matrix}$

with $P = (2, 3, 1)$

(6.18) $\det(A) = -10$
$\det(B) = 7$
$\det(C) = 51$
$\det(D) = -199$

(6.19) $\det(A) = \det(L)\det(U)$
$= (8)(8.75)(2.2)(4.8052) = 740.00$

(6.20) $\det(A^{-1}) = 1/\det(A)$, and $\det(A)$
$= \det(B)\det(C)\det(D)$

$\det(B) = 72, \quad \det(C) = -4, \quad \det(D) = 7$
So $\quad \det(A) = (72)(-4)(7) = -2016$
and $\quad \det(A^{-1}) = 1/\det(A) = -4.96 \times 10^{-4}$

(6.21) $\det(A^t) = \det \begin{bmatrix} 2 & 4 \\ 1 & 2 \end{bmatrix} = 4 - 4 = 0$

$\det(B^t) = \det \begin{bmatrix} 3 & 1 \\ 2 & -1 \end{bmatrix} = -3 - 2 = -5$

(6.23) $\det(A) = (2 - s)(-1 - s)(1 - s) + 40$
$+ 12(1 + s) - 4(1 - s)$
$= -2s^3 + 2s^2 17s + 46$

Chapter 7

(7.1) $f(\lambda) = 8 - 5\lambda + \lambda^2$

(7.2) $-1 - x + 5x^2 - x^3$

(7.4) The power series obtained becomes

$$g(x) = -20 + 33x + 8x^2 - 0.99998x^3$$

Using Bairstow for the polynomial above yields

$$x = 0.5401, -3.407, 10.86$$

(7.5) $g(x) = 182.49 + 323.406x$
$$+ 28.230x^2 - 22.5x^3 + x^4$$
Eigenvalues are $2.78, 20.86, -1.024 \pm 1.602i$

(7.8) Final results (eigenvalues of the whole matrix):

$3.07980E-01$ $6.43103E-01$ $5.04892E+00$

(7.9) Final results (eigenvalues of the whole matrix):

$2.83250E-01$ $4.26022E-01$
$9.99996E-01$ $8.29086E+00$

(7.10) Final results (eigenvalues of the whole matrix):

$-3.86748E+00$ $9.28788E-01$
$5.40035E+00$ $1.91383E+01$

(7.11) Eigenvalues (results of QR scheme):

No.	Real Part	Imaginary Part
1	20.867570	$+0.000000i$
2	-1.024574	$+1.601827i$
3	-1.024574	$-1.601827i$
4	2.781592	$+0.000000i$

(7.12) Eigenvalues (results of QR scheme):

No.	Real Part	Imaginary Part
1	19.138330	$+0.000000i$
2	-3.867481	$+0.000000i$
3	0.928787	$+0.000000i$
4	5.400349	$+0.000000i$

(7.13)

(a) -3 -4 -1.1
The roots are $-1.319 \pm 1.1429i, -0.3609$

(b) Roots are $2, -1.1196, -0.9902 \pm 0.05185i$.

Chapter 8

(8.1) Regression line: $g(x) = 0.2200 + 1.90000x$

(8.2) Regression line: $g(x) = -10.01212x + 11.0267$

(8.4) $g(x) = -3.333861E-02 + 2.632148x$
$$- .2488103x^2$$

(8.5) $g(x) = -1.667578E-02 + 2.576617x$
$$- .223822x^2 - 2.776449E-03x^3$$

(8.6) $g(x) = .1019003 + 240.21x$
$ g(x) = -9.956598E-03 + 352.0669x$
$$- 13982.13x^2$$

(8.6) $g(x) = -1.672983E-04 + 316.9891x$
$$-1745.694x^2 - 1019702x^3$$

(8.7) $g(x) = .8796641 + 37.30039x - 39.26325x^2$

(8.8) $g(x) = -.1177177 + 60.07453x$
$$- 101.6016x^2 + 41.55901x^3$$

(8.9) $g(x) = -1.857601 + 3.814397x$
$$+ 3.241863 \sin(\pi x) + 1.09415 \sin(2\pi x)$$

Chapter 9

(9.4)

x	Euler $h = 0.001$	Mod. Euler $h = 0.001$
1	0.42467 (0.08)	0.42461 (0.03)
2	-0.30287 (0.45)	-0.30405 (0.03)
5	-21.7175 (0.18)	-21.7634 (0.02)
9	-723.529 (0.28)	-725.657 (0.01)

x	Euler $h = 0.01$	Mod. Euler $h = 0.01$
1	0.42802 (0.87)	0.42741 (0.36)
2	-0.29047 (4.52)	-0.30233 (3.06)
5	-21.3809 (1.72)	-21.8364 (0.18)
9	-705.877 (2.71)	-728.118 (0.18)

(9.5) Approximately 2.51 sec.

(9.6)

x	y (Forward Euler)	y (Analytical)
1.0000	0.0434	0.0432
2.0000	0.0491	0.0491
3.0000	0.0499	0.0499
4.0000	0.0500	0.0500
5.0000	0.0500	0.0500

(9.10)

Time (hr)	N (iodine)	N (xenon)
0.0000	$1.0000E+05$	0.0000
5.0000	$5.9333E+04$	$3.3339E+04$
10.0000	$3.5204E+04$	$4.2660E+04$
15.0000	$2.0888E+04$	$4.1012E+04$
20.0000	$1.2393E+04$	$3.5108E+04$
24.9999	$7.3533E+03$	$2.8225E+04$
29.9998	$4.3629E+03$	$2.1821E+04$
34.9998	$2.5886E+03$	$1.6430E+04$
39.9997	$1.5359E+03$	$1.2138E+04$
44.9996	$9.1131E+02$	$8.8418E+03$
49.9995	$5.4071E+02$	$6.3716E+03$

(9.12) $n = 1$: $y_1 = 0.9$, $z_1 = 0.8074$
$ n = 2$: $y_2 = 1.6153$

(9.14) $y_1 = y(0) + \frac{1}{2}(0.5 + 0.375) = 0.4375$

$z_1 = z(0) + \frac{1}{2}(-0.25 - 0.2656) = 0.7422$

$y_2 = y_1 + \frac{1}{2}(0.3711 + 0.2476) = 0.7468 = y(1)$

$z_2 = z_1 + \frac{1}{2}(-0.2471 - 0.2634) = 0.4869 = y'(1)$

(9.16)

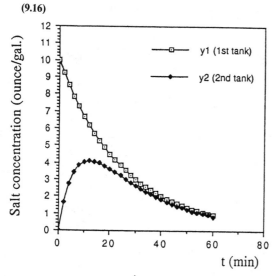

(9.19) $y(1) = y_1 = y(0) + \frac{1}{6}(k_1 + 2k_2 + 2k_3 + k_4)$

$= 1 + \frac{1}{6}[-1 + 2(-0.6666$

$- 0.7059) - 0.2707] = 0.3307$

(9.20)

(a) The local error of the second-order Runge-Kutta scheme is proportional to h^3 so we can write

$$E_h = Ah^3$$
$$[y(0.2)]_{h=0.1} + 2A(0.1)^3 = [y(0.2)]_{h=0.2} + A(0.2)^3$$

By introducing the computed solution from the table given

$$0.894672 + 0.002A = 0.8947514 + 0.008A$$

or after rewriting,

$$-0.000079 = 0.006A \quad \text{or} \quad A = -0.01316$$

Therefore, the error of $[y(0.2)]_{0.1}$ is

$$2A(0.1)^3 = 2(-0.013)(0.001) = -0.000026$$

(b) The error of $y(1)$ with $h = 0.1$ is

$$[y(1)]_{h=0.1} + 10A(0.1)^3 = [y(1)]_{h=0.2} + 5A(0.2)^3$$
$$0.3226759 + 10A(0.001) = 0.3240404 + 5A(0.2)^3$$
$$-0.0013645 = 0.04A - 0.01A = 0.03A$$
$$A = -0.04548$$

So, $\quad 10A(0.01)^3 = 0.01(-0.04548) = -0.0004548$

Estimate for the exact value is

$$0.3226759 + 10A(0.001) = 0.3226759$$
$$- 0.0004548 = 0.3222211$$

The truly exact value is 0.32219.

(9.21)

t	$h = 0.5$	$h = 1.0$
0	1	1
1	0.32233	0.32388
2	-0.59577	-0.59636

Index

Order Form
Computational Software Library in C
(CSL™)

ιe C programs in this book are available in the 5.25" PC, 3.5" PC and 3.5" Macintosh diskette
rms. Orders from U.S., Canada, and Mexico will be shipped by first class mail, and orders
ɔm overseas are shipped by surface mail with no additional cost (or airmail with an additional
st of $4.50). Please photocopy this order form and send with a check or money order in U.S.
nds to:

ɔmputational Methods, Inc.
O. Box 21230
ɔlumbus, Ohio 43221, USA

ease specify format:

_5.25" IBM PC,   ___3.5" IBM PC,   ___3.5" Macintosh
                                                           Line Total

ιantity_____   Diskette   @$15.00 ea.   _____

                             6% Sales Tax (Ohio Residents only) _____

                             Post paid (first class) to US, Canada
                             and Mexico; surface mail to other
                             countries.

                             Add $ 4.50 for oversea airmail   _____

                                         Total amount   _____

lease type or print)

ame   _____

ɔmpany _____

ɛpartment _____

reet _____

ιty _____State _____ Zip _____ Country _____

ɛlephone_____ Fax (if available)_____